AF553060

FISH
BIOLOGY

NIPA® GENX ELECTRONIC RESOURCES & SOLUTIONS P. LTD.
New Delhi-110 034

About the Author

Dr. Dina Nath Pandit (vidwan id score 9.7) is working as an Associate Professor in the Department of Zoology, Veer Kunwar Singh University, Arrah – 802 301. He did his Masters in Zoology in First Class from TM Bhagalpur University, Bhagalpur – 812 007, India and Ph. D. from the same University in 2002 on a problem related to "Live Fish Transportation." He was the beneficiary of the ICAR and the CSIR fellowship. He qualified tests for a lectureship in Life Sciences and Zoology. He taught Fish and Fisheries, Animal Taxonomy, Biostatistics, Bio-techniques, Cell and Molecular Biology, Genomics, Ethology and Ecology at UG and PG level for twenty-one years. He has supervised PhD work of sixteen research students and published around seventy research papers in the Journals of National and International repute. He has presented paper in the Science Communicators' Meet during 103rd Indian Science Congress from January 3rd to 7th of 2016 at Mysore University, Mysuru.

He is the recipient of the Bharat Shiksha Ratna Award from Global Leaders Foundation, Delhi; the Congress of Zoology Medal from the Zoological Society of India, India; the SLS Recognition Award from the Society of Life Sciences, Satna, MP, an Eminent Ichthyologist Award from Asian Biological Research Foundation, Prayagraj (India), the SKSS Best Scientist Award from Sarvhit Kalyan Seva Samiti, Meerut and recently, the Prof BB Kaliwal Award from the Zoological Society of India, India.

He is a life member of the Society of Fisheries Technologists (India), Cochin and Indian Fisheries Association, Versova, Mumbai. He is a member of the RSB, London; a Fellow member of the RSS, London; a Fellow Member of the ZSL, London; General Committee Member of the RSS, London local group in India and many prestigious societies. It signifies that he is a well qualified professional biologist subject to a rigorous code of conduct.

Earlier, he authored four textbooks entitled "ANIMAL TAXONOMY - PRINCIPLES AND PRACTICES" from Narendra Publishing House, Delhi; "LIVE FISH TRANSPORTATION" from NIPA® GENX Electronic Resources & Solutions P. Ltd. New Delhi-110 034 and "STATISTICS: A MODERN APPROACH" as well as "RESEARCH METHODOLOGY FOR Ph.D. COURSEWORK" from Hindustan Publishing Corporation (India), Delhi also available on Amazon and Flipkart.

FISH BIOLOGY

Dina Nath Pandit
M Sc, Ph D, FZSL
Department of Zoology
Veer Kunwar Singh University
Arrah - 802 301, India

NIPA® GENX ELECTRONIC RESOURCES & SOLUTIONS P. LTD.
New Delhi-110 034

NIPA• GENX ELECTRONIC
RESOURCES & SOLUTIONS P. LTD.

101,103, Vikas Surya Plaza, CU Block
L.S.C.Market, Pitam Pura, New Delhi-110 034
Ph : +91 11 27341616, 27341717, 27341718
E-mail:newindiapublishingagency@gmail.com
www: www.nipabooks.com

For customer assistance, please contact
Phone: + 91-11-27 34 17 17
Fax: + 91-11- 27 34 16 16
E-Mail: feedbacks@nipabooks.com

Print ISBN: 978-93-58872-06-4

ebook ISBN: 978-93-58877-49-6

Composed and Designed by NIPA®.

गुजरात केन्द्रीय विश्वविद्यालय
भारत की संसद के अधिनियम स. 25, के अंतर्गत स्थापित

CENTRAL UNIVERSITY OF GUJARAT
(Established by an Act of Parliament of India, No 25 of 2009)

Date: February 06, 2024

Prof. Rama Shanker Dubey
Vice Chancellor

Foreword

The 21st century ushered in an era of understanding living organisms, conservation of nature, and the technical application of biological processes for the benefit of humanity. In this context, the term "fish" encompasses comprehending past, present, and future trends, as well as their sustainability in areas where nature and human society intersect. The Roman scholar Marcus Terentius Varro (116-127 BC) recognized both the taste and recreational value of fish. A rich body of fish literature has accumulated since the dawn of civilization, resulting in several textbooks and laboratory manuals written to benefit animal science and fisheries. Without a deep understanding of their lives, habits, and behavior, effective planning, monitoring, and management of fish would be impossible.

In the current scenario, nations striving to keep pace with development across various domains must address the educational needs of the younger generation. Recently, the socioeconomic aspects of fish and fisheries have become increasingly appealing. The University Grants Commission (UGC) has introduced vocational courses such as the 'Diploma in Fisheries Sciences', 'Certificate in Fisheries and Agri-allied Science', 'Fisheries Engineering', 'Marine and Antarctic Science', and 'Freshwater Sciences' at both undergraduate and postgraduate levels. Several Indian universities have followed suit.

A comprehensive book on the domain of fish, tailored for students pursuing M.Sc. in Zoology, M.F.Sc. (Masters in Fisheries Science), Aquaculture and Fisheries,

Sector-29, Gandhinagar- 382030, Gujarat, Phone No. 079-23977402, Fax-079-23260076
Email: vc@cug.ac.in Website: www.cug.ac.in
Former Vice Chancellor: Tilka Manjhi Bhagalpur University (Bihar) and
Guru Ghasidas University, Bilaspur (Chhattisgarh)

and B.F.Sc. (Bachelor of Fisheries Science), as well as fresh aspirants preparing for UGC-NET, CSIR-NET, and other examinations present a challenging journey. This textbook aims to encompass all the accumulated information on Fish Biology from the past two decades, including insights from the COVID-19 period.

The book "Fish Biology" reflects the author's ability to meticulously organize the materials into a practical guide. It deftly gathers unexpected elements of Fish Biology using lucid methods. Dr. Dina Nath Pandit, a Professor in the Department of Zoology at Veer Kunwar Singh University, Arrah - 802 301, has made extraordinary efforts to disseminate the message of Fish Biology within the framework of current educational plans. I am confident that this textbook will significantly contribute to the overall modernization of the subject. In my opinion, it will be invaluable for students and researchers, providing them with a solid foundation and practical knowledge. I strongly recommend it as a textbook and encourage its inclusion in the libraries of various universities and colleges.

This foreword serves as a gateway to the world of fish biology, inviting readers to plunge into the depths of knowledge that reveal the secrets of these amazing creatures.

Finally, I extend my congratulations and best wishes to the writer for all his efforts.

(Rama Shanker Dubey)

Sector-29, Gandhinagar- 382030, Gujarat, Phone No. 079-23977402, Fax-079-23260076
Email: vc@cug.ac.in Website: www.cug.ac.in
Former Vice Chancellor: Tilka Manjhi Bhagalpur University (Bihar) and
Guru Ghasidas University, Bilaspur (Chhattisgarh)

Acknowledgements

First of all, I bow before the Almighty, the beneficent, the merciful, Lord of the world and owner of the Day of Judgment, who has bestowed upon my health, courage, intimates and well-wishers, whose prayers and selfless gestures have given shape to my thoughts and enabled this work to bear fruit. Words are scarce when the heart is full, but I had to delve deep into the ocean of vocabulary and raid the ranks of expression. I am ecstatic that I was able to put a few words together to express how I feel.

I thank God for weaving lovely people into the fabric of my life. Their vibrant patterns and personalities have added richness to my life. I believe the blessings and shower of God will keep my precious thread-safe and bright. As a result, it is my primary responsibility to acknowledge and thank them for their generous support and contributions.

My principal acknowledgements go to Prof. (Dr.) Rama Shanker Dubey, Vice-Chancellor, Central University of Gujarat, Gandhinagar, Gujarat; Former Vice-Chancellor, Guru Ghasidas Das University, Bilaspur, Chhattisgarh and TM Bhagalpur University, Bhagalpur, Bihar. Without her enthusiasm, support and insights into the subjects, the writing of this book would not have materialized.

I consider it an honour to express my heartfelt gratitude to Prof. (Dr.) S. K. Chaturvedi, Vice-Chancellor, Veer Kunwar Singh University, Arrah for his encouragement, undaunted support, precious suggestions and valuable recommendations in all my endeavours for this book. I am honoured to have had the opportunity to learn from your expertise and profound impact.

I would like to extend my special thanks to Prof. (Dr.) Ran Vijay Kumar, Dean of Students' Welfare, Veer Kunwar Singh University, Arrah for critically listening to my problems and extending his help. This book is a tribute to the strength and grace you embody.

The efforts made by Prof. (Dr.) Ashok Kumar Thakur, Head of the Department of Zoology and Dean Faculty of Sciences, TM Bhagalpur University, Bhagalpur in revising the contents has to be acknowledged as well. His willingness to accept this responsibility as a task is indeed worth acknowledging.

It is my proud privilege to pay my respect to Prof. (Dr.) Farida Bano, Professor and Head, Department of Zoology, Veer Kunwar Singh University, Arrah for his blessings and good wishes and also for offering me all the bits of aid and helpful recommendations for this work. Your wisdom and mentorship have been the guiding lights in the creation of this book

I acknowledge the motivational and innovative ideas extended by Dr. Uday Shankar Sinha, Dr. Dhyanendra Kumar and Dr. Suresh Prasad Srivastava, Dr. Raja Ram Singh, Dr. Anil Kumar Sinha, Dr. Ajit Kumar Sinha, Former Heads of the Department of Zoology, Veer Kunwar Singh University, Arrah as well as Dr. Ram Randhir Singh, Department of Zoology, Veer Kunwar Singh University, Arrah and Dr. Anjali Gupta, Head, P.G. Department of Zoology, H.D. Jain College, Arrah. Whenever required, they always made time from their hectic schedules which allowed me to bring this fantastic workout. This book reflects the collective knowledge and experience that you have graciously shared.

I extend my heartfelt gratitude to the diligent reviewers whose insightful feedback and constructive criticism helped shape the pages of this book. Your thoughtful assessments and valuable suggestions have been instrumental in shaping the final narrative. Your dedication to the craft of reviewing has strengthened and increased the impact of this work.

I extend special thanks to my colleague Dr. Kizar Ahmed Sumon, Associate Professor, Department of Fisheries Management, Bangladesh Agricultural University, Mymensingh-2202, Bangladesh and Ilham Zulfahmi of Department of Biology, Faculty of Science and Technology, Ar-Raniry State Islamic University, Koperma Darussalam, Banda Aceh City-23111, Indonesia.

This book is a testament to the beauty brought into my life by Dr. Mohita Sardana, Assistant Professor, Department of Zoology, Veer Kunwar Singh University, Arrah. This book is dedicated to you, my muse. This book is equally yours and mine. I have woven the threads of my thoughts into the pages of this book, but it's your unwavering support that has embroidered the fabric of inspiration. I would like to express my deepest gratitude to you for being the silent inspiration and cherished companion.

I express appreciation to Dr. Amit Priyadarshi, Assistant Professor and Dr. Rajesh Verma, Assistant Professor of the Department of Zoology, Veer Kunwar Singh University, Arrah; Dr. Anand Prakash, Assistant Professor, Department of Zoology, SVP College, Bhabhua. Their contributions to the materialisation, the spark that ignited the creative fire within, updating and enrichment of the content of this book will be remembered fondly in my heart. Your collaborative spirit and commitment to excellence were critical in bringing this book to fruition.

I am grateful to my entire family, especially my brilliant follower, Ms. Deepa Sonal, Assistant Professor, Department of Computer Science, Patna Women's College, Patna for brilliance, resilience and boundless love which inspired every word in this book. You bring me the most joy and inspiration. May the beauty you bring into my life be reflected in these pages? I consider blessed being your parent. Unique stories, cultural intersections and the beauty of Dev, Divya and Aarav as the second generation identity have enriched the narrative, making this book a celebration of diversity, heritage and the endless possibilities that arise when worlds collide.

I would like to acknowledge all the persons/institutions/organizations/ societies for their publications/work in developing software and delivering open Access Resources/External Resources to benefit of teachers/student communities and the general public. They are a beacon of learning and enlightenment, exemplifying the pursuit of knowledge and excellence. This book embodies the ideals of learning and discovery that you promote and I am grateful for the inspiration that your educational mission has provided.

I express my heartfelt gratitude to NIPA® GENX Electronic Resources & Solutions Pvt. LTD. New Delhi-110034, a visionary in the world of publishing. Your attention to detail and unwavering support have helped me to turn this manuscript into a published work. This book stands as a testament to your dedication to excellence. Your team was always to assist me whenever I required it. I also want to express my heartfelt appreciation to them. Without their assistance, it would not have been possible to publish this book.

(Dina Nath Pandit)

Preface

Fish with their amazing diversity and adaptability are more than just inhabitants of aquatic ecosystems; they are key players in the intricate web of life that sustains our planet. Fish provides both inner joy (beauty, recreation, wonder, and religious symbolism) and practical values (ecology, food, and commerce). The study of fish is a condensed and expanding record of knowledge of humanity of various fish species, incorporating discoveries, archaeological data and databases. Each species of fish from the smallest guppies to the mighty marlins, contributes to the complex mosaic of ecological balance.

The study of fish biology serves as a gateway to unravelling the mysteries that lie beneath the surface of our planet in the vast and dynamic realm of aquatic life. This book takes the reader on a journey through the intricate tapestry of fish existence, weaving together the threads of evolution, taxonomy, distribution, morphology, physiology, applied ecology, behaviour, genomics, pollution and disease impact. As we delve deeper into this enthralling discipline, we find ourselves immersed in a world where scales tell stories, fins navigate currents and gills bring ancient stories to life. Understanding the complexities of fish biology provides insights into the broader interconnectedness of life on Earth.

According to NEP-2020, the course curriculum would be modified to provide students with a diverse range of opportunities in the agricultural sector. Through educational procedures, it assists people in the fish and fishing industry in improving fishing, fish farming, and fish processing methods, increasing production efficiency and income, and improving their socioeconomic conditions. The University Grants Commission, New Delhi has updated rules to bring major changes in UG, PG and Ph.D. programmes offered across the country for the academic year 2022-2023 onward. The NEP-2020 prioritises rethinking vocational education in the primary sector, such as Aquatic Biology and Fisheries, Industrial Fisheries, Diploma in Fisheries Sciences, Diploma in Aquaculture, Diploma in Fish Product Technology, Certificate in Fisheries and Agri-allied Science, Aquaculture and Fishery Management, Fisheries and Wildlife Sciences, Fisheries Engineering, Limnology and Fisheries, Marine Biology and Fisheries, Marine and Antarctic Science, Marine Fisheries and Animal Sciences, Freshwater Sciences, Fish Production, Sustainable Management and Aquatic Ecosystems. Considering the redesign aspects of NEP-2020, Indian universities have begun to amend the Fishery Science/ Industrial Fish and Fisheries syllabus at the UG and PG levels.

This book is a celebration of the wonders of fish biology, a tribute to the pioneers who paved the way for our understanding, and an invitation to all who want to delve deeper into this fascinating field. Join us as we navigate the waters of discovery, where each chapter reveals a new facet of the extraordinary world of fish, whether you are a seasoned researcher, an aspiring biologist or a curious enthusiast.

The book has been compiled into 6 units and 47 Chapters. MCQ arranged after 47th chapter is an effective teaching and learning tool that provides prompt feedback and may assess higher cognitive abilities. A subject index and fish name index are designed to locate the important terms and names of fish. The cited references and webliography (URL list) are enlisted in this book.

The summary of the contents of the book may be pointed out as:

Unit-I Overview of Fish Diversity: The diversity of fish is astounding, ranging from primitive jawless fishes to highly evolved and specialised groups. It aims to elucidate the complexities of this group and the critical roles they play in preserving ecological balance. It delves into the fascinating world of fish diversity, investigating their geographical distribution, evolutionary history, specific features and taxonomy as well as the roles they play in global ecosystems.

Unit-II Type Study of Fish: This is a meticulous investigation into the various types that inhabit our aquatic environments. It aims to categorise, analyse and understand the vast array of fish that inhabit our oceans, rivers, lakes and streams from ancient jawless fishes to highly specialised ray-finned species. We gain a profound understanding of the diverse tapestry of fish life and the pressing need to protect their existence in the ever-changing dynamics of aquatic ecosystems through a comprehensive study.

Unit-III Fish Physiology: Fish physiology investigates the intricate mechanisms that allow these aquatic marvels to thrive in a variety of environments. This exploration sheds light on the physiological adaptations that make fish a fascinating subject of study from the molecular intricacies of their cellular functions to the physiological marvels facilitating their underwater existence. It may include a study of the fin systems, skeletal system, swimming, nutrition, circulation, excretion, osmoregulation, nervous system and sense organs. Students and researchers gain a better understanding of the adaptations that have allowed these vertebrates to conquer the aquatic realm by unravelling the complexities of fish physiology.

Unit-IV Reproduction, Development, Genomics and Behaviour of Fish: The life cycle of a fish is an enthralling journey that includes a complex interplay of reproductive strategies, developmental processes, genomic complexities and a rich tapestry of behaviours. Fish have an amazing range of reproductive strategies, from open-water external fertilisation to intricate courtship rituals and parental care. The study delves into the morphological transformations, physiological adaptations, and environmental influences that shape the development of fish during the embryonic, larval, and juvenile stages. The genomic architecture of various fish species is investigated with a focus on adaptations, genetic diversity and the role of genes in regulating key aspects of reproduction, development and behaviour. Moreover, Fish behaviour includes feeding habits, migration patterns, mating rituals, and social interactions.

Unit-V Threats to Fish Population: This investigation delves into the multifaceted factors contributing to the decline of the fish population, emphasising the importance of understanding and mitigating these threats to ensure the sustainability and imperative for aquatic biodiversity preservation. Climate change's escalating effects, such as rising temperatures, ocean acidification and altered precipitation patterns have a direct impact on fish populations. Disease and parasite spread, exacerbated by factors such as overcrowding and weakened immune systems also result in significant declines in fish populations.

Unit-VI Ecology and Importance of Fish: Understanding the ecology and importance of fish is critical for effective conservation and long-term management. Recognising the intricate roles that fish play in maintaining the balance of ecosystems is critical for ensuring their continued presence and the well-being of the diverse communities that rely on these vital resources as stewards of aquatic environments.

So, dear reader, dive into the chapters that follow with awe and curiosity. May the knowledge gained in these pages arouse awe for the wonders of fish biology and a desire to protect the delicate balance of our planet's aquatic ecosystems? As we navigate the ever-changing currents of fish biology, let us embrace the legacy of exploration and understanding.

Although great care was taken in the preparation of the book, flaws are unavoidable. As a result, I welcome suggestions, insightful comments and constructive criticism.

(Dina Nath Pandit)

Contents

Unit-IV Reproduction, Development, Genomics and Behaviour of Fish

Unit-V Threats to Fish Population

Unit-VI Ecology and Importance of Fish

1

Introduction

The term "fish" is usually a convenient description for a group of cold-blooded, ectothermic, aquatic non-tetrapod craniates that breathe with gills (Nelson, 2006). Fishes exploit every nook and niche of the domain of water. They are especially noted for their unimaginable numbers and bewildering forms. Most of them have a stream-lined body covered by scales and pose fins supported by fin rays and breathed by gills throughout life. They constitute the first group of animals that have developed biting jaws in the phylogenetic history of vertebrates. The aquatic environment has exerted a tremendous influence on the lives of fish.

The word for fish in English *fish,* German *Fisch* and Gothic *Fisks* is inherited from Proto-Germanic and is related to the Latin *Piscis*, French *Poisson*, Spanish *Pescado*, Italian *Pisce*, Hindi *Meen* or *Machli,* Sanskrit *Matsya* and Old Irish *īasc*. Some authorities reconstruct a Proto-Indo-European root *peysk**- attested only in Italic, Celtic, and Germanic. *Matsya* in Sanskrit is cognate with *maccha* in Prakrit.

1.1 Proverbs Relating to Fish

1. *Jal To Hai Doston Asli Sona, Isko Tum Kabhi Na Khona.*
   Pani Bin Hum Hai Aise, Jal Bin Machli Tadpe Jaise.
2. *Jb Pyar Kisi Se Jyada Ho Jata Hai*
   To Dil Machli Aur Woh Pani Ban Jata Hai.
3. If you give a person a fish a day, you are giving him a meal for the day, but if you teach him how to fish, you provide him meals for a lifetime. It is a quote from the Chinese philosopher Lao Tzu, the founder of Taoism.
4. Fish is considered to be rich food for poor people.
5. Fish is also known as brain food and heart food.
6. Fit, intelligent, sexy and hot: slang for FISH.
7. The symbol of two golden fish is exclusively used as a good omen, thus pointing to the general idea of good fortune.
8. Fish in the Sea is an assurance at home.
9. Do not bargain for fish that are still in the water.

10. Fish and guests smell when they are three days old.
11. Do not punish a fish by throwing it into water.

1.2 Historical Account of Ichthyology

The fish (<><), originates from the Greek word, "*ichthys/ichthus*." This word is further defined in the acrostic IXNYy:

I – Iota or Iesous (or Jesus)

X – Chi or Christos (or Christ)

N – Theta or Theou (or God)

Y – Upsilon or Yidos/Huios (or Son)

y – Sigma or Soter (or Savior).

Matsya, the first of Vishnu's ten primary avatars (Fig. 1.1), is the Hindu god *Vishnu*'s fish avatar (Bandopadhyaya, 2007). *Matsya* is said to have saved the first man, *Manu*, from a great deluge (Valborg, 2007). *Matsya* is mentioned for the first time in the *Shatapatha Brahmana* (Section 1.8.1 of the Yajur Veda), where it is not associated with any particular deity. *Matsya* is thought to represent aquatic life as the first beings on Earth.

Fig. 1.1: Matsya Avtar of Lord Vishnu (from https://www.astroved.com)

According to Bonnefoy (1993), another symbolic interpretation of *Matsya* mythology is to consider Manu's boat to represent moksha (salvation), which aids in crossing over (Fig. 1.2: A-G).

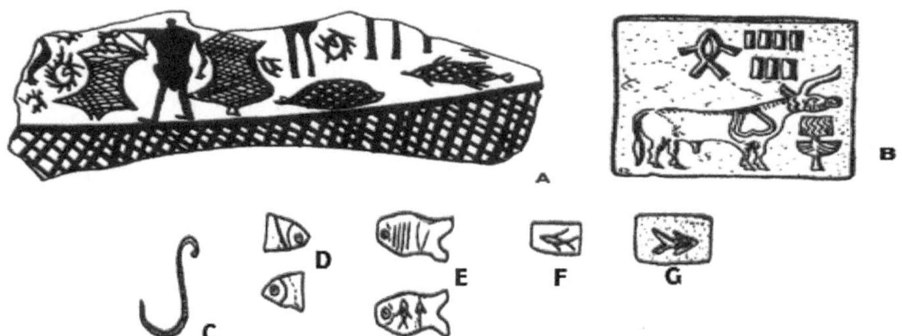

Fig. 1.2: Sherd, Fishhook and tokens from Harappa

A. sherd from Stratum VI, Mound AB (pl. LXIX: 16)

B. 'unicorn' seal of satellite with a fish sign from a Monad F (pl. LXXXIV: a)

C. copper/bronze fishhook from Monad F 'Great Granary area' Eastern extension (pl. CXXV: 8)

D. observe and reverse of the head of the fish-shaped token in steatite from an unknown province (pl. XCVII: 560)

E. observe and reverse of fish-shaped token in steatite from Mound F (pl. XCV, 428; length 15.2mm)

F. inscribed, the fish motif on one face of a three-sided steatite token from Mound F (pl. XCVIII, 590; length 5mm)

G. inscribed, the fish motif on one face of a three-sided steatite token from Mound F (pl. XCV: 428)

A-G redrew from Vats (1940) by J. M. Kenoyer (from Belcher 1991:118)

Therefore, the fish represents the phrase "Jesus Christ God's Son is Savior." Some research has suggested that the fish symbol emerged in the first century or even before Jesus died on the cross, but nothing has been confirmed as to when the symbol and its meaning first began.

1.2.1 Ancient Ichthyology (300BC to 1499AD)

The biology of the fish is so diverse that it is extremely difficult to give a concise account of the group. The branch of life sciences that deals with phylogeny, taxonomy, structure, life cycle, culture and capture of fish is known as Ichthyology.

As far as the historical account of Ichthyology is concerned, heaps of refuse of shellfish and seafish were found at the dwelling site of a man near the river or lake during the Old Stone Age (40,000 BC). The New Stone Age (10,000 BC) provided evidence of the practice of "*Salmon Smoking*".

Human beings have been interested in fish since at least 4000 BC in the Middle East. "Salting of fish" most probably started during the Bronze Age (3500 BC). The interest in Indian fish and fisheries dates back to the 3rd millennium. Freshwater fish have been considered sacred in many parts of India since the Vedic period (1750–500 BC; Nautiyal, 2014). The fishes are associated with Lord Vishnu, whose first incarnation on Earth was in the form of a fish. In this incarnation, Lord Vishnu is believed to have saved the first human on Earth by informing him of the calamitous floods that were to follow. It might be useful to consider the connections with the fish of Satyavati, the mother of Veda-Vyasa, the compiler of the Mahabharata and the arranger of the Puranas.

Aristotle (384-322 BC)

Grampus and newly-born youngs

Fig. 1.3: Aristotle as the first marine biologist (from https://slideplayer.com/)

Well-organized fishing activities were established in the Roman Empire (450-400 BC). Ichthyology remained obscure for quite a long time after Aristotle (384-322 BC). He is regarded as the first marine biologist. (Fig. 1.3). He provided the earliest taxonomic classification and described 117 fish species.

The Roman Marcus Terentius Varro (116-27 BC) wrote about freshwater ponds (dulces) and slat-water ponds (maritime or sales) in "*De re Rustica*".

Herring fisheries flourished in the Atlantic (500-1500). It is stated in Domesday Book (1086) that the Abbot of St. Edmund's had fish ponds to provide fresh food for the monastery table. The evidence of fish as food was available from the excavations of the Indian Valley Civilisations. King Somesvara (1127) composed a book entitled "*MANASOLATARA*" to record common freshwater and marine sport fishes.

1.2.2 Progress for Modern Ichthyology (1500AD to 1799AD)

Guillaume Rondelet (1507-1557) described 47 freshwater and 197 marine and 47 freshwater fishes from the Mediterranean region in "*Universae aquatilium*" and "*Libri de Piscibus marinis*". Hippopyte Salviani (1514-1572) emphasized the economy as well as the usefulness of fish and described 92 fish from Italy in "*Aquatilium animalium historia*'. Pierra Belon (1517-1575) gave a system of classification and described some 110 fish species in Europe in a publication "*De aquatilibus libri duo*".

Fig. 1.4: JT Klein
(from http://www.reformacja-pomorze.pl)

Fig. 1.5: P Artidi
(from https://prabook.com/)

Borreli (1608-1679), wrote *Do mote animalium* to explain the mechanism of swimming, and the function of the air bladder. Some 100 fishes from Brazil were described by Wilhelm Piso (1611-1678) and Georg Margrave (1610-1644) in "*Historia naturalis Braziliae*". John Ray (1628-1705) edited the work of F. Willoughby (1635-1672) and described 420 fish species from Great Britain and Germany in "*Historia Piscium*". M. Malpighi (1628-1694) examined the optic nerve of the sword-fish; Swammerdam (1637-1680) described the intestines of fishes and J. Duverney (1648-1730), entered into detailed research of the organs of respiration. Jakob Theodor Klein (1685-1759) (Fig. 1.4) developed a system of fish classification in "*Historia Piscium Naturalis*" from 1740-1744 but did not adopt the Linnaean Code.

Peter Artidi or Petrus Arctaedius (1705-1734) (Fig. 1.5) is regarded as the "Father of Ichthyology" based on the foundations of Ichthyology. The arrangement of fishes described by Artidi resembles more or less the present ones. After the death of Artidi, Carolus Von Linnaeus (1707-1778) completed the unpublished work of Artidi in a book entitled "*ARTIDI TCHTHYOLOGICA*". Linnaeus also edited the other works of Arditi as "*Bibliotheca Ichthyologica*", "*Philosophica Ichthyologica*", "*Genera Piscium*", "*Species Piscium*" and "*Synonymia Piscium*". Lorenz Theodor Gronow (1754-1780) developed a system of fish classification in "*ZOOPHYLACEUM*" in 1763 and defined genera of fish. Mark Eliezer Bloch (1723-1797) arranged and described the fishes of Germany and India. The work of Bloch is considered the standard for students working in ichthyology. Forsskål (1732—1763) worked on fishes of the Red Sea and described nearly new 200 species. The majority of the fishes described by him had a wide distribution in the Indo-pacific region. Later, Johann Gottlob Theaenus Schneider (1750-1822) edited the works of 1519 fishes of Bloch in a book "*M. E. Blochii Systema Ichthyologiae iconibus ex. Illustratum*". Georges Cuvier (1769-1832) and Achille Valenciennes (1794-1865) published a book titled "*Historie naturelles des Poissons*".

1.2.3 Modern Ichthyology (1800AD to Present)

Patrick Russell (1726—1805) published descriptions of two hundred fishes collected from Visakhapatnam on the coast of Coromandel. John Richardson (1787—1865) wrote accounts dealing with natural history and ichthyology and was the author of *Icones Piscium* (1843), Catalogue of Apodal Fish in the British Museum (1856), the second edition of Yarrell's History of British Fishes (1860). Francis Buchanan (1762—1829) studied the fishes of the Ganges.

Fig. 1.6: J. Muller (from https://en.wikipedia.org)

Fig. 1.7: A. Guenther (from https://en.wikipedia.org)

Johannes Muller (1801-1858) (Fig. 1.6) showed differences between ganoid fishes from other bony fishes. Louis Agassiz (1807-1873) expanded fish classification to include fossil forms. Peter Bleeker (1819-1878) published 500 contributions chiefly on tropical Indo-Pacific fishes.

The work of Albert Guenther (1830-1914) (Fig. 1.7), one of the most influential Ichthyologists, contain descriptions of some 6847 fish species as well as descriptions of 1682 other doubtful fish species. Modern styles of aquaria were developed during the latter half of the 19th century. Mrs. Thymne (1846) made the first attempt to keep marine fish alive and water healthy using plants. David Starr Jordan (1851-1931) published "*Guide to the Study of Fishes*" and "GENERA OF FISHES" in the following chronologies of four parts as

(a) Part First: *The Genera of Fishes* (from 1758-1833).

(b) Part Second: *The Medieval Period of Systematic Ichthyology* (from 1833-1858).

(c) Part Third: The Early Modern Period (from 1858-1880).

(d) Part Fourth: Late Modern Period in Ichthyology (from 1881-1920).

1.2.4 Indian Ichthyology

The foundation for fisheries research in India was laid by Cuvier, Valenciennes, Lacepede, Bloch, Schneider, Forsskal, Bleeker and Albert Gunther. The work of Patrick Russell, Hamilton-Buchanan, Edward Blyth, Stolizka, Sykes, J. McClelland and T.C. Jerdon contributed to our knowledge. The most outstanding contribution was that of Dr. Sir Francis Day.

Hamilton Buchanan (1822) illustrated a taxonomic account of the "Fishes of the Gangetic System". Later, Francis Day (1829-1889) (Fig. 1.8) published two books entitled "*Fishes of India*" in 1878 and "*Fauna of British India, Burma and Ceylon*" in 1889. On the other hand, James Hornell (1865-1949) was an English zoologist and ethnographer called the "Father of the Pearl and Chank Fisheries" of India and Ceylon.

After the publication of two books by F Day, BL Chaudhari (1912) presented a classical account of the fishes of India in "The Fishes of Northern India". Sunder Lal Hora (1896-1955) (Fig. 1.9) and his collaborators became immortal in the annals of Indian Ichthyology. Hora published many revisionary papers and series on fish like:

"Fauna of India"

"Game Fishes of India"

"Notes on Fishes in Indian Museum"

"Siluroid fishes of India, Burma and Ceylon"

"Angling in Ancient India" discusses fish-hook from the Indus Valley. Angling in the historical period and among the gypsies of Europe.

A genus of ricefish, *Horaichthys* ("Hora's Fish"), was created in his honor. The catfish, *Horabagrus* is named after him. He has credit for publishing 425 publications.

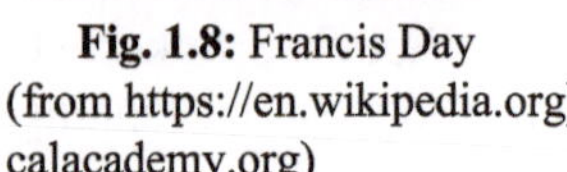

Fig. 1.8: Francis Day (from https://en.wikipedia.org)

Fig. 1.9: SL Hora (from https://www.xwhos.com/)

Fig. 1.10: DS Jordan (from https://www.calacademy.org)

David Starr Jordan (1851-1931) (Fig. 1.10) is known as the "Father of American Ichthyology". Kottore Chidambaram Jayaram (1926-2010) has been credited with publishing more than 100 publications on Siluroid fishes in India. He was considered an authority on the Indian catfishes. He initiated a series of "Contributions to the Study of Bagrid Fishes" in the following four chronological parts:

(a) Part I (1977): Fam. Bagridae

(b) Part II (1977): Fam. Siluridae, Schilbeidae, Pangasidae, Amblicipitidae and Akysidae

(c) Part III (1979): Fam. Sisoridae

(d) Part IV (1980): Fam. Claridae, Heteropneustidae, Chacidae and Olyridae

AGK Menon (1921-2002) worked for more than 20 years on fish systematic, especially on torrential fishes of the Himalayas as under:

(a) "Notes on the Fishes of Indian Museum" (1949, 1951 and 1952)

(b) "Further Studies regarding Hora's Satpura Hypothesis" (1951)

(c) "Monograph of the Cyprinid Fishes of genus Garra"

(d) "A checklist of fishes of the Himalayan and the Indo-Gangetic Plains".

Fig. 1.11: VG Jhingran (from https://insaindia.res.in/)

Fig. 1.12: Hiralal Chaudhuri (from https://en.wikipedia.org)

Fig. 1.13: Eric Godwin Silas

The earliest regional work on the fish fauna was done by Shaw and Shebbeare (1937) in the journal "Royal Asiatic Society of Bengal (Science)". The monographic work of Gopal Ji Srivastava (1968) entitled "Fishes of Eastern Uttar Pradesh and Bihar", seems to be a manual for students of ichthyology. Based on various research on ichthyology, a monograph on "Fish Culture in India" was brought out by KH Alikunhi (1957). VG Jhingran (1919-1991) (Fig. 1.11) is known as the "Father of Aquaculture" and authored a comprehensive volume on "Fish and Fisheries of India". The Commercial Sea Fishes of India by Talwar and Kacker (1984) and Fishes of the Laccadive Archipelago by Jones and Kumaran (1980) are some of the major work done in India. JS Dutta Munshi and MP Srivastava (1988) brought out a "*Natural History of Fishes and Systematics of Freshwater Fishes of India*".

Hiralal Chaudhuri (1921-2014) (Fig. 1.12) is considered the father of the Blue Revolution and also the father of induced breeding carp. It was launched in India during the seventh five-year plan (1985-1990). It aims to achieve the economic prosperity of the country and the fishers and fish farmers.

Eric Godwin Silas (1928-2018) (Fig. 1.13) is regarded as a legendary scientist in fishery and aquaculture. He has published nearly 300 scientific papers and monographs

1.3 Habitats of Fish

1. **Freshwater fish:** Fish that spend most or all of their life in freshwater, such as rivers and lakes, have a salinity of less than 0.5 ppt. Around 40% of all known species of fish are found in freshwater.
2. **Brackish water fish:** Fish tolerate a wide range of salinity (0.5 – 30.0 ppt) and live in brackish water, estuaries and coastal water. Examples: Mullet, Milkfish, Seabass, Pearlspot, Mudskipper, etc.
3. **Marine fish:** Fish spend most or all of their life in seawater, such as the seas and oceans, which have salinity above 30 ppt. There are 240 species of marine fish. Examples: Sardines, Mackerel, Ribbonfish, Anchovies, Grouper, Cobia, Tuna, etc. The important marine fishes can be grouped into the following categories:
   a) **Surface-water fish (Pelagic)**: Examples: Sardines, Anchovies, Ribbonfish, Mackerel, Seerfish, Tuna etc.
   b) **Mid-water fish (Pelagic)**: Bombay duck, Cobia, Silver Bellies, Horse Mackerel, etc.
   c) **Bottom-water Fish (Demersal)**: Perches, Catfish, Pomfrets, Flatfish, Eels etc.
4. **Coldwater fish** (5–20^{0}C): The gills are greatly reduced and the gill opening is smaller to adapt to cold temperatures. Example: Mahseer, Trout etc.,

5. **Warm water fish** (25–35°C): Example: Carps, Catfish, Snakeheads, Featherbacks etc.
6. **Shellfish**: They are aquatic invertebrates having an exoskeleton/shell. They maybe:

   a) Crustaceans: Crabs, Crayfishes, Lobsters, Prawns, Shrimps etc.

   b) Molluscs: Bivalves, Cephalopods and Gastropods.

1.4 Cultural Significance

Early literature provides us with a clear view of fishing activity by the late second or early first millennium. Thus, the *Rig Veda* refers to the method of catching fish by a net and to those people who catch fish (Das 1931). Dhaivara takes fish by netting a tank on either side, Dasa and Sauskala do so using a fish hook (badisa), Baind, Kaivarta and Mainala using a net (jala), Margara catches fish in the water with his hands, Anda by putting pegs at a ford (a type of dam) and Parnaka by putting a poisoned leaf on the water' (Das 1931).

In the *Arthasastra,* there is evidence that fishing was carried on in reservoirs and fishermen themselves were taxed; and the use of fish as manure in agriculture was recognised (Hora 1948b).

Romila Thapar commented that the edict most probably reflected the difficulty that the emperor experienced in banning the catch of fish because of the importance of fish during the Mauryan times (Thapar 1961).

Fish eating was prevalent and highly esteemed in the days of *Jataka* tales. Even ascetics enjoy fish dishes. Ajivikas, a religious order of naked ascetics, is also said to have a fish diet. Women are regarded as having a particular yearning for fish (Hora and Sarswati 1955).

In *Aranyakanda*, Rama and Laksmana are advised to cook rice and fish with salt and red pepper on reaching an *ashrama* on the west bank of the Pampa Lake (Hora 1952). References to fish in the *Sutras* and the *Smrtis* are only casual.

Apart from general allusions to fish, we have references to specific types of fish in different contexts. Mostly, it is in connection either with the *sraddhas* or with certain forbidden foods that the different types of fish are mentioned. Again, when meat does form a part of offered food, fish is often declared superior to the flesh of other animals (Hora 1953).

In the *Gautama Dharmasutra*, the *Vasistha Dharmasutra* and the *Yajnavalkya Smrti,* fish is among the items that should not be refused if offered voluntarily and the *Gautama Dharmasutra*, the *Apastamba Dharmasutra*, the *Manu Smrti* and the *Yajnavalkya Smrti* all make recommendations about fish in the *sraddha* (Hora 1953).

In one place, the *Yajnavalkya Smrti* orders three days of fasting for eating fish, but in another, it lists the fish 'fit for eating even by the *Brahmanas'*, namely, the Simhatundaka (*Bagarius bagarius (Ham.)*), the Rohita (*Labeo rohita (Ham)*), the Pathina (*Wallago attu (Bloch)*), and the Rajiva (*Mugil corsula (Ham)*) (Hora 1953).

Reports of Bengalees from just before and just after the 1914-1918 War showed that probably 80 percent of the population was fish-eaters (Das 1931).

A Hinduised pre-Dravidian tribe of Central and Western Bengal has *penkal mach* and *sal mach* as their totems. Among the Mundas, we have *aind* (an eel), *binjuar* (an eel), *area* (a fish), *dundu* (an eel), *dungdung* (a river fish), *hemram* (a fish), *jia* (a river fish), *kandru* (a fish), *machli*, *maugh* (a fish), *sal* (a fish), *sisungi* (a fish), *sohek* (a fish) and *solai* (fish). The Santals have *boar* (a fish) and *aind* (an eel) as sub-steps among them (Das 1931).

The Central Provinces (Moses 1922-1923) indicated that the Gonds believed that the souls of the dead take habitation in fish, so, after burial 'they go to the river, cry out the name of the dead man and catch a fish which they fully believe is the mortal vehicle of that soul'. This fish is then eaten in the belief that the deceased will be born again in the family (Das 1931).

1.5 Diversity of Fishes

In addition to numerous religious groups, India is home to indigenous communities, ethnic groups and regional cultures, each with its own beliefs and taboos (Sinha, 1995; Kanagavel et al, 2014). It is also likely to have contributed to the conservation of freshwater fishes, which have been associated with supernatural beings (Dandekar, 2011; Bagch and Jha, 2011; Katwate et al, 2014).

The geographical influence led to fish being an important part of our culture. Fish is considered lucky, and it signifies prosperity. In a Bengali wedding, a fish is sent for adhibasa to the bride's house on the day of the wedding as it is believed that it will bring good luck to the couple (Das 1931). This ritual is called "Tattva". The fish symbolises good luck and prosperity.

1.14: *Labeo rohita* (Rohu)

Hilsa. This fish holds a very special place in the hearts and taste buds of Bengalis. *Rohu*. Popularly known as "rui" in Bengal, this freshwater fish is widely available in various fish markets across the state (Fig. 1.2). In some districts of Bengal, *Hilsa* is placed before the goddess at the time of worship.

Goldfish are a type of carp, and carp are very lucky in Asian cultures. The story goes, that in ancient China, there was a magical carp with much power, strength, and endurance.

With 34,300 described species, fish exhibit greater species diversity than any other group of vertebrates. Approximately 95% of living fish species are ray-finned fish, belonging to the class Actinopterygii, with around 99% of those being teleosts.

1. *Rhincodon typus* (whale shark): the king of the sea and largest fish (16.5m)
2. *Schindleria brevipinguis* (stout infantfish): smallest fish (7.55mm) (Fig. 1.3)
3. *Seriphus politus* (queenfish): Queen of the sea
4. *Scomberoides pelagicus* (pola vatta): Queenfish in India
5. *Synchiropus splendidus* (Mandarin fish): the most beautiful fish
6. *Psychrolutes marcidus* (Blobfish): the ugliest fish
7. *Thunnus albacores* (yellowfin tuna): warm-blooded fish
8. *Titicaca orestias* (Lake Titicaca fish): the world's highest (3812m) lake (Fig. 1.4)
9. *Abyssocottus korotneffi* (Lake Baikal fish): the world's deepest (1642m) lake
10. *Tilapia* sps. in Africa at 44^{0}C of hot soda lake.
11. *Trematomus* live in ice.
12. *Eviota sigillata* (The Sign Eviota or tiny coral fish): fish of the shortest (8 week) lifespan
13. *Sebastes aleutianus* (Rougheye rockfish): fish with the longest (200+ years) lifespan
14. *Amatitlania siquia* (Convict cichlid): Monogamous fish
15. *Rastrelliger kanagurta* (Indian Mackerel): The National fish of India.

Fig. 1.15: *Schindleria brevipinguis* (from https://www.daviddarling.info)

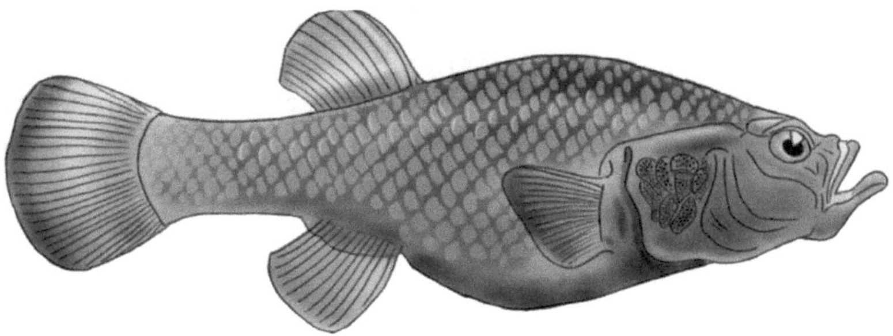

Fig. 1.16: *Titicaca orestias* (from https://twitter.com)

1.6 Ecological and Socio-political Issues

Although freshwater fishes are one of the most threatened vertebrate groups (Leidy & Moyle, 1997; Carrizo et al, 2013) they are often neglected in conservation efforts, including in countries rich in freshwater biodiversity, such as India. None of the more than 150 threatened freshwater fish species in India (IUCN, 2014) are legally protected or the focus of species-specific conservation plans. The increasing threat to freshwater fish species in India has been the subject of debate not only among scientists but also among stakeholders, including local communities (Gupta et al, 2014c). However, the role of stakeholders in freshwater biodiversity conservation is often overlooked by policymakers (Gupta et al, 2014b) as a result of the overt emphasis on centralization and adoption of a techno-centric approach to managing ecological entities (Gupta et al, 2014b).

Despite the apparent conservation benefits of sacred sites, several ecological and policy-related concerns have yet to be addressed (Dudley et al, 2009). Providing legal status to sacred sites would help ensure additional protection for these areas but could also undermine the concept of religious values and traditions associated with the sites (Dudley et al, 2009) if local communities were allowed only limited access. The success of legally protected sites is often hindered by poor management and enforcement because of a lack of human resources (Kanagavel et al, 2013) and in some cases, the transfer of site ownership to forest departments has resulted in conflict with local communities, which has adversely affected site management (Gadgil, 1991; Bhagwat and Rutte, 2006).

1.7 The Way Forward

There is a need for a greater understanding of the short and long-term socio-economic, environmental and conservation impacts of such sacred sites (Berkes, 2004). With the current dearth of conservation options for freshwater biodiversity (Strayer and Dudgeon, 2010), whether sacred sites can be supported legislatively and utilized as an additional safeguarding mechanism can be ascertained only through rigorous scientific studies that involve locally relevant stakeholders.

Deep learning models now provide tools for researchers to process large amounts of imagery data on fish behaviour in a more time-efficient and cost-effective manner as marine sciences enter the era of artificial intelligence. The most recent artificial intelligence models for detecting fish and categorising species have reached human-like accuracy (Abnagan et al, 2023).

Unit-I Overview of Fish Diversity

2

Zoogeography of Fishes

2.1 Introduction

Life exists in almost all diverse habitats, ranging from high mountain peaks of more than 20,000 feet to deep sea bottoms of up to 10,000m. Living organisms can be found in ponds, pools, ditches, sulphur springs, hot springs, the cold of the Polar Regions, the scorching heat of deserts, dense tropical forests, and anywhere else life is possible.

However, not all organisms occupy all possible life-supporting areas. Zoogeography is concerned with the horizontal distribution of animals on land and water, as well as the history of their occurrence around the world. The fish fauna of India can be divided into two groups based on distribution: freshwater and marine, with estuarine forms originating in either of the two.

The distribution of animals can be studied at the following three levels:

(A) Geographical distribution: Distribution of animals over the whole world.

(B) Regional distribution: Distribution of animals in selected segments of the world.

(C) Local distribution: Distribution of animals based on ecology and evolution.

Except for extremely hot thermal ponds and extremely salt-alkaline lakes such as Asia's Dead Sea and North America's Great Salt Lake, almost all natural water bodies support fish life.

Monitoring fish species' frequency, distribution, and abundance is necessary to inform conservation and regulatory efforts that ensure healthy ecosystems and fish stocks (Hilborn et al, 2020). Furthermore, the number and distribution of various fish species can provide important information about ecosystem health and can be used to track environmental change (Rathi et al, 2017).

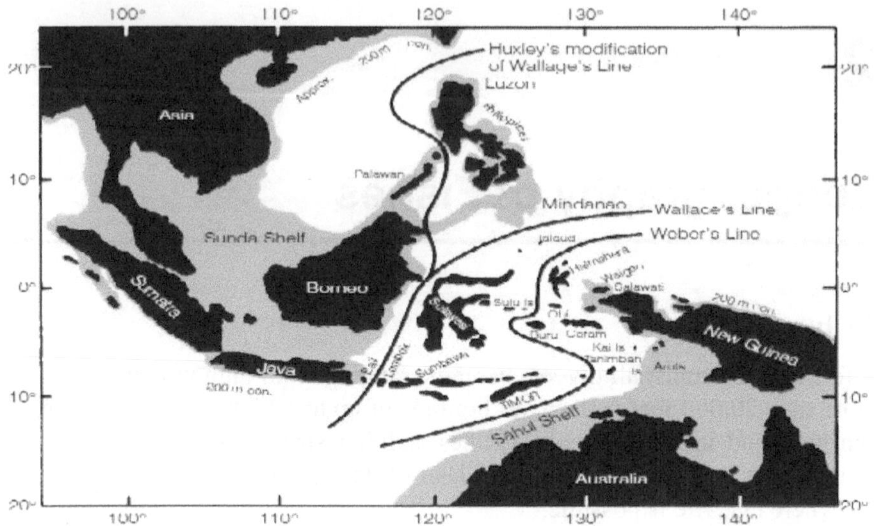

Fig. 2.1: Major freshwater fish Eco regions encompassing 1054 rivers as classified by the FishSPRICH database (Source: Brosse *et al.,* 2012).

The bathymetric distribution is the vertical distribution of animals from the highest altitude to the abyssal parts in space both on land and water can be divided into three categories:

(A) Halobiotic distribution: It is the vertical distribution of animals in marine water. e.g.:*Sardinella fimbriata, Harpodon nehereus*

(B) Limnobiotic distribution: It is the vertical distribution of animals in freshwater. e.g.: *Labeo rohita, Catla catla.*

(C) Geobiotic distribution: It is the vertical distribution of animals on land. e.g.: *Periophthalmus gracillis.*

The world's land masses can be conveniently divided into three realms based on the zoogeographical distribution of primarily freshwater fishes, particularly those of the order Cypriniformes: Megagea (Arctogaea), Neogaea and Notogaea (Fig. 2.1). These realms (Darlington, 1957) are further subdivided into the regions listed below:

2.1.1 Megagea (Arctogaea) or The Holarctic Realm

(i) **Palaearctic region:** It includes sub-regions of Europe (the Alps, the Arctic zone, the Balkans, the Black Sea, Iceland and Ireland, the Ural Mountains), Mediterranean (north of Arabia, Baluchistan, countries of the Pyrenees, north of the Sahara desert and Persia), Siberian (Northern Asia), and Manchurian (Japan and North China).

(ii) **Nearctic region:** It is made up of Canadian (the entire remainder of North America), Alleghany (the border between Canada and the United States, Colorado, Mississippi, Nova Scotia), Rocky Mountainous (the cape of St. Lucas, Rio Grande del Norte, the Gulf of California,), and Californian (British Columbia Gulf of California and Sierra Nevada) sub-regions.

(iii) **Ethiopian region:** It is divided into Malagasy (Comoro Islands, Bourbon, Madagascar and Mauritius), South African (Kalahari Desert and Limpopo Valley), West African (River Gambia and Congo), and East African (Northeast Africa, Sahara, Southern Arabia) sub-regions.

(iv) **The Oriental region:** Indo-Malayan (Borneo, Bali, Celebes, Java, Nicobar islands, Philippines and Sumatra), Indo-Chinese (Formosa, Myanmar and Southern China south of 30^0 latitudes and Thailand), Sri Lankan and Indian sub-regions are included. This region was called the Indian region by Sclater (1857) (Fig. 2.2).

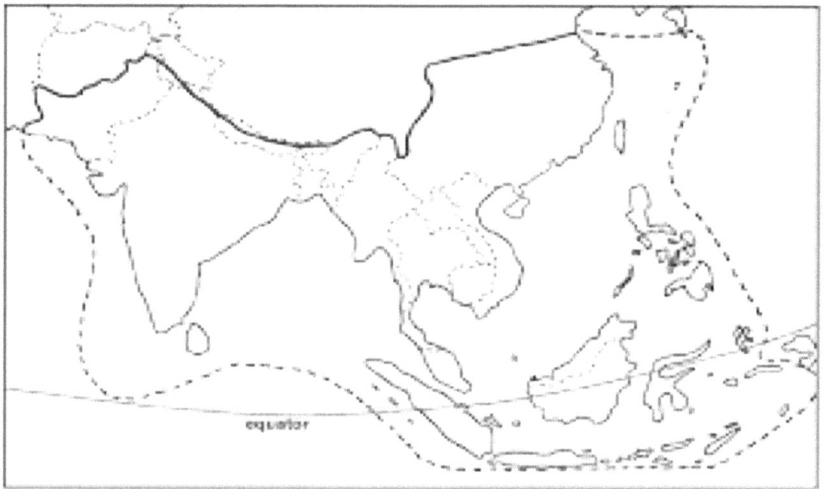

Fig. 2.2: Geographical boundaries of Oriental region (from https://www.researchgate.net/)

2.1.2 Neogaea Realm

(v) **The Neotropical region:** It is divided into Antillean (Barbuda, Grenada, Islands of Cuba, Haiti, Jamaica, Porto Rico, St. Croix), Mexican (North and South American continents), Brazilian (River Amazon and South American forest region) and Chilian (east barren plain of Patagonia, and Pampas of La Plata in the north, Tierra del Fuego swamp forest) sub-regions.

2.1.3 Notogaea Realm

(vi) **The Australian region:** It includes the Polynesian (Fiji, Islands of Sandwich, marquises, New Caledonia, New Hebrides, the Ladrone), Australian (Australia and Tasmania), and Austro-Malayan (Aru, Mysal, New Guinea and Waigiou) sub-regions.

2.2 Freshwater Fish Fauna of India

There are currently 36428 living fish species, with 18,336 valid fish species with the tag 'Habitat: freshwater.' Indian freshwater and marine water fish account for 3231 of the known species from the world. There are 2443 species of marine fish, divided into 927 genera, 230 families, and 40 orders. From 2003 to 2022, the trend of new fish species descriptions and new freshwater fish species was examined. In 2003, 330 new freshwater fish species were described compared to 119 in 222 in 2002 (Fig. 2.7). In terms of new species descriptions, the observation revealed a decreasing trend in both fish species. Inland freshwater fish are abundant in India, with approximately 940 species known from its rivers, lakes, and estuaries, of which approximately 450 are Small Indigenous Fish Species (SIFS).

The Animal Discoveries (2016) have been compiled with 313 new species and 83 new records from India. The new species comprise 27 species of fish as *Amblyceps waikhomi* Darshan et al, *Badis pancharatnaensis* Bsumatary et al, *Channa aurantipectoralis* Lalhimpura et al, *Channa pardalis* Knight, *Chaunax multilepis* Meleppura & Bineesh, *Chelidoperca stella* Matsunuma & Motomura, *Glyptothorax pasighatensis* Arunkumar, *Garra tamangi* Gurumayum & Kosygin, *Gymnothorax indicus* Mohapatra et al, *Hypselobarbus basavarajai* Arunachalam et al, *Hypselobarbus bicolor* Knight et al, *Hypselobarbus keralensis* Arunachalam et al, *Hypselobarbus kushavali* Arunachalam et al, *Hypselobarbus nilgiriensis* Arunachalam et al, *Maculabatis arabica* Manjaji-Matsumoto & Last, *Mystus catapogon* Plamoottil, *Pethia sanjaymoluri* Katwate et al, *Pseudolaguvia fucosa* Ng et al *Physoschistura chhimtuipuiensis* Lalramliana et al, *Physoschistura walongensis* Tamang & Sinha, *Puntius euspilurus* Plamoottil, *Rasbora ataenia* Plamoottil, *Schizothorax chivae* Arunkumar & Moyon, *Scomber indicus* Abdussamad, et al, *Systomus laticeps* Plamoottil, *Opistognathus ensiferus* Smith-Vaniz and *Psilorhynchus konemi* Shangningam & Vishwanath. It also includes 4 new records of Pisces: *Narcine atzi* Carvalho & Randall, *Neenchelys cheni* Chen & Weng, *Neomerinthe rotunda* Chen and *Parapercis diplospilus* Gomon.

During the year 2022, *Scomberoides pelagicus* (Abdussamad et al, 2022) was identified. The snailfish (*Pseudoliparis belyaevi*) was collected from depths greater than 8,000 meters in the Pacific Ocean south of Japan (Robert Higgs, cleveland.com, April 03, 2023).

According to Menon (1955), the fish genera endemic and exclusive to India are *Aborichthys* Chaudhuri, *Ailia* Gray, *Amphipnous* Müller, *Balitora* Gray, *Bhavania* Gray, *Ctenops* McCelland, *Horobagrus* Jayaram, *Hara* Blyth, *Horaglanis* Menon, *Jerdonia* Day, *Lepidophygopsis* Raj, *Myersglanis* Hora and Silas, *Nemachilichthys* Day, *Neotropius* Kulkarni, *Proeutropiichthys* Hora, *Parasilorhyn chus* Hora, *Somileptes* Swainson, *Sisor* Hamilton, *Silonia* Swainson and *Travancoria* Hora.

Carps and catfishes dominate the fish world (Cypriniformes). This area is home to only loaches, mullets, and mud eels. The Yangtze River in China is home to one species of Chondrostean paddlefish (*Polyodon*). Another species from this group can be found in American rivers. It is dominated by Cypriniformes, Notopteridae, Anabantidae, Osteoglossedae, cobitidae, and Nandiae. It has exclusive fauna of Mastacembelidae, Homalopteridae and Pristolepidae.

The region has 28 families of primary freshwater fishes, 12 families of catfishes, and 4 families of cypriniform ostariophysans that are endemic to the region: algae eaters (Gyrinocheilidae), loaches (Cobitidae), minnows (Cyprinidae) and river loaches (Balitoridae). Labyrinth fishes (Anabantoidei), Snakeheads (Channidae), spiny eels (Mastacembelidae) and a few cichlids are non-ostariophysan families. East of Wallace's Line, only two species of primary freshwater fish exist. All other freshwater fishes east of the line are descended from marine groups, such as catfishes and rainbow fishes.

Another review of new freshwater fish species discovered for the first time in India over the last three decades has been made. In India, 150 new freshwater fish species from 5 orders, 22 families, and 44 genera have been discovered. With 69 species representing 5 families and 17 genera, Cypriniformes was the most dominant order, followed by Siluriformes (58 species representing 8 families and 17 genera), Perciformes (19 species representing 6 families and 7 genera), Synbranchiformes (3 species representing 2 families and 2 genera), and Tetraodontiformes (1 species representing 1 family and 1 genus) (Mogaleker et al, 2015).

According to another report oriental region shares 15 orders containing 51 families of teleosts as shown in the following table:

Table 2.1: Name of orders, families and their examples of fishes of the Oriental region

Sl.No.	Name of orders	Name of families	Examples
1.	Anguilliformes	Anguillidae	*Anguilla bengalensis*
2.	Atherniformes	Neostethidae	*Ceratostethus, Neostethus*
		Phallostethidae	*Phenacostethus*
3.	Beloniformes	Belennidae	*Phenoblennis heylingeri, Xenentodon cancilla*
		Hemiramphidae	*Hemiramphus*

Sl.No.	Name of orders	Name of families	Examples
4.	Clupeiformes	Chirocentridae	*Chirocentrus dorab*
		Clupeidae	*Gadusia chapra, Gamilosa, Hilsa ilisha*
		Engraulidae	*Setipina phasa*
		Notopteridae	*Notopterus chitala, Notopterus notopterus*
5.	Cyprinodontiformes	Cyprinodintidae	*Aplocheilus*
6.	Cypriniformes	Cyprinidae	*Catla catla, Chela, Cirrhinus mrigala, Cirrhinus reba, Garra lamta, Labeo bata, Labeo gonius, Labeo rohita, Oxygaster bacaila, Tor putitora, Tor tor,*
		Cobitidae	*Botia dario, Noemacheilus chrysolaimos*
		Homalopteridae (Balitoridae)	*Homaloptera orthogoniata*
		Psilorhynchidae	*Psilorhynchus*
7.	Indostomiformes	Indostomidae	*Indostomus paradoxus*
8.	Mastacembeliformes	Mstacemebildae	*Macrognathus, Mastacembelus*
9.	Mugiliformes	Mugilidae	*Rhinomugil, Sicamugil*
10.	Myctophiformes	Harpodontidae	*Harpadon*
11.	Ophidoformes	Ophidiidae	*Ophidis*
12.	Ophiocephaliformes	Ophiocephalidae	*Channa gachua, Channa marlis, Channa punctatus, Channa striatus*
13.	Perciformes	Anabantidae	*Anabas testudineus, Colisa, Sandella*
		Belontidae	*Betta, Macropodus*
		Centropomidae	*Ambasis, Chanda*
		Chichlidae	*Tilapia, Etroplus*
		Gobiidae	*Acentrogobius, Bathogobius, Glossogobiius*
		Heleostomidae	*Heleostoma*
		Kurtidae	*Kurtus indicus*
		Luciocephalidae	*Luciocephalus pulcher*
		Nandidae	*Nandus, Badis*
		Osphronemidae	*Osphronemus gorami*
		Rhyacichthydae	*Rhyacichthys*
		Sciaenidae	*Macrospinosa, Pama, Sciama*
		Trypauchenidae	*Centrotrypauchen*
		Toxitidae	*Toxotes*

Sl.No.	Name of orders	Name of families	Examples
14.	Siluriformes	Aysidae	*Akysis*
		Amblyciptidae	*Amblyceps*
		Arlidae	*Osteogeniosus, Tachysuru*
		Bagridae	*Chandramara, Mystus, Rita,*
		Chacidae	*Chaca chaca*
		Claridae	*Clarias, Horaglanis*
		Cranoglanididae	*Cranoglania*
		Heteropneustidae	*Heteropneustes fossilis*
		Olyridae	*Olyra*
		Pangasidae	*Pangasius pangasius*
		Plotosidae	*Plotosus*
		Schilbeidae	*Ailia, Eutropichthys, Pseudotrpus, Silonia,*
		Siluridae	*Ompok, Wallago*
		Sisoridae	*Exostoma, Horaglanis, Gagata, Hara, Sisor,*
15.	Symbranchiformes	Amphipnidae	*Amphipnous cuchia*
		Chaudhuridae	*Chaudhuria caudata*
		Symbranchidae	*Monopterus*

Five of the 150 species are critically endangered maximum are vulnerable, and two are near threatened. Mizoram had the maximum number of emerging species (32) followed by Manipur (29), Kerala (28), Arunachal Pradesh (17), West Bengal (14), and Karnataka (13).

3

Ancestry of Fishes

3.1 Introduction

Fish are the only true aquatic vertebrates. These exploit every nook and cranny of the water domain. They are notable for their vast numbers and perplexing forms. In the phylogenetic history of vertebrates, this is the first group of animals to develop biting jaws. The origin, evolution and interrelationships of fishes are not fully understood due to incomplete fossil records and a scarcity of primitive types. The early appearance of bone, cartilage and enamel-like substances was crucial in the evolution of fish.

3.2 Origin

Fish most likely evolved from a coral-like sea squirt, the larvae of which are strikingly similar to early fish. Fish evolved approximately 530 million years ago during the Cambrian explosion of multicellular life. *Pikaia* (Fig. 3.1), *Haikouichthys* and *Myllokunmingia* were likely the first ancestors of fish. *Pikaia gracilens* is a transitional fossil between invertebrates and vertebrates, and it is thought to be the first known chordate. *Haikouichthys* is also a transitional species from invertebrates to vertebrates. *Myllokunmingia* is a chordate, though some argue it is a vertebrate. These allow for long periods of evolutionary divergence as well as the origin and extinction of major phyletic lines. There is no convincing evidence that gnathostomes evolved from agnathans, or that the two groups evolved independently from a common ancestor. The Devonian period is referred to as the "age of fishes."

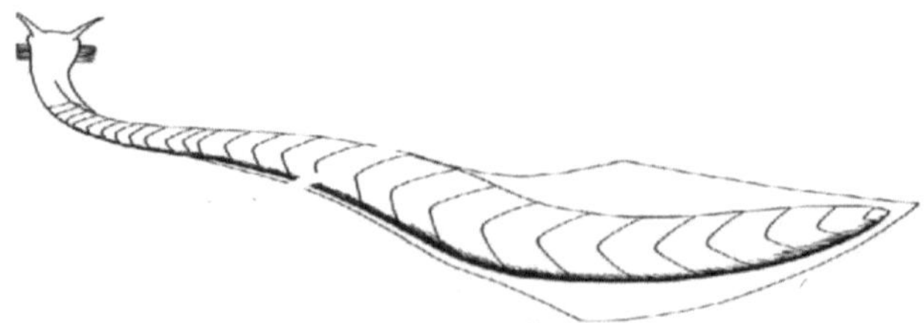

Fig. 3.1: *Pikaia* (first ancestors of fish) (from www.prehistoric-wildlife.com)

The fossil of *Jamoytius* is considered a representative of the earliest chordates. It has:

A. Length about 7 inches.
B. Unarmoured body possessing notochord and lateral fin folds.
C. No gill pouch and endoskeleton.

During the Silurian period, the following groups of fish were established:

A. Agnatha (primitive fishes)
B. Pro-gnathostomes (earlier jaw fishes)
C. Eu-gnathostomes (True jaw fishes)

Conodonts were primitive jawless eels with cone teeth. They first appeared at 520 million years ago and became extinct 200 million years later. Conodonts were once thought to be vertebrates similar to modern hagfish and lampreys, but phylogenetic analysis shows that they are more derived than either of these groups.

3.3 Agnatha

Cephalospidomorphs are the most primitive forms. Paleospidomorphs combine to form ostracoderms (bony-skinned fish). Ostracoderms first appeared in 500 million years ago in the Cambrian period, and they lived in freshwater during the Silurian and Devonian periods. According to Barrington, ostracoderms descended from urochordates. The first cephalospidomorhs were found in Osteostraci and Anaspida. Cephalospidomorhs are characterised by:

A. Finless heavy armoured and dorsoventrally flattened body.
B. Bottom dweller lacking jaws
C. A bony vertebral column was absent but the body was well protected by external bony plates.
D. They were filter feeders

Agnathans were divided into

3.3.1 Ostracoderms: The term does not appear in classification because of its paraphyletic nature and lack of phylogenetic meanings.

a. Euphanerida (e.g. *Jamoytius*): Fossil that lived 400 million years ago from the Silurian period and was discovered in England (White, 1946). The presence of myotomes in *Amphioxus* suggests that Amphioxus may be a degenerate descendent of a *Jaymoytius*-like ancestor (Fig. 3.2).

Fig. 3.2: *Jamoytius*

b. Osteostraci, Cephalaspidomorphi or Cephalaspida (e.g. *Cephalaspis* (Fig. 3.3), *Hemiccyclapsis, Ateleapsis*): They lived from the Silurian to the Devonian period. They have ten pairs of gill apertures, an endoskeleton probably cartilaginous and two semi-circular canals in the ear.

Fig. 3.3: *Cephalaspis* (from pinterest.com)

c. Coleolepida (e.g. *Coleolepis*, *Thelodus* (Fig. 3.4), *Lanarkia*): They lived during the late Silurian to the Devonian period. It seems to be the larval stages of the ostracoderms.

Fig. 3.4: *Thelodus* (from https://alchetron.com)

d. Heterostraci, Pteraspidomorphi or Pteraspida (e.g. *Pteraspis rostrata* (Fig. 3.5) and *Drepanaspis*): They appeared during the Cambrian period. Fossils are known from the Ordovician to the late Devonian period. They have prominent lateral lines, one pair of gill apertures and widely separated eyes.

Fig. 3.5: *Pteraspis rostrata* (from https://www.deviantart.com)

e. Thelodonts (Fig. 3.6) (nipple teeth): These are extinct jawless fish with distinctive scales instead of large armour plates. There is some disagreement about whether these are a monophyletic grouping or distinct stem groups to the major lines of jawless and jawed fish. The fish lived in both freshwater and marine environments, first appearing during the Ordovician and dying during the late Devonian Frasnian-Famennian extinction.

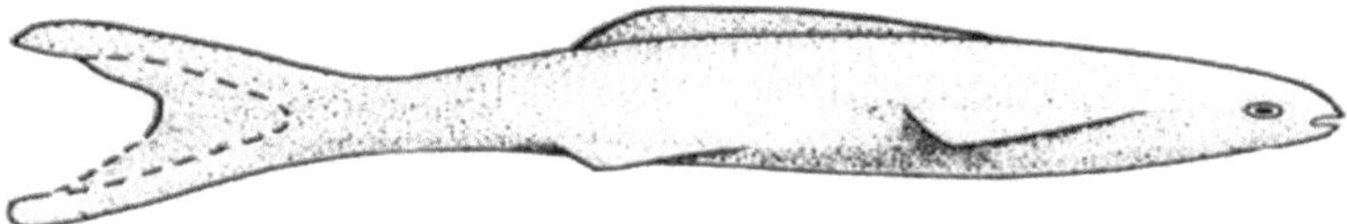

Fig. 3.6: *Thelodont* (from https://vanderbrugghenfossils.org)

f. Anaspida (e.g. *Birkenia, Pterolepis, Rhyncholepis* (Fig. 3.7)): They lived during the Silurian to the Devonian period. They were active swimmers and surface feeders, with eight pairs of gill apertures, pineal aperture between the eyes and bear longitudinal rows of plate-like bony scales or anaspid scales. They are regarded as the ancestors of lampreys.

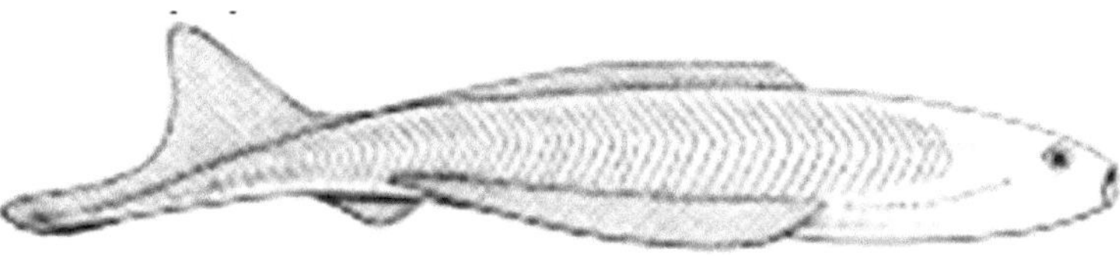

Fig. 3.7: *Rhyncholepis* (from https://slideplayer.com)

The ptyctodontids (beak-teeth) were an extinct monotypic order of unarmoured placoderms from the beginning to the late Devonian period. *Dunkleosteus* is a type of arthrodire placoderm that lived between 380 and 360 million years ago. *Titanichthys* is a massive, erroneous marine placoderm that lives in shallow seas. *Materpiscis* (mother fish) is a 380 million-year-old ptyctodontid placoderm.

3.3.2 Cyclostomata (Marsipobranchia)

Recent molecular data from ribosomal RNA (rRNA) and mitochondrial DNA (mtDNA) strongly supports the theory that cyclostomes are monophyletic. Cyclostomata (lamprey and hagfish) have degenerated descendants of the armoured agnathans. These have:

A. Eel-like body with a cartilaginous skeleton
B. Parasitic or scavenger habit and
C. No paired appendages, fins or limbs

Hagfishes differ from lamprey in having:

A. Nasal opening at the extremity of the mouth
B. The pituitary gland opens to the pharynx.
C. 3-13 pairs of gill pouches
D. Rudimentary branchial skeleton
E. One semi-circular canal
F. Direct development

According to Greenwood (1975), hagfishes are often classified with lampreys and cephalaspidomorphs. Studies by Stentio (1920) on ostracoderms show that cephalaspidomorphs are related to hagfish and lamprey.

Primitive fish have bony coverings that serve as a defensive mechanism. During evolution, the retrogressive development of the skeleton resulted in its loss, as seen in modern Cyclostomata.

When jawed vertebrates appeared during the Silurian and Devonian periods, ostracoderm became highly specialized, and there appears to be no relationship between Agnatha and Gnathostomes.

3.4 Pro-Gnathostomata

The pro-gnathostomes evolved from agnatha by extending the small terminal mouth posteriorly and converting the first visceral arch into a primitive jaw. The placoderms were the first gnathostomes to appear during the Devonian period. These were thought to be failed ancient experiments in gnathostome evolution. Except for Acanthodi, all of these became extinct by the end of the Devonian period (spiny sharks). They first appeared around 420 million years ago, during the late Silurian period, and were among the first fish to develop jaws. They have characteristics of both cartilaginous and bony fish, but they are not true sharks. *Climatius* represented the earliest Paleozoic fish, the Acanthodi. *Climatius* lived from the late Silurian to the Permian period. *Climatius* has rhombic scales, large eyes, and pectoral and pelvic fins that are paired (Fig. 3.8). Acanthodi has:

Fig. 3.8: *Climatius* (from https://prehistoric-wiki.fandom.com/wiki)

A. Superficially shark-like body
B. Spines to support the fins (except caudal)

C. Small and flat bony scales

D. Undeveloped cephalic and thoracic bony shields.

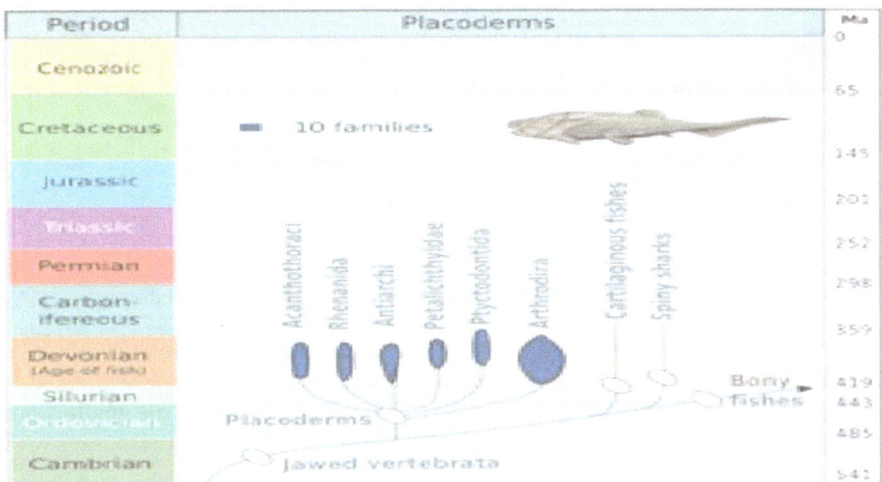

Fig. 3.9: Evolution of Placoderms (from Benton, 2005)

Acanthodian shows a mixture of primitive and specialised features. These resemble the ancestral stock that gave rise to bony fishes. Bony shields covered the heads and anterior parts of the bodies of Coccostei or Arthrodira (*Coccosteus* (Fig. 3.3), *Dinichthys* and *Titanichthys*) and Petrichthyes or Antiarchi (*Petrichthys, Bothriolrpis*). Stegoselachians (*Gemuendina*) were less armoured fish. These superficially looked like modern rays. However, none of these groups is in the main evolutionary line of higher fishes.

Fig. 3.10: *Coccosteus* (from https://www.wikidata.org/wiki)

Macropetalichthyes (*Macropetalichthyes, Lunaspis* (Fig. 3.11)) lived during the Devonian period and resembled Arthrodira.

Fig. 3.11: *Lunaspis*
(from https://en.wikipedia.org)

Fig. 3.12: *Paleospondylus*
(from https://en.wikipedia.org)

Paleospondyli (*Paleospondylus* (Fig. 3.12)) is a small mid-Devonian period fossil with a 5 cm elongated body. It had a complete hyoid arch and an ossified vertebral column with a strong centrum. According to Myo-Thomas, it is a true gnathostome and its jaw apparatus resembles that of Acanthodi.

3.5 Eu-Gnathostomata

Eugnathostomes include the following groups:

3.5.1 Chondrichthyes (cartilaginous fishes)

The Chondrichthyes are descendants of some placoderm ancestors. They appeared in fossil form in the early Devonian period, developed well, and continue up to the present time. They were characterised by steady improvements in foraging and locomotion. They have:

A. Cartilaginous skeleton

B. Marine habitat

Pituriaspida (hallucinogenic shield) lived in estuaries around 390 million years ago. It contains two bizarre species of armoured jawless fish with tremendous nose-like rostrums. They lived in estuaries around 390 million years ago (Fig. 3.13).

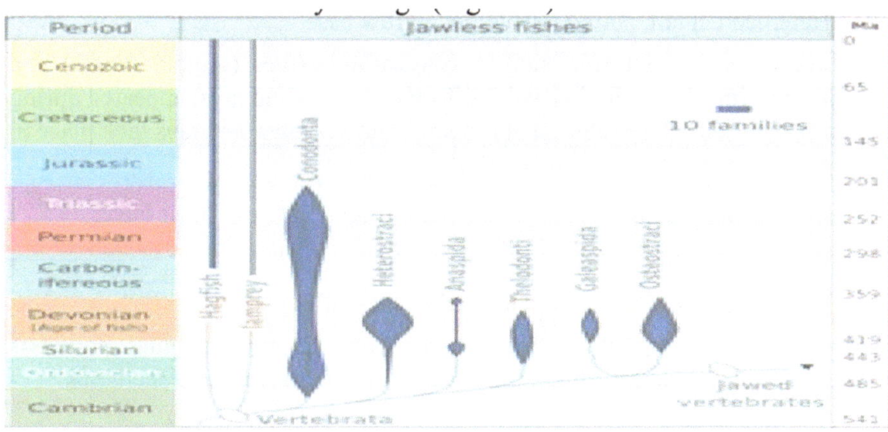

Fig. 3.13: Evolution of Jawless fishes (from Benton, 2005)

Sharks are more primitive than bony fishes, but they evolved later than the bony fishes. *Cladoselache* (Fig. 3.14) and *Xenacanthus* (*Pleurocanthus*) are the earliest sharks. *Cladoselache* was the first abundant early chondrichthyan, related to modern sharks (though probably closer to holocephalans like chimaeras) appearing about 370 million years ago. It lived during the upper Devonian period. It has:

Fig. 3.14: *Cladoselache* (from https://www.sciencephoto.com)

A. Amphistyloic skull
B. The Shark-like with a diphycercal tail
C. The long dorsal fin and may have a second dorsal fin
D. The caudal fin is externally symmetrical

The first advanced sharks appeared during the Mesozoic era, when the skull became hyostylic, allowing the jaws more freedom. Sharks evolved alongside modern sharks (e.g., *Scoliodon, Sphyrna, Carcharinus, Rhinodon*) and rays (e.g., *Pristis, Rhinobatus, Dasyatis, Torpedo*) beginning in the Mesozoic era. Modern sharks possess five lateral gill slits and hyostylic jaw suspensions. Sharks are predaceous fish that are active, fast swimmers and have the following characteristics:

A. Torpedo-like body

B. Heterocercal tail, paired fins with a narrow base.

The three orders of Selachii, namely Carcharhiniformes (ground sharks) and Lamniformes (mackerel sharks) and Rajiformes (skates and rays) appeared during the Jurassic Period. The evolution of modern sharks, skates, and rays took place in the Cretaceous Period and Cenozoic Era.

The skates and rays evolved specialized bottom-up lifestyles. These have:

A. Fused and enlarged pectoral fins

B. The narrow tail, ventrally placed whip-like gill-slits

A fossil fish, *Palaeospondylus* (Fig. 3.15), from the mid-Devonian period is probably not a placoderm, although it is sometimes classed with placoderms.

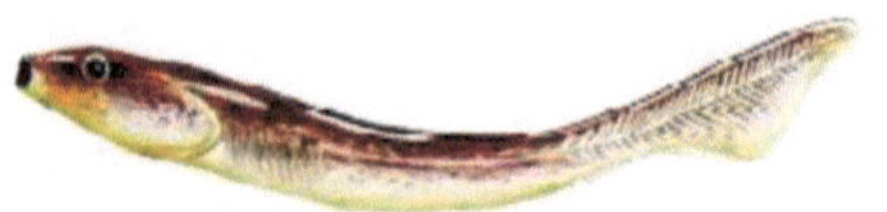

Fig. 3.15: *Palaeospondylus* (from https://allthatsinteresting.com)

Fig. 3.16: *Heterodontus* (from https://animaldiversity.org)

Some genera like *Heterodontus* (Fig. 3.16), *Chlamydoselache* occupy and an intermediate position between the primitive cladoselachians and advanced sharks. The Neoselachi (Sharks, skates and rays) probably evolved from a *Cladoselache*-like ancestor. These are regarded as "living fossils". The heterocephali (living forms *Chimaera, Callorhynchus,* and *Harriotta* and extinct forms *Squaloraja* and *Myriacanthus*) are a specialized group of deep-sea cartilaginous fishes.

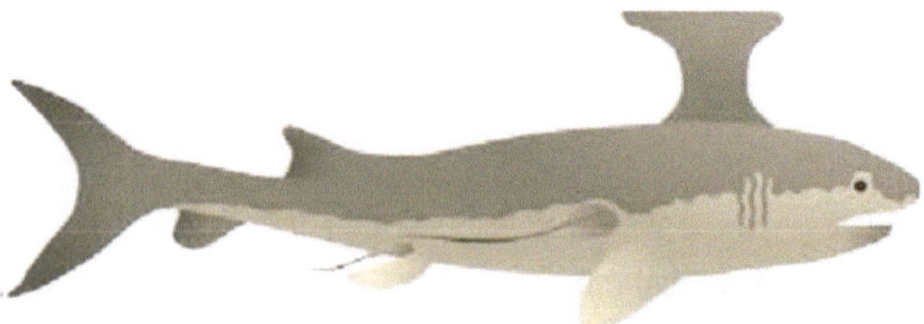

Fig. 3.17: *Stethacanthus* (from https://www.pinterest.com)

Stethacanthus (Fig. 3.17) from the Carboniferous period with unusual fins may have been used in mating rituals as a result of adaptive radiation in cartilaginous fishes. *Falcatus* is a cartilaginous fish with cladodont teeth that lived 335-318 million years ago. *Belantsea, Orodus, Chondrenchelys* and *Edestus* are examples of Carboniferous cartilaginous fishes.

Cladoselachiformes and Cladodontiformes became extinct by the end of the Permian period, while the freshwater order Xenacanthiformes lasted until the end of the Triassic period. The final Devonian order, Heterodontiformes, still has surviving members.

Fig. 3.18: *Helicoprion* (from https://blog.everythingdinosaur.com)

Helicoprion (Fig. 3.18) is related to extant chimerids, which vanished during the Permian period, with *Caseodus* and *Fadenia* surviving into the early Triassic period. *Triodus* and *Acrolepis* lived from the Carboniferous to the Triassic periods, while *Acanthodes* and *Palatinichthys* are Permian extinct fishes.

3.5.2 Oestichthes

During the mid-Devonian period, bony fishes appeared. However, some authors believe that bony fish originated around 400 million years ago during the Ordovician period. They were also characterised by steady improvements in foraging and locomotion. These have:

A. Bony endoskeleton and opercular body
B. Developed air-bladder

Romer (1959) divided the osteichthyes into two groups: Sarcopterygii and Actinopterygii. Sarcopterygii are ancient in origin; their first remains appeared in the early Devonian strata of Germany. Some authorities contend that the rhipidistians (one of the groups of sarcopterygians), gave rise to the amphibians; however, other authorities believe that tetrapods evolved from one of two other groups, the coelacanths and the dipnoans.

3.5.2.1 Dipnoi

Dipnoans originated during the mid-Devonian period and flourished well in the Permian and Triassic periods. It includes an ancestor known as *Dipterus*. *Dipterus* (two wings) is an extinct genus of lungfish from 376 to 361 million years ago. It has:

A. Body of stout cosmine or cycloid scales
B. Two small dorsal fins
C. Heterocercal tail
D. Numerous cranial teeth
E. Fusiform body

Dipnoi has certain specializations that give rise to later Dipnoi. *Ceratodus*, a fossil dipnoi of the Mesozoic era was evolved. Dipnoi is represented by *Protopterus*, *Neoceratodus* and *Lepidosiren*. According to Colbert (1969), *Neoceratodus* is a direct descendant of *Ceratodus*.

3.5.2.2 Crossopterygii

Crossopterygii (lobe-finned fish) first appeared around the mid-Devonian period. It was represented by Osteolepis. Like *Dipterus*. *Osteolepis* has:

A. Fusiform body with heterocercal tail
B. Cosmoid scale, internal nostril and ossified skeleton

Fig. 3.19: *Dipterus* (from https://www.shutterstock.com)

Fig. 3.20: *Osteolepis* (from https://en.wikipedia.org)

The *Dipterus* (Fig. 3.19) and *Osteolepis* (Fig. 3.20) had a common ancestral stock from which the Dipnoi and Crossopterygii arose. Rhipidistia (fan-webs, Osteolepitati, or Osteolepiformes) and Coelacanth evolved from pro-crossopterygians. A typical Rhipididtian (e.g., *Osteolepis, Eusthenopteron*) was a direct descendant of the first amphibians.

Crossopterygii and Dipnoi are considered independent sub-classes of Oestichthyes and Brachiopterygii is recognized as an independent sub-class (Bertin and Arambourg, 1958).

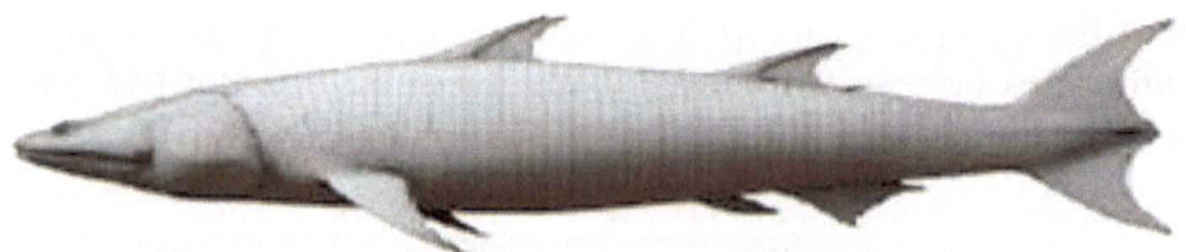

Fig. 3.21: *Megamastax* (from https://en.wikipedia.org)

Megamastax (Big Mouth) lived in China around 423 million years ago during the late Silurian period. *Guiyu oneiros,* the earliest known bony fish, has both ray-finned and lobe-finned features, though an analysis of its totality of features places it closer to lobe-finned fish. *Psarolepis* (speckled scale) existed between 397 and 418 million years ago. *Holoptychius* is a porolepiform fish that existed from 416 to 359 million years ago. *Laccognathus* (pitted jaw) was an amphibious fish that lived 398-360 million years ago. *Hyneria* was a predatory fish that lived 360 million years ago. Rhizodonts lived until the end of the Carboniferous period, 377-310 million years ago.

3.5.2.3 Brachipterygii

Palaeoniscoids gave rise to brachipterygii. These are confined to tropical Africa. They are "living fossils" possessing both primitive and specialised features. These are represented by *Polypterus* (Fig. 3.22) and *Callamoichthyes*. It has:

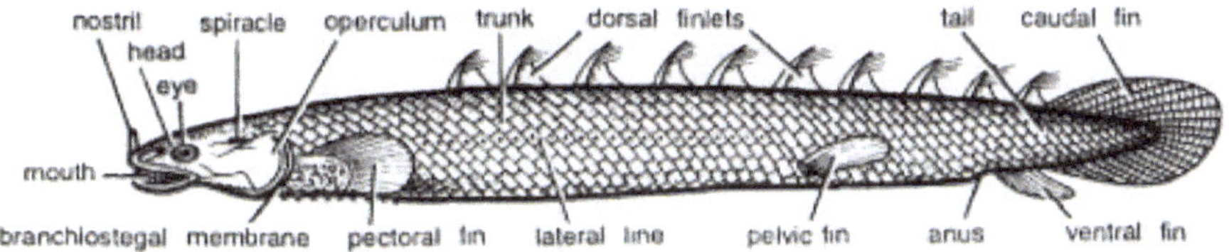

Fig. 3.22: *Polypterus* (from https://www.faunafondness.com)

A. Slender body with ganoid or rhombic scales
B. Dorsal fin with eight or more dorsal fin-lets

It is usually classified as an order of Actinopterygii. It shows some characteristics of crossopterygii and differences from actinopterygii. Therefore, it is treated as a separate suborder.

3.5.2.4 Actinopterygii

In existence for about 400 million years, since the Early Devonian, it consists of some 42 orders containing more than 480 families, at least 80 of which are known only from fossils. The extinct *Andreolepis* includes the earliest known ray-finned fish, *Andreolepis hedei*, which appeared in the late Silurian around 420 million years ago (Fig. 3.23).

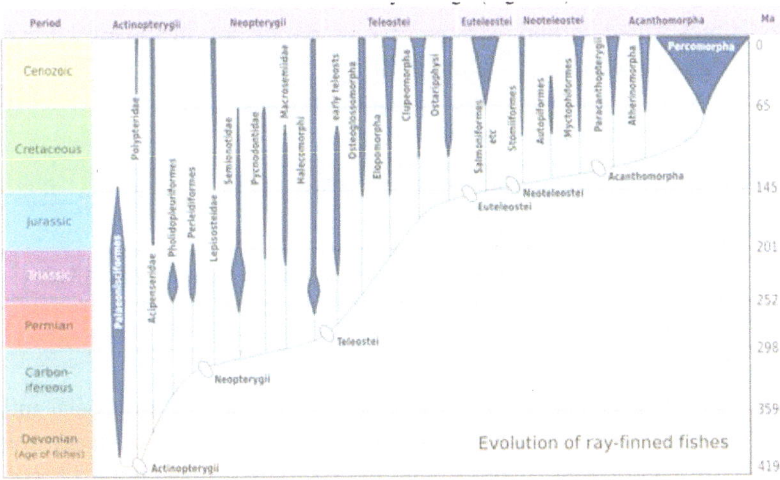

Fig. 3.23: Evolution of Bony fishes (from Benton, 2005)

There are over 50 species of *Saurichthys* in the Triassic period. Some of them show viviparity in females and gonopodia in males. Ray-finned fish such as *Perleidus, Robustichthys* (Fig. 3.24), *Semionotus* and *Pholidophorus* were also discovered in the Triassic period.

Fig. 3.24: *Robustichthys*
(from https://www.pinterest.com)

Fig. 3.25: *Chirolepis*
(from https://alchetron.com)

The fish fauna of the Early Triassic was uniform, but during the Triassic, adaptive radiation of ray-finned fishes occurred, laying the groundwork for many modern fishes. Fish and marine reptiles were the most common vertebrates in the seas during the Jurassic period.

Chirolepis (Fig. 3.25) had:

A. Small-sized body with heterocercal tail
B. Incompletely ossified vertebrae

A *Chirolepis*-like ancestor gave rise to the later Actinopterygian. These passed through three stages of development and are represented by Chondrostei or cartilaginous bony fish (e.g. *Acipenser, Polypterus*), Holostei (earlier bony fish) and Teleostei (final bony fish). The Chondrostei may have arisen as early as the Early Devonian, increased in numbers and complexity until the Permian, and thereafter became almost extinct by the mid-Cretaceous period. Several orders of the chondrosteans developed characteristics that approached the holostean level and are sometimes called sub-holosteans. Five orders of holosteans are known, with their greatest evolutionary radiation occurring during the Triassic, Jurassic, and Cretaceous periods, when the chondrosteans were declining and the teleosts were just beginning to expand.

Chirolepis also gave rise to *Acipenser. Acipneser* has:

A. No heavy scales and spiracles

During the Triassic period, *Chirolepis* ancestors gave rise to *Redfieldia*. It has:

A. Shorter jaw and a heterocercal tail
B. Less heavy scales

The mid-Triassic *Foreyia* and *Ticinepomis* also include the extant coelacanth *Latimeria*. The actinistia (Coelacanthiformes), represented by *Latimeria* was found in both fresh and marine water during the Devonian period but no fossils appeared after the Mesozoic era.

Redfieldia is regarded as a connecting link between Chondrostei and Holostei. At the end of the Triassic period, the primitive chondrostei were gradually replaced by Holosteri. Holostei includes *Amia* and *Lepidosteus*. These have:

A. Vascular lungs for reparation
B. No spiracles

The holostei were gradually replaced by Teleostei. The first teleost (*Lepiolepis*) appeared in the Jurassic period. It was the connecting link between holostei and teleostei. These have:

A. Homocercal caudal fin
B. Vascular lung as air-bladder

By the Cretaceous period, teleosts were well established. They radiated into several adaptive zones. These represent a successful group of fish. According to Greenwood et al, the Teleosts are regrouped into eight superorders in three divisions. Each division contains a distinct phyletic lineage derived from Holostei.

Bony fishes experienced extensive adaptive radiation during the Coenozoic era. Fishes account for more than half of all living vertebrate species (approximately 32,000 species), ranging from snow minnows at elevations above 4,600 meters to flatfishes in the Challenger Deep, the deepest ocean trench at approximately 11,000 meters. Fishes of various species are the primary predators in the majority of the world's freshwater and marine bodies.

Division I includes eels and eel-like fishes as well as herring-like fishes. These are descended from a primitive elopiform ancestor.

Division II contains Osteoglossiformes and Mormyriformes.

Division III bears the bulk of living teleosts. Elopids and salmonids are types of living primitive teleosts that are sufficiently generalized to serve all major teleost fish types. Teleosts are extremely diverse in terms of anatomical form and habitat. They can be divided into about 12 super orders each with distinct evolutionary significance.

3.6 Mass extinction of fishes

The Late Devonian extinctions were critical in shaping fish evolution. Throughout history, the following extinction events have had a significant impact on fish.

1. The Ordovician-Silurian extinctions: The first ones resulted in the extinction of many species.
2. The late Devonian extinction: It caused the extinction of ostracoderms, placoderms, and other fish.

3. The Permian-Triassic extinction of the spiny sharks.
4. The Triassic-Jurassic extinction of the conodonts.
5. Cretaceous-Paleocene extinction: Extinction of fish species and stocks and
6. Holocene extinction: Extinction of fish species and stocks.

4

General Characters of Fishes

While describing fish, we need complete general information about the different specimens of fish included in the particular class or grade. A flow sheet of the general characters used in this concern may be tabulated as:

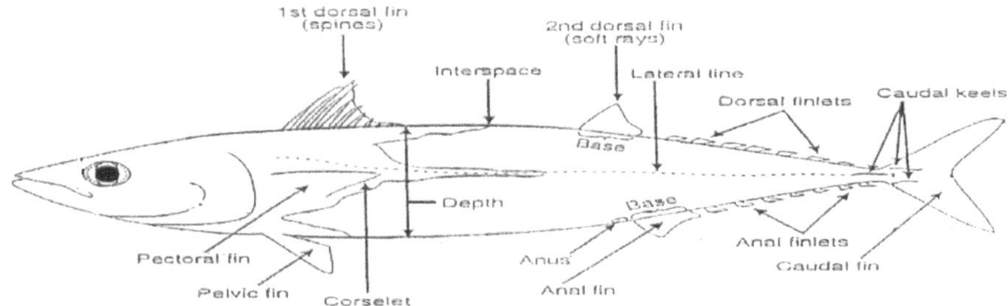

Fig. 4.1: A Hypothetical fish to show various characters

1. **Origin of the term**: The term Pisces comes from the old version of the plural of the Latin word piscis or singular word piscus meaning "fish."
2. **Literal meaning and definition**: It describes either the creature or the meat from the creature. It is a streamlined, cold-blooded aquatic vertebrate with scales and gills,
3. **Origin and evolution**: The first fish appeared around 530 million years ago during the Cambrian period and then underwent a long period of evolution so that, today, they are by far the most diverse group of vertebrates. The Devonian Period is known as the "Age of Fishes".
4. **Total number of extant and extinct species**: According to FishBase about 34,800 species of fish had been described as of February 2022, the The International Union for Conservation of Nature (IUCN) (2016) lists 65 extinct fish species, 87 possibly extinct fish species, and six extinct in the wild fish species. Some 37% of sharks and rays threatened with extinction (IUCN, 2021)
5. **Geographical distribution**: The distribution of fishes is because of the geological history and development of Earth as well as the ability of fishes to undergo evolutionary change and to adapt to the available habitats. The

pattern of distribution may be continuous, discontinuous and restricted; uniform or sporadic.

6. **Habit and Habitat**: Fish often make daily movements between feeding and resting places, seasonal movements to summer and winter habitats and annual movements to traditional spawning areas. Fish live all over the world, in the cold water of the Arctic as well as in warm tropical water around the equator; seas and oceans; rivers and streams; lakes, ponds and Reservoirs. Their habitats may include sandy sea beds, coral reefs and underwater forests. Tropical fish can be found both in freshwater like lakes and saltwater like oceans. Fish might be found in water scarcely deep enough to cover their backs, or they might swim in unfathomable depths. They may scour the bottom, roll on the surface or hover anywhere between.
7. **Shape and Symmetry**: This characteristic fusiform or boat shape is related to being quite energy efficient for swimming. They may also be filiform (eel-shaped) or vermiform (worm-shaped), compressed (laterally thin) or depressed (dorso-ventrally flat), elongated or flat. Fish show variation in shape because of elongation, compression, and depression. Fishes are generally bilaterally symmetrical but flounder is asymmetrical.
8. **Size and weight**: The length of fish ranges from less than 10 mm (0.4 inches) to more than 20 metres (60 feet) and in weight from about 1.5 grams (less than 0.06 ounce) to many thousands of kilograms.
9. **Level of Organization**: Fishes show tissue-organ-system organization.
10. **Body Plan**: The body is divided into a head, trunk and tail.
11. **Body cavity and Coelom**: The trunk contains two coelomic cavities as pericardial and abdominal cavities, separated by a transverse septum and lined with peritoneum.
12. **Integumentary (=Exoskeletal) system**: Skin is important for the health and welfare of the fish. Fish scales are part of their integumentary system, and are produced from the mesoderm layer of the dermis,
13. **Endoskeletal system**: It may be epineuralia, epicentralia, epipleuralia and myorhabdoi. It contains bones, cartilage, tendons and ligaments. Fish bones support core muscles without inhibiting their motility. The number of bones is 130 or so. It consists of:
    (a) Skull bones: Anapsid type lacking fossa and generally 22-25, Orbit, Upper jaw, a mandible (lower jaw), Clavicle, Osteichthyes have hinged jaws but hagfish and lampreys are jawless.
    (b) Girdles: In cartilaginous fishes, the pelvic girdle contains a puboischial bar and the pectoral girdle is U-shaped
    (c) Ribs and radial cartilage

(d) Fin rays: Caudal, anal, dorsal, ventral. The dorsal and the anal fin are supported by spines that may or may not be, connected to the vertebrae. The tail is supported by the caudal vertebrae (Hypurals, Epurals and Urostyle).

(e) Vertebrae: It is divided into trunk and caudal vertebrae. Caudal vertebra has haemal arch.

(f) Spines: Neural spines, haemal spines

(e) Other bones: Opercular, Clavicle,

14. Muscular system

Fish have more muscle than any other vertebrate. A male salmon or tuna can be nearly 70% muscle, the reddish meat colour found in salmon and sea trout is due to the astaxanthin.

(a) Structure and chemical composition of muscles: It is W or V-shaped and striated, consists of sarcoplasm and a number (up to 1,000 of myofibrils. The cell is surrounded by sarcolemma. It lacks tendons.

(b) Types and differences of muscles: A major white muscle, a superficial red muscle in the cross-sectional area near the middle and posterior regions and an intermediate pink muscle.

(c) Molecular basis/mechanism of muscle contraction: Muscle contraction starts when a nervous impulse sets off a release of Ca^{+2} from the sarcoplasmic reticulum to the myofibrils.

15. Movement and locomotion

(a) Organ/Organelle of movement and locomotion: The caudal fin is the primary appendage used for locomotion in many fishes. The dorsal fin, anal fin, caudal fin, pectoral fins and pelvic fins allow fish to balance and steer while swimming. Myotomes help in this process.

(b) Types of movement and locomotion (Breder, 1926):

i. Anguilliform: wave passes evenly along a long slender body; seen in highly flexible fishes.

ii. Carangiform: wave is concentrated near the tail, which oscillates rapidly as in trout, rohu, mrigal etc

iii. Ostraciiform: The body is inflexible being enclosed in a hard boxlike protective sheath, and locomotion is due to undulations of the caudal fin as in Ostracion, Tetraodon, Diodon, etc. Besides

iv. Sub-carangiform: There is a greater increase in wave amplitude along the body is done by the rear half of the fish as in trout.

v. Thuniform or autapomorphy: It is reserved for high-speed long-distance swimmers like tuna. Here, the lateral movement occurs in the region connecting the main body to the tail

(c) Mechanism of movement and locomotion: It may be swimming, walking, burrowing or flying. Most fishes generate thrust using lateral movements of their body and caudal fin, but many species move mainly using their median and paired fins. Median fin propulsion and dynamic lift are responsible for locomotion.

16. Digestion and Nutrition

(a) Feeding habit: Either herbivores, detritus feeders, carnivores or omnivores. They may be bulk, filter, deposit and fluid-feeder.

(b) Organ of digestion: It includes mouth, teeth and gill rakers, oesophagous, stomach, pylorus, pyloric caeca, pancreatic tissue, liver, gall bladder, intestine and anus.

(c) Mechanism of digestion: Fish ingest food through the mouth and break it down in the esophagus. In the stomach, food is further digested and, in many fish, processed in finger-shaped pouches called pyloric caeca, which secrete digestive enzymes and absorb nutrients.

17. Respiratory system

(a) Organ/Organelle of respiration: Gills, various air-breathing organs,

(b) Respiratory pigments: Haemoglobin (a few Antarctic ice fish lack haemoglobin)

(c) Types of respiration: Aquatic and aerial respiration

(d) Mechanism of respiration: Bimodal respiration.

18. Blood vascular system

(a) Structure and chemical composition of blood: The blood volume ranges from 1.5% to 5% of total body weight. Fish generally have hemoglobin that is responsible for the red color. Erythrocytes (20000-300000mm^3) are generally oval in *Labeo rohita* and *Labeo calbasu*. Leucocytes (20,000-150,000mm^3) may be differentiated as granulocytes or agranulocytes. Lymphocytes are 80-90% of the total white blood cells. Monocytes have either small quantities or may be absent. Neutrophils are 5-9% of the total white blood cells. Platelets are rounded, oval or spindle-shaped.

(b) Heart and blood vessels: contains the sinus venosus, atrium, ventricle, and conus or bulbus in series. The capillary arrangement is called a portal system and maybe a hepatic portal system or renal portal system.

(c) Mechanism of blood circulation: Closed blood vascular system. There are a significant number of capillary or sinusoidal systems in the arterial or venous flow of blood.

19. Lymphatic system

The lymphatic system originates from the venous system. It shows increasing complexity through Elasmobranchii to Teleostomi.

(a) Structure and chemical composition of lymph: The caudal heart is regarded as part of the lymphatic system

(b) Lymphatic organs: The thymus, kidney and spleen are the major lymphoid organs of teleosts.

(c) Mechanism of lymph circulation: A lymph heart pumps lymph in lungfishes

20. Immune system

It may be innate (non-specific) and adaptive (specific).

(a) Cellular components of Immune system: The cellular components of the fish's innate immune system consists of many different types of cells such as monocytes/macrophages, granulocytes as mast cells/eosinophilic granule cells, and neutrophils, dendritic cells, and natural killer cells (NK cells).

(b) Antigen-antibody interaction: The immune system is based on the fundamental receptors and pathways.

21. Excretory system

(a) Organ/Organelle of excretion: The primary excretory organ is the kidney. In fishes the digestive tract, skin and gills.

(b) Types of excretory products: Ammonia

(c) Mechanism of excretion: Diffusion process across the body surface through gills and into the surrounding water.

22. Osmo-regulation

(a) Osmo-regulation in marine fishes: Freshwater fishes are hypertonic so the concentration of salt is higher in their blood than in their surrounding water. They absorb a controlled amount of water through the mouth and the gill membranes.

(b) Osmo-regulation in freshwater fishes: Marine fishes are hypotonic so the concentration of salt is less in their blood than in their surrounding water. They absorb a huge amount of water through the mouth and the gill membranes.

23. Nervous system

(a) Structure of Brain (CNS)/Nerve cells and spinal cord: The skull contains a brain, which connects at the base of the skull to the spinal cord.

(b) Peripheral nervous system: It consists of 10 pairs of cranial nerves, spinal nerves and special sense organs. Maxillaries, buccalis, mandibularis and palatine.

(c) Autonomic nervous system: It controls the aperture of the iris, blood pressure and blood flow through gills for oxygenation and blood supply to various parts of the body automatically. It controls heart performance and gastric motility and controls the function of the swim bladder.

(d) Mechanism of nerve conduction: The brain receives information from sense organs that monitor conditions both within and around the body.

24. Synaptic transmission

(a) Structure of synapses: Synapses may be symmetric or asymmetric. Asymmetric synapses are characterized by rounded vesicles in the presynaptic cell and a prominent postsynaptic density

(b) Types and chemical composition of neurotransmitters: Glutamate and gamma-aminobutyric acid (GABA)

(c) Mechanism of synaptic transmission: Both chemical and electrical transmission can occur at synapses.

25. Sense organs

(a) Photoreceptors: The fish retina has rod cells that provide high visual sensitivity in low light conditions and cone cells that provide higher temporal and spatial resolution.

(b) Phonoreceptors: They have labyrinths with maculae and cristae but have no auditory papillae

(c) Chemoreceptors: It helps in feeding, prey detection, predator avoidance, species and sex recognition, sexual behaviour and migration

(d) Gustatoreceptors: It detects various organic acids, nucleotides and bile salts.

26. Endocrine system

(a) Structure and types of endocrine glands: Pituitary glands, interrenal tissue, chromaffin tissue, corpuscles of stannous, ultimobranchial glands, islets of Langerhans, thyroid gland, gastrointestinal glands, pineal glands and urophysis.

27. Neuro-secretory System

(a) Types of neuro-secretory cells: It contains magnocellular neuroendocrine Dahlgren cells.

(b) Functions of neuro-secretory cells: They help in homeostatic roles in osmoregulation and reproduction.

28. **Ethological responses**

(a) Innate behavior: Innate ('built-in') behaviour includes patterns of maturation (developmental changes).

(b) Learned behavior: Learning processes (imprinting and trial-and-error learning)

29. Reproductive System

(a) Type of puberty (Protogyny/Protandry):

Promiscuous: Both sexes with multiple partners.

Polygamy: Either male or female will have multiple partners in a given breeding season

Polygynous: Males with multiple mates (cichlids)

Polyandry: Females with multiple mates – few (Anglerfish, males "parasitize" females).

Monogamy: Mating pair remains together over time with long gestation of young (some cichlids, seahorses and pipefish)

(b) Bisexual/Unisexual: Bisexual fishes: *Perca, Stezostedon, Micropterus.*

(i) Simultaneous hermaphrodites function as male and female at the same time (23 families; ex. Anguilliformes, eels; Atheriniformes, killifish)

(ii) Sequential hermaphrodites start life as one sex, change sex after maturity

Protandrous: male first, female later

Protogynous: female first, male later (most common, Wrasses)

(c) Parthenogenesis:

Parthenogenetic (asexual): (Amazon molly *Poecilia formosa).*

In parthenogenesis, unfertilized eggs develop into embryos.

Gynogenetic: Sperm needed for egg development, but mating without fertilization, the result is daughters are genetic clones of mothers

Hybridogenetic: Egg development with fertilization by males of other species, but male genes discarded at the next generation

(d) Development: Direct or Indirect

30. Fertilization

(a) Types of breeding (Continuous/Seasonal)

(b) Site of fertilization (Internal/External): Most fishes use external fertilization. Internal fertilization occurs in sharks, rays, skates, ratfishes, guppies, mollies, surfperches

31. Stages of Development

(a) Cleavage: It occurs only in blastodiscs.

(b) Blastula: Mid-blastula transition is observed around the tenth cell division

(c) Gastrula: It is the epiboly of the blastoderm cells over the yolk.

32. Post-gastrular Development

(a) Histogenesis: It is a series of organized, integrated processes by which cells of the primary germ layers of an embryo differentiate and assume the characteristics of the tissues into which they will develop.

(b) Organogenesis: The primary organ rudiments continue to give rise to the rudiments of the various organs of the fully developed is known as organogenesis.

(c) Metamorphosis: These larval metamorphoses include the formation of pectoral fins and ossification of fin rays, maturation of internal organs and sensory systems, formation of scales, modifications to pigment patterns, and allometric changes in body proportions.

33. Foetal Membranes

Fishes lack amnion (anamniota)

34. Larval Stages

Ammocoete larva, Leptocephalus, Kasidoron larva, Lantern fish larva, Sea bass larva, Exterilium larva, Squirre fish larva, Scutatus larva, Flagelloserranus larva etc.

35. Cytogenetic Approach

(a) Organization of genome: The genome structure of major fishes in the world fisheries have been sequenced.

(b) Karyotype and DNA barcode: DNA barcoding is used to identify groups of fish based on DNA sequences within selected regions of a genome. These methods can be used to study fish, environmental DNA (eDNA) or cells freely diffused in the water.

(c) 16S rRNA analysis: This study evaluated the 16S and 12S rRNA genes for fish species

36. Special Features

(a) Trophic level: In marine ecosystems, the trophic level of most fish takes a value between 2.0 and 5.0. The trophic level of freshwater fish species ranged from 2.0 to 4.5,

(b) IUCN status: IUCN is leading the assessment of freshwater fish species. IUCN listed 1,000 critically endangered fish species

5

Taxonomy of Fishes

The term "fish" most commonly refers to Agnatha (jawless fishes), Chondrichthyes (sharks and rays), Sarcopterygii (lobe-finned fishes) and Actinopterygii (ray-finned fishes). Although the ichthyologist Regan (1910) defined a fish species as a product of interrelated communities with common "morphospecies", in the early 20th century, it should be emphasized that the concept of species classification varied across scientists.

Aristotle (335-332 BC) provided the earliest classification of fish and described 117 Mediterranean fish species obtained from the Aegean Sea close to Greece.

While the majority of researchers are concerned with fish as a food source and their studies entail adding to the body of aquaculture knowledge, others are interested in their diversity, distribution patterns, ecology and functional physiology. Since humans learned how to hunt fish, species were identified based on some simple anatomical features. Observation of specimen anatomy and differentiating fish species based on their morphological features is the most practical, rapid and low-cost method. Besides experienced local fishermen and fishmongers, people who lived by the river or wetland would learn to identify fish at a young age. This is due to knowledge and memory acquired from long-term observation or through oral tradition maintained by elders.

Such traditional knowledge has been interwoven into modern ichthyology by many researchers (Calamia, 1999; Drew, 2005; Stacey *et al,* 2008; Ferreira *et al,* 2014), and the term for it is "traditional ecological knowledge" (TEK) (Berkes *et al,* 2000). Recently, there has also been a surge of interest in the molecular constitution of fish (Wong et al, 2011; Pereira et al, 2013; Quraishia et al, 2015) and their use as bio-indicators to monitor water body pollution (Fonge et al, 2011; Khodadoust et al, 2013; Authman et al, 2015). Correspondingly, the interest in fish has grown exponentially, and ichthyology is frequently sought to contribute to many other fields of study (Padilla and Williams, 2004; Feist and Longshaw, 2008). In retrospect, the advent of molecular, computational and statistical approaches has led many people to believe that "species" can be simply classified using nucleotide sequences, software and mathematical calculations.

Recently, for the categorization of fish-species, the PatternNet technique has been employed (Prasenan and Suriyakala, 2022). Colour pattern, anatomical traits, meristic and morphometric measurements are used for the identification of a fish species. Measurement of various parts of the body and different fins are considered in morphometric parameters. Meristic characteristics include counting scales following various parameters. In this identification, the restriction fragment length polymorphism (RFLP), amplified fragment length polymorphism (AFLP), random amplified polymorphic DNA (RAPD), microsatellites or simple sequence repeat (SSR), single nucleotide polymorphism (SNP), DNA barcoding and expressed sequence tag (EST) markers (O'Reilly and Wright, 1995; Khoo *et al,* 2003; Chauhan and Rajiv, 2010; Khoo *et al,* 2011; Kress *et al,* 2015) DNA analyses are applied. PCR-RFLP, PCR-FINS, PCR-specific primers, real-time PCR and microarray technology are also most used.

A typical fish genome contains roughly a billion nucleotide pairs (Stepien and Kocher, 1997), and analysing all of them would be too daunting and the results would cause an information overload. Since Herbert et al. (2003) discovered a technique to amplify the mitochondrial cytochrome c oxidase subunit 1 (COI) gene, there is now a consensus to analyse a 648-base pair (bp) region of COI to rapidly identify a fish species (Cawthorn *et al,* 2012). Such an approach is now known as DNA barcoding and it has gained widespread acceptance as a fast, cost-effective and standardised technique. So far, the Fish Barcode of Life Initiative has barcoded 7,882 fish species, which is only approximately 25% of the estimated 31,220 species, present globally (Jinbo *et al,* 2011).

In recent years, DNA sequencing techniques have made substantial progress and they can analyse 30 to 1,500 nucleotides for hundreds of thousands to millions of DNA molecules in a single process within a complex or degraded DNA source (Davey and Blaxter, 2011; Mehinto *et al,* 2012). These are classified as next-generation sequencing (NGS) technologies (Tillmar *et al,* 2013) and are sometimes termed "DNA metabarcoding" to refer to their ability to automatically identify multiple species from a single bulk sample (Taberlet *et al,* 2012). Because NGS technologies can extract massive numbers of reads, the results increase the chances of finding and annotating matches (Hemmer-Hansen *et al,* 2014).

There are already attempts to digitalize specimens using these technologies. For example, Berquist et al. (2012) have progressively scanned specimens with magnetic resonance imaging (MRI) technology and created an online digital archive called the Digital Fish Library (DFL, http://www.digitalfishlibrary.org) to share high-resolution and high-contrast visual data.

FishBase (www.fishbase.org.in latest version 02/2023) is a global species database that contains a database of 35100 fish species. It is the web's largest and most widely used online database of adult finfish. It has "evolved into a dynamic and versatile ecological tool" that has been widely cited in scholarly publications over time.

5.1 Earlier classifications

The classification of fish is a debatable/problematic issue and there is no agreement amongst the Ichthyologists regarding this issue. The controversy is due to huge numbers, diversity in shape, size, habit and habitats; variations in structural and functional organization; incomplete fossil records; ever-changing tools for identification; purpose and basis of classification of fishes.

Fishes may be classified as Marine (M); Marine, Estuarine (M, E); Estuarine (E); Estuarine, Fresh (E, F) and Marine, Estuarine, Fresh (M, E, F) groups based on their habitats.

Parental care is a behavioural and evolutionary strategy adopted by some animals, involving a parental investment being made to the evolutionary fitness of offspring. In fishes, it may be biparental, maternal or paternal and vary from spawning to internal fertilization and viviparity (such as *Sebastus alutus, Scoliodon*). Among fishes, 77% do not show parental care, 17% care for eggs and 6% care for eggs and hatch young ones.

The differences in classification are caused by variations in morphometric, meristic, chromatic, karyotypic, anatomical, biochemical, physiological and ethological characteristics.

I. According to Peter Arditi (1705-1734): Order 1. Malacpterygii

2. Acanthopterygii

3. Brachiostegi

4. Chondropterygii

5. Plagiuri

He developed standard methods for making counts and measurements of anatomical features.

II. According to Cuvier: Group 1. Poissons Osseusx (Bony fishes)

2. Cartilagineux (Cartilaginous fishes)

III. According to Agassiz (1820), based on the nature of scales and dermal appendages: Order 1. Placoids

2. Ganoids

3. Ctenoids and

4. Cycloids

IV. According to Agassiz (1857): 1. Myzontes (lampreys and hagfishes)

2. Pisces

3. Ganoids and

4. Selachii

V. According to Muller (1844):

Subclass 1. Dipnoi (Lung Fishes)

2. Teleostei (Bony Fishes)

3. Ganoidei (*Polypterus, Amia, Lepidos*)

4. Selachii or Elasmobranchii (Sharks, Rays and Rat fishes)

5. Cyclostomi or Marsipobranchi (Cyclostomes)

6. Leptocardii (*Amphioxus*)

He identified anatomical peculiarities of living types and reviewed sharks and rays to create major groupings of bony fishes.

VI. According to Gunther (1859-1870) Subclass 1. Palaeichthys

2. Teleostei

3. Cyclostomata

4. Leptocardii

He described some 6800 fish species.

VII. According to Jordan (1923) Subclass 1. Leptocardii

2. Marsipobranchi

3. Ostracophora

4. Arthrodora

5. Elasmobranchii

6. Pisces

VIII. According to Berg (1940): Class 1. Pterichthys

2. Coccostei (Arthrodira)

3. Acanthodii (Antiarchi)

4. Elasmobranchii

5. Holocephali

6. Dipnoi

7. Teleostei: Subclasses A. Crossopterygii

B. Actinopterygii

Berg followed the scheme of classification of Regan and Stensio to consider resemblances and differences between both fossil and extant species. He included all gnathostomes bearing respiratory organs and fins in the series Pisces. However, many workers considered the inclusion of Crossopterygii in Teleostomi as an error. He applied the –iformes uniform endings to orders of fishes.

IX. According to Bertin and Arambourg (1958):

Class 1. Acanthodii

2. Placodermi: Subclasses A. Antiarchi

B. Arthrodira

3. Chondrichthyes: Subclasses A. Selachii

B. Bradyodonti

4. Osteichthyes Subclasses A. Actinopterygii

B. Brachiopterygii

C. Dipneusti

D. Crossopterygii

The scheme of classification of Bertin and Arambourg (1958) follows the discussion of the International Committee on Fish Classification. In this scheme class Teleostomi was abolished and a subclass Brachiopterygii recognized.

X. According to Romer (1959):

Class 1. Agnatha: Subclass A. Monorhina

B. Diplorhina

2. Placodermi: Subclass A. Pterichthys

B. Coccostei

C. Acanthodii

3. Chondrichthyes: Subclasses A. Elasmobranchii

B. Holocephali

4. Osteichthyes: Subclasses A. Sarcopterygii: Infraclass a. Crossopterygii

b. Dipnoi

B. Actinopterygii

It seems that the classificatory schemes of Berg, Bertin and Arambourg and Romer are essentially similar. But, Common association of crossopterygian and dipnoan done by Romer appears to be misleading.

XI, According to Greenwood et al, (1966) teleostean fishes may be classified as

Division I: Superorder 1. Elopomorpha

2. Clupeomorpha

II: Superorder 3. Osteoglossopmorpha

III: Superorder 4. Protacanthopterygii

5. Ostariophysii

6. Paracanthopterygii

7. Atherinomorpha

8. Acanthopterygii

XII. According to Parker and Haswell (1967): Class 1. Olacodermi (Aphetohyoidea): Subclasses

A. Acanthodii

B. Arthrodira

C. Petalichthyida

D. Antiarchi

E. Rhenanida

F. Paleospondylia

2. Chondrichthyes: Subclasses A. Selachii

B. Bradyodonti

3. Osteichthyes: Subclasses A. Crossopterygii

B. Actinopterygii

4. Dipnoi

XIII. Combined classification (after Romer 1959 and Parker and Haswell 1967):

Class 1. Placodermi: Subclasses A. Acanthodii

B. Arthrodira

C. Petalichthyida

D. Antiarchi

E. Rhenanida

F. Paleospondylia

2. Chondrichthyes: Subclasses A. Selachii

B. Bradyodonti

3. Osteichthyes: Subclasses A. Crossopterygii

B. Actinopterygii

XIV. Based on DNA content per cell, Hinegrander and Rosen (1972) divided fishes into two groups:

1. Highly specialized fishes: less DNA content per cell

2. Less specialized fishes: more DNA content per cell

They established cladistic relationships between major groups of fish based on DNA content. This finding was applied by Nelson (1976) in the classification of fishes.

XIV. According to Nelson (1976):

Phylum Chordata:

Subphylum 1. Calcichordata

2. Hemichordata

3. Pogonophora

4. Urochordata

5. Cephalochordata

6. Vertebrata: Superclass A. Agnatha Class: (a) Cephalospidomorpha

(b) Pteraspidomorpha

B. Gnathostomata Grade (a) Placodermi

(b) Chondrichthyes

(c) Acanthodi

(d) Osteichthyes

XV. According to Romer and Parsons (1979):

1. Class Agnatha (jawless fish)
   - A. Subclass Cyclostomata (hagfish and lampreys)
   - B. Subclass Ostracodermi (armoured jawless fish)
2. Class Chondrichthyes (cartilaginous fish)
   - A. Subclass Elasmobranchii (sharks and rays)
   - B. Subclass Holocephali (chimaeras and extinct relatives)
3. Class Placodermi (armoured fish)
4. Class Acanthodii («spiny sharks», sometimes classified under bony fishes)
5. Class Osteichthyes (bony fish)
   - A. Subclass Actinopterygii (ray-finned fishes)
   - B. Subclass Sarcopterygii (fleshy finned fishes, ancestors of tetrapods)

XVI. According to Lauder and Liem (1983) the classification of teleostean fishes may be arranged as

Infra division I. Osteoglossopmorpha: Order 1. Osteoglossiformes

2. Mormyriformes

II. Elopomorpha: Order 1. Elopiformes

2. Anguillformes

3. Notacanthiformes

III. Clupeomorpha: Order 1. Clupeiformes

IV. Euteleostei: Super order 1. Protacanthopterygii

2. Ostariophysi

3. Paracanthpterygii

4. Acanthopterygii

XVII. According to Pough et al. (1989) teleostean fishes may be classified as

Division I. Osteoglossopmorpha:

II. Elopomorpha:

III. Clupeomorpha:

IV. Euteleostei: Super order 1. Protacanthopterygii

2. Ostariophysi

3. Paracanthpterygii

4. Acanthopterygii

5. Scopelomorpha

XVIII. According to Nelson (1994):

Super class A. Agnatha Class: (a) Cephalospidomorphii

(b) Paraspidomorphii

B. Gnathostomata Grade (a) Pisces: Sub grade (i) Elasmobranchiomorphi

Class (I) Placodermi

(II) Chondrichthyes

Sub class (A) Elasmobranchi

(B) Holocephali

(ii) Teleostomi

Class (I) Acnathodii

(II) Osteichthys

(b) Tetrapoda Class: (i) Amphibia

(ii) Reptilia

(iii) Aves

(iv) Mammals

5.2 Recent classifications

There are established conventions in fish taxonomy for expressing taxonomic ranking. The "Family-group Names of Recent Fishes" published by Van der Laan et al. (2014) appears to be the most updated for family naming. According to ICZN regulations, the name of a family must end with the suffix "-idae" (e.g., Cyprinidae) while the subfamily with the "-inae" (e.g., Cyprininae).

I. Mohanty et al, (2016) classified fish based on their fat content as:

White or lean fishes (<2% fat): Example *Rita rita.*

Low-fat fishes (2-4% fat): For example *Catla catla, Labeo rohita, Notopterus notopterus, Heteropneustes fiossilis, Clarias batrachus* etc.

Medium fat fish (4-8% fat): Example *Anabas testudineus.*

Blue or high fat fish (8% fat): Example *Tenualosa ilisha.*

II. According to Nelson, 2016

The classification proposed by Muller, Jordan, Regan, Berg, Romer and Nelson considered to be the recent classification of bony fishes. Nelson (2016) and his book Fishes of the World became a central clearinghouse for fish classification. According to Nelson (2016), the classification of fish may be given as:

5.2.1 Class Myxini

An ancient and unique lineage of primitive eel-like marine fishes, The skeleton is made up of cartilage, produces slime, lacks true jaws and scales, degenerate eyes and many gill openings.

1. Order **MYXINIFORMES**

Nostril at the tip of the snout, tongue with 2 rows of horny teeth, mouth surrounded by barbells.

   1. Family **Myxinidae** (Hagfishes)

5.2.2 Class Petromyzontida or Hyperoartia

They show the beginnings of a vertebral column as neural arches protruding from the notochord, possess unpaired fins and eggs hatch into ammocoete larvae.

2. Order **PETROMYZONTIFORMES** (Berg, 1940) (38 living and 5 extinct species)

   2. Family Petromyzontidae (Northern lampreys or Vampire fish)
   3. Family Geotriidae (Southern lampreys)
   4. Family Mordaciidae (Southern top-eyed lampreys)

5.2.2.1 Grade Pisces

The jawed aquatic vertebrates in this grade have gills throughout life and paired limbs, if any, in the shape of fins that are not polydactylous.

The living forms of the grade Pisces are divided into two classes, Chondrichthyes and Osteichthyes which may be distinguished by the following characters.

5.2.3 Class Placodermi

They have characteristic armour of dermal or skin bones. The armour formed a head and trunk shield. Their jaws evolved from 1st pair of gills.

5.2.4 Class Chondrichthyes (Huxley, 1880)

The jawed fish possessing a cartilaginous skeleton is often calcified but never ossified, paired nares, a heart with its chambers in series and the skull lacks sutures in living forms.

3. Order **CHIMAERIFORMES** (Chimaeras)
   5. Family Callorhinchidae (Ploughnosed chimaeras)
   6. Family Rhinochimaeridae (Long-nosed chimaeras)
   7. Family Chimaeridae (Shortnose chimaeras or ratfishes)
4. Order **HETERODONTIFORMES** (Bullhead sharks)
   8. Family Heterodontidae (Bullhead sharks)
5. Order **ORECTOLOBIFORMES** (Carpet sharks)
   9. Family Parascyllidae (Collared carpet sharks)
   10. Family Brachaeluridae (Blind sharks)
   11. Family Orectolobidae (Wobbegong sharks)
   12. Family Hemiscyllidae (Bamboo sharks)
   13. Family Ginglymostomatidae (Nurse Sharks)
   14. Family Stegostomatidae (Zebra shark)
   15. Family Rhincodontidae (Whale sharks)
6. Order **LAMNIFORMES** (Berg, 1958) (Mackerel sharks)

The dorsal fins (one dorsal fin in the scyliorhinid *Pentanchus*), without spines; anal fin present; five gill-slits; gill rakers usually absent; spiracle usually present.

   16. Family Mitsukurinidae (Goblin sharks)
   17. Family Odontaspididae (Sand tiger sharks)
   18. Family Pseudocarchariidae (Crocodile sharks)
   19. Family Alopiidae (Thresher sharks)
   20. Family Megachasmidae (Megamouth sharks)
   21. Family Cetorhinidae (Basking sharks)
   22. Family Lamnidae (Threshers and Mackerel or mako sharks)
7. Order **CARCHARHINIFORMES** (Ground sharks)
   23. Family Scyliorhinidae (Catsharks)
   24. Family Proscyllidae (Ground sharks)
   25. Family Pseudotriakidae (False catshark)
   26. Family Leptochariidae (Barbeled hound sharks)
   27. Family Triakidae (Hound sharks)
   28. Family Hemigaleidae (Weasel sharks)
   29. Family Carcharhinidae (Requiem sharks)
   30. Family Sphyrnidae (Hammerhead sharks)

8. Order **HEXANCHIFORMES** (deBuen, 1926) (Six-gill sharks)

One dorsal fin, without spine, anal fin present; 6 gill slits; eyes without a nictitating fold; spiracle present

31. Family Chlamydoselachidae (Frilled sharks)
32. Family Hexanchidae (Cow sharks)

9. Order **SQUALIFORMES** (Goodrich, 1909) (Dogfish sharks)

Two dorsal fins, with or without spines; anal fin absent; 5 or 6 gills slits.

33. Family Centrophoridae (Gulper sharks)
34. Family Etmopteridae (Lantern sharks)
35. Family Somniosidae (Sleeper sharks)
36. Family Oxynotidae (Rough sharks)
37. Family Dalatiidae (Kitefin sharks)
38. Family Squalidae (Dogfish or dogsharks)

10. Order **ECHINORHINIFORMES** (Bramble sharks)

39. Family Echinorhinidae (Bramble sharks)

11. Order **SQUATINIFORMES** (Angel sharks)

40. Family Squatinidae (Angel sharks)

12. Order **PRISTIOPHORIFORMES** (Sawsharks)

41. Family Pristiophoridae (Sawsharks)

13. Order **TORPEDINIFORMES** (Electric rays)

42. Family Torpedinidae (Electric rays)
43. Family Narcinidae (Numbfishes)

14. Order **PRISTIFORMES** (Sawfishes)

44. Family Pristidae (Sawfishes or Carpenter sharks)

15. Order **RAJIFORMES** (Skates)

45. Family Rajidae (Skates)

16. Order **PRISTIFORMES** (Guitarfishes and Sawfishes)

46. Family Rhinobatidae (Guitarfishes)
47. Family Rhinidae (Bowmouth guitarfishes)
48. Family Rhynchobatidae (Wedgefishes)

17. Order **MYLIOBATIFORMES** (Stingrays)

49. Family Platyrhinidae (Thornbacks)

50. Family Zanobatidae (Panrays)
51. Family Plesiobatidae (Deepwater stingrays)
52. Family Urolophidae (Round stingrays)
53. Family Hexatrygonidae (Sixgill stingrays)
54. Family Dasyatidae (Whiptail stingrays)
55. Family Potamotrygonidae (River stingrays)
56. Family Gymnuridae (Butterfly rays)
57. Family Urotrygonidae (American round stingrays)
58. Family Myliobatididae (Eagle rays)
59. Family Rhinopteridae
60. Family Mobulidae (Manta and Devi rays)

5.2.5 Class Osteichthyes

5.2.5.1 Subclass Sarcopterygii

18. Order **COELACANTHIFORMES** (Coelacanths)
    61. Family Latimeriidae (Gombessas or coelacanths)
19. Order **CERATODONTIFORMES** (Living lungfishes or Dipnoi)
    62. Family Neoceratodontidae (Australian lungfishes)
    63. Family Lepidosirenidae (South American lungfishes)
    64. Family Protopteridae (African Lungfishes)

5.2.5.2 Subclass Actinopterygii

20. Order **POLYPTERIFORMES** (Bleeker, 1859) (Bichirs)
    65. Family Polypteridae (Bichirs)
21. Order **ACIPENSERIFORMES** (Sturgeons and Paddlefishes)
    66. Family Polyodontidae (Paddlefishes)
    67. Family Acipenseridae (Sturgeons)
22. Order **LEPISOSTEIFORMES** (Gars)
    68. Family Lepisosteidae or Lepidosteidae (Gars)
    69. Family Amiidae (Bowfins)
23. Order **ELOPIFORMES** (Tenpounders)
    70. Family Elopidae (Tenpounders, ladyfish, skipjack, jack-rashes)
    71. Family Megalopidae (Tarpons)
24. Order **ALBULIFORMES** (Bonefishes)

72. Family Albulidae (Bonefishes)

25. Order **NOTACANTHIFORMES** (Halosaurs and Deep-sea spiny eels)

73. Family Halosauridae (Halosaurs)
74. Family Notacanthidae (Deep-sea spiny eels)

26. Order **ANGUILLIFORMES** (Eels)

75. Family Protanguillidae (Primitive cave eels or living fossils)
76. Family Synaphobranchidae (Cutthroat eels)
77. Family Heterenchelyidae (Mud eels or eels)
78. Family Myrocongridae (Myroconger eels or thin eels)
79. Family Muraenidae (Moray eel)
80. Family Chlopsidae (False morays)
81. Family Derichthyidae longneck or narrowneck and shorttail eels
82. Family Ophichthidae (Snake eels and worm eels)
83. Family Muraenesocidae (Pike congers)
84. Family Nettastomatidae (Duckbill eels)
85. Family Congridae (Conger and garden eels)
86. Family Moringuidae (Spaghetti eels, Worm eels)
87. Family Cyematidae (Bobtail snipe eels)
88. Family Monognathidae (One jaw gulpers)
89. Family Saccopharyngidae (Swallowers)
90. Family Eurypharyngidae (Gulpers or Pelican eels)
91. Family Nemichthyidae (Snipe eels)
92. Family Serrivomeridae (Sawtooth eels)
93. Family Anguillidae (Freshwater eels)

27. Order **HIODONTIFORMES**

94. Family Hiodontidae (Mooneyes)

28. Order **OSTEOGLOSSIFORMES** (Bonytongues)

95. Family Pantodontidae (Butterflyfishes)
96. Family Osteoglossidae (Bonytongues)
97. Family Notopteridae (Featherback and knifefishes)
98. Family Mormyridae (Freshwater elephant fishes)
99. Family Gymnarchidae (Abas, freshwater rat-tail, Poisson-cheval)

29. Order **CLUPEIFORMES** (Herrings)

100. Family Denticipitidae (Denticle herrings)
101. Family Pristigasteridae (Longish herrings)
102. Family Engraulidae (Anchovies)
103. Family Chirocentridae (Wolf herrings)
104. Family Clupeidae (Herrings)

30. Order **ALEPOCEPHALIFORMES** (Slickheads and Tubeshoulders)

105. Family Platytroctidae (Tubeshoulders)
106. Family Bathylaconidae (Bathylaconids)
107. Family Alepocephalidae (Slickheads, Smoothheads)

31. Order **GONORYNCHIFORMES** (Milkfishes)

108. Family Chanidae (Milkfishes)
109. Family Gonorhynchidae (Beaked sandfishes or Beaked salmon)
110. Family Kneriidae (Knerias and Snake mudheads)

32. Order **CYPRINIFORMES** (Carps)

Parietal, symplectic, subopercular, and intermuscular bones present; vomerine teeth absent; heavy long plates never present on the body; branchiostegal rays 5-8. In terms of species, it is the second-largest order of fishes. It contains some 4250 species.

111. Family Cyprinidae (Minnows, Carps, and Loaches) (largest and most diverse fish family including 376 genera and 3160 species)
112. Family Psilorhynchidae (Mountain carps)
113. Family Gyrinocheilidae (Algae eaters)
114. Family Catostomidae (Suckers)
115. Family Botiidae (Botiid loaches)
116. Family Vaillantellidae (Long-fin loaches)
117. Family Cobitidae (True loaches)
118. Family Balitoridae (Hillstream or River loaches, flossensaugers, Lizardfish)
119. Family Gastromyzontidae (Gastromyzontid or Sucker loaches)
120. Family Nemacheilidae (Stone loaches)
121. Family Barbuccidae (Fire-eyed loaches)
122. Family Ellopostomatidae (Sturgeon-mouthed loaches)
123. Family Serpenticobitidae (Serpent loaches)

33. Order **CHARACIFORMES** (Regan, 1911) (Characins)

124. Family Distichodontidae (Distichodontids)

125. Family Citharinidae (Citharinids or Lutefishes)
126. Family Crenuchidae (South American darters)
127. Family Alestidae (African tetras)
128. Family Hepsetidae (African pikes characins)
129. Family Erythrinidae (tiaras)
130. Family Parodontidae (Parodontids)
131. Family Cynodontidae (Cynodontids)
132. Family Serrasalmidae (pacus, silver dollars, serrated salmon and piranhas)
133. Family Hemiodontidae (Hemiodontids)
134. Family Anostomidae (Toothed headstanders)
135. Family Chilodontidae (Headstanders)
136. Family Curimatidae (Toothless characins)
137. Family Prochilodontidae (Flannel-mouth characiforms)
138. Family Lebiasinidae (Pencil fishes)
139. Family Ctenoluciidae (Pike-characids)
140. Family Acestrorhynchidae (Acestrorhynchids)
141. Family Characidae (Characins)
142. Family Gasteropelecidae (Freshwater hatchetfishes)

34. Order **SILURIFORMES** (Catfishes)

143. Family Diplomystidae (Primitive velvet catfishes)
144. Family Cetopsidae (whale catfishes)
145. Family Trichomycteridae (Pencil or parasitic catfishes)
146. Family Nematogenyidae (Mountain catfishes)
147. Family Callichthyidae (Callichthyid armored catfishes)
148. Family Scoloplacidae (Spiny dwarf catfishes)
149. Family Astroblepidae (Climbing catfishes)
150. Family Loricariidae (suckermouth armored catfishes)
151. Family Siluridae (sheatfishes)
152. Family Austroglanidae (Rock catfishes)
153. Family Pangasiidae (Shark catfishes)
154. Family Chacidae (Squarehead, angler or frogmouth catfishes)
155. Family Plotosidae (Eeltail catfishes)
156. Family Ritidae (Ritas)
157. Family Ailiidae (Asian schilbeids)

158. Family Horabagridae (imperial or sun catfishes)
159. Family Bagridae (naked or Bagrid catfishes)
160. Family Akysidae (Stream catfishes)
161. Family Sisoridae (Sisorid catfishes)
162. Family Erethistidae (Erethistid catfishes)
163. Family Amphiliidae (Loach catfishes)
164. Family Malapteruridae (Electric catfishes)
165. Family Mochokidae (Squeakers or upside-down catfishes)
166. Family Schilbeidae (Schilbeid catfishes)
167. Family Auchenoglanididae (Auchenoglandids)
168. Family Claroteidae claroteids
169. Family Lacantuniidae (Chiapas catfishes)
170. Family Clariidae (Airbreathing catfishes)
171. Family Heteropneustidae (Airsac catfishes)
172. Family Anchariidae (Malagasy catfishes)
173. Family Ariidae (Sea catfishes)
174. Family Aspredinidae (Banjo catfishes)
175. Family Doradidae (Thorny catfishes)
176. Family Auchenipteridae (Driftwood catfishes)
177. Family Cranoglanididae (Armourhead catfishes)
178. Family Ictaluridae (North American catfishes)
179. Family Heptapteridae (Heptapterids)
180. Family Pimelodidae (Long-whiskered catfishes)
181. Family Pseudopimelodidae (Bumblebee catfishes)

35. Order **GYMNOTIFORMES**

182. Family Gymnotidae (Nakedback knifefishes)
183. Family Rhamphichthyidae (Sand knifefishes)
184. Family Hypopomidae (Bluntnose knifefishes)
185. Family Sternopygidae (Glass knifefishes)
186. Family Apteronotidae (Ghost knifefishes)

36. Order **LEPIDOGALAXIIFORMES** (Salamanderfishes)

187. Family Lepidogalaxiidae (Salamanderfishes)

37. Order **SALMONIFORMES** (Bleeker, 1859)

188. Family Salmonidae (Trout, Salmon and whitefish)

38. Order **ESOCIFORMES** (Pikes and Mudminnows)

189. Family Esocidae (Pikes)
190. Family Umbridae (Mudminnows)

39. Order ARGENTINIFORMES (Marine smelts)

191. Family Argentinidae (Argentines or herring smelts)
192. Family Opisthoproctidae (Barreleyes or spookfishes)
193. Family Microstomatidae (Pencilsmelts)
194. Family Bathylagidae (Deepsea smelts)

40. Order **GALAXIIFORMES**

195. Family Galaxiidae (Galaxiids)

41. Order **OSMERIFORMES** (Freshwater smelts and allies)

196. Family Osmeridae (Northern Hemisphere smelts)
197. Family Plecoglossidae (Ayu or sweetfish)
198. Family Salangidae (Icefishes or noodlefishes)
199. Family Prototroctidae (Southern graylings)
200. Family Retropinnidae (Southern smelts)

42. Order **STOMIIFORMES** (Regan, 1909) (Dragonfishes)

201. Family Gonostomatidae (Bristlemouths)
202. Family Sternoptychidae (Marine hatchetfishes)
203. Family Phosichthyidae (Lightfishes)
204. Family Stomiidae (Barbeled dragonfishes)

43. Order **ATELEOPODIFORMES** (Jellynose fishes)

205. Family Ateleopodidae (Jellynose or tadpole fishes)

44. Order **AULOPIFORMES** (Rosen, 1973) (Lizardfishes)

206. Family Synodontidae (Typical lizardfishes)
207. Family Aulopidae (Flagfins or Aulopiform fishes)
208. Family Pseudotrichonotidae (Sand-diving lizardfishes)
209. Family Paraulopidae (Cucumber fishes)
210. Family Ipnopidae (Deepsea tripod fishes)
211. Family Giganturidae (Telescopefishes)
212. Family Bathysauroididae (Bathysauroidids)
213. Family Bathysauridae (Deepsea lizardfishes)
214. Family Chlorophthalmidae (Greeneyes)

215. Family Notosudidae (Waryfishes)
216. Family Scopelarchidae (Pearleyes)
217. Family Evermannellidae (Sabertooth fishes)
218. Family Paralepididae (Barracudinas)
219. Family Alepisauridae (Lancetfishes)
220. Family Lestidiidae (Naked barracudas)

45. Order **MYCTOPHIFORMES** (Regan, 1911) (Lanternfishes)

221. Family Neoscopelidae (Blackchins)
222. Family Myctophidae (Lanternfishes)

46. Order **LAMPRIDIFORMES** (Opahs)

223. Family Veliferidae (Velifers)
224. Family Lamprididae (Opahs)
225. Family Lophotidae (Crestfishes)
226. Family Radiicephalidae (Tapertails)
227. Family Trachipteridae (Ribbonfishes)
228. Family Regalecidae (Oarfishes)

47. Order **POLYMIXIIFORMES**

229. Family Polymixiidae (Beardfishes)

48. Order **PERCOPSIFORMES** (Berg, 1940)

230. Family Percopsidae (Trout-perches)
231. Family Aphredoderidae (Pirate perches)
232. Family Amblyopsidae (Cavefishes)

49. Order **ZEIFORMES**

233. Family Cyttidae (Lookdown dories)
234. Family Oreosomatidae (Oreos)
235. Family Parazenidae (Smooth dories)
236. Family Zeniontidae (Armoreye dories)
237. Family Grammicolepididae (Tinselfishes)
238. Family Zeidae (True dories)

50. Order **STYLEPHORIFORMES** (Miya et al. 2007)

239. Family Stylephoridae (Tube-eyes or thread-tails)

51. Order **GADIFORMES** (Goodrich, 1909) (Cods and Hakes)

240. Family Melanonidae (Pelagic cods)

241. Family Steindachneriidae (Luminous and southern hakes)
242. Family Bathygadidae (Rattails)
243. Family Macrouridae (Grenadiers)
244. Family Trachyrincidae (Whiptails and trachyrincines)
245. Family Euclichthyidae (Eucla cod)
246. Family Moridae (Deep-sea cods)
247. Family Macruronidae (Southern hakes)
248. Family Merlucciidae (Merluccid hakes)
249. Family Ranicipitidae (Tadpole cods)
250. Family Bregmacerotidae (Codlets)
251. Family Muraenolepididae (Eel cods and moray cods)
252. Family Gadidae (Cods)

52. Order **HOLOCENTRIFORMES** (Betancur et al, 2013)

253. Family Holocentridae (Squirrelfishes)

53. Order **TRACHICHTHYIFORMES** (Roughies)

254. Family Anoplogastridae (Fangtooths)
255. Family Diretmidae (Spinyfins)
256. Family Anomalopidae (Flashlight or lantern eye fishes)
257. Family Monocentrididae (Pinecone fishes)
258. Family Trachichthyidae (Roughies)

54. Order **BERYCIFORMES** (Regan, 1909)

259. Family Gibberichthyidae (Gibberfishes)
260. Family Stephanoberycidae (Pricklefishes)
261. Family Hispidoberycidae (Hispidoberycids)
262. Family Rondeletiidae (Redmouth whalefishes)
263. Family Barbourisiidae (Red whalefishes)
264. Family Cetomimidae (Flabby whalefishes)
265. Family Melamphaidae (Bigscale fishes)
266. Family Berycidae (Alfonsinos)

55. Order **OPHIDIIFORMES** (Berg, 1937) (Cusk-eels)

267. Family Carapidae (Pearlfishes)
268. Family Ophidiidae (Cusk-eels)
269. Family Bythitidae (Viviparous brotulas)

270. Family Aphyonidae (Aphyonids)
271. Family Parabrotulidae (False brotulas)

56. Order **BATRACHOIDIFORMES** (Toadfishes)

272. Family Batrachoididae (Frogfishes, toadfishes)

57. Order **KURTIFORMES** (Jordan, 1923)

273. Family Kurtidae (Nurseryfishes)
274. Family Apogonidae (Cardinalfishes)

58. Order **GOBIIFORMES** (Günther, 1880) (Gobies)

275. Family Rhyacichthyidae (Loach gobies)
276. Family Odontobutidae (Freshwater sleepers)
277. Family Milyeringidae (Blind cave gobies)
278. Family Eleotridae (Sleepers)
279. Family Butidae (Butid sleepers)
280. Family Thalasseleotrididae (Ocean sleepers)
281. Family Oxudercidae (Gobionellus-like and mudskipper gobies)
282. Family Gobiidae (Gobies)
283. Family Ambassidae (Asiatic glassfishes)
284. Family Embiotocidae (Surfperches)
285. Family Grammatidae (Basslets)
286. Family Plesiopidae (Roundheads)
287. Family Polycentridae (South American leaffishes)
288. Family Pomacentridae (Damselfishes)
289. Family Pseudochromidae (Dottybacks)
290. Family Opisthognathidae (Jawfishes)

59. Order **MUGILIFORMES**

291. Family Mugilidae (Mullets or Grey mullets)

60. Order **CICHLIDIFORMES**

292. Family Cichlidae (Cichlids)
293. Family Pholidichthyidae (Convict blenny)

61. Order **BLENNIFORMES** (Blennies)

294. Family Tripterygiidae (Triplefin blennies)
295. Family Dactyloscopidae (Sand stargazers)
296. Family Blenniidae (Combtooth blennies)

297. Family Clinidae (Kelp blennies)
298. Family Labrisomidae (Labrisomid blennies)
299. Family Chaenopsidae (Tube blennies)

62. Order **GOBIESOCIFORMES**

300. Family Gobiesocidae (Clingfishes)

63. Order **ATHERINIFORMES** (Silversides)

301. Family Atherinopsidae (New World silversides)
302. Family Notocheiridae (Surf sardines)
303. Family Isonidae (Surf sardines)
304. Family Melanotaeniidae (Rainbowfishes and blue eyes)
305. Family Atherionidae (Pricklenose silversides)
306. Family Dentatherinidae (Mercer's tusked silverside)
307. Family Phallostethidae (Priapiumfishes)
308. Family Atherinidae (Old World silversides)

64. Order **BELONIFORMES** (Berg, 1937) (Needlefishes)

309. Family Adrianichthyidae (Ricefish and medakas)
310. Family Exocoetidae (Flyingfishes)
311. Family Hemiramphidae (Halfbeaks)
312. Family Zenarchopteridae (Viviparous halfbeaks)
313. Family Belonidae (Needlefishes)
314. Family Scomberesocidae (Sauries)

65. Order **CYPRINODONTIFORMES** (Berg, 1940) (Killifishes)

315. Family Aplocheilidae (Asian rivulines)
316. Family Nothobranchiidae (African rivulines)
317. Family Rivulidae (New World rivulines)
318. Family Profundulidae (Middle American killifishes)
319. Family Goodeidae (Goodeids)
320. Family Fundulidae (Topminnows)
321. Family Valenciidae (Valencia toothcarp)
322. Family Cyprinodontidae (Pupfishes)
323. Family Anablepidae (Four-eyed fishes)
324. Family Poeciliidae (Livebearers)

66. Order **SYNBRANCHIFORMES** (Berg, 1940)

325. Family Synbranchidae (Swamp eels)
326. Family Chaudhuriidae (Earthworm eels)
327. Family Mastacembelidae (Freshwater spiny eels)

67. Order **CARANGIFORMES** (Jordan, 1923) (Jacks)

328. Family Nematistiidae (Roosterfishes)
329. Family Coryphaenidae (Dolphinfishes)
330. Family Rachycentridae (Cobias)
331. Family Echeneididae (Remoras) (Sharksuckers)
332. Family Carangidae (Jacks and pompanos)
333. Family Menidae (Moonfishes)

68. Order **ISTIOPHORIFORMES** (Betancur et al, 2013)

334. Family Sphyraenidae (Barracudas)
335. Family Xiphiidae (Swordfishes)
336. Family Istiophoridae (Billfishes)

69. Order **ANABANTIFORMES** (Britz, 1995) (Labyrinth fishes)

337. Family Anabantidae (Climbing gouramies)
338. Family Helostomatidae (Kissing gouramis)
339. Family Osphronemidae (Gouramies and fighting fishes)
340. Family Channidae (Snakeheads)
341. Family Nandidae (Asian Leaffishes)
342. Family Badidae (Chameleonfishes)
343. Family Pristolepidadae (Malayan leaffishes)

70. Order **PLEURONECTIFORMES** (Flatfishes)

344. Family Psettodidae (Spiny turbots)
345. Family Citharidae (Largescale flounders)
346. Family Scophthalmidae (Turbots)
347. Family Paralichthyidae (Sand flounders)
348. Family Pleuronectidae (Righteye flounders)
349. Family Bothidae (Lefteye flounders)
350. Family Paralichthodidae (Measles flounders)
351. Family Poecilopsettidae (Bigeye flounders)
352. Family Rhombosoleidae (Rhombosoleids)
353. Family Achiropsettidae (Southern flounders)

354. Family Samaridae (Crested flounders)
355. Family Achiridae (American soles)
356. Family Soleidae (True soles)
357. Family Cynoglossidae (Tonguefishes)

71. Order **SYNGNATHIFORMES**

358. Family Pegasidae (Seamoths)
359. Family Solenostomidae (Ghost pipefishes)
360. Family Syngnathidae (Pipefishes and Seahorses)
361. Family Aulostomidae (Trumpetfishes)
362. Family Fistulariidae (Cornetfishes)
363. Family Macrorhamphosidae (Snipefishes)
364. Family Centriscidae (Shrimpfishes)
365. Family Dactylopteridae (Flying gurnards)

72. Order **ICOSTEIFORMES** (Ragfishes)

366. Family Icosteidae (Ragfishes)

73. Order **CALLIONYMIFORMES** (Dragonets)

367. Family Callionymidae (Dragonets)
368. Family Draconettidae (Slope draconets)

74. Order **SCOMBROLABRACIFORMES** (Longfin escolars)

369. Family Scombrolabracidae (Longfin escolars)

75. Order **SCOMBRIFORMES** (Mackerals)

370. Family Gempylidae (Snake mackerels)
371. Family Trichiuridae (Cutlassfishes)
372. Family Scombridae (Mackerels and Tunas)
373. Family Amarsipidae (Amarsipas)
374. Family Centrolophidae (Medusafishes)
375. Family Nomeidae (Driftfishes)
376. Family Ariommatidae (Ariommatids)
377. Family Tetragonuridae (Squaretails)
378. Family Stromateidae (Butterfishes)

76. Order **TRACHINIFORMES**

379. Family Chiasmodontidae (Swallowers)
380. Family Champsodontidae (Gapers)

381. Family Pinguipedidae (Sandperches)
382. Family Cheimarrhichthyidae (New Zealand torrentfishes)
383. Family Trichonotidae (Sanddivers)
384. Family Creediidae (Sandburrowers)
385. Family Percophidae (Duckbills)
386. Family Leptoscopidae (Southern sandfishes)
387. Family Ammodytidae (Sand lances)
388. Family Trachinidae (Weeverfishes)
389. Family Uranoscopidae (Stargazers)

77. Order **LABRIFORMES** (Kaufman & Liem, 1982) (Wrasses)

390. Family Labridae (Wrasses)
391. Family Odacidae (Cales)
392. Family Scaridae (Parrotfishes)

78. Order **PERCIFORMES** (Bleeker, 1863) (Perches) The largest order of the class Teleostomi contains 6000 species and 150 families.

393. Family Centropomidae (Snooks)
394. Family Latidae (Lates perches)
395. Family Gerreidae (Mojarras)
396. Family Centrogeniidae (False scorpionfishes)
397. Family Perciliidae (Southern basses)
398. Family Howellidae (Oceanic basslets)
399. Family Acropomatidae (Lanternbellies)
400. Family Epigonidae (Deepwater cardinalfishes)
401. Family Polyprionidae (Wreckfishes)
402. Family Lateolabracidae (Asian seaperches)
403. Family Mullidae (Goatfishes)
404. Family Glaucosomatidae (Pearl perches)
405. Family Pempherididae (Sweepers)
406. Family Oplegnathidae (Knifejaws)
407. Family Kuhliidae (Flagtails)
408. Family Bathyclupeidae (Bathyclupeids)
409. Family Toxotidae (Archerfishes)
410. Family Arripidae (Australian salmon or kahawai)
411. Family Dichistiidae (Galjoen fishes)

412. Family Kyphosidae (Sea chubs)
413. Family Terapontidae (Grunters or tigerperches)
414. Family Percichthyidae (Temperate perches)
415. Family Sinipercidae (Chinese perches)
416. Family Enoplosidae (Oldwives)
417. Family Pentacerotidae (Armorheads)
418. Family Dinopercidae (Cavebasses)
419. Family Banjosidae (Banjofishes)
420. Family Centrarchidae (Sunfishes)
421. Family Serranidae (Sea basses)
422. Family Percidae (Perches)
423. Family Lactariidae (False trevallies)
424. Family Dinolestidae (Long-finned pikes)
425. Family Scombropidae (Gnomefishes)
426. Family Pomatomidae (Bluefishes)
427. Family Bramidae (Pomfrets)
428. Family Caristiidae (Manefishes)
429. Family Monodactylidae (Moonfishes or fingerfishes)
430. Family Priacanthidae (Bigeyes or Catalufas)
431. Family Leiognathidae (Ponyfishes, slimys or slipmouths)
432. Family Chaetodontidae (Butterflyfish)
433. Family Pomacanthidae (Angelfishes)
434. Family Malacanthidae (Tilefishes)
435. Family Haemulidae (Grunts)
436. Family Hapalogeniidae (Barbeled grunters)
437. Family Lutjanidae (Snappers)
438. Family Caesionidae (Fusiliers)
439. Family Cirrhitidae (Hawkfishes)
440. Family Chironemidae (Kelpfishes)
441. Family Aplodactylidae (Marblefishes)
442. Family Cheilodactylidae (Morwongs)
443. Family Latridae (Trumpeters)
444. Family Cepolidae (Bandfishes)
445. Family Scatophagidae (Scats)

446. Family Siganidae (Rabbitfishes)
447. Family Bovichthyidae (Temperate icefishes)
448. Family Pseudaphritidae (Catadromous icefishes)
449. Family Eleginopidae (Patagonian blennies)
450. Family Nototheniidae (Cod icefishes)
451. Family Harpagiferidae (Spiny plunderfishes)
452. Family Artedidraconidae (Barbeled plunderfishes)
453. Family Bathydraconidae (Antarctic dragonfishes)
454. Family Channichthyidae (Crocodile icefishes)

79. Order **SCORPAENIFORMES** (Greenwood et al, 1966) (Mail-cheeked fishes)

455. Family Scorpaenidae (Scorpionfishes or rockfishes)
456. Family Aploactinidae (Velvetfishes)
458. Family Pataecidae (Australian prowfishes)
459. Family Gnathanacanthidae (Red velvetfishes)
460. Family Congiopodidae (Racehorses or pigfishes or horsefishes)
461. Family Triglidae (Searobins or gurnards)
462. Family Peristediidae (Armored searobins)
463. Family Bembridae (Deepwater flatheads)
464. Family Platycephalidae (Flatheads)
465. Family Hoplichthyidae (Ghost flatheads)
466. Family Normanichthyidae (Barehead scorpionfishes or mote scorpionfishes)
467. Family Bathymasteridae (Ronquils)
468. Family Eulophiidae (Eulophiids)
469. Family Zoarcidae (Eelpouts)
470. Family Stichaeidae (Pricklebacks)
471. Family Cryptacanthodidae (Wrymouths)
472. Family Pholidae (Gunnels)
473. Family Anarhichadidae (Wolffishes)
474. Family Ptilichthyidae (Quillfishes)
475. Family Zaproridae (Prowfishes)
476. Family Scytalinidae (Graveldivers)
477. Family Hypoptychidae (Sand eel)
478. Family Aulorhynchidae (Tubesnouts)
479. Family Gasterosteidae (Sticklebacks)

480. Family Indostomidae (Armored sticklebacks)
481. Family Anoplopomatidae (Sablefishes)
482. Family Zaniolepididae (Combfishes)
483. Family Hexagrammidae (Greenlings)
484. Family Trichodontidae (Sandfishes)
485. Family Jordaniidae (Longfin sculpin)
486. Family Rhamphocottidae (Grunt sculpins)
487. Family Scorpaenichthyidae (Cabezon)
488. Family Agonidae (Poachers and searavens)
489. Family Cottidae (Sculpins)
490. Family Psychrolutidae (Fathead sculpins)
491. Family Bathylutichthyidae (Antarctic sculpins)
492. Family Cyclopteridae (Lumpfishes or lumpsuckers)
493. Family Liparidae (Snailfishes)
494. Family Eschmeyeridae

80. Order **MORONIFORMES** (Temperate basses)

495. Family Moronidae (Temperate basses)
496. Family Drepaneidae (Sicklefishes)
497. Family Ephippidae (Spadefishes)

81. Order **ACANTHURIFORMES** (Jordan, 1923) (Surgeonfishes)

498. Family Emmelichthyidae (Rovers)
499. Family Sciaenidae (Drums croakers)
500. Family Luvaridae (Louvar)
501. Family Zanclidae (Moorish Idols)
502. Family Acanthuridae (Surgeonfishes)

82. Order **SPARIFORMES** (Breams and Porgies)

503. Family Callanthiidae (Splendid perches or groppos)
504. Family Sillaginidae (Sillagos whitings or smelt-whitings)
505. Family Lobotidae (Tripletails)
506. Family Nemipteridae (Threadfin breams)
507. Family Lethrinidae (Emperors or emperor breams)
508. Family Sparidae (Porgies)

83. Order **CAPROIFORMES**

509. Family Caproidae (Boarfishes)

84. Order **LOPHIIFORMES** (Anglerfishes)

510. Family Lophiidae (Goosefishes)
511. Family Antennariidae (Frogfishes)
512. Family Tetrabrachiidae (Tetrabrachiid frogfishes)
513. Family Lopichthyidae (Lophichthyid frogfishes)
514. Family Brachionichthyidae (Handfishes warty anglers)
515. Family Chaunacidae (Coffinfishes or sea toads)
516. Family Ogcocephalidae (Batfishes)
517. Family Caulophrynidae (Fanfins)
518. Family Neoceratiidae (Spiny seadevils)
519. Family Melanocetidae (Black seadevils)
520. Family Himantolophidae (Footballfishes)
521. Family Diceratiidae (Double anglers)
522. Family Oneirodidae (Dreamers)
523. Family Thaumatichthyidae (Wolftrap anglers)
524. Family Centrophrynidae (Prickly seadevils)
525. Family Ceratiidae (Warty seadevils)
526. Family Gigantactinidae (Whipnoses anglers)
527. Family Linophrynidae (Leftvents)

85. Order **TETRAODONTIFORMES** (Plectognaths)

No parietal, nasal, infraorbital, or lower ribs; post-temporal, if present, simple and fused with the pterotic of the skull; hyomandibular and palatine firmly attached to the skull; gill openings restricted; maxillae usually firmly united or fused with premaxillae; scales usually modified as spines, shields or plates; lateral line present or absent, sometimes multiple.

528. Family Triodontidae (Threetooth puffers)
529. Family Triacanthodidae (Spikefishes)
530. Family Triacanthidae (Triplespines)
531. Family Aracanidae (Deepwater boxfishes)
532. Family Ostraciidae (Boxfishes)
533. Family Balistidae (Triggerfishes)
534. Family Monacanthidae (Filefishes)
535. Family Molidae (Molas ocean sunfishes)

536. Family Tetraodontidae (Puffers)
537. Family Diodontidae (Porcupinefishes)

Fish classification has undergone significant revisions in recent years, and further changes are likely in the future. Ichthyologists frequently differ on both major and minor concepts of phyletic relationships. There is still much to discover about both living and fossil fishes.

Zhang et al, (2021) applied deep learning to understand various activities like identification and classification of fish species, feeding, behaviour, estimation of size and prediction of water quality.

Kandimalia et al, (2022) explained automated detection, counting of fish in its passages as well as classification by applying deep learning.

The classification of a fishery on the list of fisheries (LOF) for 2022 as published by The National Marine Fisheries Service (NMFS) determines whether participants in that fishery are subject to certain provisions of the Marine Mammal Protection Act (MMPA), such as registration, observer coverage, and take reduction plan (TRP) requirements (NOAA, 2022).

Saleh et al, (2022) conducted a survey on fish classification in underwater habitats using computer vision and deep learning.

The Catalog of Fishes by Eschmeyer (2022) is the standard reference for taxonomic fish names, with a searchable online database. The Catalog of Fishes contains about 61,700 species and subspecies, over 11,000 genera and subgenera, and over 34,000 bibliographic references. According to Eschmeyer (2022), fish are divided into the following eight classes:

1. Myxini
2. Petromyzonti
3. Elasmobranchii
4. Holocephali
5. Cladistii
6. Actinpteri
7. Coelacanthi
8. Dipneusti

Gong et al, (2023) proposed a novel method for multi-water fish classification (Fish-TViT) based on transfer learning and visual transformers. Fish-TViT uses a label smoothing loss function to solve the problem of overfitting and overconfidence of the classifier.

Unit-II Type Study of Fish

6

Agnatha

Agnatha consists of cyclostomes, conodonts and ostracoderms. Among recent animals, cyclostomes are sister to all jawed vertebrates known as gnathostomes. They have little economic importance. It includes the oldest known craniate fossils and has many primitive characteristics.

The oldest fossil of agnathans appeared in the Cambrian period and two groups the lampreys (Hyperoartii) and the hagfish (Myxini or Hyperotreti) comprising about 120 species still survive today (Fig. 6.1).

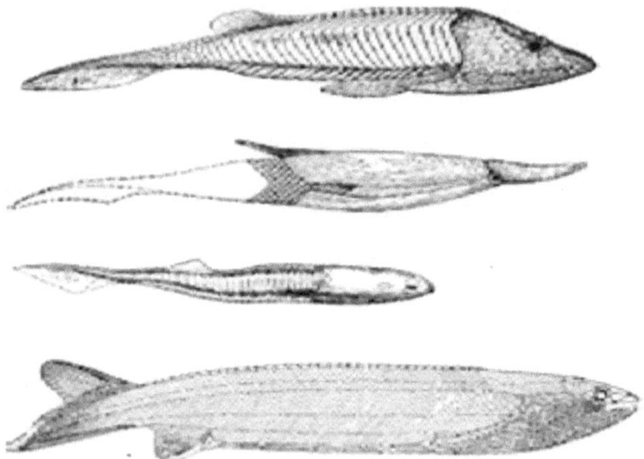

Fig. 6.1: An overview of Agnatha (from https://www.sciencedirect.com)

Hagfishes are considered members of the subphylum *Vertebrata*, because they secondarily lost vertebrae; before this event was inferred from molecular and developmental data, the group Craniata was created by Linnaeus to reference hagfish and vertebrates. Recent studies from rRNA, mtDNA and embryology strongly supports the hypothesis that living agnathans, the cyclostomes, are monophyletic.

While a few scientists still regard the living agnathans as only superficially similar and argue that many of these similarities are probably shared fundamental characteristics of ancient vertebrates. Recent taxonomic studies clearly place hagfish with the lampreys as being more closely related to each other than either is to the jawed fishes. Kent (1967) believed that agnathan have lost their bony structure due to mutation.

6.1 Features

1. Freshwater or marine.
2. Scavengers and parasites
3. Fish-like or eel-like body
4. Absence of jaws, exoskeleton and generally paired fins.
5. Early extinct species had heavy bony scales and plates in their skin.
6. Endoskeleton mostly cartilaginous.
7. The embryonic continuous encapsulated notochord persists in the adult.
8. Seven or more paired gill pouches are present.
9. Segmental myotomes but little modified from head to tail.
10. Ciliated alimentary tract straight and without much regional specialisation.
11. Presence of suctorial mouth, single nostril and unpaired fins.
12. Lampreys have a light-sensitive pineal eye.
13. The middle Ordovician period, flourished during the Silurian and lower Devonian periods became rarer in the Devonian period and some survive to present.

6.2 Ancestry

Although a minor element of modern marine fauna, agnathans were prominent in the early Paleozoic era. Two types of early animals having fins, vertebrate musculature and gills are known from the early Cambrian Maotianshan shales of China: *Haikouichthys* and *Myllokunmingia*. They have been tentatively assigned to Agnatha by Janvier. A third possible agnathid from the same region is *Haikouella* which has not been formally described and was reported by Simonetti from the Middle Cambrian Burgess Shale of British Columbia.

Many Ordovician, Silurian and Devonian agnathans were armoured with heavy bony-spiky plates. The first armoured agnathans (the Ostracoderms), precursors to the bony fish and the tetrapods are known from the middle Ordovician period and by the Late Silurian period the agnathans had reached the high point of their evolution. Most of the ostracoderms, such as thelodonts, osteostracans, and galeaspids were more closely related to the gnathostomes than to the surviving agnathans, known as cyclostomes. Cyclostomes split from other agnathans before the evolution of dentine and bone, which are present in many fossil agnathans, including conodonts. According to Kent (1967), cyclostomes have lost the bony armour due to mutation. Agnathans declined in the Devonian period and never recovered.

Approximately 500 million years ago, two types of recombinatorial adaptive immune systems (AISs) arose in vertebrates. The jawed vertebrates diversify their repertoire of immunoglobulin domain-based T and B cell antigen receptors through the rearrangement of V(D)J gene segments and somatic hypermutation, but none of the fundamental AIS recognition elements in jawed vertebrates have been found in jawless vertebrates. Instead, the AIS of jawless vertebrates is based on variable lymphocyte receptors (VLRs) that are generated through recombinatorial usage of a large panel of highly diverse leucine-rich-repeat (LRR) sequences. Three VLR genes (VLRA, VLRB, and VLRC) have been identified in lampreys and hagfish, and are expressed in three distinct lymphocyte lineages. VLRA+ and VLRC+ cells are T-cell-like and develop in a thymus-like lymphoepithelial structure, termed thymoids. VLRB+ cells are B-cell-like, develop in hematopoietic organs, and differentiate into "VLRB antibody"-secreting plasma cells.

6.3 Organisation of Agnatha

6.3.1 Morphology

There is the absence of jaws, modern agnathans are characterised by the absence of paired fins; the presence of a notochord both in larvae and adults; and seven or more paired gill pouches. Lampreys have a light-sensitive pineal eye. All living and most extinct Agnatha do not have an identifiable stomach or any appendages. The Agnathans have a cartilaginous skeleton, and the heart contains 2 chambers.

6.3.2 Body covering

In modern agnathans, the body is covered in skin, with neither dermal nor epidermal scales. The skin of hagfish has copious slime glands for defense mechanisms. The slime can sometimes clog up enemy fishes' gills, causing them to die. In direct contrast, many extinct agnathans sported extensive exoskeletons composed of either massive, heavy dermal armour or small mineralised scales.

6.3.3 Appendages

Almost all agnathans lack paired appendages, although most do have a dorsal or a caudal fin. However, osteostracans and pituriaspids did have paired fins, a trait inherited in their jawed descendants.

6.3.4 Metabolism

Agnathans are ectothermic or cold-blooded, so their metabolism is slow in cold water and therefore they do not have to eat very much. They have no distinct stomach but rather a long gut, more or less homogeneous throughout its length. Lampreys feed on other fish and mammals. Anticoagulant fluids preventing blood clotting are injected into the host, causing the host to yield more blood. Hagfish are scavengers, eating mostly dead animals. They use a row of sharp teeth to break down the animal. The fact that Agnathan teeth are unable to move up and down limits their possible food types.

6.3.5 Reproduction

Fertilization in lampreys is external but in hagfishes is not known. Development in both groups probably is external. There is no known parental care. Not much is known about the hagfish reproductive process. It is believed that hagfish only have 30 eggs over a lifetime. Most species are hermaphrodites. There is very little of the larval stage that characterizes the lamprey. Lamprey is only able to reproduce once. After external fertilization, the lamprey's cloacas remain open, allowing a fungus to enter their intestines, killing them. Lampreys reproduce in freshwater riverbeds, working in pairs to build a nest and burying their eggs about an inch beneath the sediment. The resulting hatchlings go through four years of larval development before becoming adults.

6.4 Classification

Phylum **Chordata**

Subphylum **Vertebrata**

Super class **AGNATHA**

Agnatha relates to Marsipobranchii or Jordan (1923) and Regan (1929) or Cyclostomi of Stensio (1968).

Class **MYXINI**

An ancient and unique lineage of primitive eel-like marine fishes, the Skeleton is made up of cartilage, produce slime, lack true jaws and scales, degenerate eyes and many gill openings.

1. Order MYXINIFORMES

Nostril at the tip of the snout, eyes vestigial, tongue with 2 rows of horny teeth, mouth surrounded by barbels, slime glands along the body, 1–16 pairs of gills and single fin extending around the posterior end of the body (Fig. 6.2). They arose during Pennsylvanian sub-period.

1. Family **Myxinidae** (Hagfishes)(Rafinesque, 1815)(7 genera, about 70 species)

Myxine glutinosa (Linnaeus, 1758) *M. limosa* (Girard, 1859)

M. formosana (Mok & Kuo, 2001)
M. capensis (Regan, 1913)
M. greggi (Mincarone et al, 2021)
Notomyxine tridentiger (Garman 1899)
Nemamyxine elongata Richardson 1958
Eptatretus strickrotti Møller & Jones, 2007.
M. mcmillanae (Hensely, 1991)
M. australis (Jenyns, 1842)
M. phantasma (Mincarone et al, 2021)
Neomyxine biniplicata Richardson 1953
Eptatretus cirrhatus J. R. Forster, 1801

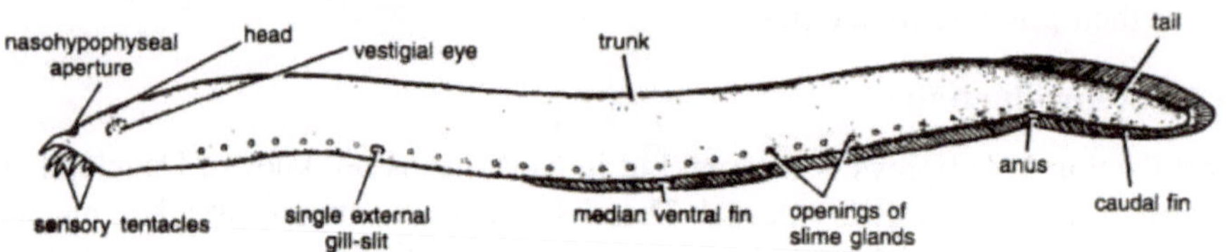

Fig. 6.2: *Myxine* (from https://www.faunafondness.com)

Class **PETROMYZONTIDA or HYPEROARTIA**

They show the beginnings of a vertebral column as neural arches protruding from the notochord, possesses unpaired fins and eggs hatch into ammocoete larvae.

2. Order PETROMYZONTIFORMES (Berg, 1940) (10 genera and 43 species)

7 pairs of gills open through pores, laterally placed eyes, single nostril dorsal, horny teeth on an oral sucker, the horny teeth on the tongue. 1 or 2 dorsal fins present. They arose during the Pennsylvanian sub-period.

2. Family **Petromyzontidae** (Northern lampreys or Vampire fish) (Risso, 1827) (8 genera and 39 species).

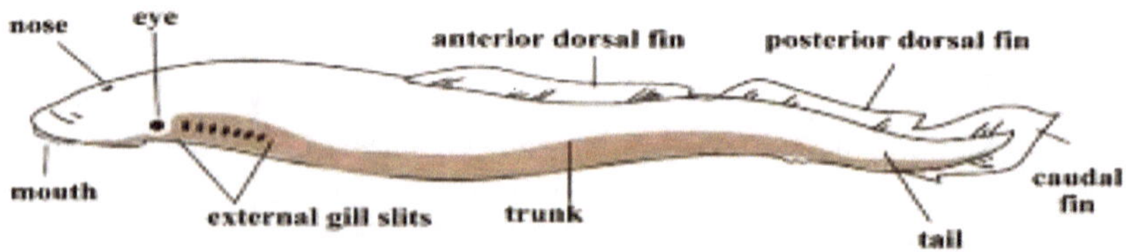

Fig. 6.3: *Petromyzon* (After *Forey and Janvier 1993*)

Single horny tooth plate present above the mouth and carry pointed or rounded teeth. Found in Eurasia and North America (Fig. 6.3).

Petromyzon marinus (Linnaeus, 1758)
Entosphenus tridentatus (Richardisn, 1836)
E. macrostomus (Beamish, 1982)
Ichthyomyzon castaneus (Girard, 1858)
I. unicuspis (Hubbs & Trautman, 1937)
Lampetra fluviatilis (Linnaeus, 1758)

Caspiomyzon wagneri (Kessler, 1870)
Eudontomyzon danfordi (Regan, 1911)
E. graecus (Renaud & Economidis, 2010)
Tetrapleurodon spadiceus (Bean, 1887)
L. auremensis (Mateus et al, 2013)
Lethenteron camtschaticum (Tilesius, 1811)
L. ninae (Naseka, Tunjyev & Renaud, 2009)

3. Family **Geotriidae** (Southern lampreys) (Jordan, 1923) (1 genus and 2 species)

A single tooth-plate is present above the mouth and carries 4 teeth. From southern Australia, New Zealand, South America.

Geotria australis (Gray, 1851) *G. macrostoma* (Burmeister, 1868)

4. Family **Mordaciidae** (Southern top eyed lampreys)(Gill, 1893) (1 genus and 3 species)

Tooth plate above mouth paired, each being tricuspid. They are reported from Eastern Australia to Western South America.

Mordacia mordax (Richardson, 1846) *M. praecox* (Potter, 1968)

Class **OSTRACODERMI**

Ostracoderms are jawless vertebrates commonly known as bohy-skinned fishes or armoured fishes lacking a bony vertebral column. They are supposed to be first the first vertebrates that appeared on the earth. They include all extinct agnathans.

3. Order HETEROSTRACI (Pterapsida): Marine or freshwater, Head covered with 2 large bony plates and separated by a variable number of smaller plates. A slit-like mouth (as rostrum) is adapted for bottom-feeding. Eves widely separated. These plates lack true bone cells. Their size ranges from 25cm up to 1.5 meter. Single gill opening on each side, body covered with large scales, possibly paired nostrils. Bone lacks enclosed bone cells. A prominent lateral line was present. Small tail laterally flattened and hypocercal. The nasal aperture is absent but the pineal apparatus is present on the head. According to Parker and Haswell (1962), the dorsal shield contains 1-rostral, 1-pineal, 1-dorsal, 2- orbital, 2-branchial, 2-cornual, 1-ventrla and many oral plates. The brain is primitive and long containing an unspecialised medulla and uncompressed telencephalon. Found during the Cambrian to late Devonian period. Example: *Pteraspis rostrata* (Fig. 6.4), *Coelolepsis, Drepanaspis, Thelodus*.

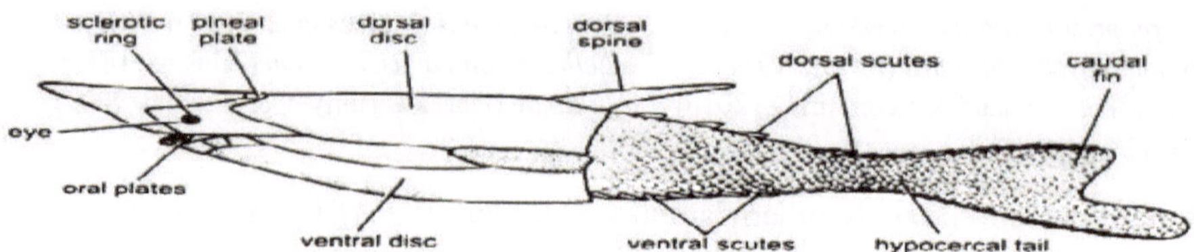

Fig. 6.4: *Pteraspis rostrata* (from https://www.notesonzoology.com)

4. Order ARANDASPIDA (Fig. 6.5): The oldest craniates that contain all chordates with a cartilage-derived skull. Body fusiform with a flat dorsal shield and a bulging ventral shield is present. In the anterior part of the dorsal shield, a paired pineal opening is likely the external opening of the endolymphatic ducts. The eyes are surrounded by a sclerotic ring present at the anterior end of the dorsal shield. The minute gill openings are probably more than 15. The body is covered with rod-shaped scales arranged in chevrons, and the tail is probably pad-shaped and diphycercal. The dermal bones consist of aspidine (acellular bone) and are ornamented with oakleaf-shaped tubercles which seem to contain no dentine. Found during the Middle to Late Ordovician Period. Example: *Aseraspis*(Fig. 6.6), *Andinaspis*, *Apedolepis, Pircanchaspis, Sacabambaspis* .

Fig. 6.5: *Arandaspida* (from https://en.wikipedia.org)

Fig. 6.6: *Astrapsis* (from https://animals.fandom.com)

5. Order ASTRASPIDA: Head covered with small mushroom-shaped plates, gill openings separate. Found the Late Ordovician to Early Silurian Period in North America. They have a dermal ornamentation of large, mushroom-shaped tubercles of fine-tubuled dentine (“astraspidine”), covered with a thick, glassy cap of enameloid. Example: *Astraspis*,

6. Order OSTEOSTRACI (Cephalaspida): Head was protected by the broad bony shield; the under surface of the head was covered with tiny scales, gill openings on the undersurface, eyes dorsal close to the top of the head, nostril median and placed on top of the head, areas of a bony shield covered with tiny tesserae. Their cartilaginous body was fish-like and bears a median dorsal fin, paired pectoral fins, and heterocercal tail. The well-marked area of the head seems to bear a kind of electric or sense organs. Found during the Silurian to Late Devonian Period. Examples: *Hemicyclaspis, Cephalapsis* (Fig. 6.7).

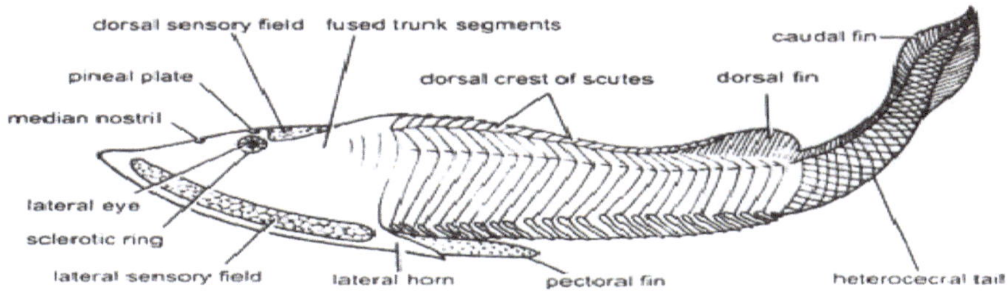

Fig. 6.7: *Cephalapsis* (from https://www.notesonzoology.com)

7. Order GALEASPIDA: Head covered in broad semicircular head shield that is sometimes drawn out to a pointed snout, eyes dorsal, medium nostril very large, gills on the undersurface. The large opening on the dorsal surface of the head shield connected to the pharynx and gill chamber and scalloped sensory lines. They had up to 45 gill openings. The body is covered with minute scales arranged in oblique rows and there is no other fin besides the caudal fin. The mouth is the ventral side of the head and suggests bottom-dwellers. Found during Silurian to Late Devonian Period. Example: *Hanyangaspis* (Fig. 6.8).

Fig. 6.8: *Hanyangaspis*
(from spinops.blogspot.com)

Fig. 6.9: *Birkenia*
(from https://alchetron.com/Birkenia)

8. Order ANASPIDA: Small streamlined fishes having habits of active swimming and surface feedings, the body lacks a shield and head covered with elongated anaspid scales, mouth terminal, nostril between laterally placed eyes, gill openings lateral and arranged in slanting line, hypocercal tail bent downward. Paired fins absent but paired pectoral spines and many median spines are present. Found during the Silurian to Late Devonian Period. Example: *Birkenia* (Fig. 6.9).

9. Order THELODONTI (Coleolepids): A little-known monophyletic group of unknown affinities. The body is covered in tiny scales; in some, the body is flattened from top to bottom and in others from side to side. They resemble classes Heterostraci and Anaspida Found during the Silurian to Late Devonian Period. Example: *Lanarkia* (Fig. 6.10), *Loganellia*.

Fig. 6.10 *Lanarkia* (from spinops.blogspot.com)

6.5 Degenerate, primitive and specialised characters of agnathans

6.5.1 Degenerate characters

1. Body elongated and eel-like
2. Exoskeleton absent
3. Endoskeleton cartilaginous
4. Paired fins and girls absent
5. Eyes rudimentary
6. Liver reduced with bile duct and gall bladder in the adult

6.5.2 Primitive characters

(I) Characters resembling Cephalochordates

1. Unconstricted notochord due to rudimentary vertebral column.
2. Segmental musculature, myotomes extending from head to tail.
3. Relatively large numbers of gill-slits
4. No Gonoducts
5. Straight and ciliated alimentary canal without stomach but has less regional modifications.
6. Presence of endostyle in Ammocoete larva.

(II) Characters primitive than fishes

1. No hinged jaws, paired appendages, medullated nerves and true teeth
2. Median fin-fold continuous with diphycercal caudal fin.
3. Fin-rays without muscular attachments
4. Incomplete cranium
5. Single and median nostril

6. Vertebrae are either absent or poorly developed.
7. The alimentary canal lacks a spiral valve
8. Rudimentary pancreas
9. 'S' shaped Heart lacks conus arteriosus.
10. Brain primitive with small cerebellum and possesses IX and X cranial nerves. Sympathetic nervous system ill developed. Nerves are non-medullated and dorsal and ventral nerves are present in each segment.
11. Hypophyseal duct is large but not connected to the pituitary
12. Semicircular canals either 1 or 2.

6.5.3 Specialised characters

(I) Unique Specialised Characters

1. Mouth suctorial with a strong, muscular, denticulate rasping tongue and horny teeth.
2. Anticoagulant present in saliva.
3. Gills present in gill pouches, gill openings posteriorly placed
4. Alimentary canal with an oesophagus and a pharynx.
5. A cartilaginous buccal basket supports the pharynx and helps in respiration.
6. Mucous glands secrete mucus.
7. Presence of single nasal aperture

(II) Characters advanced than Cephalochordata

1. Distinct head and skull present.
2. Cranium encloses the brain.
3. Kidneys are either pronephric or mesonephric.
4. Well-developed heart and liver.
5. Multilayered epidermis with unicellular mucous glands
6. Spinal nerves with dual roots and ganglia.
7. Vertebrae are either absent or poorly developed.
8. Sympathetic nervous system ill developed.
9. Lateral line organs present.
10. Myotomes 'W' shaped.

6.6 Special features of Cyclostomata

1. Vermiform appearance
2. Marine or freshwater, ascends freshwater to breed
3. Free-living, ectoparasites, sanguivorous or scavengers

4. Absence of tongue, scales, paired fins, true fin rays, genital ducts, cloaca, conus arteriosus, renal portal system, exoskeleton, sympathetic nervous system
5. Dorsal fin may be present
6. Presence of soft and slimy skin, cartilaginous skeleton, unpaired fins, a single nostril, lateral line organ, single gonad, 8 pairs of cranial nerves,
7. Labial cartilage well developed
8. Mouth suctorial, buccal cavity with epidermal teeth
9. Paired gill pouch as Marsipobranchs, internal gills 5-16 pairs
10. Heart two-chambered, aortic arches many
11. Fertilisation external
12. Development direct or indirect, indirect development with Ammocoete larva
13. Ammocoete larva is microphagous
14. Olfactory lobe large but hollow or solid
15. Median olfactory organ opens in the pituitary sac
16. Cerebellum very small
17. Nerves non-medullated
18. Internal ear with 1 or 2 semi-circular canal

7

Placodermi

7.1 Introduction

Placoderms (plate-skinned fishes) (Fig. 7.1) flourished and evolved in divergent lines during the Devonian period and became extinct by the end of the Permian period. They were bottom-dwelling, mud-sucking Palaeozoic fishes that may have been nectonic. Their bodies were divided into three parts: the head, the trunk and the tail. They ranged in size from 10 cm to 9 meters. The head and trunk are surely covered by bony plates. The bony plates were made up of laminar and trabecular bones. Their dorso-ventrally flattened body showed a primitive type of autostylic or aphetohyoidean jaw condition. They have hyoid gill slits, a heterocercal or diphycercal tail and parried fins but are devoid of spiracles. They had developed internal ears with semicircular canals. Males bear pelvic claspers. According to Romer (1959), Placoderms originated and lived exclusively in freshwater in the early stages but later forms invaded the seas during the Devonian period.

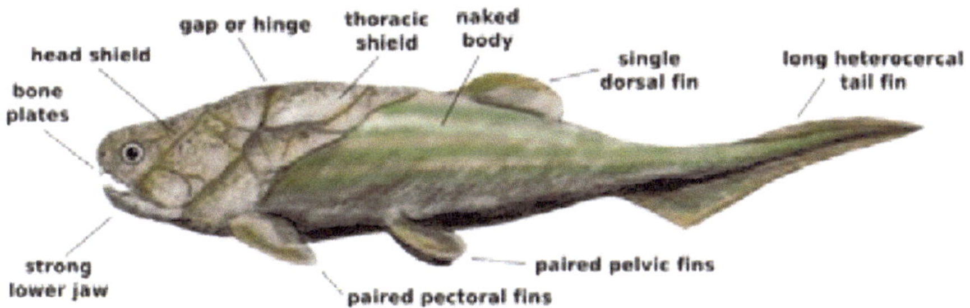

Fig. 7.1: A typical placoderm (from https://en.wikipedia.org)

Placoderms are thought to have evolved from some types of ostracoderms. However, their fossil record does not show any connecting link between ostracoderms and placoderms. They were polymorphic and closely related to cartilaginous fishes. Their fragmentary remains have been reported from the late Carboniferous and early Silurian periods.

The placoderms present a wide range of adaptive radiation. They possess many peculiar and specialised characteristics and are regarded to be the progenitors of the modern fishes.

7.2 Taxonomic position

Phylum: Chordata

Subphylum: Vertebrata

Division: Gnathostomata

Superclass: Pisces

Class: Placodermi

7.3 Specific features

Presence of primitive aphetohyoidean jaw suspension as the hyoid arch did not support the jaw.

1. Hyoid gills were present.
2. Hyoid is unspecialized and does not take part in jaw suspension.
3. Males with pelvic claspers.
4. Paired fins were present.
5. Pineal opening situated between the eyes.
6. Spiracles were absent.
7. The body is divided into a head, trunk and tail.
8. The body is protected by heavy bony armour.
9. The eyes are ventrolateral.
10. The upper jaw fused with the skull.
11. They have five pairs of gills.
12. They possess internal ears and semicircular canals.
13. They possess paired fins and heterocercal tails.
14. They were close to cartilaginous fishes.
15. Unossified palatoquadrate and mandibular fused head shield.

Placoderms only lived for a short time and were regarded as "unsuccessful" in the evolution of gnathostomes. They are regarded as ostracoderm derivatives. Placoderms either became completely extinct or gave rise to osteichthyes and chondrichthyes.

The evolution of jaws significantly altered their way of life. This led to the evolution of efficient muscular, sensory and nervous systems.

Class 1: Coccostei

Order 1. Arthrodira (Coccostei, Arthrodiriformes): It is the largest order containing some 60% of placoderms. It is an armoured, joint-necked, jawed fish that flourished in the Devonian period before its sudden extinction, surviving for about 50 million years in freshwater to brackish water and penetrating most marine ecological niches and having a very efficient tearing mechanism. The head is composed of many dermal paired and unpaired median bones. The occipital is short and has only one pair of paranuchal plates; the medulla oblongata is elongated and poorly differentiated from the spinal cord. The nostrils are small and located near the midline. Due to the presence of a condyle and socket, they have a movable joint between the armour surrounding their heads and bodies. They lack distinct teeth but have bony cusps and use the sharpened edges of a bony plate as a biting surface. The eye sockets are protected by a bony ring. *Arctolepis* were well-armoured fish with flattened bodies. The upper jaw bears 2 pairs of tooth plates. The movable mandible was made of a single bone. The operculum is formed by the posterolateral parts of the head shield. The lateral line sensory canal system is well developed. The vertebral center, ribs and hyoid were lacking. Gills were perhaps protected by posteroventrolateral extensions of bony plates. They had a notochord at the floor of the neurocranium with neural and haemal arches. A heterocercal tail, a single dorsal fin, paired but small pelvic fins and paired pectorals as an immovable bony structure. The dorsal fins are supported by 2 rows of skeletal elements. The hypophysis is anteriorly placed and pineal organs are two. The largest member of this group, *Dunkleosteus*, measures 6 m in length and *Titanichthys* up to 10 m. In contrast, the long-nosed *Rolfosteus* measured just 15 cm. Example: *Arctolepis, Coccosteus* (Fig. 7.2), *Dunkleosteus (=Dinichthyes).*

Fig. 7.2: *Coccosteus* (from https://prehistopedia.fandom.com)

Class 2: Pterichthyes

Order 2. Antiarchi (Pterichthyes, Antiarchiformes): They form the second-most successful group of placoderms. The orbital fenestra was the opening for the mouth (oral siphon) and the opening for the anal siphon was on the other side of the body as opposed to having both the oral and anal siphons together at one end. Their cephalic plates had a different arrangement. The pair of pectoral fins was modified into a pair of caliper or arthropod-like limbs. In *Yunnanolepis*, the limbs were thick and short, while in *Bothriolepis*, the limbs were long and had elbow-like joints. The function of the limbs is still not perfectly understood, but, most hypothesize that they helped their owners pull themselves across the substrate, as well as bury themselves in it. A median dorsal fin was present and a heterocercal tail was present. Scales were present on the trunk and tail of *Petrichthys*. A pineal opening between the eyes was present. Example: *Pterichthys* (Fig. 7.3), *Bothriolepis*.

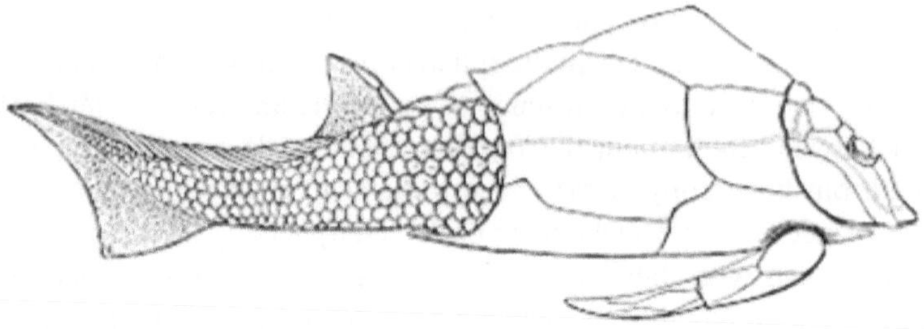

Fig. 7.3: *Pterichthys* **(from https://etc.usf.edu)**

Order 3. Brindabellaspida: According to Janvier, anatomical similarities of *Brindabellaspis stensioi's* (Fig. 7.4) brain and those of Osteostraci and the Galeaspida, strongly suggests that *B. stensioi* and also the antiarchs, are basal placoderms closest to the ancestral placoderm. *B. stensioi* is a placoderm with a flat, platypus-like snout from the Early Devonian of the Taemas-Wee Jasper reef in Australia. When it was first discovered in 1980, it was originally regarded as a Weejasperaspid acanthothoracid due to anatomical similarities with the other species found at the reef.

Fig. 7.4: *Brindabellaspis* (from https://en.wikipedia.org)

Order 4. Phyllolepida (Phylolepidiformes): They were bottom-dwelling predators that ambushed prey and inhabitants of freshwater environments. The armour of the phyllolepids was made of whole plates, rather than numerous tubercles and scales, They are considered to have been blind, as the orbits for the eyes are extremely small, so much to suggest that the eyes were vestigial and that they were placed on the sides of the head, as opposed to visually-oriented bottom-dwelling predators, like, say stargazers or flatfish, which have their eyes placed high on top of the head. Most fossils consist of fragments of their thoracic armour, *Phyllolepis* (Fig. 7.5) and *Austrophyllolepis* have been thoroughly studied.

Fig. 7.5: *Phyllolepis* (from https://www.wikiwand.com)

Order 5. Ptyctodontida (Ptycodontiformes): Ptyctodontids (folded teeth placoderms) have a superficial resemblance to chimaeras because of their big heads, big eyes, reduced armour and long bodies. Their armour was reduced to a pattern of small plates around the head and neck. Most ptyctodontids lived near the sea bottom and preyed on shellfish and other molluscs. They lacked shagreen on their skin, armoured plates and scales were made of bone, the anatomy of their craniums was more similar to those of other placoderms and they had beak-like tooth-plates. They were sexually dimorphic, the males had hook-like growths on their pelvic fins analogous to the clasping organs. Except for the ptyctodontids, the claspers were lost in the evolutionary process. Examples: *Australoptyctodus* (Fig. 7.6), *Campbellodus, Chelyophorus, Deinodus, Destnoporella, Eczematolepis, Goniosteus, Kimbryanodus, Materpiscis, Ptyctodontidsis, Rhamphodopsis, Rhynchodus.*

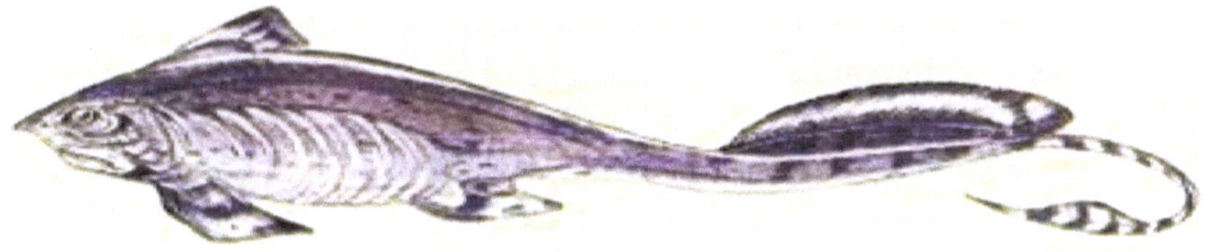

Fig. 7.6: *Australoptyctodus* (from https://www.deviantart.com)

Order 6. Rhenanida (Stegoselachii, Rhenaniformes): The rhenanids' armour was made up of a mosaic of unfused scales and tubercles. All rhenanids were flat, ray-like, bottom-dwelling predators that lived in marine environments. Paired fins present as large pectoral and small pelvic fins but unpaired fins present as a median dorsal supported by spines. Gills were covered by a large operculum but the spiracle opening was absent. Examples: *Asterosteus, Nefudina, Gemuendina, Jagorina, Bolivosteus* (Fig. 7.7).

Fig. 7.7: *Bolivosteus* (from https://www.deviantart.com)

Class 3: Acanthodii

Order 7. ACANTHOTHORACI: Acanthothoraci (spine chests) are chimaera-like placoderms and closely related to the rhenanids. They were distinguished from other placoderms due to differences in the anatomy of their skulls, and due to patterns on the skull plates and thoracic plates that are unique to this order. Example: *Arabosteus variabilis*.

Order 8. Petalichthyida (Petalichthyformes): It is a small, flattened placoderm fish. They are typified by their splayed pectoral fins, exaggerated lateral spines, flattened bodies, and numerous tubercles that decorate all of the plates and scales of their armor. They reached a peak in diversity during the Early Devonian and were found throughout the world, particularly in Europe (especially in Germany), North America, Asia, South America, and Australia. The most primitive petalichthyid is *Diandongpetalichthys* from the earliest Devonian-aged strata of Yunnan. The presence of *Diandongpetalichthys*, along with other primitive petalichthyids including *Neopetalichthys* and *Quasipetalichthys*, and more advanced petalichthyids, suggests that the order may have arisen in China, possibly during the late Silurian. Examples: *Lunaspis* (Fig. 7.8). *Macropetalichthys, Wijdeaspis.*

Fig 7.8: *Lunaspis* (from https://en.wikipedia.org)

Order 9. Pseudopetalichthyida: They are lightly armored placoderms known only from rare fossils in Lower Devonian strata in Hunsrück, Germany. Like *Stensioella heintzi*, and the Rhenanida, the Pseudopetalichthids had armor made up of a mosaic of tubercles. Like *Stensioella heintzi*, the Pseudopetalichthids' placement within Placodermi is suspect. Some suggested that *Paraplesiobatis heinrichsi* is the same species as *P. problematica*. The anatomy of the specimens of *Nessariostoma granulosum* is sufficiently different from the other two pseudopetalichthyids to be recognized as a separate species. *N. granulosum* bears a superficial resemblance to *Stensioella heintzi*, and it was this resemblance that originally led to the pseudopetalichthyids' grouping within Stensioellidae. Example: *Pseudopetalicthys problematica.*

Order 10. Stensioellida: *Stensioella heintzi* is an enigmatic placoderm of arcane affinity. It is only known from the Lower Devonian Hunsrück slate of Germany. *Stensioella heintzi* has an elongated body, a whip-like tail, and long, wing-like pectoral fins. In life, the animal would have looked vaguely like an elongated ratfish. Like the sympatric *Gemuendina*, *S. heintzi* had armor made up of a complex mosaic of small, scale-like tubercles. Example: *Stensioella heintzi.*

8

Elasmobranchii

8.1 Introduction

Cartilaginous fishes have ventrally placed mouths with teeth in several series. They possess rigid dorsal fins and placoid scales. They have five pairs of gill clefts opening individually to the exterior, a lower jaw articulated with the upper jaw and upper jaw not fused to the skull, no swim bladder, an operculum, an endolymphatic duct and a heterocercal or homocercal tail. It shows internal fertilization, meroblastic cleavage and direct development.

Elasmobranchii includes a wide range of cartilaginous fishes. Dogfish sharks are commonly studied all over the world probably because of their suitable size for dissection. The common example of the group, the dogfish or dog shark, is chosen as an introductory type to give a general idea of the class. There are several dogfish genera found in various parts of the world. *Scoliodon*, a typical Indian genus, is described below. The genus contains about nine species, four of which are very common in the Indian Ocean. They are the black shark (*Scoliodon sorrakowah*), *Scoliodon dumerilii, Scoliodon palasorrah* and *Scoliodon walbeehmi*. *Scoliodon sorrakowah* was first worked out and published by EM Thillayampallam (1928) in Indian Biological Memories.

8.2 Systematic Position

Kingdom: Animalia

Phylum: Chordara

Subphylum: Vertebrata (= Craniata)

Superclass: Gnathostomata

Class: Elasmobranchii (Chondrichthyes)

Order: Carcharhiniformes

Family: Carcharhiniidae

Genus: *Scoliodon*

Species: *Scoliodon sorrakowah* Bleeker, 1853

= *Scoliodon walbeehmi*, Bleeker, 1856

Scoliodon differs from other dogfish genera by having an elongated snout, a depressed head, and a compressed body. The teeth in both jaws are similar. The caudal pit and subcaudal lobe are visible and distinct.

8.3 Organisation of Elasmobranchs

8.3.1 Habit and Habitat

Scoliodon is a marine carnivorous fish. It relies on its quick movement to catch prey such as crabs, lobsters and worms. It is a fast swimmer and primarily a fish eater, catching prey with its sharp teeth. It feeds on small pelagic schooling and bottom-living bony fishes, codlet (Bregmacerotidae), burrowing gobies (Tripauchenidae), Bombay ducks (Harpadontidae), shrimp and cuttlefish.

8.3.2 Distribution

From Zanzibar to Ceylon (Sri Lanka), Ceylon to the Malay Archipelago of the Indian Ocean, Bay to Bengal, Pacific West India, Eastern Pacific (from Mexico to Panama), the Atlantic Ocean (from Labrador to Brazil), the West Indies, and the eastern coasts of South America are all home to *Scoliodon. Scoliodon fossils* have been discovered in geological strata ranging from the lower Eocene to later periods. These are marine creatures that can be found in all open seas.

8.3.3 External structures

Scoliodon has a laterally compressed spindle-shaped and elongated body tapered at the ends, making it a very fast swimmer. A fully developed specimen of the genus grows 30—60 cm long. *Scoliodon* has a dark grey dorsal side and a pale white underside, while the caudal fin is more or less dark.

The body is divided into three parts: the head, the trunk, and the tail. The head is flattened dorsoventrally and ends anteriorly in a compressed snout (rostrum). In the transverse section, the trunk is more or less oval. The middle region has the most thickness, and the body gradually tapers posteriorly into a long tail (Fig. 8.1A). The tail is oval in cross-section as well (Fig. 5.2B) and has a heterocercal caudal fin, which means that the posterior end of the vertebral column is bent upwards and lies in the dorsal or epichordal lobe.

The mouth is a large crescentic aperture on the ventral side of the head near the front. It is bounded by the upper and lower jaws, each of which has one or two rows of sharply pointed and backwardly directed teeth to catch the slippery prey. If a tooth is broken, it is replaced. *Scoliodon* teeth are modified scales. Scales cover their body and extend inside their jaws to act as teeth (Fig. 8.1C). The transition from placoid scales to teeth is well documented in the jaw regions.

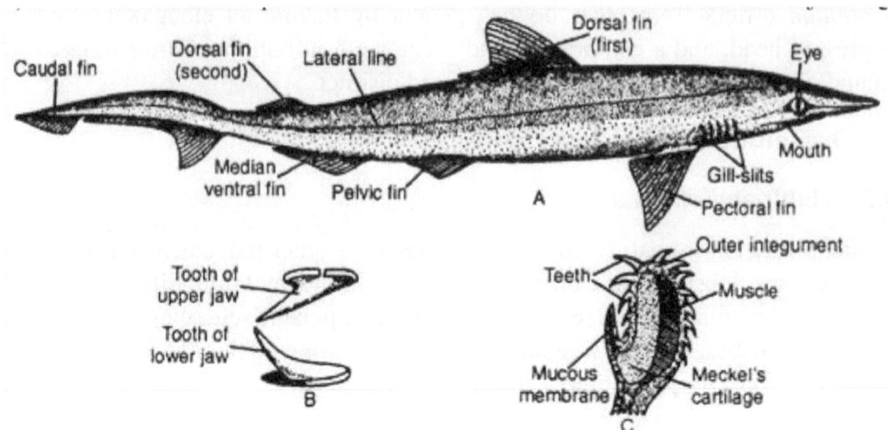

Fig. 8.1: A. Morphological aspects of a female *Scoliodon.* B. Disposition of teeth in the lower and upper jaw and C. Lower jaw showing the transformation of placoid scale into teeth.(from https://www.notesonzoology.com)

There are two prominent circular eyes. The upper and lower eyelids are movable in each eye. In an emergency, the nictitating membrane, the third eyelid, can cover the entire eye. The pupil is a slit-like vertical aperture. One nostril is located at each angle of the mouth. These are solely olfactory and have no connection to the oral cavity. A small fold of skin covers part of each nostril. Each side has five vertical slits posterior to the eyes. They are known as gill or branchial slits. Branchial slits lead to gill pouches, which open into the pharyngeal cavity. The cloaca is accessible to the outside world via a cloacal aperture located between the two pelvic fins. The cloacal aperture is a long and narrow opening. The cloaca is a common chamber that contains anus, urinary, and genital apertures. The abdominal pores are located on either side of the cloaca. The abdominal pores are paired structures located on elevated papillae that communicate with the coelom. There is a faint lateral line. A canal runs beneath this line. At regular intervals, minute pores in the canal open to the outside. There are also numerous ampullary pores on the head.

8.3.4 Fins

Scoliodon, like other fishes, has unpaired and paired fins that are flap-like integumentary extensions of the body. These are flexible, with cartilaginous rods or horny fin rays to stiffen them. All of the fins are angled backward, which helps with fast-forward movement in the water.

8.3.4.1 Median unpaired fins

It includes two dorsal, one caudal, one ventral (anal) fin and a caudal fin. The dorsal fins are triangular in outline. The anterior or first dorsal is larger and triangular located near the centre of the body. The posterior or second dorsal fin is relatively small and sits in the middle between the first dorsal and the tip of the tail. The caudal fin has one well-developed ventral lobe (hypochordal) and one poorly developed dorsal lobe (epichordal). The genus is distinguished by the presence of two shallow depressions known as caudal pits. There are two of these at the base of the tail, one on the dorsal surface and one on the ventral. The median ventral fin is located just anterior to the caudal fin in the midventral line about 5.0cm in front of the caudal fin.

8.3.4.2 Lateral paired fins

The lateral paired fins are made up of two pectoral and two pelvic fins. The pectoral fins are large and located behind the gill clefts. The pelvic fins are significantly smaller. These are simple in females, but in males, each is linked to a copulatory organ called the muxipterygium or clasper. Clasper resembles a rod and has a dorsal groove that leads to a siphon at its base.

8.3.5 Skin

The skin or integument is slimy in fresh specimens and is made up of two layers: the epidermis on the outside and the dermis on the inside (Fig. 8.2).

The epidermis is made up of many layers of epithelial cells and numerous unicellular mucous glands. The epithelial cells are ciliated in the early stages. However, cilia are lost in adults. The stratum germinativum or innermost layer of cells rests on a basement membrane. Just below the epidermis, melanophores (chromatophores) are present.

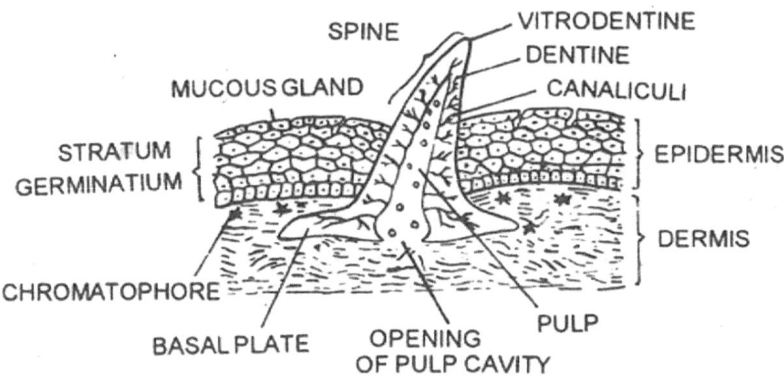

Fig. 8.2: V.S. of the skin of *Scoliodon* (from https://www.bioscience.com.pk)

The dermis is firmly attached to the underlying muscles and is made up of three layers:

(i) Stratum laxum: The outer layer with no fibres.

(ii) Stratum compactum: This is the fibrous median layer of the stratum compactum. Fibres connect the placoid scale's basal plate to this layer.

(iii) Subcutaneous layer: It has a variable thickness and fine fibres arranged in a reticular pattern.

Functions of skin

(i) Skin protects viscera against mechanical injuries.

(ii) Secretion of mucus makes the body surface slippery.

(iii) Its colouration imparts camouflage from predators.

(iv) Its receptors help to changes in the swimming.

8.3.6 Muscular system

Muscles are densely packed between the endoskeleton and the skin. The muscles in the trunk and tail are well-developed and have a distinct segmental arrangement. Numerous myotomes are separated from one another by myocommata, which are partitions made of tough connective tissue. In each myotome, muscle fibres run parallel to the length of the body. The muscles in the head region show no signs of segmentation and have become specialised to control the movement of the jaws, pharynx, and eyes. In the trunk region, the muscles are greatly thickened on the dorsal side on each side of the vertebral column, whereas in the tail region, the muscles are equally developed around the vertebral column.

8.3.7 Skeletal structures

The exoskeletal and endoskeletal structures of *Scoliodon* are well-developed. The exoskeletal system primarily consists of scales that cover the entire body. The axial and appendicular skeletons are included in the endoskeletal system.

8.3.7.1 Exoskeleton

The entire body is covered in oblique rows of placoid scales or odontoids (Fig. 8.3A to D). A typical placoid scale has a basal plate made of calcified tissue that remains embedded in the skin and a spine that projects backward out of the skin.

Sharpey's and other fibres anchor the diamond-shaped basal plate to the dermis (Fig. 8.3E). A perforation at the base of the spine connects the pulp cavity of the basal plate to the pulp cavity of the spine. During life, the pulp cavity is filled with pulp containing odontoblasts, blood vessels, nerves and lymph channels. The spine is made of dentine that has been enamelled on the outside. The dentine is coated externally with vitrodentine and traversed by parallel canaliculi. The composition of enamel is debatable. The name fibro dentine comes from the calcification of fibrous material between the dentine and the enamel organ. The scales that cover

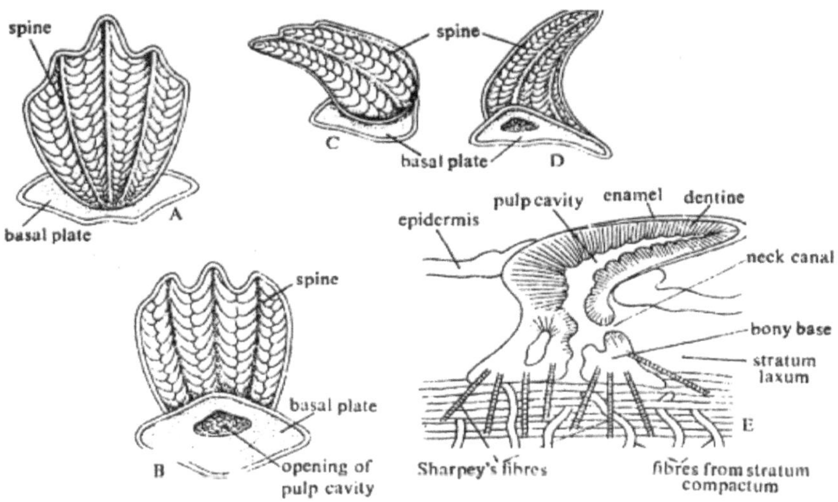

Fig. 8.3: Placoid scale of *Scoliodon.* A. Dorsal view, B. Ventral view, C. Dorsolateral view, D. Ventrolateral view and E. Section of integument to show placoid scale. (From https://www.notesonzoology.com)

the body extends inside the jaws and serves as teeth. The placoid scale has a dual origin and develops from both the epidermis and the dermis. The basal plate and dentine of the spine are both mesoderm derivatives. The enamel organ is formed by the ectodermal enamel organ.

8.3.7.2 Endoskeleton

The endoskeleton is entirely made of cartilage, which can be hardened by calcification in some places. It can be divided into two categories:

(1) The axial skeleton: It includes the skull and the vertebral column and

(2) The appendicular skeleton: It includes the pectoral and pelvic girdles as well as the skeleton of the fins.

8.3.7.3 Axial skeleton

The skull is a simply cartilaginous casket. The cranium, four sense capsules enclosing the auditory and olfactory organs and the visceral skeleton which forms the jaws and supports the pharynx with gills, make up the skeleton. These three parts of the skull are still inextricably linked (Fig. 8.4A to C).

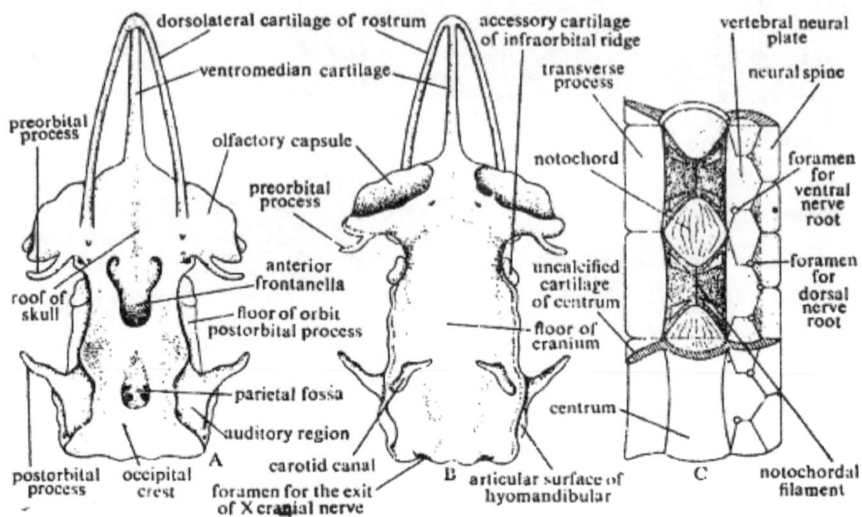

Fig 8.4: A. Dorsal view and B. Ventral view of the axial skeleton and C. Vertebral column of trunk region of *Scoliodon***.** (from https://www.notesonzoology.com)

The cranium is divided into four distinct regions:

(1) The occipital region: It is located at the back of the skull and contains a large foramen magnum through which the spinal cord travels. On either side of the foramen magnum, there is an occipital condyle. Above this foramen, there is a median ridge called the occipital crest.

(2) The auditory region: It is composed of auditory capsules that remain firmly attached to the cranium in adults. The parietal fossa is an oval depression between the two capsules. On the dorsal side of the cranium's roof, there are two large fontanelles. It possesses three ridges that mark the position of three semicircular canals.

(3) The orbital region: It is located in the middle of the skull. The supraorbital ridge (=crest) delineates the boundary between the cranium's roof and the orbit. It is ventrally bordered by an infraorbital (suborbital) ridge. The orbit is partially encircled anteriorly by a slender cartilage called the preorbital process. Similarly, a postorbital process emerges from the side of each postorbital cartilage and runs along the orbit's upper margin.

(4) The ethmoidal region: It bears the olfactory capsules and the rostrum is located in the cranium anteriorly. The posterolateral surface of each capsule has a short ethmopalatine ridge. The internasal septum separates the two olfactory capsules.

8.3.7.4 Visceral skeleton (splanchnocranium):

Seven half-hooped cartilaginous U-shaped visceral arches encircle the buccal cavity and pharynx.

A. The first (mandibular) pair of visceral arches or palato-pterygoquadrate: It forms the upper jaw (palatoquadrate) and the lower part is known as Meckel's cartilage, which forms the lower jaw.

B. The second or the hyoid arch made up of five cartilages and bears three parts:

(i) a ventral or lower median basihyal in the floor of the pharynx,

(ii) A lateral ceratohyal in the buccal floor and

(iii) A dorsal hyomandibular that bears gill rays.

The jaws are suspended from the cranium via the hyomandibular, and this type of suspensorium is known as hyostylic.

C. The branchial arches: The remaining visceral arches support the pharynx and gills. Typically each branchial arch is made up of nine pieces of cartilage, a lower median rod-like basibranchial to which is attached on each side of a ventral hypobranchial, a lateral ceratobrnachial and epibranchial and dorsal pharyngobranchial cartilages.

8.3.7.5 Vertebral column: A chain of 130 cartilaginous vertebrae makes up the vertebral column. The vertebrae form around the notochord, which remains stable. The vertebrae vary slightly in length. Two types of vertebrae named trunk and caudal are found in *Scoliodon*.

A. Trunk vertebra: These vertebrae are mesospondylous. The notochord is enclosed by the centrum of a trunk vertebra. Above the centrum, there is a neural canal through which the spinal cord travels. The neural canal is covered by a neural arch with a short blunt neural spine. A pair of transverse processes project ventrolaterally from the amphicoelous centrum. In the transverse section, the centrum shows four wedge-shaped calcified fibrocartilages or the Maltese cross. Within the centrum, the notochord is very narrow, but it becomes much dilated in the intervertebral spaces. The centra are reinforced by calcified fibrocartilage, which forms four wedges and forms a cross across the body of the centrum. The posterior vertebrae have a haemal arch on the ventral side of the centrum. The haemal arch protects the haemal canal, which contains the caudal artery and vein. The haemal arch provides a haemal spine to support the caudal fin's ventral lobe. The vertebral column's posterior end is bent dorsally.

B. Caudal vertebrae: These vertebrae are diplospondylous. The caudal vertebrae are slightly modified in comparison to the trunk vertebrae. These vertebrae have short neural arches, large and neural spines and backwardly pointed haemal spines.

8.3.7.6 Appendicular skeleton

The appendicular skeleton is made up of the fins' supporting skeleton. Pterygiophores or somactidia are cartilaginous rod-like structures found on the two dorsal fins and the ventral fin. The somactidia's distal ends have a double series of ceratotrichia or horny fin-rays. The caudal fin is devoid of somactidia. The neural and haemal spine extensions support the caudal fin.

A. The pectoral girdle: It is located posterior to the last branchial arch and is made up of two semicircular cartilages connected along the midventral line. Each half is made up of a thick and rod-like scapula on the dorsal side and a thin and flattened coracoid on the ventral side. Between the scapula and coracoid is a glenoid region with three foramina for the branchial artery with the pectoral fin by a triplet facet. Many radial cartilages (radials) support the pectoral fin on the basal cartilages. The endoskeleton has pteygiophores cartilages to which muscles are attached for moving the fin. The proximal pterygiophores are basals and the distal ones are radials. There are three thick basals, an outer propterygium, a middle mesopterygium and an inner metapterygium (Fig. 8.5A).

B. The pelvic girdle: It is a flattened cartilaginous rod located in front of the cloaca along the transverse plane. It is a flat transverse rod of cartilage known as an ischiopubic bar. On each side of the girdle, there is an acetabular region for the attachment of basal cartilage. The pelvic fin's radials are supported by a curved basal cartilage called the basipterygium (Fig. 5.3B). The basipterygium is attached to the pelvic girdle anteriorly. The distal ends of the radials have small cartilages that bear the ceratotrichia (Fig. 8.5B).

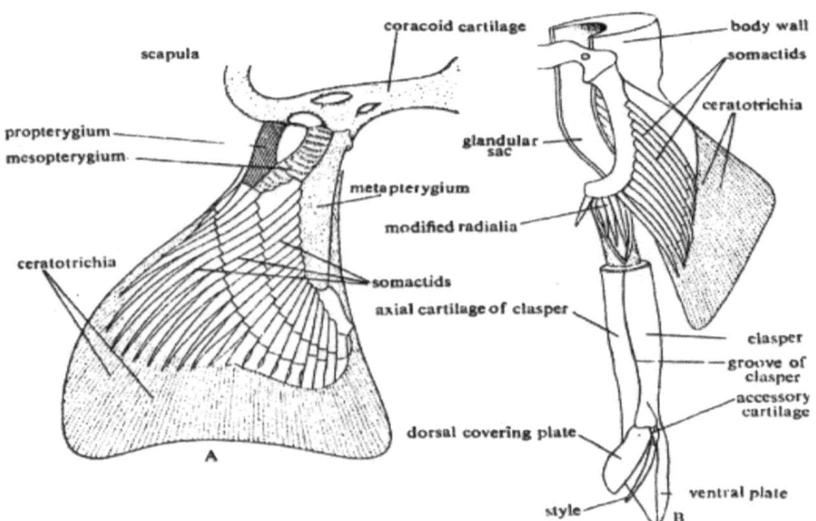

Fig. 8.5: Appendicular skeleton of *Scoliodon*. One-half of A. pectoral and B. pelvic girdle. (from https://www.notesonzoology.com)

Each clasper in males is a tubular cartilage that is grooved dorsally. The groove ends proximally in a sharp style surrounded by two sheathing plates. A small accessory cartilage is located at the top of the style.

8.3.8 Locomotion

The movement of *Scoliodon* is caused by myotomal longitudinal muscle fibre activity and is aided by fin movement. Throughout the phylogenetic history of fishes, the fins were primarily used to lift the body off the bottom, but they later became secondary in swimming by producing undulatory movements. The locomotion in fish has been explained by Brder (1926) and Gray (1933).

Swimming relies heavily on the longitudinal muscle fibres that comprise the myotomes. The myotomes are located on either side of the incompressible vertebral column, which serves as a lever for the myotomes to operate. The contractility of the myotomes causes the body to bend. During forward progression, myotomes contract in an anteroposterior direction, causing waves of curvature to pass down each side of the body alternately from the head to the tail. This type of contraction is known as a metachronal contraction. During steady swimming, it has been calculated that approximately 54 waves/minute are produced. The factors that govern muscle contraction are not fully understood. The transection of the spinal cord behind the medulla oblongata in *Scoliodon* can produce swimming movements for several days.

The dorsal fin prevents rolling and keeps the fish already in a horizontal plane. Equilibrium in a vertical plane is controlled by the movement of pectoral fins. Turning is produced by either a wave of myotomal contraction only on one side of the body or by asymmetrical braking with the pectoral fins.

This phenomenon suggests that the rhythm of contraction is largely governed by the spinal cord, but the role of the brain in the entire process remains unknown.

The body and fins are specially designed to maintain balance in the water while swimming. The dorsal fin is well-developed and aids in the restoration of stability and equilibrium if there is any deviation along the vertical axis of the body. The pectoral fins play an important role in fish turning by unilateral breaking with the pectoral fins. The pectoral fins also contribute to vertical plane stability. The movable pectoral fins raise the head, which is compensated for by the heterocercal tail. The hypo-chordal lobe is flexible, whereas the epichordal lobe is rigid. The hypo-chordal lobe's flexibility results in a vertical lift of the tail. The pelvic fins serve no purpose in locomotion and are typically used to aid in reproductive function, particularly in males.

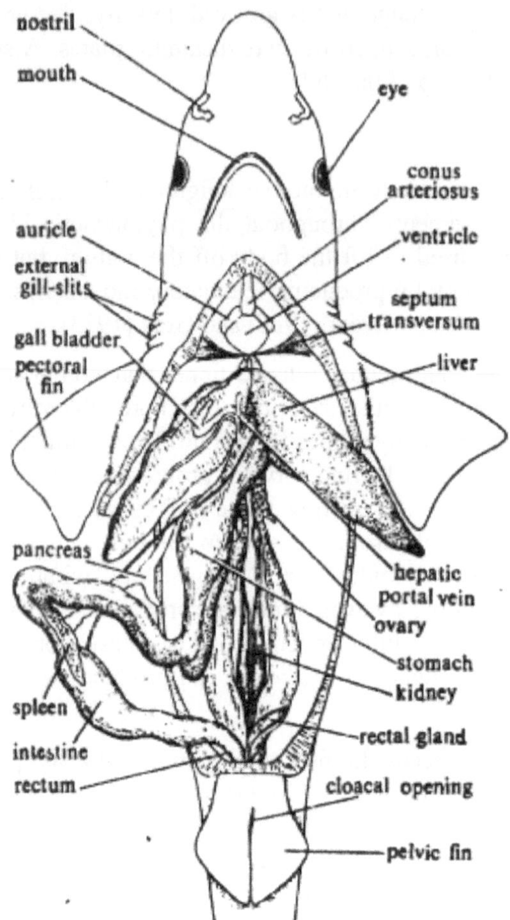

Fig 8.6: Coelom and viscera of *Scoliodon.* (from https://www.notesonzoology.com)

8.3.9 Coelom

The coelom is large and divided into two unequal cavities a smaller pericardial cavity and a large abdominal cavity. These two cavities are separated by the septum transversum and communicate with one another via the septum's pericardioperitoneal canal. The viscera are housed in the abdominal cavity, which is accessible to the outside via a pair of abdominal pores (Fig. 8.6). The pericardial coelom houses the heart.

8.3.10 Digestive System

The alimentary canal and digestive glands make up the digestive system. The alimentary canal begins in the mouth and ends in the anus.

The mouth is located on the ventral side, bound on both sides by jaws.

The mouth opens into a large buccal cavity lined with mucous membranes. The floor of the buccal cavity folds to form a non-muscular, non-granular tongue. Because of the presence of dermal denticles or teeth, the mucous membrane is very thick and rough. The teeth are extremely sharp and obliquely placed (See Fig. 8.2C). There are homodont, polyphyodont and lyodont teeth.

The buccal cavity connects to the pharynx. The internal openings of the spiracles and five branchial clefts are located on either side of the pharynx. The pharyngeal mucous membrane contains numerous dermal denticles.

The pharynx connects to a narrow oesophagus. The pharyngeal inner mucous membrane is raised to form longitudinal folds. To form a large stomach, the oesophagus dilates posteriorly.

The stomach is extremely muscular and bent inward to form a J-shaped configuration. The stomach's long limb connects to the oesophagus, while the shorter one enters the intestine. The oesophagus entrance into the stomach is provided with a crescentic fold that serves as a valve. The long anterior limb is known as the cardiac stomach, and the short posterior limb is known as the pyloric stomach. A small outgrowth known as a "blind sac" can be found at the junction of the cardiac and pyloric limbs. The inner lining of the cardiac stomach, like that of the oesophagus, is folded longitudinally (Fig. 8.7A). The internal lining of the pyloric stomach is mostly smooth, with some slight folding at the distal end. The pyloric valve guards the pylorus's entrance into the bursa entiana, a small chamber with thick walls.

The bursa entiana is immediately followed by a broad tubular intestine that narrows posteriorly to form the rectum. The rectum connects to the cloaca. A tubular caecal, rectal or digitiform gland protrudes from the rectum.

The intestine's inner surface folds to form an anticlockwise spiral of about two and a half turns. The scroll valve (Fig. 8.7B) increases the absorptive surface of the intestine while also limiting the rapid flow of digested food through the intestine. In *Scylliorhinus* and *Squalus*, the spiral valve takes about 14—15 turns arranged like a spiral staircase of a tower.

The liver, a massive yellowish gland with two lobes, is the main digestive gland. The lobes join anteriorly. In the anterior part of the right lobe of the liver, there is a thin-walled V-shaped gall bladder. The bile duct receives a few smaller ducts from the two lobes of the liver and opens into the anterior end of the intestine near the commencement of the scroll valve. The pancreas is a pale compact irregular body with a dorsal lobe parallel to the posterior part of the cardiac stomach and a ventral lobe tightly attached to the pyloric stomach. The pancreatic duct, which is located opposite the bile duct's aperture, drains pancreatic juice into the intestine.

The functional significance of the rectal (caecal) gland is unknown. The central cavity of the rectal gland is lined with cuboidal cells. It has a high vascularity and is made up of lymphoid tissue. It excretes a fluid into the intestine lumen, but its exact function is unknown. There are no such glands in the buccal cavity that can be compared to the salivary glands of higher vertebrates.

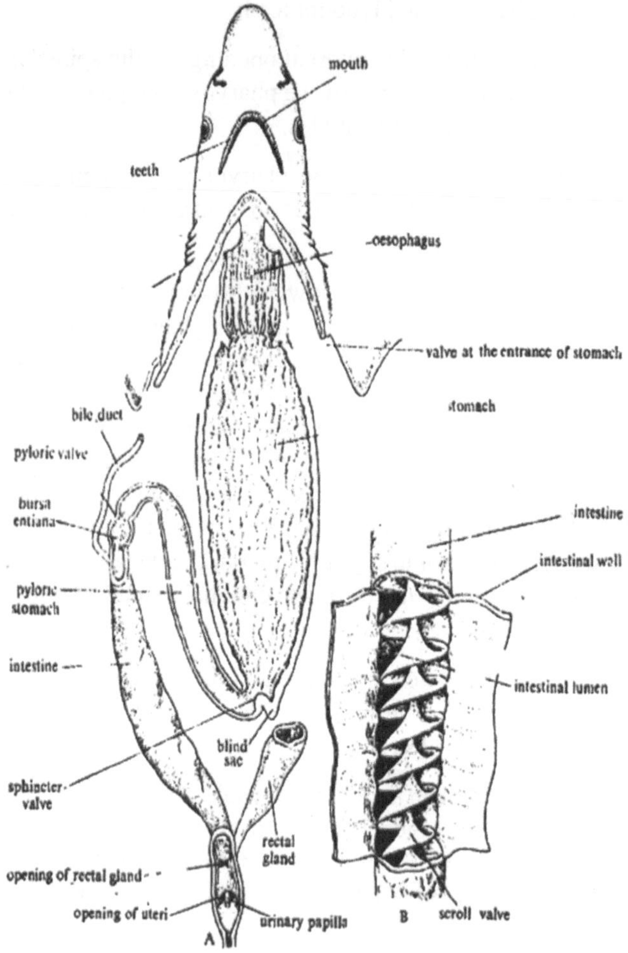

Fig. 8.7: Digestive system of *Scoliodon* (from www.old.amu.ac.in)

The spleen is located dorsal to the distal end of the stomach's body. The spleen is functionally linked to the circulatory system but morphologically linked to the alimentary canal.

Physiology of digestion: *Scoliodon* feeds mainly on other fishes, but may also include rock crabs, lobsters and spider crabs. Food is swallowed without mastication. Since there is no salivary gland, there is no digestion in the buccal cavity. The wall of the pharynx possesses numerous mucous glands.

The digestion starts only in the cardiac stomach. The mucous membrane of the cardiac stomach secretes the gastric juice which contains pepsin and HCl that converts proteins into syntonin, proteases and peptones. The pyloric stomach and spiral valve have no digestive activity in the pancreas. The pancreas secretes trypsin, amylopsin and lipase. The semi-digested food enters the intestine; it is acted upon by the bile and the pancreatic juice. The bile makes the food alkaline and thus helps the action of pancreatic juice. The trypsin acts on the remaining proteins, the amylopsin converts starch into sugar and lipase from fats into fatty acids. The digested food is absorbed into the blood over the extensive surfaces of the intestine and spiral valve.

8.3.11 Respiratory System

The gills, which are carried in gill pouches, are the respiratory organs. The structure of gill pouches varies between dogfish species. The organisation of the gill pouch of *Brachalurus*, a related genus of *Scoliodon*, is depicted in Fig. 8.8B. There are five pairs of gill pouches, each with a large internal branchial aperture that communicates with the pharyngeal cavity and opens to the outside via exterior gill slits (Fig. 8.8A).

The mucous membrane that lines the gill pouches has horizontal branchial lamellae. Branchial lamellae are extremely vascularized structures. Each gill pouch has two sets of branchial lamellae, one anterior and one posterior. The interbranchial septum, which projects beyond the branchial lamellae, separates the gill pouches (Fig. 8.8C). A visceral arch supports the pharyngeal end of each interbranchial septum. Each arch supports the anterior lamellae of the one gill pouch and the posterior lamellae of the next-gill pouch behind it. The first gill pouch is located between the hyoid and the first branchial arches, while the last is located between the fourth and fifth branchial arches.

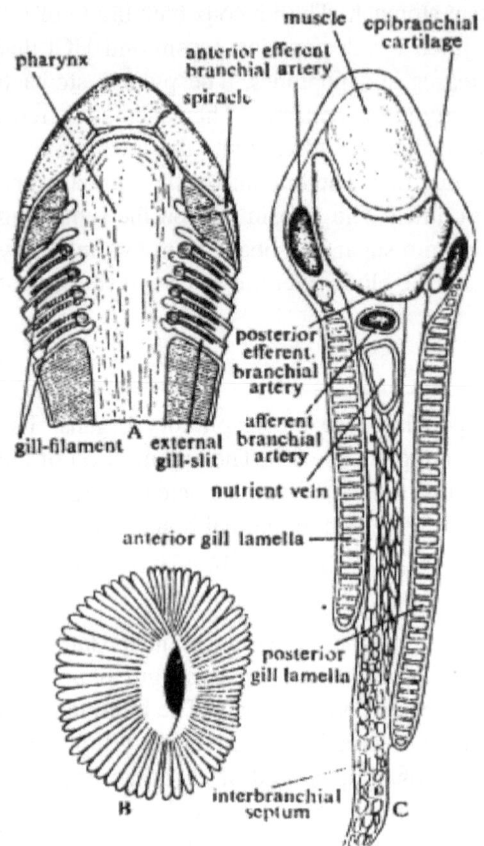

Fig 8.8: Respiratory system of *Scoiodon.* A. Gill apparatus. B. Gill pouch and C. Holobranch of *Scoliodon.* (From https://www.notesonzoology.com)

There are two types of gills:

(i) When a branchial arch bears two sets of gill lamellae, this is referred to as a holo-branch or complete gill.

(ii) Demibranch, hemibranch, or half-gill when there is only one set of gill lamellae. The hyoid arch can only support a demibranch, whereas the first four branchial arches can support holobranchs. The final branchial arch lacks gills.

Mechanism of respiration

(A) Inspiration: Contraction of hypobranchial muscles lowers the floor of the buccopharyngeal cavity, causing the visceral arches to expand throughout the pharynx. An anterior fold of skin on each gill pouch prevents water from entering external branchial apertures.

(B) Expiration: The mouth closes during expiration due to the action of the adductor muscle. The floor of the buccopharyngeal cavity is now raised, and the contraction of the pharyngeal wall forces water into the internal branchial apertures, closing the oesophagus, and then into the gill clefts, where it washes the branchial lamellae and exits through the external branchial aperture.

When the mouth is otherwise occupied, the spiracles are occasionally used as secondary pathways for water entry for respiration.

Fresh seawater entering the gill pouches with the respiratory current contains oxygen dissolved in it. The water is separated from the blood contained within the capillaries. The oxygen of the water passes by the water passes by endosmosis through the thin capillary walls into the blood and at the same time, the carbon dioxide of the blood passes into the water by exosmosis. As blood makes a complete circuit in the capillaries of the gills in a very short time, it is evident that the exchange of gases also takes place very quickly.

8.3.12 Circulatory System

The circulatory system comprised of

(a) The circulatory fluid called the blood,
(b) The heart,
(c) The arteries and
(d) The veins.

8.3.12.1 Blood

The blood is made up of colourless plasma and corpuscles suspended in it. There are two types of corpuscles encountered: R.B.C. (or erythrocytes) and W.B.C. (or leucocytes). Erythrocytes are oval bodies with a nucleus. Haemoglobin can be found in erythrocytes. Leucocytes are amoeboid in shape.

8.3.12.2 Heart

The heart is a dorso-ventrally bent S-shaped muscular tube that consists of receiving parts (a sinus venosus and a dorsally placed auricle or atrium) and forwarding parts (a ventricle and a conus arteriosus) (Fig. 8.9A). The heart of *Scoiodon* contains only impure blood; hence, it is called venous or branchial heart. The pericardium surrounds and protects the heart. The basibranchial cartilage supports the dorsal part of the pericardium. The heart is located on the body's ventral side, between two series of gill pouches.

Receiving parts of the heart: The sinus venosus is a tubular chamber with thin walls. The sinus venosus is a highly contractile part of the heart that causes the heart to beat. On each lateral side, two great veins, the ductus Cuvieri, open into the sinus venosus. The sinus venosus is entered by two hepatic sinuses from the back. The sinus venosus enters the auricle through a sinu-auricular aperture guarded by two valves. The auricle is a large, triangular, thin-walled chamber dorsal to the ventricle but anterior to the sinus venosus. Through a slit-like auriculo-ventricular aperture guarded by two-lipped valves, the auricle communicates with the ventricle. The receiving chambers (sinus venosus and auricle) take in venous blood from all over the body.

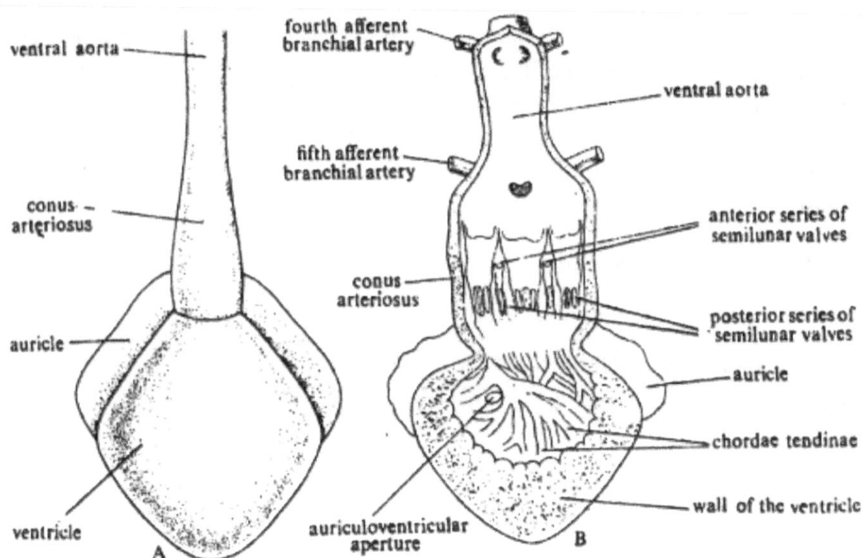

Fig 8.9: Heart of *Scoliodon.* A. Position of different chambers. B. L.S. of Heart. (from https://www.notesonzoology.com)

Forwarding parts of the heart: The ventricle has a thick muscular wall with many muscular strands on the inner surface, giving it a spongy texture (Fig. 8.9B). It is an oval chamber that is the most visible part of the heart. The conus arteriosus is a muscular tube that arises from the ventricle. Two transverse rows of semilunar valves line the lumen of the conus arteriosus. The free ends of the valves are attached to the ventricular wall by fine tendinous threads known as chordae tendinae to keep them in place. The conus arteriosus continues as the ventral aorta.

The heart's function is to receive deoxygenated blood from all parts of the body and pump it to the gills for aeration. Because only deoxygenated blood circulates through its various parts, this type of heart is known as the venous or branchial heart.

8.3.12.3 Arterial System

Scoliodon's arterial system is divided into two distinct types of arteries:

(1) The afferent branchial arteries that arise from the ventral aorta and transport deoxygenated blood to the gills for oxygenation.

(2) The efferent branchial arteries, which arise from the gills and transport oxygenated blood to various parts of the body (Fig. 8.10).

Afferent Branchial Arteries: The ventral aorta runs from the ventral surface of the pharynx to the posterior border of the hyoid arch. The ventral aorta splits into two branches known as innominate arteries, which then split into the first and second afferent branchial arteries. The ventral aorta gives rise to the third, fourth, and fifth afferent arteries. Except for the anteriormost pairs, all afferent branchial arteries arise from the ventral aorta via independent openings.

Efferent branchial arteries: In the gills, the afferent branchial arteries divide into capillaries. Blood is collected from the gills by efferent branchial arteries. There are nine pairs of efferent branchial arteries on each side, which are evenly distributed. The first eight arteries form a series of four complete loops around the first four gill slits, and the ninth efferent branchial artery collects blood from the fifth-gill pouch's demibranch, from which blood is poured into the fourth loop. The four loops are linked together by a network of longitudinal commissural vessels known as the lateral hypobranchial chain, in addition to short longitudinal connectives. An epibranchial artery arises from each efferent branchial loop. The dorsal aorta is formed by four pairs of epibranchials joining in the mid-dorsal line. The ninth efferent branchial artery does not have an epibranchial branch but connects to the eighth efferent branchial artery.

Anterior arteries: The first efferent branchial artery and the proximal end of the dorsal aorta both supply blood to the head region. The following arteries arise from the first efferent branchial (hyoidean efferent):

(a) The external carotid artery: It arises from the first collector loop and divides into (i) a ventral mandibular artery that supplies the muscles of the lower jaw and (ii) a superficial hyoid artery that supplies the second ventral constrictor muscle, the skin and the subcutaneous tissue beneath the hyoid.

(b) The afferent spiracular artery: It emerges from the middle of the hyoidean efferent and enters the cranial cavity as the spiracular epibranchial artery. It sends a great ophthalmic artery to the eyeball just before entering the cranial cavity. Immediately after entering the cranium, it connects with an internal carotid artery branch to form the cerebral artery. The cerebral artery immediately divides into two branches that supply the brain, one anterior and one posterior.

(c) The hyoidean epibranchial artery: It gets a branch from the dorsal aorta. The hyoidean epibranchial artery runs forward and inward to the orbit's posterior border, where it receives an anterior branch from the dorsal aorta. It immediately divides into:

(i) The stapedial artery: It originates as the inferior orbital artery and continues as the superior orbital artery, supplying six eye muscles as well as the superficial tissue above the auditory capsule. The superior orbital artery gives rise to a large buccal artery that becomes the maxillonasal artery. The maxillonasal artery supplies several arteries to the upper jaw muscles, the olfactory sac, and the rostrum.

(ii) The internal carotid artery: It travels inward and enters the cranium, where it divides into two branches. One of the branches connects with its counterpart on the opposite side, while the other connects with the stapedial.

Dorsal aorta and its branches: The dorsal aorta is formed by the union of epibranchial arteries. It runs posteriorly and is located ventral to the vertebral column. It continues as the caudal artery to the tip of the tail. The dorsal aorta gives rise to the following arteries in the anteroposterior direction:

(1) Several buccal and vertebral arteries branch anteriorly.

(2) A pair of small subclavian arteries arises near the origin of the fourth epibranchial arteries. The subclavian artery carries the apicoracoid artery and divides into three branches:

(i) A branchial artery to the pectoral girdle and pectoral fin,

(ii) An anterolateral artery to the musculature of the body and

(iii) A dorsolateral artery to the dorsal musculature.

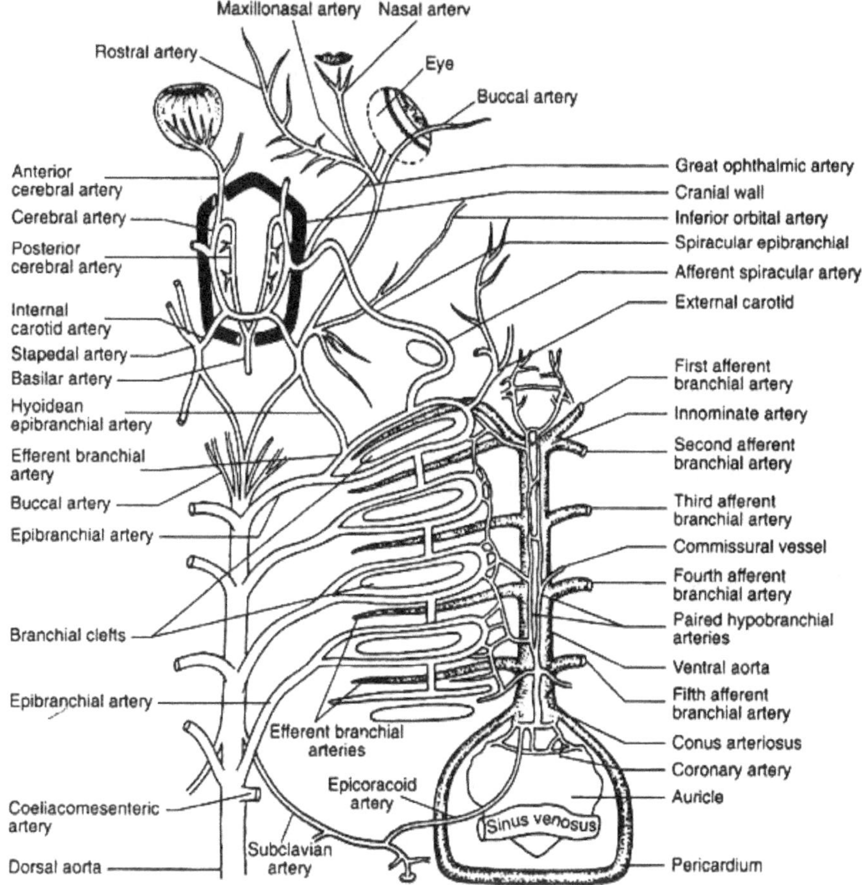

Fig. 8.10: Arterial system of Scoliodon. (from https://www.notesonzoology.com)

(3) A large coeliacomesenteric artery originates just behind the origin of the fourth epibranchial artery. It gives rise to a smaller coeliac artery and a larger anterior mesenteric artery.

(4) A lienogastric artery gives rise posterior to the coeliacomesenteric artery and gives off

(i) An ovarian (in females) or spermatic artery (in males) to the gonad,

(ii) A posterior intestinal artery to the posterior part of the intestine,

(iii) A posterior gastric to the posterior part of the cardiac stomach and

(iv) A splenic artery to the spleen.

(5) Behind the subclavian artery, a series of paired parietal arteries emerge.

Each parietal produces a dorsal and a ventral parietal artery. The dorsolateral musculature, the vertebral column, the spinal cord, and the dorsal fin are all supplied by the dorsal parietal artery. The ventral parietal artery provides blood to the ventral muscles and peritoneum. The kidneys receive renal branches from the ventral parietal.

(6) A pair of iliac arteries extend to the pelvic fin as femoral arteries.

Hypobranchial blood plexus: A network of slender arteries arising from the ventral ends of the loop of the efferent branchial arteries forms a lateral hypobranchial chain. Four commissural vessels arise from the lateral hypobranchial chair, which joins to form a pair of median hypobranchials on the ventral wall of the ventral aorta and communicate with one another via transverse vessels. The median hypobranchials connect posteriorly to form the median coracoid artery, which gives rise to the coronary artery and the pericardial artery. The pericardial artery gives rise to the common epicoracoid artery, which divides into left and right epicoracoid arteries, each of which connects to one subclavian artery.

8.3.12.4 Venous System

Deoxygenated blood from various parts of the body is returned to the heart via veins that form irregular blood sinuses along their paths (Fig. 8.11). The presence of numerous blood sinuses is a distinguishing feature of *Scoliodon*'s venous system. The venous system is extremely complex and is divided into the following sections:

(A) Cardinal System

Blood is returned to the heart from the anterior region of the body by the paired jugular and anterior cardinal sinuses. A pair of posterior cardinal sinuses collects blood from the posterior region. The anterior and posterior cardinals join on each side to form the ductus Cuvieri, a transverse sinus.

(1) **Anterior cardinal system:** The venous system, which consists of a pair of internal jugular veins, returns blood from the head region. Each internal jugular vein is made up of two parts: the olfactory sinus and the anterior cardinal sinus. The anterior facial vein drains blood from the rostral region to the olfactory sinus and then to the orbital sinus. The postorbital sinus connects the orbital sinus to the anterior cardinal sinus. The anterior cardinal sinus enters the Cuvieri duct. The hyoidean sinus and five dorsal nutrient branchial sinuses are received by the anterior cardinal sinus from the gills,

(2) **Posterior cardinal system:** The caudal vein takes blood from the tail region and travels through the haemal canal. The caudal vein divides the abdominal cavity into the left and right

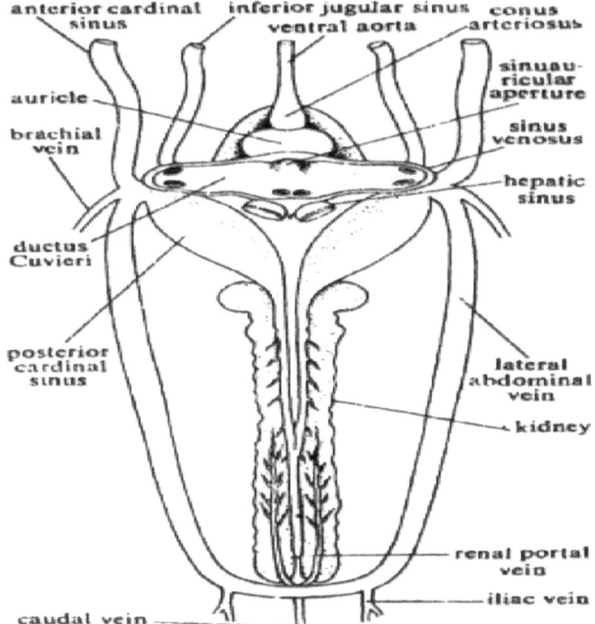

Fig. 8.11: Venous system of Scoliodon (from https://www.notesonzoology.com)

renal portal veins, which break up into sinusoid capillaries in the kidney substance. The renal portal vein receives small parietal veins along its length. The renal veins transport blood from the kidneys to the posterior cardinal sinuses. The ducts are served by two posterior cardinal sinuses or ductus Cuvieri.

(B) Hepatic Portal System

The hepatic portal vein is made up of many small veins that carry blood from the alimentary canal and its associated glands. The lienogastric vein, as well as the anterior and posterior gastric veins, enters the hepatic portal vein. In reality, the hepatic portal vein is formed by the union of the anterior and posterior intestinal veins. In the liver, the hepatic portal vein divides into capillaries. Blood is collected from the liver by another set of capillaries, which join to form two large hepatic sinuses that open into the sinus venosus.

(C) Cutaneous System

This system consists of a dorsal, a ventral and two paired lateral cutaneous veins. The inferior lateral cutaneous vein joins with the lateral cutaneous vein near the anterior end of the pectoral fin. Each lateral cutaneous vein ultimately opens into the brachial vein.

(D) Ventral System

This system is made up of two groups of veins: anterior central veins that drain into the ductus Cuvieri via the inferior jugular sinuses and posterior veins that drain into the subclavian vein. Each inferior jugular sinus is formed by the union of the lower jaw's submental sinus, the hyoidean sinus, and the ventral nutrients from the gills. The ductus Cuvieri is reached by each inferior jugular vein. The subclavian vein connects to the ductus Cuvieri on both sides.

A small caudal vein and two iliac veins join to form two large lateral abdominal veins. A commissural vein connects the lateral abdominal veins posteriorly. The lateral abdominal vein joins the brachial vein anteriorly to form the subclavian vein, which opens to the ductus Cuvieri.

8.3.13 Nervous System

The nervous system of *Scoliodon* comprises of

(i) The central nervous system,
(ii) The peripheral nervous system and
(iii) The autonomous nervous system.

8.3.13.1 Central Nervous System

The brain and spinal cord are part of the central nervous system.

BRAIN: The brain is highly organised and far superior to that of agnathans. The brain is divided into three major sections:

(a) **The forebrain or the prosencephalon**: It is a massive undivided cerebral hemisphere. The cerebral hemisphere of other fishes is relatively larger. Two stout olfactory peduncles emerge from the anterior end of the cerebral hemisphere, each terminating in a large bilobed olfactory lobe (Fig. 8.12). The olfactory lobes are located near the olfactory capsules. Each olfactory nerve is made up of numerous bundles of nerve fibres. The cerebrum has a smooth surface and thick walls. The neuropore is a small opening on the midventral surface of the cerebrum. The diencephalon (the posterior part of the forebrain) is very short. The diencephalon's roof is thin and non-nervous, and it houses the anterior choroid plexus. The thalami are two thickened bodies formed by the diencephalon's lateral walls. The pineal organ, also known as the epiphysis cerebri, is a long and slender tube that projects from the roof of the diencephalon up to the membrane that covers the anterior frontanella. The diencephalon (or hypothalamus) floor is well-formed. The floor of the diencephalon emits a hollow infundibulum. The infundibulum dilates to form two oval thick-walled bodies known as the lobi inferiores, the distal ends of which produce two thin-walled glandular sacs known as the sacci vasculosi. The inferior lobi are the sites of gestation and smell. The sacci vasculosi are thought to be a centre for receiving cerebrospinal fluid pressure while also producing cerebrospinal fluid. The hypophysis connects to the infundibulum. In front of the infundibulum is the optic chiasma. The decussation of the nerve fibres of the two optic nerves forms optic chiasma (Fig. 8.12B).

(b) The midbrain or mesencephalon: The midbrain is large, with two round optic lobes. Behind the diencephalon are the optic lobes. The floor and side walls are significantly thicker. The midbrain is thought to be the centre of coordination.

(c) The hindbrain or the rhombencephalon: The hindbrain is made up of a well-developed cerebellum and medulla oblongata. The cerebellum's dorsal surface has many irregular convolutions. There is a small cavity in the cerebellum. The cerebellum is also a coordination centre. Two distinct transverse furrows divide the cerebellum into three lobes. In adults, the medulla oblongata is triangular, with a pair of hollow corpara restiformia and traces of convolutions at the anterior end. A transverse nerve band connects two corpora restiformia. The medulla oblongata roof is non-nervous and contains the posterior choroid plexus. Swimming movements are controlled by the hindbrain.

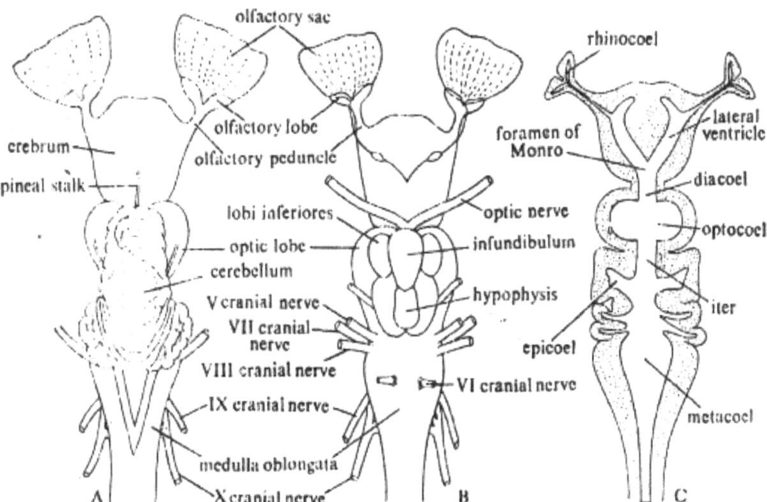

Fig 8.12: A. Dorsal, B. Ventral and C. Cavities of the brain of *Scoliodon.* (from https://www.notesonzoology.com)

The cavities or ventricles of the brain: These are moderately developed (Fig. 8.12C). The lateral ventricles (paracoels of I and II ventricles) in the cerebral hemispheres are narrow. The foramen of Monro opens between the I and II ventricles of the cerebrum and the III ventricle. The third ventricle (diocoel or cavity of the diencephalon) is about half the length of the cerebral hemispheres forward. The fourth ventricle's floor is significantly thickened. The metacoel (cavity of medulla oblongata) or fourth ventricle is large and extends dorsally into the cavity of the cerebellum (epicoel), continuing behind the spinal cord cavity. The iter (cavity of the midbrain) or aqueducts of Sylvii (communication duct between the third and fourth ventricles) is larger and communicates together with III and IV ventricles, epicoel and optocoels (cavity of optic lobes). Although the cerebrum is undivided, two lateral ventricles continue to the rhinocoels (cavities of olfactory sacs).

SPINAL CORD: Scoliodon spinal cord advances clearly towards the plan of higher vertebrates. The spinal cord receives only pia matter. The grey matter is divided into two horns: dorsal and ventral. The dorsal horns are joined to form a single broad region, giving the grey fatter an inverted 'T' shape.

8.3.13.2 Peripheral Nervous System

The cranial nerves and spinal nerves are part of the peripheral nervous system.

Cranial Nerves

In all fishes, there are ten pairs of cranial nerves. They are identified by their serial Roman numbers and names. *Scoliodon* has an extra pair of anterior terminal nerves. The cranial nerves in fishes are suited to an aquatic mode of life and are related to gills, lateral-line receptors and ampullae of Lorenzini.

Terminal pair or pre-olfactory nerve (nerve 0): It is sensory and discovered in 1894 and connects the anterior ends of the cerebral hemispheres. It was first discovered in *Protopterus* and it has since been found in all gnathostomes except birds. To avoid confusion caused by changes to the long-established nomenclature and symbols of other cranial nerves, this newly reported cranial nerve '0'. Some workers believe nerve '0' is a ganglionated remnant of the first branchial nerve. This nerve does not appear to be related to the olfactory nerve. It exits the cerebral hemisphere's anteroventral region and enters the olfactory organ's mucous membrane. It is a sensory nerve with one or more ganglia. Elasmobranchs have the most developed terminal nerve. The terminal nerve's functional significance is not fully understood. It has been proposed that it is the remains of an anterior branchial nerve that has lost its significance over time.

I. **The first pair or olfactory nerves**: It is a sensory nerve to taste and is produced by the olfactory lobes and innervates the olfactory sacs. These nerves emerge from the telencephalon and bear a ganglion near the origin called ganglion terminale. The nasal septum and the external nostril are supplied by these nerves.

II. The second pair or optic nerves: After the origin from the optic thalami lying on the ventral side of the diencephalon. It forms the optic chiasma and supplies the eyes. It is a sensory nerve to sight.

III. The third pair or oculomotor nerve: It is a motor nerve that arises from the ventral surface of the mesencephalon and supplies the anterior, superior, and inferior recti muscles, as well as the inferior oblique muscles of each eyeball.

IV. The fourth pair or trochlear or pathetic nerve: It is a motor nerve that arises from the anterolateral side of the medulla oblongata. The surface of the midbrain and supplies the superior oblique eye muscle.

V. The fifth pair or trigeminal nerve: It is a mixed nerve that arose from the dorsolateral surface of the medulla oblongata along with the VII and VIII nerves. It bears a Gaserian ganglion within the cranium. It has four branches:

(i) Ophthalmicus superficialis: It is sensory and supplies the skin of the snout. It enters the orbit along with the ophthalmicus superficialis of the VII nerve.

(ii) The maxillaries: It is sensory and divided into (a) the maxillaris superior runs along the floor of the orbit along with the buccal branch of the VII nerve. It supplies nerves to the skin of the upper jaw and (b) the maxillaris inferior innervating the posterior part of the upper jaw.

(iii) The mandibularis: It is mixed and innervates the muscles of the lower jaw.

(iv) The ophthalmicus profundus: It is sensory and becomes secondarily associated with the trigeminal to supply nerves to the eyeball and the dorsal surface of the snout (Fig. 8.13).

VI. The sixth pair or abducens nerve: It is a motor nerve that arises mid-ventrally from the medulla oblongata and supplies the posterior rectus muscle of the eyeball.

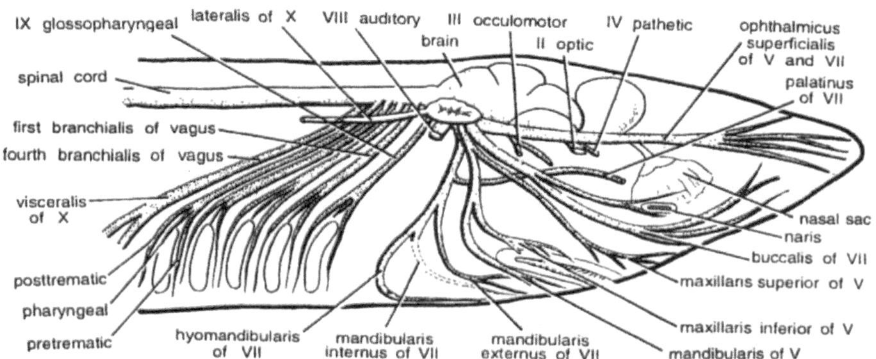

Fig. 8.13: Cranial nerves of *Scoliodon.* (from https://www.faunadondness.com)

VII. The seventh pair or facial nerve: It emerges from the side of the medulla oblongata. It bears a ganglion and is closely associated with the V nerve in its distribution. It is mixed in nature and divides into two major branches:

(1) The ophthalmicus superficialis branch like that of the V cranial nerve and

(2) A bundle of mixed nerves which subdivides into three routes:

(a) A ramus buccalis innervating the infraorbital canal of the snout along with the maxillaries of the V nerve.

(b) A ramus hyomandibularis supplying nerves to the lower jaw and the throat.

(i) Mandibularis externus: It innervates the mandibular canal.

(ii) Mandibularis internus: It innervates the mucous membrane of the buccal floor.

(iii) Hyoidean or Hyomandibular: It innervates the muscles of the throat.

(c) A ramus palatinus gives nerve supply to the roof of the buccal cavity and the pharynx.

VIII. The eighth pair or auditory or statoacoustic or vestibulocochlear nerve: It is a sensory nerve and arises close to the origin of the V and VII nerves from the side of the medulla oblongata. It gives the vestibular and saccular branches to the internal ear.

IX. The ninth pair or glossopharyngeal nerve: It is a mixed nerve that arises from the ventrolateral surface of the medulla oblongata behind the origin of the VI nerve. In the region of the first-gill cleft, it divides into a small pretrematic nerve and a large posttrematic nerve. These nerves supply branches to the pharynx, pharyngeal muscles and the mucous membrane surrounding the first-gill slit.

X. The tenth pair or vagus (pneumogastric) nerve: It arises from many roots from the postero-lateral side of the medulla oblongata behind the IX nerve. It arises from multiple roots and gives off the following branches.

(i) The branchial nerves supplying the gills,

(ii) The visceral nerves supply the viscera including the heart, liver, etc.,

(iii) The lateralis supplies the lateral line sense organs and gives numerous branches along its course.

An occipital nerve arises behind the vagus and joins the first two spinal nerves to form a hypobranchial nerve which goes to muscles in the floor of the buccal cavity.

8.3.13.3 Spinal Nerves

The spinal nerves arise from the spinal cord. Each has one dorsal (sensory) and one ventral (motor) root. The dorsal root bears a ganglionic swelling of somatic sensory, visceral sensory and visceral motor fibres, while the ventral root has somatic motor and visceral motor fibres. After emerging out through the vertebral column, the dorsal and the ventral roots unite to form a commonly mixed nerve.

Each spinal nerve has three branches

(a) ramus dorsalis,

(b) ramus ventralis and

(c) ramus communicans to join the autonomic nervous system.

In the region of the pectoral fin, 3—6 spinal nerves join each other near the pectoral girdle to form the cervicobrnachial plexus. Then 9 or 10 spinal nerves go to the pelvic fin and some of them join to form a lumbosacral plexus.

8.3.13.4 Autonomic Nervous System

This system consists of a series of paired ganglia arranged irregularly on the kidney's dorsal wall and the posterior cardinal sinuses. The largest ganglion is the gastric ganglion, which sends nerves to the viscera. Each segment usually has one ganglion. The successive ganglia in *Scoliodon* do not form a distinct continuous chain.

8.3.14 Sense Organs or Receptors

The nervous system is associated with highly developed sense organs or receptors, viz. eyes (photoreceptors), nose, ear (phonoreceptors) and many others.

8.3.14.1 Eyes

The circular eyes work on the same principle as a photographic camera. Each eye has two poorly developed immovable eyelids and an anteroventrally located nictitating membrane. The eyeball is made up of three layers: sclerotic, choroid, and retina.

The sclerotic is made of cartilage. The exposed part of the sclerotic is transparent and is known as the cornea. The cornea is flat and fused with an outer conjunctiva that is continuous with the immovable eyelids. The pupil is a vertical slit that can neither dilate nor constrict. Rods and cones make up the retinal layer.

The sclerotic is lined internally by a connective tissue coat or choroid. It is continued in the front as a strongly circular curtain, the iris. The inner surface of the choroid known as the tapetum lucidum, contains a large number of guanine plates. It serves as a reflector. The crystalline lens has a spherical shape. The lens is held in place by the suspensory ligament, which runs from the lens's margin to the ciliary processes. The ciliary processes are choroid layer longitudinal folds.

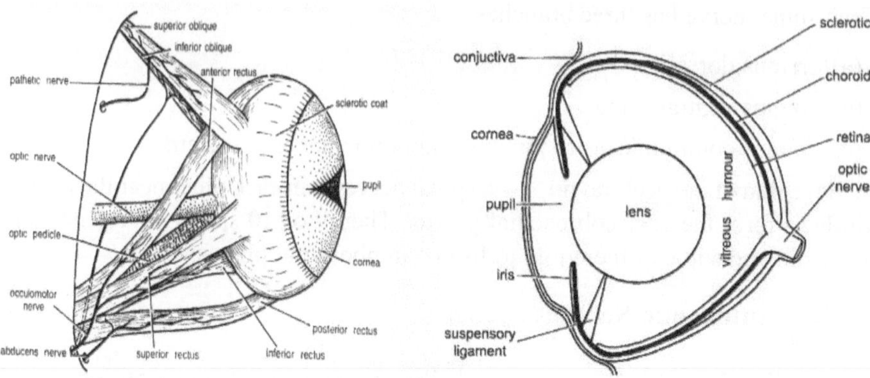

Fig 8.14: Eye muscles of *Scoliodon* (from https://www.faunadondness.com)

Fig 8.15: V.S. of the eye of *Scoliodon* (from https://www.notesonzoology.com)

The chamber in front of the lens is filled with a saline liquid or aqueous humour, while the chamber behind the lane contains a denser fluid or vitreous humour.

Six extrinsic eye muscles hold the eye in place in the orbit. There is an optic nerve and a cartilaginous optic pedicle in addition to the eye muscles (Figs. 8.14 and 8.15). Two groups of eye muscles are connected to the eyeball. The first group consists of:

(i) Superior rectus: It runs outwards and upwards and is inserted on the dorsal side of the eyeball.

(ii) Inferior rectus: It extends outwards and downwards to insert on the ventral side of the eyeball.

(iii) Anterior rectus: It runs outwards and downwards and is attached to the ventral surface of the eyeball.

(iv) Posterior rectus: It runs backward and is attached to the posterior surface of the eyeball.

The second group includes

(i) Superior oblique: It is attached to the dorsal surface of the orbit just in front of the insertion region of the superior rectus muscle.

(ii) Inferior oblique: It is attached to the ventral surface of the eyeball just in front of the inferior rectus muscle.

Scoliodon eyes are quite prominent and proportionately larger. *Scoliodon* has monocular vision because the eyes are laterally placed and each has its range of vision. *Scoliodon* is colorblind due to the lack of cones in the retina. However, the eye is adapted for near vision in dim light and fish can see their prey.

8.3.14.2 Olfactory Organs

In front of the mouth, there are two crescentic apertures, the nares or blind sac-like olfactory organs. Each olfactory sac is contained within a cartilaginous capsule and has no communication with the buccal cavity. The olfactory sac's mucous membrane is folded into two series of folds known as Schneiderian folds, which are formed by the median raphe. Schneiderian folds are made up of olfactory sense cells as well as supporting cells. *Scoliodon* has highly developed olfactory sense organs. Physiologically, three muscular nasal valves divide each nasal opening into a median excurrent siphon and a lateral incurrent siphon.

8.3.14.3 Internal Ear

The internal ear (the membranous labyrinth) is a closed ectodermal sac. It has three semicircular canals and a central body otosaccus. The otosaccus is laterally compressed and contains an anterior or dorsal utriculus and a posterior or ventral sacculus (Fig. 8.16A). The three semicircular canals are arranged in three planes of the body, with both ends opening into the central body. However, one end of each canal dilates and forms an ampulla. The utriculus produces an invagination beneath the ampullae of the anterior vertical and horizontal semicircular canals known as the recessus utriculi. The sacculus produces a posterior outgrowth known as the lagena cochlae. A cartilaginous auditory capsule houses the membranous labyrinth. The inner cavity is filled with a fluid called endolymph. Endolymph is a type of seawater that contains numerous calcareous bodies (otoliths). The perilymph fills the space between the membranous labyrinth and the auditory capsule. Through a small opening, a long tube called the endolymphatic duct (ductus endolymphaticus) connects the sacculus cavity to the outside world.

The membranous labyrinth is innervated by the auditory nerve. The group of receptor cells located in the utriculus and sacculus are called maculae and those found in ampullae are known as cristae.

Scoliodon's membranous labyrinth serves the following purposes:

(1) It helps in orientation to gravity,
(2) It accelerates in changing direction during swimming and
(3) It helps in hearing.
(4) It maintains muscle tone.
(5) It may detect a low frequency of vibrations of water.

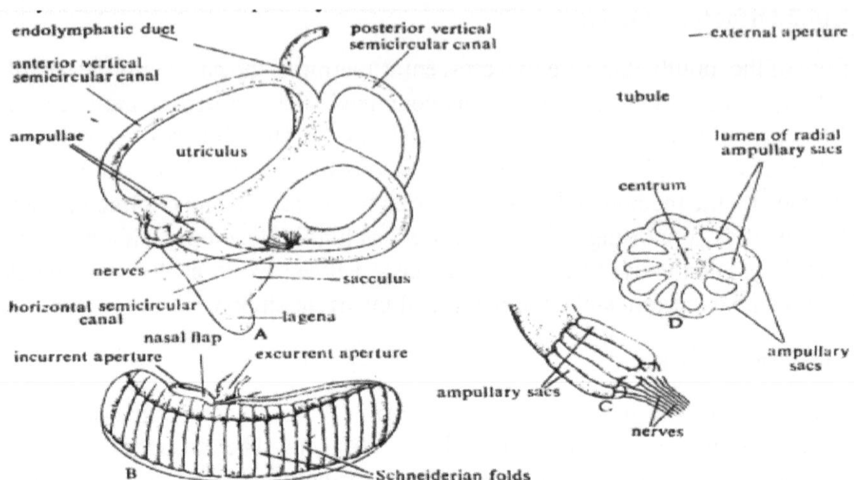

Fig 8.16: A. Internal ear of *Scoliodon.* B. Olfactory sac, C. Ampulla of Lorenzini and D. Ampullary sac. (From https://www.notesonzoology.com)

The utriculus together with the semicircular canals is responsible for the orientation and acceleration, while the sacculus is meant for hearing. The internal ears are the statoacoustic organs in *Scoliodon.*

8.3.14.4 Neuromast Organs

These organs comprise of

(i) **The lateral line receptors**: Inside the lateral line canal are the lateral line sense organs. These sensory organs, known as neuromasts, are epidermal derivatives that remain embedded in the lateral line canal wall. This canal communicates with the outside world via minute pores. Each lateral line canal is continued into the cephalic canal on the head.

In the head region, two lateral canals are joined by a transverse commissural occipital canal above the head and then each lateral line canal runs forward as a post-orbital canal below the orbit, both run up to the snout. The dorsal branch continues as the ventrolateral branch and joins with the infra-orbital canal. Arising from the infra-orbital canal is a jugal canal below the eye which runs back up to the first-gill cleft. The jugal canal gives rise to the mandibular canal to the lower jaw.

(ii) **The pit organs (Rheoreceptors):** Pit organs are scattered individual neuromasts found in all fishes. Pit organs are present on the lateral and dorsal sides of the head. They are ectodermal pits linked together by sense cell groups.

(iii) **The neuromast organs (Rheoreceptors):** They are scattered ectodermal sensory receptor organs. Each receptor has a stiff sensory hair projecting into the canal and a nerve fibre at the other end. The hairs are tipped with a heavy gelatinous substance. They aid in orienting the body to currents and waves.

(iv) **Ampulla of Lorenzini:** Numerous pores on the head's dorsal and ventral sides lead into a long tube that terminates in radially arranged ampullary sacs (Fig. 5.14C). The ampullae are arranged in clusters of eight or nine chambers arranged radially around a centrum (Fig. 5.14D).

Each ampulla has a pore opening on the surface; the pore leads into an elongated mucous-filled tubule which ends in a radially septate ampullary sac lying deep beneath the skin. The ampullae contain two types of cells pear-shaped glandular cells and sensory (pyramidal) hair cells.

The names of the ampullae are determined by their location:

(a) Supra-ophthalmic group: It lies around the supra-orbital canal.
(b) Outer buccal group: It lies between the supra-orbital and infraorbital canal.
(c) Inner buccal group: It lies around the infraorbital canal.

These sense organs are thermoreceptors (Sand, 1938) as well as pressure receptors.

8.3.15 Urinogenital System

In *Scoliodon*, the sexes are separate and sexual dimorphism is conspicuous. The male *Scoliodon* only possesses two claspers as the modification of pelvic fins.

The excretory system: It consists of two elongated kidneys, ureters, and a single urinogenital sinus. Graham Kerr classifies functional adult kidneys as opisthonephros. The anterior portion of the kidney (cranial mesonephros) becomes non-functional as the genital kidney, whereas the posterior renal kidney (caudal mesonephros) develops greatly.

The kidneys are made up of coiled glandular uriniferous tubules, also known as nephrons. Each tubule is made up of a double-walled cup (or Bowman's capsule) that houses the glomerulus and a tightly coiled renal tubule. A few renal tubules connect to a single collecting tubule. The kidney tubules have the unusual ability to reabsorb urea. The collecting tubules of the kidney's anterior non-renal portion open to the Wolffian duct (epididymis), while the posterior tubules open to the ureter, which then opens to the urinogenital sinus.

The mesonephric duct is divided into two parts. The dorsal duct is known as the Wolffian duct, and the ventral duct is known as the Mullerian duct. In males, the Wolffian duct becomes the vas deferentia, which connects to the vasa efferentia from the testis. In females, the Mullerian duct becomes the oviduct.

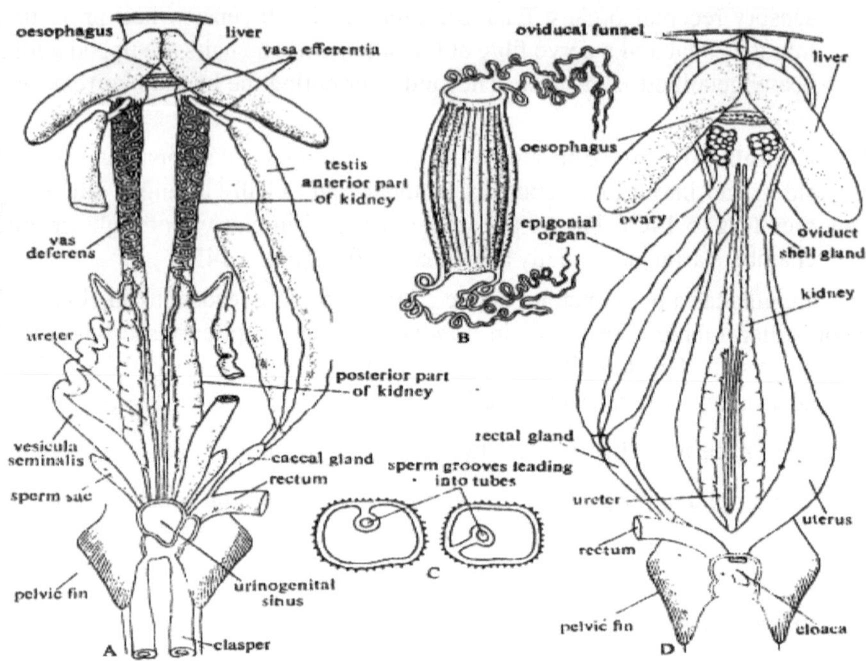

Fig 8.17: A. Male and D. Female urinogenital system of *Scoliodon*. B. Eggs and C. Sperm groove of *Scoliodon* (from www.old.amu.ac.in)

The kidneys are extremely long structures that extend from the liver's root to the cloacal region. The kidney proper is the kidney's posterior portion, which is highly developed. The anterior portion narrows and aids in the transport of genital products. This section is known as the Leydig organ or epididymis.

Sexual dimorphism: The male bears rod-shaped copulatory organs called claspers (myxopterygia) on the inner margins of the pelvic fins, each clasper leads into a siphon beginning at the base of the clasper.

Male reproductive system: The testes are elongated paired organs (Fig. 8.17A). Each testis is connected to the dorsal body wall by a peritoneal membrane called the mesorchium and to the caecal gland by ordinary tissue. Sperm cells escape through the vasa efferentia into the vas deferens, which become extremely coiled in the kidney's anterior portion. The vas deferens dilates greatly posteriorly to form the seminal vesicle (vesicular seminalis). The seminal vesicles connect to the urinogenital sinus, which connects to the cloaca. The urinogenital sinus wall evaginates, forming a club-shaped sperm sac.

A pair of sacs known as siphons is present. The siphons are found beneath the skin on the ventral side of the body. These are continued posteriorly to the sacs as siphon tubes that open to the groove of the clasper of the respective sides. The siphons are unrelated to the genital system. They contain sea water and aid in sperm expulsion through the clasper grooves. Each clasper has a closed groove with an anterior opening, an apopyle lying near the cloaca and a posterior opening as a hypopyle. Sperms enter the apopyle from the urinogenital papilla and are then transferred to the cloaca of the female during coitus.

Female reproductive system: There is no connection between the kidneys and the genital organs in females. The kidneys are typical, except that the ureters join posteriorly and open into the urinary sinus through a single urinary aperture.

The ovaries are two in number and are held in place by peritoneal folds known as the mesovarium. The shape, size, and colour of the ovaries vary greatly depending on the individual's age. Between the ovary and the rectal gland are two epigonial organs. The oviducts are extremely long tubes that stay connected both posteriorly and anteriorly (Fig. 8.17D). Two oviducts connect posteriorly to form the vagina, which opens into the cloaca. The oviducts converge anteriorly and open into the coelomic cavity via a longitudinal slit-like opening known as the oviducal funnel. A dilated shell (oviducal) gland at the anterior portion of each oviduct has little significance in *Scoliodon*.

Because *Scoliodon* is viviparous, the posterior oviduct dilates to form the uterus, allowing the young to develop. The number of embryos contained within a single uterus varies greatly between *Scoliodon* species. Fertilisation is internal and takes place in the oviduct between the oviducal funnel and the shell gland. The fertilised eggs enter the uterus. *Scoliodon laticuadus, Scoliodon sorrakowah* and *Scoliodon palasorrah* produce 8—19, 7 and 3 embryos respectively, when they conceive. The uterine mucous membrane is divided into several compartments, each containing an embryo. The number of compartments increases in direct proportion to the number of embryos.

8.3.16 Development

the eggs are large and yolked heavily. During its journey down the oviduct, each egg is coated with albumen. The egg is surrounded by a horny shell that is oblong, especially in oviparous forms. The shell's angles are extended into four coiled elongated filamentous processes (Fig. 5.17B). This condition is not seen in *Scoliodon*, but it is seen in oviparous sharks. Development of eggs occurs in the uteri and gives birth to young ones.

Scoliodon is ovoviviparous, meaning it gives birth to living offspring. It is claimed that during copulation, claspers are introduced into the female's cloacal aperture for spermatozoa transmission.

Scoliodon eggs are highly telolecithal. The cleavage is limited to a small germinal disc floating on top of the yolk mass (Fig. 5.18A). The cleavages are meroblastic, and the first cleavage plane is simply a furrow in the germinal disc's surface.

The second cleavage occurs at an angle to the first (Fig. 5.18B). The cleavage plane becomes irregular after this stage. A blastodisc is separated from a layer of periblast cells using this method of cleavage. A syncytial layer of periblast cells covers the yolk mass.

A blastocoel forms between the blastodisc and the periblast's central region. The blastodisc becomes multi-layered as the blastocoel expands. The blastodisc's posterior region grows faster than the other regions and is raised from the yolk mass to form a double-layered germ ring. A blastopore is formed when this germ ring is raised posteriorly (Fig. 5.18 D).

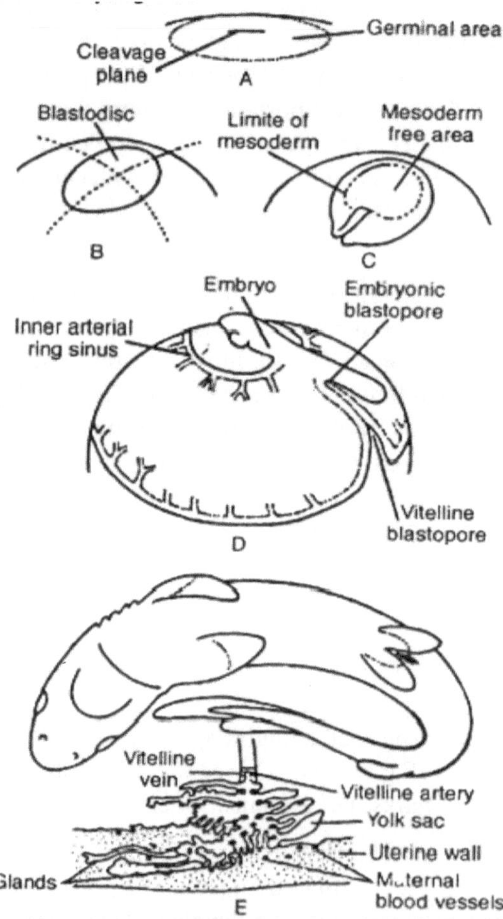

Fig 8.18: Development of *Scoliodon*. (from https://www.notesonzoology.com)

An upper layer (epiblast) and a lower hypoblast are separated at this stage. Between the ectoderm epiblast and the inner hypoblast is the blastocoel (mesendoderm). The formation of the dorsal lip of the blastopore initiates the gastrulation process.

Cellular proliferation from the hypoblast's underside differentiates the endoderm. A prechordal plate is formed by the mesoderm.

As the embryo lengthens along the anteroposterior axis, a head fold separates from the blastodisc and rises to form the neural folds. The blastodisc constricts the embryo's body, and the tail fold differentiates and extends backward. The fusion of the yolk sac edges divides the blastopore into two embryonic and vitelline blastopores. The circulatory system first appears in the mesoderm as blood islets that connect to form vascular networks. Because the embryonic blastopore closes, the mouth forms from stomoedeal invagination and the anus forms from proctodeal invagination as the embryo grows.

A tubular yolk stalk is provided to the developing embryo. This yolk stalk connects the embryo's intestine to the yolk sac. The yolk sac contains yolk material, which nourishes the developing embryo. When the yolk sac is fully exhausted, it folds and becomes anchored to the mother's uterine wall as the yolk-sac placenta. The yolk stalk is lost during the formation of the placenta, and the blood vessels of the yolk stalk form the placental cord. The placental cord connects the embryo to the placenta in the yolk sac (Fig. 5.18E). Appendicularia are finger-like processes that form on the placental cord. Each appendicularium is composed of a loose connective tissue core surrounded by several layers of epithelial cells. The location of the appendicularia in the placental cord varies by region. The appendicularia aids in the absorption of nutrients secreted by the mother's uterine wall.

9

Holocephali

9.1 Introduction

Holocephalians are a group of highly specialised cartilaginous fishes that live in the sea. It is structurally similar to Elasmobranchii and Teleostomi. It exists in both extinct and living forms. Previously considered a class of Pisces, Nelson (1984) classified Holocephali as a sub-class.

9.2 Systematic Position

Phylum: Chordata

Subphylum: Vertebrata

Division: Gnathostomata

Superclass: Pisces

Class: Chondrichthyes (=Elasmobranchii)

Subclass: Holocephali

Order: Iniopterygiformes (Chondrenchelyformes)

Chimaeriformes

Genus: *Squaloraja, Myriacanthus, Iniopteryx* (all extinct)

Helodus, Cochiodus, Menaspis, Deltoptcychius, Myriacanthus, Chimaeropsis, Squaloraja (all extinct)

Callorhynchus, Harriotta, Neoharriotta, Chimaera (=Hydrolagus), Rhinochimaera

Species: *Callorhynchus antarticus*

Harriotta ralcighana

Chimaera affinis, Chimaera monstrosa

9.3 Organisation

9.3.1 Origin and evolution

During the Carboniferous period, Holocephali diverged from the main Chondricthian stock. They originated during the Carboniferous period and evolved from the lower Jurassic to the recent period. They represent the Chondrichthyes' second major line of evolution.

9.3.2 Geographical distribution

Chimaera monstrosa (ratfish, rabbitfish, King of Herrings, Ghostfish, elephant fish) lives on the Pacific coasts of Europe, Japan, Australia, New Zealand, North America, Africa (=Cape of Good Hope), and India. The South Pacific oceans are teeming with *Callorhynchus antarticus. Harriota* is a North Atlantic Ocean deep-sea fish.

9.3.3 Habits and Habitat

They are marine fishes that live in water less than 80 metres deep but move to shallow areas to lay eggs.

9.3.4 Shape and Size

Holocephali has a variety of morphological forms. *Chimaera monstrosa* has an elongated shark-like body and a blunt conical snout. As a tactile organ, *Callorhynchus antarticus* has a rostrum produced forward and ending in a ventrally directed flap. *Harriota* is a long, elongated fish with a tapering, depressed rostrum. Holocephalians range in size from 60 to 150cm.

9.3.5 Morphological features

Holocephalians have a large and compressed head with a small mouth giving an appearance like that of parrot. The operculum is formed by a fold of skin, and there is only one branchial aperture. They are deficient in the spiracle and cloaca. The urinogenital aperture differs from the anus. There are two dorsal and one ventral fin. As in *Chimaera* (Fig. 9.1), the heterocercal tail transforms into a whip. A shallow glandular pouch near the pelvic fins is present in males. The pouch has unusual anterior claspers covered in denticles. Males have frontal (= cephalic) claspers or tentaculum on the dorsal surface of the head, as well as posterior claspers behind the pelvic fins. The median and paired fins are both well-developed. *Chimaera* has a lateral line groove, but *Callorhynchus* has a closed tube.

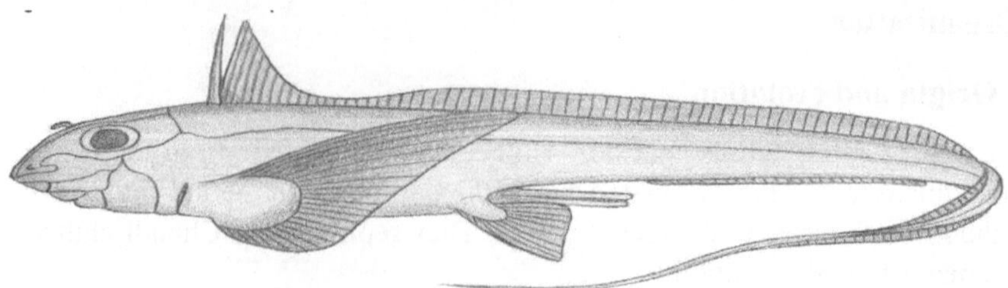

Fig 9.1 Morphology of Holocephali (from https://alchetron.com)

9.3.6 Exoskeleton system

Skin is smooth, silvery and without scales, but placoid scales are present during developmental stages.

9.3.7 Endoskeleton system

The vertebral column has a poorly developed notochord with cartilaginous arches. The vertebrae are reduced as nodules (Fig. 9.2). *Chimaera* has calcified rings embedded in its notochordal sheath. The neural arches of the first few vertebrae fuse to form a plate that connects to the first dorsal fin. The cranium is fused with a holostylic or autostylic skull and palate-quadrate. It has been reported that there is specific flexibility between the skull and the vertebral column. The hyomandibular is not involved in jaw suspension. As a result, jaw suspension is holostylic. Lateral cartilage is mature. The tail is filamentous and diphycercal in appearance. Three cartilaginous rods grow forward from the cranium to support the snout in *Callorhynchus*.

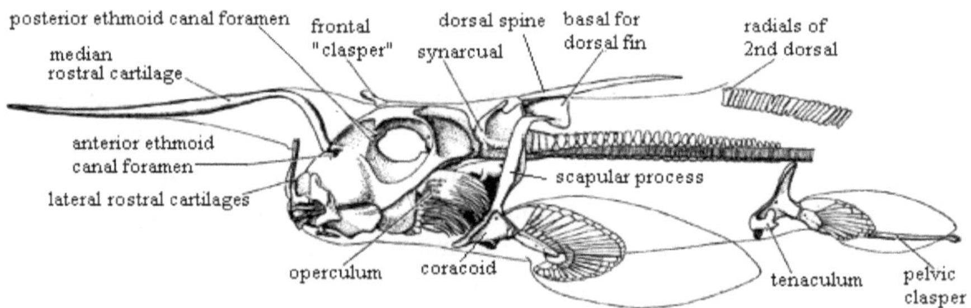

Fig 9.2: Endoskeleton of Holocephali (from https://alchetron.com)

These rods are shortened in *Chimaera*. First dorsal fin pterygiophores are fused into a single plate. Their iliac region is narrow and their pube-ischial region is broad. Branchial rays support a fleshy (spinny) operculum. The operculum is attached to the hyoid arch and forms gill opening covers on both sides. The operculum is not related to the teleost operculum. The occipital condyles are well-developed and superior to those of sharks.

9.3.8 Locomotion

The first of two dorsal fins bears a strong spine. The anal fin is small. The tail may be isocercal or heterocercal. Pelvic fins are positioned abdominally. These fish can swim due to the flapping movement of their pectoral fins.

9.3.9 Digestive system

These fish are omnivorous. The mouth is small, bounded by three folds of lips, and equipped with tooth plates with cutting edges. Teeth joined together to form crushing plates, with enamel replaced by vasodentine. Teeth may be vomerine, palatine, or mandibular in shape. It evolved to crush molluscs, crustaceans, and sea urchins. The gut is a simple, straight tube. The short oesophagus connects to the broad intestine, which then connects to the short rectum. They do not have a stomach, but their intestine has spiral valves.

9.3.10 Respiratory System

There is no mandibular pseudoarch and air bladder. There are four-gill openings without a spiracular cleft. Holobracnhs live in the first three branchial arches, while hemibranchs live in the fourth and hyoid arches. The fifth arch is gill less, and the gill's interbranchial septum is reduced. All of the gills are enclosed in a large branchial cavity that opens to the outside through a single external branchial aperture. This is due to the formation of a peculiar skin fold over the gill clefts.

9.3.11 Blood vascular system

As in a dogfish, the heart is made up of the sinus venosus, atrium, ventricle, and conus arteriosus with three rows of valves. They keep urea in their bloodstream for osmoregulation.

9.3.12 Excretory system

It consists of the opisthonephric kidneys, which are lobed structures with numerous urinifeorus tubules. They do not have peritoneal funnels but do have abdominal pores.

9.3.13 Nervous system

Small, spindle-shaped cerebral hemispheres connect to the olfactory bulb via a narrow peduncle in the brain. Diencephalon is a long, trough-shaped structure. The cerebellum is small, and the medulla oblongata is produced laterally into large frill-like bodies known as restiform bodies. The pineal is a small, rounded body found at the tip of the pineal stalk. The hypophysis is divided into intracranial and extracranial sections.

9.3.14 Sense organs

In comparison to body size, the eyes are large. On the cheeks and snout, there are small Ampullae of Lorenzini openings. Three semicircular canals run through the membranous labyrinth. The lateral line system is well developed, particularly in the head region.

9.3.15 Reproductive system

Unisexuality and sexual dimorphism are prominent. Males are bigger than females. Females have two ovaries, as well as shell glands and uteri. Males have two large oval testes but only immature sperm. The vas deferens coils to form the epididymis, which is connected to the kidneys. Sperm migrate to the epididymis, where they mature and form spermatophores. Enlargement of the lower epididymis forms the vesicula seminalis that opens into the urinogenital sinus. Transverse septa divide the vesicula seminalis into several compartments which serve to store spermatophores for later transfer into the urinogenital sinus. They don't have a cloaca.

9.3.16 Fertilisation

They are oviparous. Fertilisation is internal. *Callorhynchus* has a lengthy incubation period of 9-12 months.

9.3.17 Development

Eggs are capsulated and horny. Capsule secreted by shell glands. Capsules range in size from 13 to 25 cm. Cleavage is holoblastic.

9.4 Affinities of Holocephali

Holocephali occupies a position between the chondrichthyes and teleosts as well as certain features of placoderms. They also possess many well-defined peculiar characteristics that entitle them to a separate class.

9.4.1 Affinities with Placoderms

Some placoderms have a close relationship with holocephalians. *Rhamphodopsis* and *Ctenurella* have the following characteristics in common:

1. Males have large labial cartilage, large tooth plates, a long rat tail, a short platoquadrate, paired rostral processes to support a fleshy snout, and prepelvic claspers.
2. Menaspids have a series of large tubercles over their bodies and heads that resemble modern holocephalians' reduced dermal denticles.

When holocephalians are traced back through the geological record from living

Chimaeroids to Palaeozoic Helodontids, it is clear that they were very different at first. As a result, similarities between ptycodontid placoderms and living chimaeroids appear to be the result of parallel evolution due to the adoption of a similar benthic mode of life.

9.4.2 Affinities with Chondrichthyes

The holocephalians show resemblances with chondrichthyes in the following characters:

1. Exoskeleton: Smooth and silvery skin. Placoid scales are scattered throughout the claspers. Powerful denticles are found in *Squaloraja* and rows on the heads of *Callorhynchus* young.
2. Endoskeleton: Cartilaginous. The notochord is tenacious. The pectoral girdle fused in the midventral line. Chondrichthyes limbs and girdles Acentrous and ribless vertebral column. There are no bony jaws.
3. Fins: As in Cladoselache, the first dorsal fin has a spine in the front. A metapterygium is attached to a pectoral fin with many fan-like rays. Pelvic fins are smaller than pectoral fins. As in *Callorhynchus*, the caudal fin is heterocercal.
4. Alimentary canal: Ventral mouth. The nose leads to the mouth. A spiral valve is used to close the gut.
5. Swim bladder: They lack a swim bladder.
6. The heart has three rows of valves and a contractile conus.
7. Urinogenital organs: Opisthonephric kidneys. Ovaries are paired. There is a shell gland and uteri present. The epididymis was formed by the coiling of the vas deferens. Urea is kept in the body for osmoregulation.
8. Brain: The brain is shark-like, with highly developed restiform bodies. The olfactory lobes are relatively small, and the diencephalon is greatly elongated. Chondrocranium development in general.
9. Sense organs: Differentiated lateral line organs and ampullae of Lorenzini are present in the head region.
10. Eggs: Eggs are encased in horny capsules.

The holocephalians differ from chondrichthyes in the following characters:

1. Holostylic jaw suspension.
2. Unconstructed notochord.
3. Presence of crushing tooth plates.

Therefore, it is possible to conclude that holocephalians are descended from a primitive form of a shark with an unconstricted notochord that diverged from the main stem during the pre-Devonian period.

9.4.3 Affinities with Teleosteans

Holocephalins relationship with teleosteans shares the following common points:

1. Absence of spiracles and cloaca.
2. Branchial slits reduced to 4.
3. Gills with septum between the gill-lamellae.
4. An operculum is present so that the gills do not open directly to the exterior.
5. A single external branchial chamber is present.
6. Reduced interbranchial septum.
7. Urinogenital aperture lies in the anus.

9.4.4 Specialised characters of Holocephali

The holocephalians are the descendants of some primitive Selachians who deviated from the main line of evolution during the pre-Devonian period and became a specialised group.

1. Absence of denticles, spiracles and cloaca.
2. The palatoquadrate fused with the cranium. Jaw suspension holostylic. Neither the skull nor the jaw is attached to the hyoid arch.
3. Claspers on the head as well as in front of pelvic fins. Presence of Frontal clasper or tentaculum.
4. The large dental plate forms a parrot-like beak.
5. Teeth are devoid of enamel but bear vasodentine.
6. Operculum as a flap of skin.
7. Head produced in front to form a rostrum.
8. Placoid scales are lost from a large part of the body.
9. Presence of vertebral ring.

The main line of holocephalian evolution can be traced back to living chimaeroids, then to Mesozoic Squalorajas and Myriacanthoids, and then to Palaeozoic Menaspids, Helodontis, and Iniopterygiformes.

According to Young (1981), based on resemblances to chondrichthyes, holocephlians may be considered an aberrant group that diverged from chondrichthyes' ancestors. Holocephalians could be an imperfect extinct order of cladoselachiformes.

As a result, there is little agreement on the phylogeny of the holocephalians. They are a group of fishes that exist between the chondrichthyes and the teleosts, retaining some primitive characteristics of placoderms while developing some specialised characteristics to justify their status as a separate subclass among the chondrichthyes.

10

Teleostomi

10.1 Introduction

Bony fishes (perfect mouth fishes) have a distinct mechanism for opening their mouths by lowering the mandible through the hyoid apparatus, skull either hyostylic or amphistylic. They have a single pair of respiratory openings, a movable maxilla and premaxilla, four-gill arches, a swim bladder, an operculum and a homocercal tail. They are both freshwater and marine. They vary greatly in shape but are built on the same basic plan.

Labea rohita, also known as the *Rohu* fish, is perhaps the most common freshwater bony fish in India. Many Indian universities use *Labeo, Lates calcarifer* and *Mugil* as the type specimen of bony fish. *Channa (=Ophiocephalus) punctatus* is widely used in practical classes in different universities and other states of India.

The following description follows *Labeo rohita* which is perhaps the most common of Indian freshwater bony fishes. Meroblastic cleavage occurs during embryonic development and has a diploid chromosome number 52. It is most esteemed for eating purposes. *Labeo rohita* belongs to:

10.2 Systematic Position

Phylum: Chordata

Subphylum: Vertebrata (= Craniata)

Grade: Pisces

Class: Teleostomi (Osteichthyes)

Subclass: Actinopterygii

Superorder: Telestoi

Order: Cypriiformes

Family: Cyprinidae

Subfamily: Labeoninae

Scientific Name: *Labeo rohita*

Common name: Rohu

10.3 Organisation of Teleostomi

10.3.1 Habit and Habitat

It is abundantly found in freshwater ponds, lakes, rivers and reservoirs. It is chiefly herbivorous and bottom feeders but young fry feed on zooplankton. It is oviparous and breeds in June and July in running water. It shows external fertilization.

10.3.2 Geographical Distribution

Labeo rohita is found in tropical and temperate regions. It is the most common carp in the plains of India, except in the south. It is also common in Bangladesh and Myanmar (Burma). *Labeo rohita* and *Labeo calbasu* occur almost throughout India, Pakistan and Bangladesh.

10.3.3 External Structures

The *Rohu* fish has a spindly body that can grow to be 1 m long and weighs 20-25 kg. The dorsal side of the body is blackish, while the ventral and lateral sides are silvery. The body is distinguished by a prominent head, trunk and postanal tail (Fig. 10.1). They have a depressed head which forms an obtuse, short and blunt snout. Its dorsal profile is more arched than the ventral profile. Lips are thick and fringed with a distinct inner fold.

The head extends from the snout to the operculum's posterior margin. The snout is depressed and extends past the jaws. The lack of lateral lobes in the snout distinguishes *Labeo rohita*. On the dorsal side of the snout, there are two nostrils. The mouth is a transverse crescentic opening surrounded by thick fringed lips. Teeth are missing from the jaws. The eyes are prominent and without lids. On the dorsolateral sides of the mouth, one or two pairs of barbels are present. The maxillary barbels are shorter and more delicate than the rostral barbels.

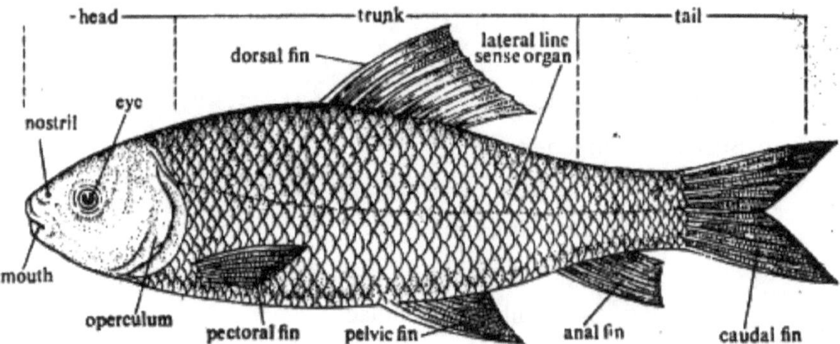

Fig. 10.1: Morphology of *Labeo rohita*

In cross-section, the trunk is elongated and oval. It is covered in cycloid scales that are thin and overlapping. The lateral line runs along the body's lateral sides. The scales along the lateral line have pores that are linked by a tubular canal. The vent is located ventrally, just ahead of the anal fin.

10.3.4 Fins

The paired and unpaired fins are well-developed. The pectoral and pelvic girdles support the respective fins. The pectorals are located behind the operculum on the anterolateral side of the trunk. 19 fin rays support each pectoral fin. The pelvis is located behind the pectorals on the ventral side. Each pelvic fin has nine fin rays. *Rohu* has only one dorsal fin, which emerges from the trunk's mid-dorsal line halfway between the snout and the base of the tail. The anal fin is located behind the anus. The dorsal fin has 13 fin rays, while the anal fin has 4-6 fin rays. The tail fin has two symmetrical lobes and is homocercal. The tail fin is supported by several fin rays (Fig. 10.1).

10.3.5 Skin

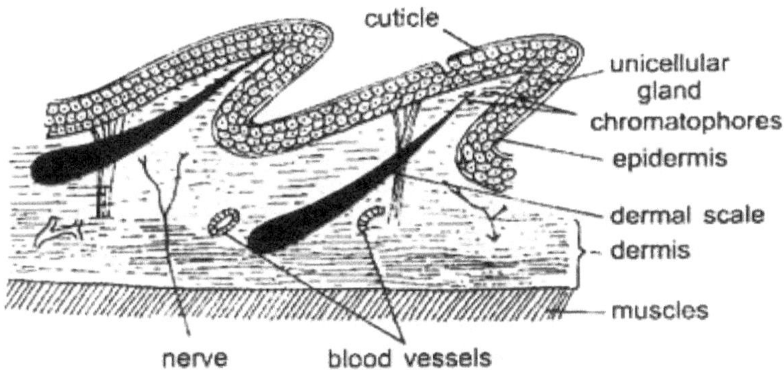

Fig. 10.2: VS of Skin of bony fish (From https://www.notesonzoology.com)

An outer soft epidermis and an inner dermis cover the integument. The epidermis is made up of epithelial cells and many unicellular mucous glands. The musculature of the body wall is located beneath the integument. The disposition of the body wall musculature is similar to that of other fishes (Fig 10.2).

10.3.6 Skeletal Structures

The scales and fin rays constitute the exoskeleton of *Rohu* fish. The endoskeleton is ossified and consists of

(1) The axial skeleton and
(2) The appendicular skeleton

The endoskeleton of *Rohu* was described by Sarbahi (1932).

10.3.6.1 Axial skeleton

The axial skeleton is composed of the skull, vertebral column with the ribs and the skeletal elements supporting the median fins.

10.3.6.2 Skull

Rohu's skull has a very complicated structural organisation. The formation of the skull involves many investing and replacing bones (Fig. 10.3). It is made up of a cranium, sense capsules, and visceral arches. The cranium and sense capsules are inextricably linked, whereas the visceral arches are only loosely connected to the skull. The skull is made up of two parts: a posterior basal plate and an anterior trabecular region. The auditory capsules connect to the basal plate, while the nasal capsules connect to the trabecular region. The lateral walls are derived from the orbital cartilage and are mostly incomplete. This cartilage connects to the auditory capsule posteriorly and the nasal capsule anteriorly. The palate-quadrate (visceral arch) articulates with the trabecular region of the skull anteriorly via a basal process and with the auditory capsule posteriorly via an otic process.

The skull in adults assumes an elongated shape. It is broadly divided into

(1) A dorsal roof: The dorsal side is more or less convex.

(2) A posterior occipital region: On its posterodorsal side, there is a shallow supratemporal groove. This groove runs posterolaterally to the main occipital spine. The posterior wall of the skull has three apertures, a median foramen magnum, and two large oval fenestrae. Fenestra is characteristic of cyprinoid skulls. Each fenestra penetrates the exoccipital bone and is a distinguishing feature of the cyprinoid skull. The occipital region is made up of three bones: the supraoccipital, the basioccipital, and a pair of exoccipital. The dorsal portion of the supraoccipital has a median vertical occipital spine. The occipital spine makes up the posteroinferior portion of the supraoccipital (or keel).

The exoccipitals are large bones that are made up of

(a) A basal plate: The basal plate is a component of the cranial cavity's floor.

(b) A paroccipital process: The paroccipital process is the side wall of the cranial cavity that forms the posterior boundary of the auditory capsule.

(c) A small dorsal process: The foramen magnum is encased by the dorsal process.

    The basioccipital is a large bony structure that is covered by the occipital condyle. The posterior surface of the occipital condyle has a deep depression. The ventral surface of the basioccipital bears a large oval masticatory process.

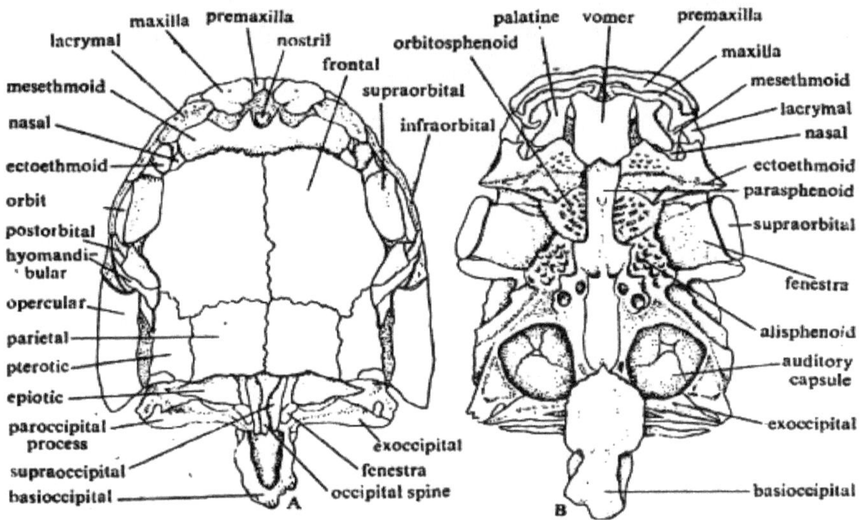

Fig 10.3: Skeletal system of *Labeo rohita* (From https://www.notesonzoology.com)

(3) The otic region: It consists of the auditory capsule bones: The otic region is represented by the paired auditory capsules on side-prootic, epiotic, sphenotic and pterotic, which are located on the back of the skull between the VII and IX cranial nerves. The anterior prootic unites in front of the basioccipital on the opposite side of the floor of the brain. The epiotic wedges between supraoccipital, exoccipital and pterotic form the epiotic process. Opisthotic lies in the posterior part of the boundary of the orbit to the exoccipital. Sphenotic lies above the prootic and forms part of the boundary of the orbit. Pterotic lies above the exoccipital and opisthotic to form the pterotic process. The opisthotic bone is absent in *Rohu* and the other four bones form a compact inverted cup-like structure. A small bean-like supra-temporal is present between the pterotic and its posterior process.

Each auditory capsule develops from otic cartilage that grows around the internal ear.

(4) A spatiotemporal (orbitotemporal) region: The temporal (=sphenoidal) region and the orbit make up the orbitotemporal region of the skull. The orbit contains anterior preorbital; posterior postorbital, supraorbital and postorbital and postfrontal are of the dorsal side bones and infraorbital froms the anteroventral part. The anterior boundary of each orbit is formed by the ectoethmoid, the posterior boundary is formed by the sphenotic. An inter-orbital septum separates two cavities from one another. The temporal region is divided into three sections:

(a) The parietal region: It includes parietal bones that overlap the wider anterior end of the dorsal portion. It also contains parasphenoids below and paired alisphenoids on the sides. It lacks a basisphenoid.

(b) The frontal region. The frontal region includes the frontals above, orbitosphenoids on the sides and parasphenoids below. Frontals articulate anteriorly with mesethmoid; laterally with nasals and ectoethmoids; anterolaterally with supraorbital and posterolaterally with postfrontals.

(c) The supratemporal region: It is located at the posterolateral angle of the skull.

(5) An anterior nasal (ethmoidal) region: The nasal (or ethmoidal) region is made up of the bones that form the nostrils and snout. The paired nasals, ectoethmoids and lacrymal, a median mesethmoid, a vomer, and a rostral are involved. The nasals, lacrymal, and vomer are all investing bones, while the mesethmoid, ectoethmoid and rostral are all replacing bones. The mesethmoid is transversely placed in front of the frontal. The rostral fits into the notch of the parasphenoid ventrally. The vomer lies beneath the mesethmoid in front of the parasphenoid ventrally. The nasal articulates with the posterolateral side of the mesethmoid and forms the roof of the olfactory capsule along with ectoethmoids. The lacrymal are attached to the anterolateral borders of the mesethmoid.

10.3.6.3 Visceral skeleton

The visceral skeleton is made up of seven half-hoops that surround the pharyngeal wall. The half-hoops of two sides connect along the midventral line to form seven visceral arches. The entire network of visceral arches arced together midventrally to form a basket-like visceral skeleton.

(A) The first visceral or mandibular arch: The primary lower jaw is formed by the mandibular arch, which is divided into a dorsal palate-pterygoquadrate bar and a ventral Meckel's cartilage. The palatoquadrate is homologous to the upper jaw of *Scoliodon*. The palatopterygoquadrate attaches to the cranium and forms the primary upper jaw. The primary upper jaw is ossified by the palatine, meta-pterygoid, and quadrate replacing bones. The anterior margin of the mouth is supported by two investing bones, the premaxilla and maxilla, which together form the secondary upper jaw. Each lower jaw half is made up of a small articular, a large dentary, and a small angular. In the middle line, two dentaries come together.

(B) The second or hyoid arch: The hyoid arch is also divided into two sections: upper and lower hyomandibular. The hyomandibular form is the suspensorium, which holds the jaws to the cranium. Many investing bones like interhyal, epihyal, ceratohyal and basihyal are linked to the hyoid arch and help to support the operculum (Fig. 10.4A).

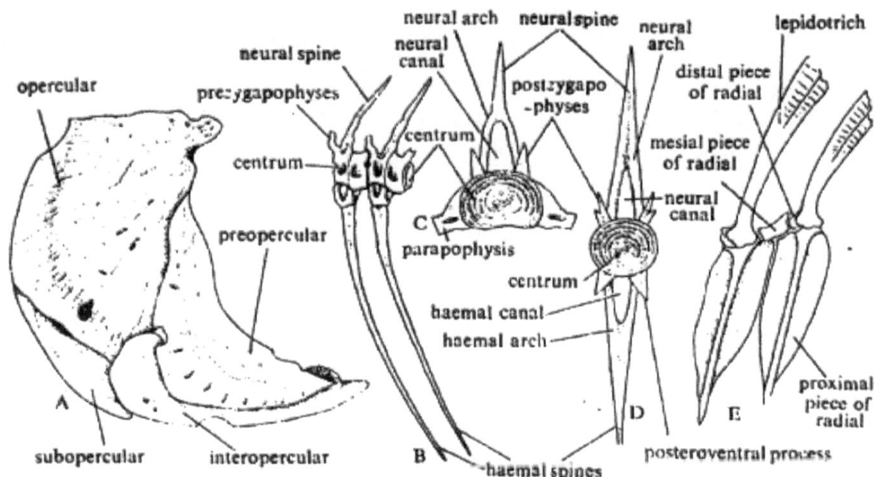

Fig 10.4: A. Opercular and B-E vertebrae of *Labeo rohita.* (From https://www.notesonzoology.com)

The operculum is made up of four bones: opercular, preopercular, subopercular, and interopercular. Four replacing bones, pharyngobranchial, epibranchial, ceratobranchial, and hypobranchial, ossify each branchial arch.

(C) Rest five or branchial arches: The gills are supported by four of the five branchial arches, while the fifth forms the inferior pharyngeal bones. Masticating plates with large teeth develop from the inferior pharyngeal bones.

10.3.6.4 Vertebral Column

The vertebral column is a fully ossified structure made up of 37-38 simple and comparatively similar vertebrae (Fig. 10.4B). The vertebrae are amphicoelous. The vertebral column is divided into the following sections:

(1) An anterior trunk region consisting of 21 trunk vertebrae (Fig. 10.4C) and bearing movable ribs: The first four trunk vertebrae are greatly altered since these vertebrae connect the swim bladder with the internal ear.

(2) The last 3–4 trunk vertebrae bear posteroventral processes.

Typical trunk vertebra: A typical trunk vertebra has a deeply biconcave centrum (Fig. 10.4C). During the embryonic stage, the concavities are linked by a narrow notochordal canal that perforates the centrum's body. In adults, the notochord canal closes. The centre's edges are joined by connective tissue ligaments, and the spaces between the vertebrae are filled with notochordal remnants. A pair of backwardly directed processes emerge from the centrum's anterolateral borders, encircle the spinal cord, and join above to form the neural arch. The neural arch produces a long dorsal spine that is backwardly directed. The prezygapophyses are a pair of small blunt processes found anteriorly at the base of the neural arch. Another pair of postzygapophyses emerges from the vertebra's posterolateral edges. Postzygapophyses point upwards and backward. A pair of short paraphyses extends downward from the ventrolateral surfaces of the centrum. Ligaments connect the ribs to the parapophyses.

Typical caudal vertebra: The caudal vertebral column is made up of 16-17 caudal vertebrae. A typical caudal vertebra, like a trunk vertebra (Fig. 10.4D) has:

(1) An amphicoelous centrum with a median dorsal, a medium ventral, and two lateral depressions,

(2) In the same position, there is a neural arch with a long backwardly directed neural spine and

(3) Articulating processes such as prezygapophyses and postzygapophyses.

A pair of backwardly directed processes emerge from the centrum's anterolateral margin and meet in the midventral line to form a canal for the placement of the caudal blood vessels. The haemal arch is the name given to this process. This arch results in a haemal spine that is oriented backward. At the bases of each haemal arch, there are two small blunt anteroventral processes. Similar processes can be found on the centrum's posterolateral side. As seen in posterior trunk vertebrae, these posteroventral processes are directed downwards and backward.

Median fin skeleton: The skeleton that supports the median fin is made up of

(1) Somactidia or endoskeletal radials: Somactids are parallel bony rods that are embedded within the muscles of the body. Each somactid is divided into three segments: proximal, mesial, and distal.

(2) Dermotrichia or dermal fin-rays: The dermotrichia help to keep the fin folded. Dermotrichia are branched and jointed in *Rohu* and are commonly referred to as lepidotrichia.

In addition to these fin rays, delicate horny rays (actinotrichia) can be found at the free edges of the fins.

Lepidotrichia supports the dorsal fin. On fourteen radials, there are fifteen or sixteen lepidotrichia. Each radial's proximal segment is enlarged and dagger-shaped. These are referred to as the interspinous bone or axonost. The distal sector is greatly reduced and the median segment is short. The anal fin has eight fin rays supported by seven radials. The first six are well-developed. Many flattened bony rods support the caudal fin. The urostyle has two epiurals and a radial on the dorsal side and nine hypourals on the ventral side. The fin-rays are connected in two symmetrical halves with the hypourals and epiurals.

Ribs

Pleural ribs are paired with slender bony rods that are segmentally arranged. The ribs are joined to the parapophyses at their distal ends. The ribs are divided into seventeen pairs, the first of which is connected to the parapophyses of the fifth trunk vertebrae. The ribs surround the abdominal cavity and are located between the muscles and the peritoneum. In addition to the ribs, there are a series of Y-shaped intermuscular bones that support the connective tissue septa or myocommas. The neural arch of all the vertebrae that make up the vertebral column gives rise to the intermuscular bones.

10.3.6.5 Appendicular Skeleton

The appendicular skeleton consists of the supporting structures of the paired fins and their girdles.

Pectoral girdle

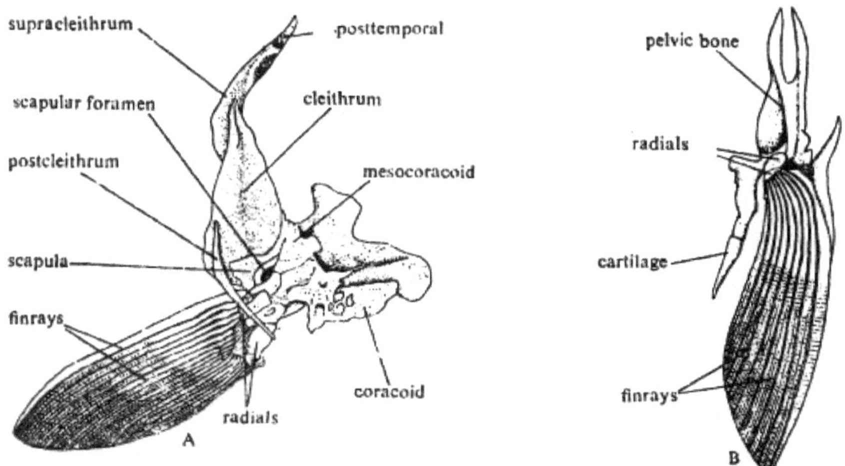

Fig 10.5: A. Pectoral girdle and B. Pelvic girdle of *Labeo rohita.* (From https://www.notesonzoology.com)

The pectoral girdle is located directly behind the final branchial arch. It has a reduced 'primary' endoskeletal structure and a well-formed secondary dermal girdle. The primary girdle is divided into two lateral halves, each of which is divided into three replacing bones, the scapula, coracoid, and mesocoracoid (Fig. 10.5A). The girdle's two halves do not meet in the midventral line. Investing bones make up the secondary dermal girdle. This girdle is made up of cleithrum (or clavicle) tips, supracleithrums, posttemporals, and postcleithrums. The post-temporal ligament connects the dermal girdle to the pterotic process of the skull.

The scapula is a bony ring with a large scapular foramen for the branchial artery and nerve to pass through. The coracoid is a triangular bone that is internal to the scapula but ventral to the mesocoracoid. The mesocoracoid bone is shaped like an inverted Y. The coracoid and scapula help to form the glenoid articulation, to which three of the four pectoral fin radials are movably attached. The largest bone in the secondary pectoral girdle is the cleithrum, which completely covers the primary pectoral girdle. The cleithrum's posterior inner surface is joined to a stout curved bone called the postcleithrum. The supracleithrum is an elongated dagger-shaped bone that covers the cleithrum's dorsal end. The supracleithrum articulates with a minute conical posttemporal bone at its dorsal end, which remains attached to the supratemporal bone.

Pectoral fin

Nineteen lepidotrichia are attached with four radials to support the pectoral fin. The radials are articulated with the scapula directly.

Pelvic girdle

The anal fin is located anterior to the pelvic girdle. It is divided into two identical halves. Each half is mostly made up of a large osseous pelvic girdle with a small cartilaginous rod attached to the pelvic bone's posterior end (Fig. 10.5B). The pelvic bone is divided into two parts: an anterior elongated portion and a posterior stout rod-like portion. The anterior part has a deep ventral groove and a forked frontal end. Ligaments connect the forked end to the ribs of the twelfth trunk vertebra. The middle line connects the posterior rod-like parts of the pelvic girdle's two halves.

Pelvic fin

Nine fin rays and three radials support each pelvic fin. The fin rays are proximally attached to the radials, and the radials articulate with the posterior border of the pelvic bone. The first two radials carry two fin-rays each, while the third radial carries the rest.

10.3.7 Coelom

The coelom is lined with peritoneum and divided into two cavities: an anterior pericardial cavity that houses the heart and a posterior perivisceral cavity that houses the major viscera.

10.3.8 Digestive System

The digestive system is made up of a very long alimentary canal and digestive glands (Fig. 10.6A and B). The alimentary canal is divided into the mouth, buccal cavity, pharynx, oesophagus, intestinal bulb, intestine, and rectum, which has an external opening known as the anus.

Upper and lower soft lips surround the mouth. The lips' free edges are broad, with four or five rows of blackish conical papillae. The buccal cavity is a small dorsoventral cavity with a flat floor and an arched roof. The buccal cavity's mucous membrane is lined with minute papillae. In Rohu, there is no distinct tongue, but the mucous membrane lining the floor of the buccal cavity has highly developed thick muscles.

The buccal cavity leads into a pharynx that is dorsoventrally flattened. The pharynx is divided into an anterior respiratory part and a posterior narrow masticatory part by gill arches. The anterior portion is narrower and has gill slits perforating it laterally. The ventrolateral walls of the pharynx bear closely set pharyngeal teeth, and the ventral wall is highly folded transversely. The pharyngeal teeth aid in the crushing of solid foods. These teeth are homodont, and they are arranged in three rows, one after the other. Each tooth has a cylindrical projecting crown and a narrow basal root. The root remains embedded in the mucous membrane, and the crown is compressed laterally.

The pharynx's posterior portion connects to a very short tube called the oesophagus. The oesophageal mucous membrane is divided into several longitudinal folds. The anterior respiratory part is perforated laterally by 4 pairs of gill slits. From gill arches project into the pharyngeal cavity small spiny gall rakers prevent the passage of food through gill slits. The swim-ductus bladder's pneumaticus enters the oesophagus.

There is no true stomach in *Rohu*. However, it is difficult to explain the absence of a stomach in *Rohu* and related forms such as *Labeo gonius, Cirrhina mrigala, Catla catla* and many other cyprinids. As suggested by Barrington (1957), a lack of stomach is not related to feeding habits but may be a case of neoteny.

The intestine is extremely long (about 7.5m) and thin-walled. The intestine has a relatively uniform diameter and many coils. The intestine's length and extensive coiling are related to its herbivorous diet. The intestine lacks intestinal villi and spiral valves. Absorptive cells with a free striated margin line the intestine. The intestine's muscular layer is thin. The intestine's anterior portion swells into a sac just behind the oesophagus. This sac is known as an intestinal swelling or an intestinal bulb and stores food. The intestinal bulb has an anterior broader cardiac part that opens dorsally to the pancreatic and bile duct, while a posterior narrower pyloric part without pyloric caeca is common in teleosts. The intestinal bulb lacks gastric glands and is histologically similar to the intestine. The intestinal bulb mucous lining of the cardiac part is honey-comb folds made up of absorptive and mucous cells. The pyloric part contains longitudinal folds. The intestine's mucous membrane folds in a variety of ways. The anterior portion of the intestine has oblique transverse folds, whereas the posterior portion of the intestine has distinct longitudinal folds.

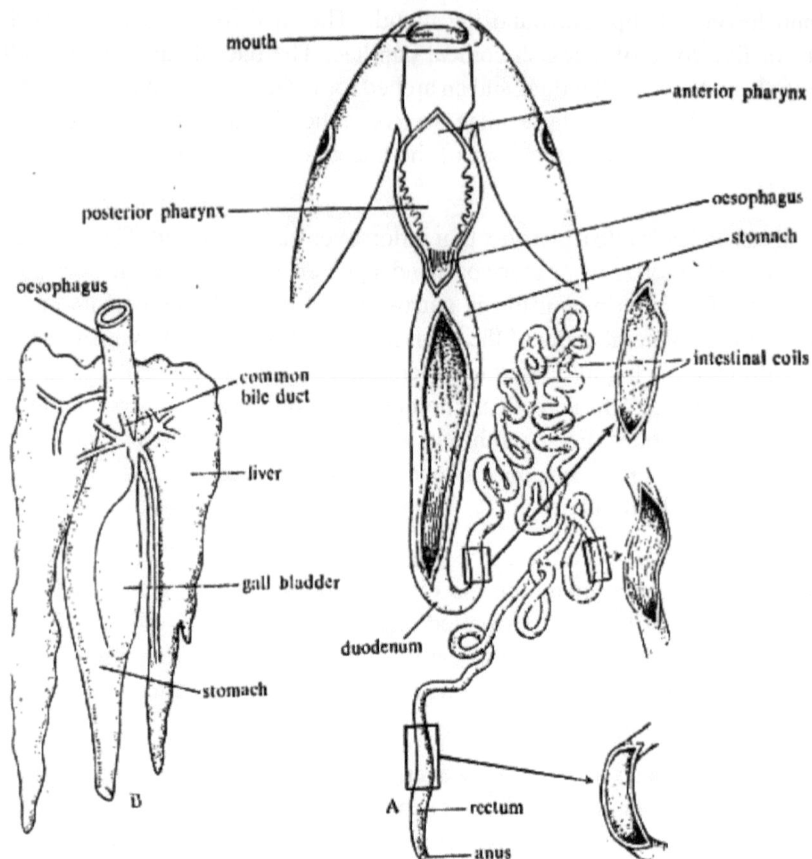

Fig 10.6: A and B Digestive system of *Labeo rohita.* (From https://www.notesonzoology.com)

The intestine's terminal portion is slightly dilated and forms a thin-walled sac known as the rectum. The length of the rectum is around 1 metre and there are no rectal glands. The mucous lining of the rectum, like that of the anterior part of the intestine, has indistinct oblique transverse folds.

The anus, which is located just in front of the urinogenital opening, connects the rectum to the outside world.

The digestive glands include the liver and pancreas (Fig. 10.6B). The liver is a massive, dark-brown gland. The liver is divided into two lobes: the narrow right lobe and the larger left lobe. The liver's two lobes are linked in three places: anteriorly by a median lobe, medially by a median connective lobe, and posteriorly by a median mass. The gall bladder is an elongated sac about 8 cm long and 2.5 cm wide. The gall bladder is situated between the liver's right and left lobes. A cystic duct receives three hepatic ducts after emerging from the anteroventral end of the gallbladder. The pancreas remains diffused along the intestine's coils. It penetrates the liver. Acini can form in the exocrine pancreas. The acini's cells are quite large. Each cell is columnar, with a prominent nucleus and two parts: basal and apical. The basal end has homogeneous cytoplasm, whereas the apical end has large zymogen granules. The presence of exocrine pancreatic cells in the spleen tissue of *Rohu, Catla Catla* and *Mrigal* is unusual.

The mechanism of digestion is unknown. The production of pancreatic trypsin and erepsin, as well as enterokinase from the intestinal mucosa, compensates for the absence of the stomach. Amylase is manufactured by pancreatic cells. Lipase and maltase have also been found in intestinal extracts, though the location of their secretion has not been determined. Pepsin and HCl are not present in this genus of fish because it lacks a stomach. Because this fish is herbivorous, the concentration of carbohydrate-splitting enzymes is highest, while the concentration of protein-splitting enzymes is lowest.

10.3.9 Hydrostatic Organ

The swim bladder is a gas-filled perivisceral sac with thin walls. It is located dorsal to the gut in the cavity of the coelom. It is connected to the roof of oesophagus by a slender pneumatic duct. The swim bladder is of physostomous type and is divided into two chambers: anterior and posterior. A slender ductus pneumaticus connects the anterior chamber to the oesophagus.

10.3.10 Weberian Ossicles

Labeo has a unique bony chain that connects the anterior end of the swim bladder to the internal labyrinth. Weberian ossicles are made up of four bony pieces. Tripus, intercalarium, scaphium, and claustrum are the pieces (Fig. 10.7). The tripus is the ossicles' largest piece. It consists of three processes:

(i) The anterior process connects to the inter-ossicular ligament,

(ii) The medial process connects to the second and third vertebrae and

(iii) The posterior process is curved and connects to the anterior end of the swim bladder to act as a transformator. The intercalarium is a rod-like piece that extends up to the second vertebral centrum. Scaphium is a large piece. An inter-ossicular ligament connects the scaphium and intercalarium. The claustrum is the smallest piece that articulates into the first vertebra's neural arch.

Weber (1820) considered the Weberian ossicles to be homologous to the mammalian ear ossicles. These ossicles are thought to be non-homologous to mammalian ear ossicles due to significant differences in structure, function, and development. Weberian ossicles serve many functions in the lives of fish. The functions are as follows:

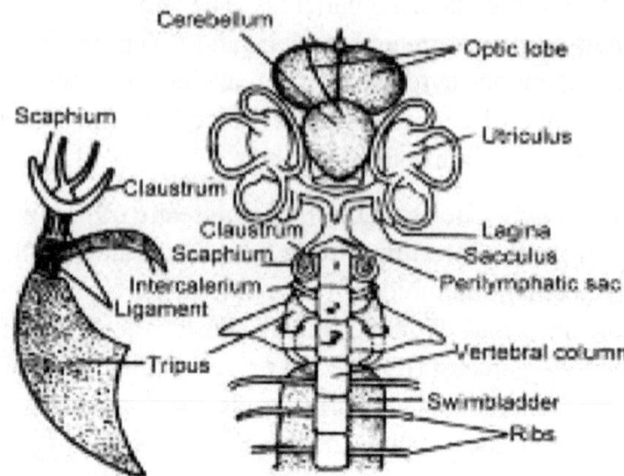

Fig 10.7: Relationship between Weberain ossicles and swimbladder of *Labeo rohita.* (From https://biologyeducare.com/)

(i) Serve as a pressure gauge. The Weberian ossicles carry any change in gas volume in the swim bladder caused by changes in hydrostatic pressure to the perilymph. This change is transmitted to the endolymph by the perilymph.

(ii) Serve as a barometer, detecting changes in atmospheric pressure and providing warning of impending weather changes.

(iii) May help with auditory function. Fish with Weberian ossicles are more sensitive to sounds with higher frequencies and lower intensities than fish without such ossicles. The ossicles also aid in sound localization.

10.3.11 Respiratory System

In the branchial chambers, there are four pairs of gills and the gills are holobrach types. The gills of *Labeo* are known as filiform or pectinate. The operculum is supported by bony plates and covers each branchial chamber. The chamber opens outside by a large crescentic branchial or gill aperture behind the operculum and in front of the pectoral fin. A branchiostegal membrane is attached to the posterior margin of the operculum. Each gill arch bears one afferent and two efferent branchial vessels (Fig. 10.8A to C). The hyoid arch pseudobranch is made up of a comb-like body. Each pseudobranch is made up of a single row of gill filaments on the operculum's inner surface. Therefore, four holobranchs are only attached to the first four branchial arches and fifth gill slits on either side beneath an operculum. Each gill is supported by a gill arch with gill rakers and has a double row of gill filaments. The two rows of gill lamellae are separated by the inter-branchial septum which is short and compact.

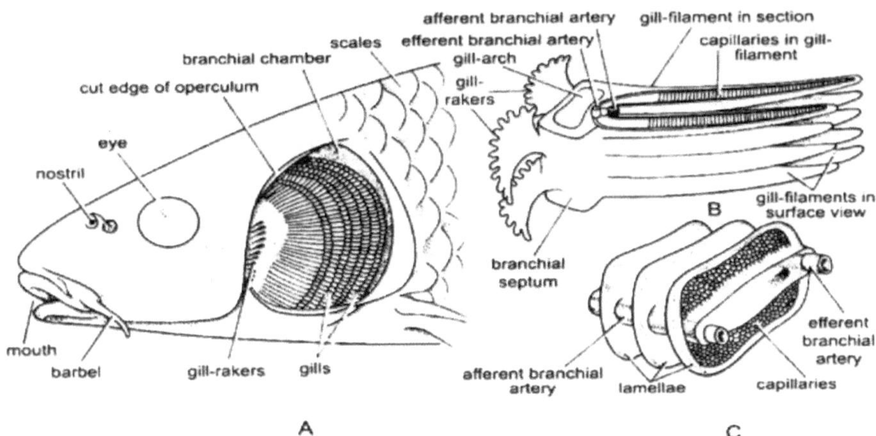

Fig 10.8: Gills of bony fish. A. Gills in left gill chamber. B. Part gill showing gill chamber and gill raker and C. Blood supply to the gill. (From https://www.notesonzoology.com)

Physiology of respiration

Rohu utilizes the oxygen dissolved in water. The physical mechanism of respiration can be described under two sequences.

Inspiration

During inspiration, the outer opening of the gill chamber remains tightly closed to the body wall by the branchiostegal membrane and the two opercula bulge out to increase the accommodating capacity of the pharyngeal and buccal cavities. As a consequence, water from the exterior rushes inside through the opened mouth and fills in the buccopharyngeal cavity (Fig. 10.9A).

Expiration

Immediately with the entry of water, the pharyngeal and the buccal cavities contract and exert pressure on the contained water. As the mouth, by this time, becomes closed by oral valves, the contained water finds its way out through the gill slits (Fig. 10.9B).

The operculum as well as the branchiostegal membrane is lifted by this time and the water from the gill chambers goes out through the opening of the gill chamber. The dilatation and the contraction of the pharyngeal cavity are caused by the alternate retraction and protraction of the hyoid arch supporting the buccopharyngeal cavity.

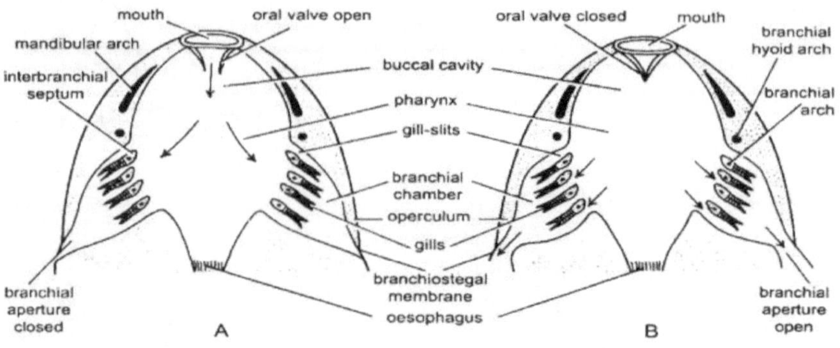

Fig 10.9: A. Inspiration and B. Expiration in *Labeo rohita*. (From https://www.notesonzoology.com)

Physiology of gaseous exchange

The gills are highly vascular structures and are supplied by afferent and efferent branchial arteries. The afferent branchial artery carrying the deoxygenated blood is situated very superficially on the outer edge of the gill. The afferent brachial artery breaks up into capillaries in the substance of the gill.

During the transit of water through the gill slits, the deoxygenated blood in the capillaries of the gill filaments takes oxygen dissolved in water and gives out carbon dioxide by diffusion. The blood thus aerated, is collected by efferent branchial arteries and is conveyed to the different parts of the body.

10.3.12 Circulatory System

The blood vascular system is of single circulation type and contains Blood, Heart, Arteries and Veins.

10.3.12.1 Blood

It contains plasma and corpuscles. Erythrocytes are elliptical and nucleated with haemoglobin.

10.3.12.2 Heart

Rohu's circulatory system is teleostean in design. The heart is 2 chambered and contains the sinus venosus, the auricle, and the ventricle. The sinus venosus is thin-walled and has two lateral appendages and is larger in proportion. The texture of the heart is spongy. Sinus venosus receives deoxygenated blood by ductus Cuvieri. The auricle is thin-walled and lies ventral to the sinus venosus. The auricle receives blood from the sinus venosus through the sinu-auricular aperture. In absence of conus arteriosus, bulbus arteriosus is represented by a pair of valves (Fig. 10.10).

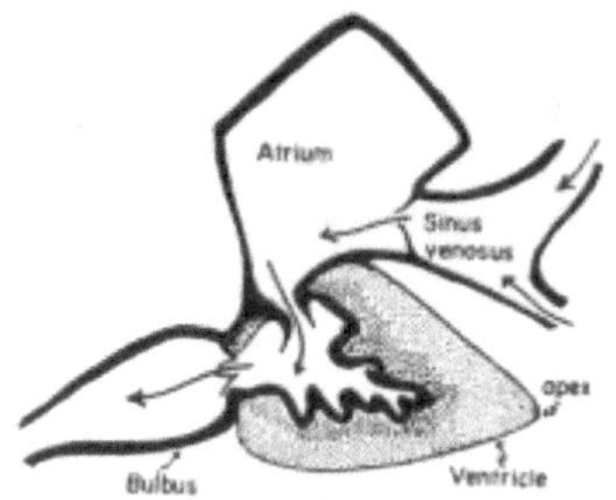

Fig. 10.10: Heart of Bony fish (From https://biologyeducare.com)

10.3.12.3 Arterial System

(A) Afferent branchial arteries: These transport deoxygenated blood to the associated gills. Anteriorly, the ventral aorta divides into the first two afferent branchial arteries. The proximal part of the ventral aorta dilated to form the bulbus arteriosus. The second, third, and fourth pairs of afferent branchial arteries arise independently from the ventral aorta.

(B) Efferent branchial arteries: These transport oxygenated blood from the gills to the heart. The first and second efferent branchial arteries unite to form the anterior epibranchial artery. The third and fourth efferent branchial arteries unite to form the posterior epibranchial artery. The first efferent branchial artery gives off a cephalic artery which further divides into the external carotid artery and internal carotid artery. Along its route, the dorsal aorta gives the following branches:

  (i) A coelico-mesentric artery to provide blood to the swim bladder and parts of the alimentary canal.
  (ii) A few pairs of renal arteries to the kidneys.
  (iii) Several paired segmental arteries for segmental muscles.
  (iv) A pair of genital arteries to the gonads.
  (v) A pair of subclavian arteries, each dividing into pectoral and pelvic arteries.

10.3.12.4 Venous System

The venous system is made up of systemic veins and portal veins. These veins carry deoxygenated blood from various parts of the body to the heart, either directly or indirectly. It contains paired anterior and posterior cardinal veins, unpaired hepatic and renal portal veins and paired subclavian veins. The renal portal vein is visible.

10.3.12.5 Systemic venous system

Right and left ductus Cuvieri transport blood to the sinus venosus. Each ductus

Cuvieri is made up of three major veins: an anterior cardinal sinus, a jugular sinus, and a posterior cardinal sinus.

The anterior cardinal sinus transports blood from the anterior part of the body to the posterior part of the body. The posterior cardinal veins both receive segmental veins, renal veins, genital veins, and other veins.

Aside from the three major veins mentioned above, the pectoral and pelvic veins form the pectoral and pelvic fins, respectively, and the slender hepatic vein opens into the ductus Cuvieri.

A caudal vein transports blood from the tail region to the trunk, where it splits into two branches. The right posterior cardinal sinus enters the right ductus Cuvieri after passing through the substance of the right kidney. The capillaries of the renal portal vein give rise to the left posterior cardinal vein.

10.3.12.6 Portal system

The portal system is made up of a special vein that begins and ends in capillaries, and the blood from these veins passes through some intermediate organs before reaching the heart. When the intermediate organ is the kidney, the system is referred to as the renal portal system; when the organ is the liver, the system is referred to as the hepatic portal system.

(A) Renal portal system

After entering the left kidney, the left branch of the caudal vein divides into capillaries, forming the renal portal vein. The left posterior cardinal vein is formed when these capillaries join.

(B) Hepatic portal system

The capillaries of the alimentary canal and its associated structures join to form the hepatic portal vein, which enters the liver substance and divides into capillaries. The capillaries join to form the hepatic vein, which connects to the ductus arteriosus.

10.3.13 Nervous System

10.3.13.1 Central nervous system

Brain: *Rohu*'s brain is typically built on the piscine plan. However, the brain of *Labeo* is more developed than *Scoliodon*. Olfactory lobes, cerebrum and diencephalon are somewhat smaller while the optic lobe and cerebellum are larger than *Scoliodon*.

The small and undivided prosencephalon comes in front of the midbrain. The cerebral hemispheres are small and separate. Although the corpora striata are visible, the pallium is thin and non-nervous. The olfactory lobes are developed to a moderate degree. Behind the pituitary body, there is a saccus vasculosus. A dorsal pineal body and a ventral pituitary body are present in the reduced diencephalon. The optic lobes are large, with two large ventral lobi inferiores. The optic chiasma is missing and the optic nerves simply cross. The cerebellum is prominent, and it extends anteriorly to form the valvula cerebella as a characteristic of teleosts.

Cranial nerves: *Rohu* has ten pairs of cranial nerves. The origin and branching of the cranial nerves are also located in this manner.

Spinal nerves: These are paired and arise a dorsal sensory root and ventral motor root.

10.3.13.2 Autonomic nervous system

The autonomic nervous system in teleosts is more developed than in cartilaginous fishes. In teleosts, a paired symmetrical chain of sympathetic ganglia begins at the terminal level and runs back to the end of the tail. The ganglion is connected with each cranial dorsal root. The sympathetic trunk is connected with the mixed spinal nerves by rami communicanates.

10.3.13.3 Sense Organs

Special sense organs are well-developed and built on a typical teleostean plan.

Olfactory pits: They open to the external using two apertures but they do not communicate with the buccal cavity.

The sense of taste (chemoreceptors) is extremely refined. There are numerous taste buds in the lips, the epithelium lining the first three gill slits, and on the barbels.

Tactile receptors are abundant throughout the body, particularly on the lips and barbels.

The organisation of the lateral line system, eyes, and ears is strikingly similar to that of previously described bony fishes.

The eye is monocular and has three layers: (a) Sclerotic, (b) choroid and (c) retina. Between sclerotic and choroid, there is a silvery layer or argentea which gives colour. The eye lacks an eyelid with a flat cornea a large pupil and the lens almost spherical. A vesicular stickle-shaped fold of choroid or falciform process extends from the black spot to the back of the lens where it meets a knob-like muscle from the lens or campanula Halleri (retractor lentis). The falciform process and Campanula Halleri takes an important part in accommodation. Further, between choroid and argentea, there is a choroid gland that surrounds the optic nerve. The choroid gland is a complex network of rete mirabile. Together they form the accommodation apparatus which helps in focusing the eye on distant or near objects by shifting the lens (Fig. 10.11).

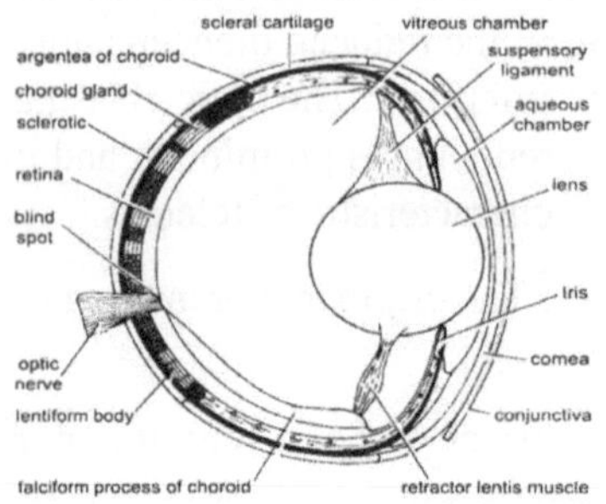

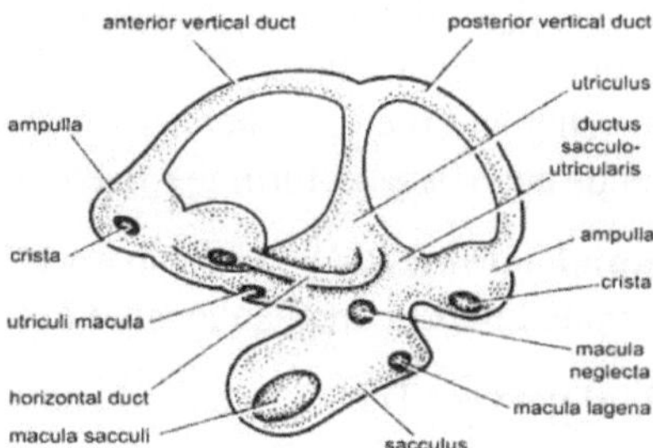

Fig 10.11: V.S. of eye of the *Labeo rohita.* **Fig. 10.12:** Internal ear of *Labeo rohita*
(From https://www.notesonzoology.com)

The ear is made up of four parts: the utriculus, sacculus, lagena and three semicircular canals (Fig. 10.12). The otoliths are three in number and are remarkably large:

(1) Sagitta is found within the sacculus,

(2) Asterisus is found within the lagena and

(3) Lapillus is found within the utriculus.

10.3.14 Urinogenital System

The excretory and reproductive organs are quite independent in *Labeo*. Fishes are ammonotelic.

The kidneys are extremely long bodies that extend the entire length of the visceral cavity. These are located on the dorsal side of the body wall, above the swim bladder and are distinct anteriorly but partially fused in the middle. The kidneys are mesonephric. The free anterior end of the kidney has knob-like a non-renal pronephric head kidney. Nitrogenous waste is carried posteriorly from each kidney in a slender, tubular ureter. The ureter opens into a thin-walled urinary bladder ventral to the cloaca, one from each kidney. The urinogenital sinus is formed by the opening of the urinary bladder (Fig. 10.13 A and B). Excretory fluid is temporarily stored in the bladder and then expelled through the midventral urinary aperture situated behind the anus.

During the breeding season, the gonads grow significantly larger. The testes in males extend the entire length of the abdominal cavity. A vas deferens emerges from the posterior end of each testis and eventually opens into the urinogenital sinus. In females, the ovaries are also paired structures that grow larger than the testes. There are no oviducts. The eggs are released into the body cavity and emerge through a pair of genital pores formed temporarily from the anterior wall of the urinogenital sinus.

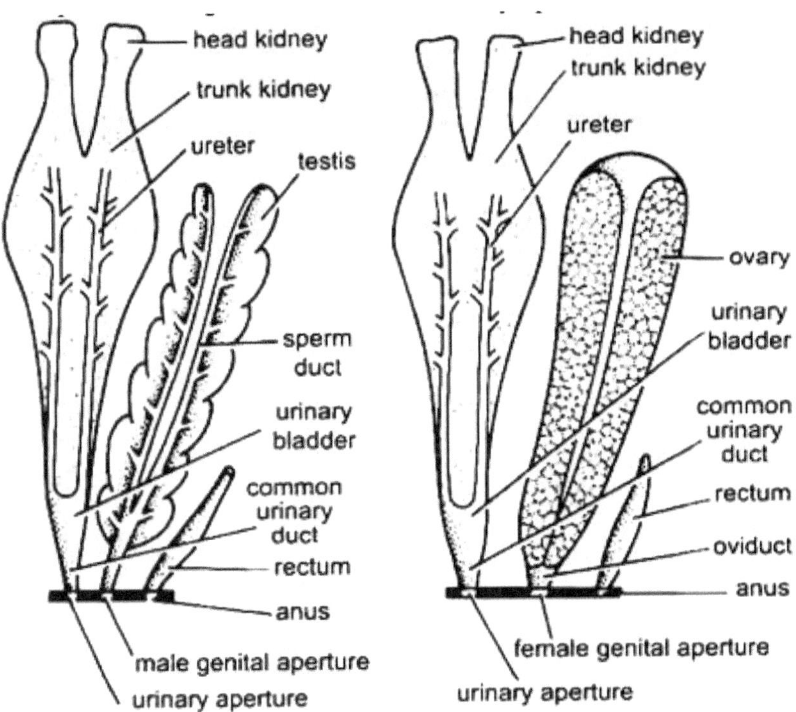

Fig 10.13: Urinogenital system of A. Male and B. Female *Labeo rohita***.** (From https://www.notesonzoology.com)

10.3.15 Development

In *Rohu,* a large number of eggs are laid at once, and the eggs sink to the bottom. The eggs come into contact with the spermatic fluid (milt) immediately after discharge, and fertilisation occurs externally. In *Rohu*, the cleavage is meroblastic and developmental direct. It is anticipated that the development will be similar to that of other bony fishes. *Labeo rohita* is a rapidly expanding carp. The eggs hatch in 2 to 15 hours. Hatchlings are newly hatched young with a prominent yolk sac attached to the ventral side of the body. It takes 5-7 days for the yolk sac to be absorbed, after which the young (called fry) begin to feed. The fry grows to be about 2-3 mm long and has fringed lips as well as a prominent vertical dark spot at the base of the tail that disappears as it grows. When the fry reaches 5 mm in length, it is classified as a fingerling, which can range in length from 5 to 15 mm. Sexual maturity takes about two years to achieve (Fig. 10.14).

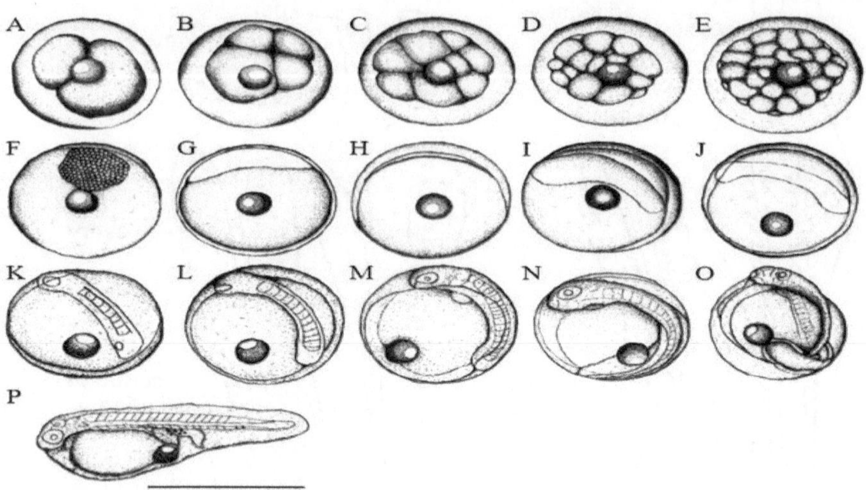

Fig. 10.14: Egg development stages of *Epinephelus septemfasciatus*. (A) 2 cell stage; (B) 4 cell stage; (C) 8 cell stage; (D) 16 cell stage; (E) 32 cell stage; (F) Morula stage; (G) Blastula stage; (H) Gastrula stage; (I)-(J) Formation of embryo; (K) Kuffer's vesicles appearance; (L) 12~13 mytomes stage; (M) Formation of eyes; (N) Heartbeat; (O) Hatching; (P) Hatched larvae. Scale bar = 1.0 mm.

11

Crossopterygii

11.1 Introduction

The crossopterygians, primitive, lobe-finned, bony fishes, represent a very interesting group of fish showing many features of amphibians and other tetrapods. They possess paired fins with a large median lobe, fins covered by scales, a cosmine layer, swim bladder as lungs, nostrils that open into the mouth cavity and autostylic jaw suspension.

11.2 Taxonomic position

Phylum: Chordata

Subphylum: Vertebrata

Superphylum: Gnathostomata

Grade: Pisces

Subgrade: Teleostomi

Class: Oestichthyes

Subclass: Crossopterygii (Lobe-finned fishes, Tassel-finned fishes or Fringe-finned fishes)

Superorder: 1. Osteolepimorpha (Osteolepida or Rhipidistia) All extinct

Order: 1. Porolepiformes (Holoptychiiformes): e.g. *Glyptolepis, Porolepis.*

2. Osteolepiformes: e.g. *Gyroptychius, Glyptopomus, Eusthenopteron, Latvius, Panderichthyes, Thursius, Megaligthyes, Ectosteorhachis, Lohsania, Platycephalichthys, Tristichopterus.*
3. Rhizodontiformes (root teeth fish): e.g. *Rhizodon, Rhizodus, Callistiopterus.*
4. Struniiformes (Onychodontiformes) (dagger-tooted fish): e.g. *Onychodus, Strunius.*

Superorder: 2. Coelacanthiformes e.g. *Coelacanthus, Nesides, Rhabdoderma, Latimeria chalumnae* (Fig. 11.1), *Macropoma, Undina, Laugia, Chagrinia, Dictyonosteus, Diplocercides, Euporosteus, Spermatodus, Synaptotylus.*

11.3 Organisation

11.3.1 Origin and evolution

Osteolepis existed from the mid-Devonian to late-Permian periods (416–251 million years ago). Coelacanths first appeared in mid-Devonian marine deposits and have survived to the present day. The crossopterygians underwent a rapid decline and then almost became extinct after the end of the Triassic Period (about 200 million years ago). Coelacanths probably originated from *Osteolepis*. Among the primitive crossopterygians are *Youngolepis* and *Powichthys*.

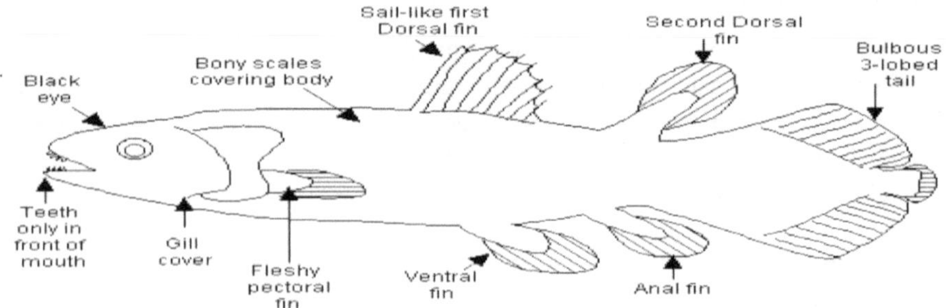

Fig 11.1: Morphology of *Latimeria chalumnae* (from https://www.enchantedlearning.com)

11.3.2. Habits and habitats

They are freshwater or marine, free-swimming, shallow or deep water forms, large predators with a strong sense of smell and small eyes. Freshwater forms existed in the Carboniferous period.

11.3.3 Geographical distribution

Osteolepids occurred in the old red sandstone of Scotland and also in North America, Antarctica, Asia and Europe. The coelacanth was caught off South Africa in 1838. The rocky bottom of the Comoro Islands, which lie between Madagascar and Mozambique, has yielded 80 specimens. They are distributed in the Ethiopian region only.

11.3.4. Shape, size and colour

Living coelacanths are enormous. Their bodies are shorter and stockier than those of Rhipidistians, but larger than those of their forefathers. They can weigh up to 80 kg and have a length of 2.5 meters. Their colour is steel-blue fusiform.

11.3.5 Morphology

Paired pedunculate fins act as active swimmers and are used for movement at the seafloor. These fins are provided by a large median lobe and covered by scales.

Anterior dorsal fins are supported by a bony plate and a joined median axis while the posterior ones are almost similar to paired fins. The tail is diphycercal with a small median lobe. Eyes are highly reflective and golden coloured.

11.3.6 Exoskeleton

Scales of the cosmoid type; rhombic, quite thick or thin and overlapping; containing an outer layer underlined by a layer of vascular tissue, the inner isopendine tissue and a more superficial enamel-like layer.

11.3.7 Endoskeleton

The endoskeleton is mostly bony. The skull is divided into two units: an anterior or ethmo-sphenoidal and a posterior or oto-occipital. Bones periodically lose their cosmine layer. The vertebral column is acentrous, cyclocentrous or amphicentrous. The cranium is divided into two regions: antero-ethno-sphenoid and postero-occipital. The mandible has infradentaries and splenials. The operculum is made up of opercular, subopercular and interopercular. Branchiostegal rays 4-13 or absent. The skull has a well-developed joint between the condyle and cavity. The joint allows the anterior portion of the skull to move, which aids in prey capture. There is autostyling of the jaw suspension. The pineal foramen is absent.

11.3.8. Digestive system

The nostril opens into the mouth cavity. Teeth fused to the mandible with or without a folded (labyrinthine) dentine and sigmoid whorls. The buccal cavity is large and opens into a strong muscular oesophagus. The intestine is subfusiform and contains a complicated spiral valve. A median nodular (homologous to the rectal) organ pours its secretion into the cloaca. The cloacal pouch bears a urinogenital papilla. The liver is bi-lobed with an elongated gall bladder. The pancreas is well developed. A spiral valve takes 20 turns.

11.3.9 Swim bladder

It arises as a tube of 3-8cm tube from the ventral side of the oesophagus and continues backward to occupy the dorsal side of the abdominal cavity. Its cavity is greatly reduced and contains about 90% fatty tissues. It is neither hydrostatic nor respiratory but calcified. The swim bladder is modified as the lung.

11.3.10 Respiratory System

The deep spiracular pouch lacks the mandible, psedobranch and external opening. Gills are made up of four holobranchs and a small hyoidean hemibranch, or the posterior surface of the hyoid arch. The fifth branchial arch is greatly reduced and devoid of gills. Its spiracles were almost as large as those of *Tiktaalik* and *Acanthostega*.

11.3.11 Blood vascular system

The heart is very simple, primitive and S-shaped. The heart contains the sinus venosus and auricle located behind the ventricle and sinus arteiosus. The large RBC looked like Dipnoi.

11.3.12 Urino-genital system

The paired kidneys fuse behind to form a characteristic 'T' shape. The kidneys are attached to the abdominal cavity's ventral wall. Two ureters dilate into an expanded bladder. The ureter opens into a urinogenital papilla. Urea is retained in the blood for osmoregulation.

11.3.13 Nervous system

The brain is small and occupies about 1.5% of the part of the cranial cavity. The rest contains fatty substances. A large corpus striatum is present in the forebrain and its roof is very thin.

11.3.14 Sense organs

Besides the eyes, internal ears and lateral line sense, a peculiar sense organ is also present. The rostral sense organ is filled with gelatinous substances rather than olfaction. This organ resembles the ampulla of Lorenzini and is innervated by the superficial ophthalmic nerve. This organ has electroreceptive functions.

11.3.15 Development

Fertilisation internal. Eggs are very large (about 9cm). Development occurs in the right oviduct only. Ovoviviparous. The embryo is about 32cm. The gestation period is one year.

11.4 Affinities

Romer (1971) classified crossopterygii as a distinct subclass of sarcopterygii with two distinct orders. On the other hand, analysis of endoskeleton characteristics in crossopterygians is supposed to be based on the direct line of descent between fish and amphibians. As a result, it is worthwhile to discuss crossopterygii affinities with at least dipnoans and amphibians. There are similarities, particularly in the osmoregulation mechanisms of *Latimeria* and Chondrichthyes.

11.4.1 Affinities with Vertebrates

Crossopterygians resemble vertebrates in the following respects:

1. Similar bony jaw and bony pattern
2. A pair of bones between eyes as parietals of vertebrates

3. Pineal opening in the suture between two parietals
4. Presence of frontal, rostral and temporal a component of the skull
5. Circumorbital around the eyes
6. Anatomy of paired fins equated with the limbs of tetrapods
7. Proximal represents the upper bone (e.g. humerus or femurs) of tetrapod limbs

The agreement of bones in paired fins is regarded as the starting point for the evolution of limb bones in land vertebrates.

11.4.2 Affinities with Amphibians

A comparison of *Osteolepis* and *Eusthenopteron* with the early amphibians can be made as follows:

1. Two large bones on the top of the skull between the eyes of *Osteolepis* may be homologised as the parietals of other vertebrates.
2. The skull of 'Ichthyostegids' is comparable to the pattern seen in the advanced crossopterygians.
3. The sclerotic plates around the orbit are comparable with the circumorbital bone of early amphibians.
4. The large unconstructed notochord with vertebrae closely resembled a rachitomous type
5. The enamel with dentine is highly folded as in 'labyrinthodont amphibians'
6. The presence of nostrils on the palatal surface between the vomer and palatine bone is common
7. Presence of single proximal bone in paired fins

The relationship of crossopterygians with amphibians has been considered so much closer, some authors have preferred to include these two groups in a single group 'Choanata' to emphasize their affinity.

11.4.3 Affinities with Rhipidistians

The rhipidistians are fairly large and voracious fishes, differing from crossopterygians in the number and arrangement of dermal bones in the head, having branched lepidotrichia and the presence of internal nostrils. The rhipidistians are plump, fat-bodied fish with acutely lobate paired fins but the crossopterygians are slender-bodied fusiform fishes. They have cosmoid scales and poorly ossified vertebral columns.

A marked reduction of bony tissue and absence of choana exclude coelacanth from the ancestry of tetrapods. They seem to have constituted a lateral branch of Rhipidistians that has been retained without notable changes up to the present.

11.4.4 Affinities with Dipnoi

Like dipnoans, *Osteolepis* had a heterocercal tail, lobate archipterygial paired fins, two dorsal fins and heavy rhombic scales of the cosmoid type.

But while in crossopterygians there was no trend towards the reduction of bony elements, in dipnoans the bony pattern of the former was well-defined. Crossopterygians, therefore, occupy a position on the direct line of descent between fishes and amphibians.

1. The closeness of crossopterygians with Dipnoi becomes evident from the following characters:
2. Presence of two dorsal fins, smooth cosmine layer on dermal bones and scales
3. Presence of rhombic scales, basibranchial with thin tooth-plates
4. Presence of an interclavicle and a scapulocoracoid attached to the cleithrum of the pectoral girdle.
5. Opercular very large relative to subopercular
6. Vertebrae with distinct inter and pleurocentra
7. Heterocercal tail with an epichordal lobe

11.4.5 Affinities with Actinopterygii

Crossopterygii and Actinopterygii share the following resemblances:

Rhipidistian and Actinistian had separate anterior and posterior nasal openings on the side of the snout.

Dermal bones of the skull roof and check are developed

Onuchodus and Strunius resemble Plaleoniscoid

Based on characters and classification of crossopterygian, it has become obvious that once thought to be represented by only extinct forms, this group is also represented by *Latimeria*, the sole survivor of this very old group.

The classification also reveals that all known and studied fossils and living forms can be grouped into Osteolepimorhs and Coelacanthimorphs.

12

Dipnoi

12.1 Introduction

Dipnoans are commonly referred to as 'lungfish'. Dipnoi owes its name to the presence of two internal nostrils. Dipnoans, which stand on the basic piscine platform, have many interesting features that the early fishes went through to become land vertebrates, particularly amphibians. As a result, the group's systematic position becomes contentious. Dipnoans evolved during the middle Devonian period and thrived during the Permian and Triassic periods. They became scarce after the Triassic period and are now represented by three specialised genera. According to Young (1981), dipnoans have acquired special characteristics through paedomrphosis.

12.2 Taxonomic Position

Phylum: Chordata

Subphylum: Vertebrata (= Craniata)

Series: Pisces

Class: Teleostomi

Subclass: Sarcopterygii

Order: Dipnoi

Sub-order:	Dipneumona	Monopneumona
Family:	Lepidosirenidae	Ceratodontidae
Scientific Name:	*Lepidosiren paradoxa*	*Neoceratodus forsteri*
	Protopterus aethiopicus (in Nile basin)	
	Protopterus amphibius (in East Africa)	
	Protopterus annectans (in West Africa)	
	Protopterus dollei (in Congo)	

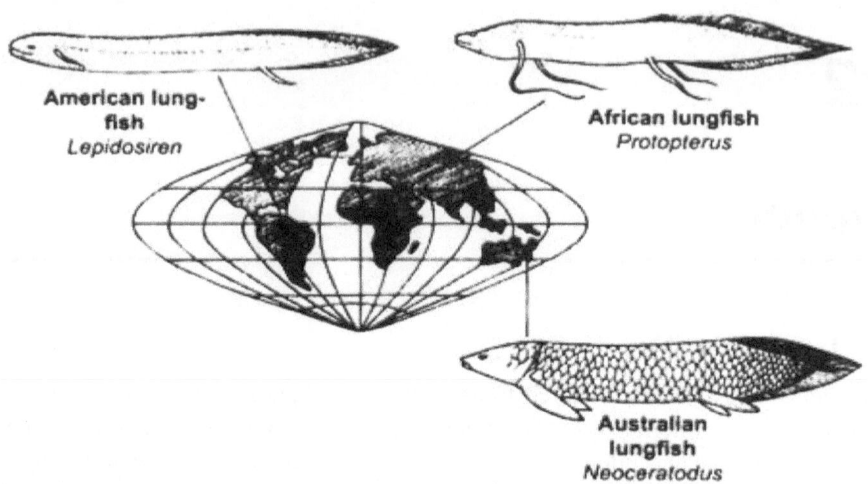

Fig 12.1: Distribution of Dipnoi (from https://www.notesonzoology.com)

12.3 Geographical Distribution

The fact that closely related species exist in widely separated places without having any form in the intermediate territories presents problems that are difficult to interpret. The same is true in the case of the dipnoans living today. The dipnoans are represented by three surviving genera (Fig. 12.1). *Neoceratodus* (=*Epiceratodus*) *forsteri* (Burnett Salmon or Australian lung-fish) in Australia of the Burnett and Mary Rivers of Queensland, *Protopterus* (Nile lungfish or African lungfish) in tropical Africa (ranging from the river Senegal, While Nile, Lake Tanganyika and Zambesi River) and *Lepidosiren* (South American lungfish) in the tropical South America (in River Amazon and its tributaries) *Protopterus* and *Lepidosiren* belong to the same family, Lepidosirenidae and *Neoceratodus* is the only genus of the family, Ceratodontidae. *Neoceratodus* is considered a living fossil.

It is apparent that despite the oceanic barriers separating them, the living dipnoans have a special relationship. The palaeontological records show that the dipnoans once enjoyed almost worldwide distribution. Their persistence in the Southern parts of the globe and their complete absence from Asia can be explained by the assumption that the southern parts became separated from Asia before the evolution of the dipnoans.

12.4 Organisation

12.4.1 Habit and Habitat

The dipnoans inhabit the rivers and are also capable of breathing air by the 'lungs'. They are sluggish and bottom dwellers. *Protopterus* lives in large lakes and rivers of Tropical Africa. It can survive even when the rivers become completely dried up in summer. It can dig into the mud and aestivate there for at least six months in a 'cocoon' made out of clay and mucus (Fig. 12.2). The mouth of the cocoon is closed by a lid which is perforated by a minute pore for the entry and exit of air. During the period of aestivation, nutrition is derived from the stored fat. Transfitment of the cocoon is possible even when the fish is enclosed in a cocoon. This habit of aestivation has been adopted since the Permian period. This fact is attested by the remains of cylindrical burrows associated with fossilised dipnoan bones.

Lepidosiren has also the property of aestivation. It lives in rivers that become shallow and stagnant in up completely. It lives mostly at the bottom of the river.

During summer when the water level becomes low and toxic due to the decomposition of the organic materials, *Neoceratodus* thrives well by switching over to pulmonary respiration. When the water is plenty and oxygen is easily available, *Neoceratodus* adopts gill respiration. *Neoceratodus* dies if it is taken out of the water.

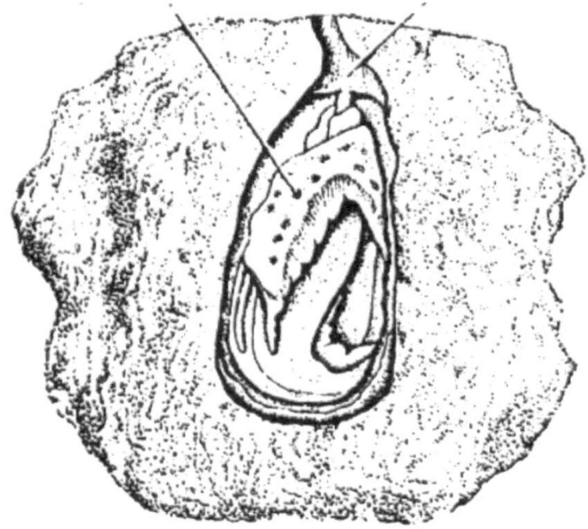

Fig 12.2: Habit of Dipnoi

All the surviving genera of the lungfish are sluggish and are bottom dwellers. They are carnivorous, although *Lepidosiren* sometimes feeds on plant materials. It lives mainly on snails. *Protopterus* is predaceous and feeds on worms, crustaceans, insects, frogs and many other small animals. These fishes are often found to attack their species and bite off portions of the limbs, tail and other parts of the body. The lost parts are regenerated by the animals. *Neoceratodus* is also a carnivorous fish that devours molluscs, crustaceans and other worms.

In *Lepidosiren* and *Protopterus*, parental care is present. The eggs are laid in muddy nests. The nest is simply a tunnel of about 30 cm deep. The males usually guard the nest till the development is complete. Parental care is highly specialised in *Lepidosiren* because its pelvic fins become specially modified during the breeding season. The pelvic fins become greatly enlarged and highly vascularised. The significance of such modification is either to release excess of oxygen to the surrounding water for the respiration of the developing young or possibly to act as an accessory respiratory structure to compensate for aerial respiration while remaining around the nest.

12.4.2 External Structures

The three extant dipnoans have eel-like bodies covered by overlapping cycloid scales. The mouth is subterminal or ventral. The scales are thin but in the fossil genus *Dipterus*, the scales were thick and covered by cosmine. The dorsal, anal and tail fins are continuous and are supported by partly calcified fibre like rays called the camptotrichia. Two small dorsal fins were present in *Dipterus*. The pectoral and pelvic fins are usually designated as the 'limbs'. These are extremely elongated, filamentous structures and are devoid of fin rays. These are highly mobile and help in 'walking' along the bottom by using the so-called limbs as the 'legs'. The paired fins in all of them are typical of 'archipterygial' types, each having an axis with two rows of radials. The tail is diphycercal but in *Dipterus* it is heterocercal with a small epichordal lobe. The operculum and a slit-like branchial opening are present on either side. The spiracles are absent. The external nostrils are enclosed within the upper lip and two internal nostrils open into the mouth cavity. The lateral line sensory system is well-developed. The cloacal aperture lies at the root of the tail. Two abdominal pores usually open into the cloaca.

12.4.3 Skeletal Structures

The internal skeleton in the dipnoans is largely cartilaginous. It is composed of an axial and appendicular skeleton.

12.4.3.1 Axial Skeleton: The notochord persists as an unconstricted rod and remains ensheathed by a tough fibrous covering. The vertebrae are represented by paired neural or basidorsal and basiventral cartilages which form the neural arches in the trunk and both the neural and haemal arches in the tail region respectively. The vertebrae lack centra. Distinct joint between the skull and vertebral column is lacking. The pleural ribs, like that of other teleosts, are present in the body wall.

The autostylic skull shows the tendency towards reduction of bones in the roof. The anterior part of the roof is incomplete. The roof and walls of the cranium are mostly composed of parietals and frontals. The cranial floor is formed by the parasphenoid. The parasphenoid is expanded and assumes a rhombic shape. The premaxillae, maxillae and nasals are absent. The vomer is narrow. The jaw suspension is autostylic because the mandible is attached to the skull by a palatoquadrate. The lower jaw is composed of paired Meckel's cartilages, tooth-bearing coronoid and supraangulars. In *Neoecratodus*, vestigial dentaries are present. The hyoid and branchial arches are all cartilaginous. The branchial arches bear the gill rakers.

12.4.3.2 Appendicular Skeleton: The pectoral girdle consists of a stout cartilage with a pair of investing bones, the cleithra and the infraclavicle. The pectoral girdle is attached to the skull by ligament. The pelvic girdle is composed of a cartilaginous plate with a long epipubic and prepubic process. Each pectoral fin consists of a long basal cartilage and a central axis with rows of jointed cartilaginous radials. The central axis is made up of small cartilaginous pieces with preaxial and postaxial radials on the sides. The pelvic fin has a similar structure except that the preaxial radials are lacking.

12.4.4 Digestive System

The food of the lungfish consists largely of invertebrates (especially the molluscs) and decomposed vegetation. The teeth form characteristic tooth plates for crushing the molluscan shells. The tooth plates are formed by the fusion of many small denticles. The tooth plates are borne on the bony plates. Each tooth plate is produced into three major cusps. The shape and disposition of the tooth plates depend on the degree of adaptation to crushing and shearing the molluscan body.

The alimentary canal is a simple tube. The pharynx leads into the oesophagus. The lungfish lack distinct stomachs. The posterior region of the oesophagus shows slight dilatation as a part of the stomach (Fig. 12.3A). The cavities of the stomach and intestine are separated by a flap-like pyloric valve. The intestine bears bursa entiana and pylorus (Fig. 12.3B). The intestine is ciliated and contains a spiral valve running along the entire length of the intestine and makes about six and a half turns (Fig. 12.3C). The spiral valve ends a short distance ahead of the cloacal aperture. This portion is designated as the rectum which opens into a small cloaca. The cloaca extends upward to the urinogenital papilla and then forwards as a dosed bladder (rectal gland). The cloacal aperture is situated on the left side of the median fin.

The liver is a single massive gland that is slightly divided into two unequal lobes. The liver lies ventrally to the right side of the stomach. The gallbladder is large and situated on the left margin of the liver. The pancreas remains embedded within the walls of the gut. The islets of Langerhans have not been seen in the pancreas of the dipnoans. The spleen is composed or vascular tissue and is attached to the right dorsolateral wall of the stomach. The alimentary canal exhibits little histological difference and the whole of the gut is lined by columnar, ciliated and goblet cells. The gut, particularly its posterior portion is kept in position by a ventral mesentery which is attached to the ventral body wall.

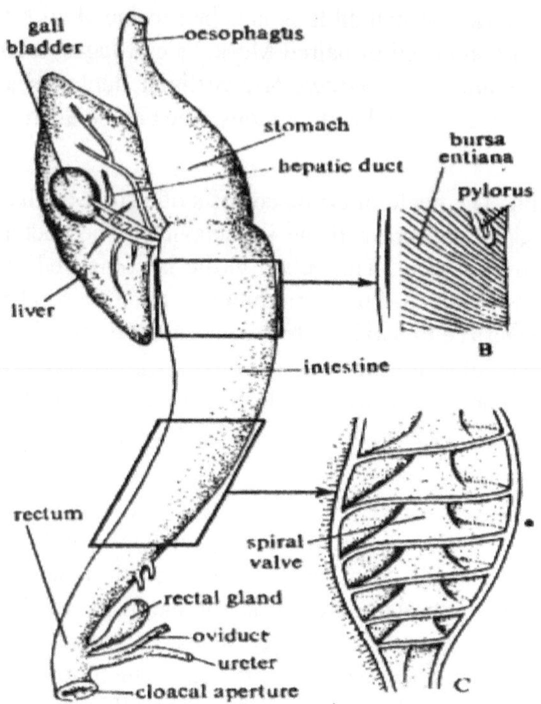

Fig 12.3: Digestive system of *Protopterus* (from www.lkouniv.ac.in)

12.4.5 Respiratory System

Both the gill and pulmonary respiration take place in the lung-fishes. Although the dipnoans possess the gills as well as lungs, they use mostly the lungs. The nostrils help in aerial respiration. The external nostrils lie at the margin of the mouth and the internal nostrils open into the buccal cavity. A slit-like glottis is present in the floor of the oesophagus which opens into a short *trachea.* The trachea passes into the lungs around the right side of the oesophagus. The glottis is provided with a fibrocartilaginous plate which resembles the epiglottis.

The swim bladder is modified into the 'lung' which is similar to that of other tetrapods in structure and function. The main difference lies in its topography. The lung is placed dorsal to the gut while in the tetrapods it is ventral. The dorsal position of the lung in the dipnoans is regarded to be the result of shifting from its original ventral to the dorsal side. The walls of the lungs contain muscle fibres and the internal cavity produces numerous *alveoli* which lead to minute alveolar sacs. In *Protopterus* and *Lepidosiren,* the supply of blood to the lungs is elaborate. The blood is supplied to the lungs by the sixth embryonic right afferent branchial arch as seen in amphibians. The blood from the lungs is returned by a special pulmonary vein.

Aquatic respiration takes place through the gills. The gill structure is similar in many respects to that of other crossopterygians except for the absence of a mandibular pseudobranch and having only a vestigial mandibular pouch. *Neoceratodus*, the most aquatic of the dipnoans, possesses a hyoidean hemibranch and four holobranchs. In the lungfish which depend more on well-developed aerial respiration, the gills help in the excretion of carbon dioxide. *Protopterus* and *Lepidosiren* obtain 98% of their oxygen from the air.

12.4.6 Circulatory System

The circulatory system is well-developed. The blood is composed of all the cellular entities observed in the higher vertebrates. The introduction of pulmonary respiration in the lungfish results in the attainment of much complicacy in the cardiac structure. The heart is enclosed in a stiff pericardium. The heart is situated somewhat posterior to the gills. The heart of *Protopterus* and *Lepidosiren* is typically built on the same plan, but in *Neoceratodus,* the heart is slightly different (Fig. 12.4A and B).

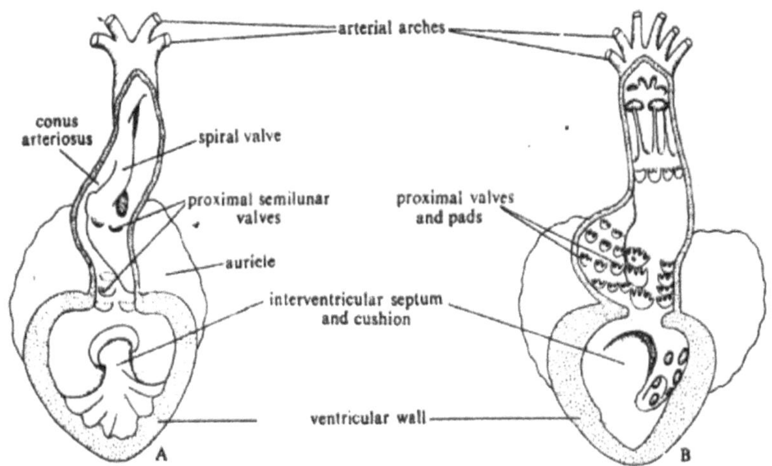

Fig 12.4: Heart of A. *Protopterus* and B. *Neoceratodus* (from www.lkouniv.ac.in)

The heart of the lung-fishes consists of four chambers, the sinus venosus, auricles, ventricle and conus arteriosus. The sinus venosus is incompletely divided into two parts. It opens into the auricle by a broad sinuauricular opening. The auricle becomes dilated on either side of a thin and perforated interauricular septum, i.e., the cavity of the auricle is almost divided. As the septum contains pores, the blood of the two auricular cavities becomes mixed up. The right cavity of the auricle receives venous blood from the sinus venosus while the left cavity gets oxygenated blood from the pulmonary vein. The pulmonary vein from the lungs passes through the sinus venosus. The auricles are communicated with the ventricle by a large auriculoventricular aperture. This aperture is plugged by a large fibrous cushion called the auriculoventricular cushion. It is continued into the ventricular cavity as an incomplete interventricular septum. Although the ventricle appears to be divided into two parts by the presence of a septum, the ventricular cavity is single and lies anterior to the so-called interauricular septum. The presence of an auriculoventricular cushion is peculiar and the auriculoventricular aperture may be opened or closed by raising or lowering the cushion. The conus arteriosus becomes spirally twisted and the cavity becomes complicated by the presence of valves.

A spiral valve is present. It begins ventrally and extends forward to the anterior end of the conus. Three rows of proximal valves are present in the conus of *Protopterus* and *Lepidosirren*. But in *Neoceratodus* the conus lacks the spiral valve and a series of semilunar valves marks the course. There are a few rows of proximal and two rows of distal valves in the conus of *Neoceratodus*. The valves in the conus are so arranged that the blood from the right side of the auricle is directed into the last two branchial arches and from the left side into the first two. In this way, a mechanism for the separation of systemic and pulmonary circulations is achieved.

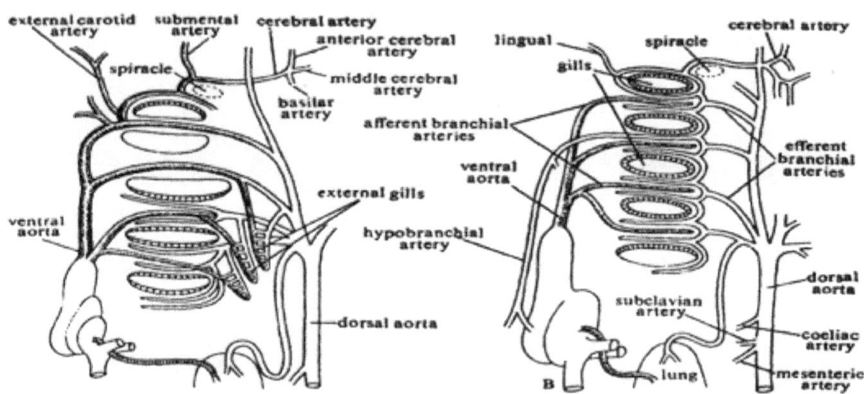

Fig. 12.5: Aortic arch and Blood supply in Dipnoi

The lung-fishes possess inconspicuous ventral aorta and the four afferent branchial arteries arise from the anterior end of the conus immediately outside the pericardium. The afferent artery carrying blood to the hyoidean hernibranch originates from the first one (Fig. 12.5A and B).

The arches bearing gills are provided with afferent and efferent branchial divisions. Two efferent branchial arteries from each gill-bearing arch join to form four epibranchial arteries. These four epibranchial arteries on either side join to form a single median dorsal aorta. The pulmonary arteries carry the blood to the lungs. *Neoceratodus* is peculiar by lacking the second efferent branchial vessel. In all the lungfish, a coronary artery is present which arises from the anterior efferent branchial arches.

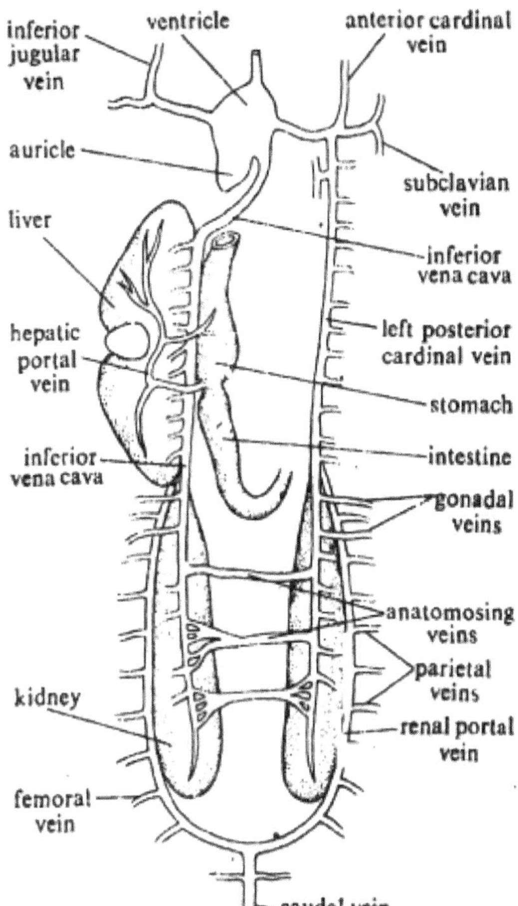

Fig. 12.6: Venous system of Dipnoi

The venous system presents certain advanced features that are not observed in other fishes (Fig. 12.6). The blood from the anterior part of the body is returned to the heart by two precavals or ductus Cuvieri. Each precaval is formed of an inferior jugular, anterior cardinal and subclavian veins. But the left Precaval, in addition to the three veins, receives the posterior cardinal vein. The posterior parts of both the posterior cardinal veins constitute the renal portal veins and are connected by interconnecting cross channels. The presence of an inferior vena cava is a characteristic feature in the venous system of lungfish. This vessel collects the majority of the blood from the posterior part of the body and conveys it directly to the sinus venosus. The inferior vena cava enters into the liver and also receives the hepatic veins. The hepatic portal drains the intestine by the sub-intestinal and infra-intestinal veins. Both the inferior vena cava and the left posterior cardinal vein have been formed by the renal veins draining blood from the kidneys. Both these veins receive segmental and genital veins on their way to the heart. The blood from the tail region is carried by a caudal vein which is bifurcated into two renal portal veins. The renal portal veins break up into capillaries inside the kidneys. In *Neoceratodus*, the caudal vein opens into the inferior vena cava and not into the renal portal veins as seen in the other two genera. In *Neoceratodus*, the renal portal vein is connected by a ventral abdominal vein and the caudal vein by the lateral cutaneous veins.

The pulmonary veins carrying blood from the lungs unite to form a common pulmonary vein and open into the left auricle. The venous system presents many peculiarities and the posterior portion shows an intermediate stage between that of Chondrichtthyes and Amphibia.

12.4.7 Nervous System

The brain of lungfishes (Fig. 12.7A to C) shows many characteristics resembling that of amphibians. The telencephalon becomes evaginated into a pair of well-marked cerebral hemispheres. These are elongated structures and united at the posterior ends but are distinct anteriorly. The olfactory lobes are sessile and lie dorsal to the anterior ends of the cerebral hemispheres. The roof of the cerebral hemispheres is thin but nervous. The diencephalon is relatively small and its roof is formed of a large mass of choroid tissue, the saccus dorsalis. A pineal body is present on the saccus dorsalis and its stem extends back towards the posterior commissure. The hypothalamus bears small inferior lobes. The optic lobes are slightly developed and become fused to form a single oval mass in front of the cerebellum. The cerebellum is small and forms a narrow transverse ridge. The roof of the diencephalon is covered over by the anterior plexus which projects into the third ventricle while the fourth ventricle is formed of the choroid plexus. The brain of *Neoceratodus* is slightly different and the cerebral hemispheres are smaller in size and the roof is membranous. In *Neoceratodus* the optic lobes are slightly separated and form paired lobes.

A peculiarly lobed saccus endolymphaticus or endolymphatic sac lies above the medulla oblongata. This is formed by the backward extension of the internal ear and the significance of this structure is not known.

The cranial nerves are similar to those of other teleosts. A sympathetic nervous system is associated with the vagus nerve. An ill-developed lateral line sense organ is present. In the living dipnoans, this system resembles closely that of amphibian larvae. The sensory organs are placed in grooves in *Protopterus*, while in others these are lodged inside the canals in the skin. However, the pattern of development of this lateral line sensory system is the same in the lung-fishes.

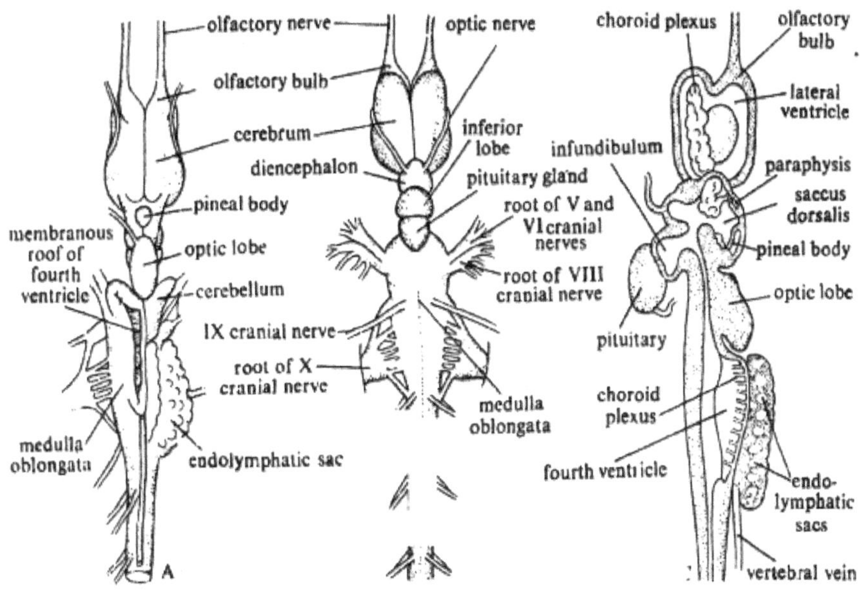

Fig 12.7: Brain of *Protopterus*. A. Dorsal, B. Ventral and C. Longitudinal sectional view. (from www.lkouniv.ac.in)

12.4.8 Endocrine System

'The adrenals of dipnoans are represented by interrenal and chromaffin tissues. These tissues remain intermingled and are situated along with the venous channels on the ventral aspect of the kidneys. These tissues function like the adrenals of the higher forms, especially that of mammals.

12.4.9 Excretory System

The excretory system comprises of a pair of elongated kidneys which are separate anteriorly but are usually fused on the posterior ends. But in *Lepidosiren*, the posterior portions of the kidneys remain separate. The kidneys are of mesonephric type and remain in intimate contact with the gonads. As the spermatozoa pass out through the kidney tubules, this particular type of kidney is designated by Greham Kerr as the opisthonephros. The kidneys extend throughout the greater part of the visceral cavity. Two thick-walled ducts, one from each kidney, may unite in *Neoceratodus* or may remain separate in *Protopterus* and *Lepidosiren* before opening into the cloaca. A cloacal bladder is present on the dorsal side. The lung-fishes normally excrete 30-70% of nitrogenous waste products through the gills in the form of ammonia. But during aestivation, the nitrogenous waste products are stored in the die form of non-toxic urea because the cloacal aperture becomes closed by the cocoon.

12.4.10 Reproductive System

The sexes are separate. Sexual dimorphism is absent except *Lepidosiren* where the males develop vascular papillae on the pelvic fins during the breeding season.

12.4.10.1 Female Reproductive Organs

The ovaries are also paired and elongated bodies (Fig. 12.8A). The ovaries are typically like that of other fishes and are kept in position in *Protopterus* by mesovarium but in *Neoceratodus* these are attached to the dorsal body wall. The oviducts are located on the lateral side of the ovaries. Each oviduct (Mullerian duct) opens anteriorly into the body cavity by a fringed slit-like opening. The anterior part of the oviduct is greatly convoluted and the posterior part is slightly expanded to form the uterus. The uteri join posteriorly to form a median vagina which opens into the urinogenital sinus. In the non-breeding stage, the oviducts are nearly straight but in breeding season these become highly convoluted. The eggs are shed free into the body cavity and carried out by the oviducts.

12.4.10.2 Male Reproductive Organs

There are two elongated testes in lungfish. (Fig. 12.8B). In *Lepidosiren* and *Protopterus*, the testes are narrow bodies and appear round in cross-section. But in *Ncoceratodus* the testes are thick and triangular in cross-section.

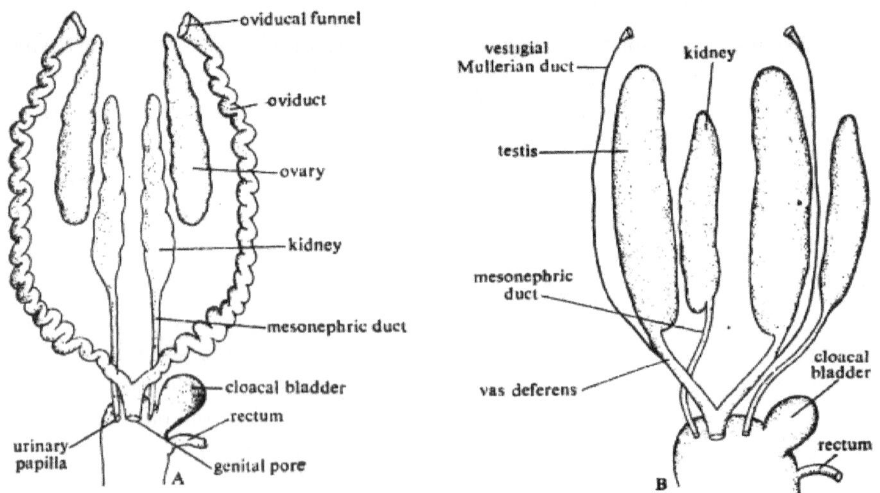

Fig. 12.8: Urinogenital system of dipnoi. A. Female and B. Male.

The testes are enclosed by the fatty tissue and lie on the ventrolateral sides of the kidneys. In *Protopterus* the anterior half of the testis remains in contact with the body wall while its posterior half is suspended by mesorchium. The testes of *Protopterus* are proportionately larger than that of *Lepidosiren* and extend the whole length of the body cavity. The testes are attached to the kidneys dorsally. The anterior end of the right testis in the dipnoans is attached to the liver while the left one extends far forward.

The vasa deferentia (numbering about six in *Lepidosiren* and *Neoceratodus*) opens into the kidney ducts. But in *Protopterus* two vasa deferentia unite at the posterior end. The Mullerian ducts are usually reduced. In *Lepidosiren* and *Protopterus* the anterior end of the Mullerian duct bears an opening in the larval stage while in adults it is closed. These ducts do not open into the urinary sinus. In *Neoceratodus*, the Mullerian ducts of the two sides fuse to form a common sinus and end blindly in the cloaca.

12.4.11 Development

The eggs are minute (3.5-4 mm in *Protopterus aethiopicus*) and shellless. In *Lepidosiren* the first three cleavage planes are vertical (Fig. 12.9) and are restricted to the animal pole which subsequently extends around the vegetal pole. Gradually the egg gets divided into cells. The cells on the vegetal side are megameres and contain yolk and those on the animal end are micromeres. A blastula is produced which contains a blastocoel inside. During the onset of gastrulation, the megameres invaginate through the blastopore and the micromeres spread over the entire embryo. The gastrula elongates, neural folds are produced and the developing embryo transforms into the larva. Each larva after hatching, attaches itself to the substratum by the ventral sucker.

The larva is a non-feeding form and derives its nutrition from the yolk. In lung-fishes the egg contains a limited quantity of yolk; as a result fully-formed larva is very small. Four pairs of external gills are visible in the larvae of *Protopterus* and *Lepidosiren*, while in *Neoceratodus* the opercular folds develop before the formation of the external gills. The external gills degenerate with the development of the lung. Vestiges of the external gills may persist in the adult stage. The lungs in the dipnoans, like that of other tetrapods, develop as an outgrowth from the floor of the pharynx which becomes eventually shifted to the dorsal side of the gut. In *Protopterus* and *Lepidosiren,* a peculiar structure of unknown function called the Pinku's organ is present. This structure is derived from the spiracular rudiment. Development of such a structure is not found in *Neoceratodus*.

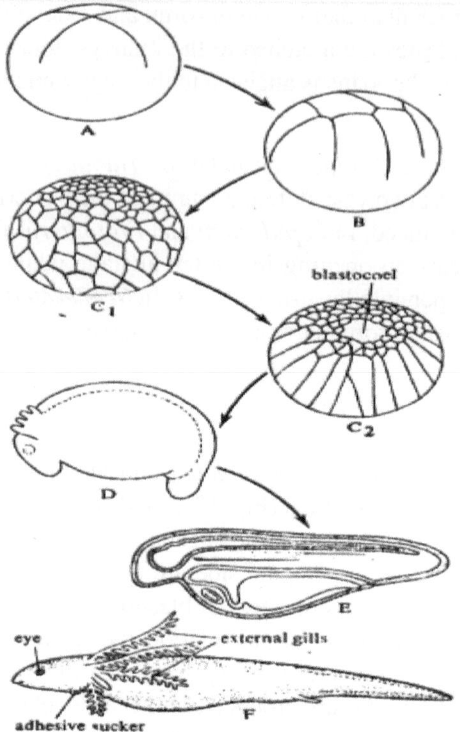

Fig. 12.9: Development of Dipnoi

12.5 Comparison Of *Protopterus*, *Lepidosiren* and *Neoceratodus*

Protopterus, *Lepidosiren* and *Neoceratodus* have similar structural construction. These forms differ in minor points which are primarily due to the different modes of adaptations. The differences existing between them are tabulated below:

Features	*Protopterus*	*Lepidosiren*	*Neoceratodus*
1. Distribution	Tropical Africa	Tropical South America	Australia
2. Habitat	Inhibits the river and large lakes. It aestivates when the river becomes completely dried up in summer.	Like *Protopterus, Lepidosiren* also aestivates during summer.	Lives in the river which becomes shallow in summer, but never dry up completely. It lives in the toxic water by adopting aerial respiration. Aestivation is absent. The lungs work in conjunction with the gills.
3. Feeding habit	Carnivorous and eats worms, crustaceans, insects, frogs and many other animals. They also feed on their species.	Mostly carnivorous feeding on *Ampullaria* and sometimes on plants.	Carnivorous which devours molluscus, crustaceans and various worms.
4. Length (m)	1.80-2.10	1.20	1.50
5. Paired appendages	Elongated and slender with extremely reduced radials.	Like that of *Protopterus*.	Broad and leaf-like with highly developed radials.
6. Eyes	Smaller and ill-developed.	Moderate in size and well-developed.	Like that of *Lepidosiern*
7. Skull	Lacks frontanelles.	Frontanelles are present.	Frontanelles absent.
8. Lower Jaw	True splenials and dentaries are present.	True splenials and dentaries are present.	The splenials and dentaries are vestigial.
9. Lungs	Two lungs are present.	Two lungs are present.	Only one lung possibly due to the suppression of the original left lobe.
10. Hyoidean gill	As hemibranch.	A hemibranch like that of *Protopterus*.	As a pseudobranch supplied by an artery from the first efferent branch.
11. External gills and suckers	Four pairs of external gills and suckers in the larva. Vestiges of these external larval gills may persist in the adult.	Four pairs of external gills and suckers. These are absent in the adult.	As the opercula develop earlier than the formation of the gills, external gills are absent in the larval stage.
12. Accessory respiratory organ	The gills and the lungs are the main respiratory structures. No accessory respiratory structure.	During the breeding season, vascular filaments are developed in the pelvic fins of males which serve as accessory respiration structures besides the gills and lungs.	No accessory respiratory organ.

Features	*Protopterus*	*Lepidosiren*	*Neoceratodus*
13. Abdominal vein	Absent.	Absent.	Present.
14. Conus arteriosus	A spiral valve and three rows of proximal valves present.	Exactly similar to that of *Protopterus*.	No spiral valve. Both distal and proximal rows of valves present.
15. Optic lobes	Unpaired	Unpaired.	Slightly separated and form paired lobes.
16. Lateral line sense organs	The sense organs are lodged in grooves.	The sense organs placed inside the canals in the skin.	Similar to that of *Lepidosiren*.
17. Kidneys	Extremely elongated structures and the posterior ends are fused.	Like that of *Protopterus* but the posterior ends remain separate.	Similar to that of *Protopterus*, but the kidneys are not so elongated.
18. Ureter	Two ureters, one from each kidney, remain separate.	Like that of *Protopterus*.	Two ureters may unite before opening into the cloaca.
19. Testes	The testes elongated narrow bodies and round in cross-section.	Similar to *Protopterus*, but extremely elongated extending the entire length of the body cavity.	The testes are thick and triangular in CS.
20. Sexual dimorphism	Absent.	Seen only during the breeding season. Oviparous.	Absent.
21. Pinku's organ	Present.	Present.	Absent.

12.6 Affinities

From the anatomical standpoint, the dipnoans form a sort of structural bridge between the fishes and the amphibians. Like that of other fishes, the dipnoans possess gills and spend life in water. But unlike fishes, the lung-fishes can survive long dry spells by respiring through the air-breathing lungs like that of amphibians. The paired fins are also used much the same way the amphibians employ their limbs. Because of this fact, many workers postulate that the dipnoans hold the ancestry of the amphibian. Before going into the details of the phylogenetic consideration, the affinities with the different groups of fishes and also with the amphibians are considered first.

12.6.1 Relationship with fishes

The lung-fishes show close similarity with the fishes and their affinity is well established. The characteristics by which the lungfishes resemble the fishes are:

1. Body stream-lined.
2. Locomotory appendages fin.
3. The notochord is persistent.
4. Largely ossified, slender dermal fins rays.
5. The vertebrae lack centra.
6. Branchial arches 4—6 pairs present.
7. The skull shows little ossification and is with many investing bones.
8. The dermal fin rays are slender and are highly ossified.
9. The body is covered by overlapping cycloid scales.
10. The tail fin is diphycercal.
11. The gills constitute the respiratory organs to absorb the oxygen which remains dissolved in water.
12. The lateral line sense organs are present.

But the relative position of the lungfish amongst the bony fishes is controversial because they bear many primitive features by which they resemble the elasmobranchs.

12.6.2 Relationship with Elasmobranches

The lungfishes resemble the elasmobranches by having

1. Similar conus arteriosus,
2. Spiral valve in the intestine,
3. Each gill with two efferent arteries.
4. Similar female reproductive system,

5. Identical diencephalon
6. No nephrostome in the kidney tubules.

However, the above similarities are not sufficient enough to establish any relationship. The presence of lungs, diphycercal tail and opercula and lack of claspers as well as external fertilisation put difficulty in such contention. However, the similarities speak about the primitiveness of the dipnoans.

12.6.3 Relationship with Holocephalians

Jarvik (1964, 1967) has demonstrated that the dipnoans remarkably agree with the holocephalians, especially in the structural upper and lower jaw. The mandibular and palatine toothplates are the main biting elements in both groups. The other similarities are:

1. Similar histological picture of teeth.
2. Gills covered by operculum.
3. Similar cranial muscles.
4. Jaw suspension autostylic.
5. Fusion of the operculo-gular membranes of both sides.
6. Similar kidneys, gonads and ducts.
7. Lack of stomach and the presence of spiral valves.
8. Two efferent arteries in the gill arches.
9. Shifting of excurrent nostril into the mouth cavity.

Despite such similarities, it is extremely difficult to establish any relationship between dipnoans and holocephalians. Some of the features are shared by other groups. The dipnoans differ from holocephalians by having lungs and the absence of claspers.

12.6.4 Relationship with Rhipidistians

The dipnoans and rhipidistians show close similarities and it is regarded by many workers that the two groups should be placed under a common systematic status (Goodrich, 1909; Holmgren and Stensio, 1936). The alleged similarities are:

(1) Westoll (1949) has shown that the dermal bones and scales have similar histological organisation, especially having cosmine.

(2) Watson and Gill (1923) established resemblances between Middle Devonian members particularly in their body form and median fins. The presence of two separate dorsal fins, a single anal fin, and a heterocercal tail, etc., are some of the notable similarities.

(3) Romer (1955) suggested that they have fleshy lobate fins with scales and well-developed endoskeleton.

(4) Save-Soderbergh (1984) gave the support based on the similarities in the number and disposition of dermal bones on the skull.
(5) Similar opercular and gular bones as advocated by Watson and Gill (1923).
(6) Similar general plan of laterosensory system (Westoll, 1949).
(7) Presence of nostril in the roof of the mouth cavity.
(8) Comparable lower jaw.

But Jarvik (1968) and many authors offer doubt on the resemblances between the two groups. Most of the similarities, according to him are either insignificant or non-existent. The structural organisation of the vertebral column, neural endocranium, visceral endoskeleton, and snout in the dipnoans is fundamentally different from that of the rhipidistians. So the question of establishing a close relationship is untenable.

12.6.5 Relationship with Struniiformes

Jessen (1966) tried to establish the relationship between the dipnoans with the group struniiformes. But the fundamental plan of the operculo-gular series and lower jaw establishes its relation with the rhipidistians more than others. The dipnoans differ widely from the struniiformes and they are not closely allied.

12.6.6 Relationship with Teleostomi

The lung-fishes and ray-finned bony fishes show close affinities. But as regards the affinities of the group, the relationship with the subclass Crosspterygii is quite close. The lung-fishes resemble the subclass Actinopterygii by having:

(1) Similar lobate paired fins.
(2) Presence of cosmine covering the cycloid scales and skull bones in *Dipterus*.
(3) Paired inferior jugular veins are present.
(4) Presence of powerful palatine and splenial teeth.
(5) Presence of blunt snout with ventral nostril.
(6) Presence of swim bladder, although modified into 'lungs' in dipnoans.
(7) Presence of operculum. But the similarities with other crossoptergians are very close. The common features are:
    (1) Presence of internal nostrils.
    (2) The tail fin is diphycercal.
    (3) Existence of cosmine covering the scales.
    (4) Similar skull bones.
    (5) Presence of swim-bladder functioning as the lung.
    (6) The intestine contains spiral valves.

(7) The conus arteriosus is contractile.

(8) External gills are present in the larvae.

Such a close similarity between the groups suggests the contention that the lung-fishes are phylogenetically related to the other crossopterygians.

12.6.7 Relationship with Amphibian

The lung-fishes resemble the amphibians by many characteristics. The common features are:

1. The presence of vomerine teeth.
2. Semiaquatic or marshy habitat.
3. Internal nostril piercing of the roof of the mouth cavity.
4. The lungs are capable of pulmonary respiration. A pulmonary artery and a pulmonary vein are present.
5. Spiral valve lacking.
6. The circulatory system of dipnoans resembles that of amphibians by many points, viz., the spirally twisted conus arteriosus is divided into two by a longitudinal partition.
7. The auricle and the sinus venosus are imperfectly divided into two parts.
8. The pericardium is thin-walled.
9. The Ventral aorta is short.
10. Both have similar inferior vena cava and anterior abdominal veins.
11. The skin glands are multicellular.
12. Presence of dermal scales in Gymnophiona.
13. The brain of the lung-fishes resembles that of amphibians, especially in the structure of the cerebrum and cerebellum.
14. The spermatozoa are carried through the excretory part of the mesonephros.
15. The structure of the egg and its development are similar.
16. The larvae of *Lepidosiren* and *Protopterus* possess the external gills and sucker. In some species of *Protopterus* external gills may be persistent.
17. The jaw suspension is autostylic.

According to Dolo, These resemblances are probably due to convergent evolution on account of similar habits and habitats. Although the dipnoans resemble the amphibians by several features, there are many individual specialised characteristics by which they are separated from the amphibians. These are:

1. The skull is largely cartilaginous with little ossification.
2. The maxillae and premaxillae are absent.

3. Some anterior vertebrae have become fused with the skull.
4. Presence of characteristic crushing tooth plates.
5. The lungs are located dorsal to the gut.
6. Lobate fins instead of limbs for locomotion.
7. Urinary bladder develops from the dorsal wall of the cloaca in Dipnoi but from the ventral wall in Amphibia.

The common characteristics present in lungfish and amphibians are due to their evolutionary convergence and have possibly resulted from adaptive modifications to a similar environmental condition.

12.7 Phylogeny of Dipnoans

The affinities and origin of the dipnoans are still an unsolved problem on which diverse opinions exist. The similarities with the amphibians are due to evolutionary parallelism. The dipnoans exhibit similarities with other fishes, particularly with the elasmobranchs and holocephalians. There are reasons to believe such a relationship, but their mutual interrelationship becomes difficult to establish. So in the present state of knowledge which of the surviving vertebrate group is the sister group of the dipnoans remains far from the solution.

The lung-fishes possess real lungs and *Neoceratodus* even 'walks' across the muddy bottoms of the river by using the paired fins like the 'legs'. This fact tempts one to think that these fishes have some direct connection with the forms that led to the origin of land vertebrates. From the anatomical standpoint, the dipnoans exhibit close similarities with the bony fishes, especially with the rhipiditians on one hand and the amphibians on the other. This transitional status of the dipnoans makes this group of fishes interesting. The contention of holding the direct ancestry of the land vertebrates is proven to be untrue nowadays. But their relation with the ancestral stock of the land vertebrates is supported by numerous evidences. The dipnoans, although survived by only three genera living in widely separated parts of the earth, have left behind a complete geological succession of the extinct dipnoans leading to the modern forms. All the dipnoans are united under a common order by several similar characteristics. These features are:

(1) The presence of internal nostrils.
(2) The scales and bones are covered by cosmine.
(3) The autostylic mode of jaw suspension.
(4) The peculiar pattern of teeth.

The presence of these features also testifies to the biological truth that the lungfishes have evolved from a common ancestor with the other crossopterygians. The earliest known fossil dipnoan is represented by the genus, *Dipterus* (*Dipnorhynchus* is supposed by many to be more primitive than *Dipterus*). *Dipterus* possesses the following primitive features:

(1) The body is covered by stout cosmine-covered cycloid scales.
(2) Two small dorsal fins are present.
(3) The tail fin is of heterocercal type with a small epichordal lobe.
(4) Numerous conical teeth are present.

During evolution modern lungfish have witnessed the following changes:

(1) Fusion of median fins to form a continuous one.
(2) Reduction in the number of dermal plates in the skull.
(3) Change over from the heterocercal to the diphycercal tail.
(4) Fusion of the conical teeth to form characteristic tooth-plates with ridges.
(5) Reduction in the number of opercular bones.

Dipterus is fundamentally akin to the most primitive crossopterygian, *Osteolepis*. The scales and the head bones of *Dipterus* are strikingly similar to those the contemporary crossopterygians. The resemblances in some cases are so close that their relative position within the subclass becomes sometimes confusing.

The amphibians have evolved from the crossopterygian ancestor. Although crossopterygians are on the verge of their racial passing, they give us an insight into the emergence of the land vertebrates. A survey of their past geological history suggests that the rhipidistian line and the dipnoan line converge back as we follow them further backward. This is evidenced by the presence of close similarity in the earlier representatives of both the above groups. The dipnoans constitute a very primitive race and possess many primitive characteristics in comparison to the then crossopterygians. This primitive organisation nullifies the concept of the derivation of the dipnoans from the rhipidistian stock. However, both groups after emerging from a common ancestral stock diverged widely over time.

The phylogenetic significance of the dipnoans, as regards the holding of the direct ancestry of the amphibian, is questionable. This view is confronted with serious objections from an anatomical standpoint, and is difficult to interpret. The most important obstacle is the derivation of the terrestrial limb from the peculiar and specialised archipterygial paired appendages of the dipnoans. The development of the cloacal bladder is different. In the dipnoan, the cloacal bladder develops from the dorsal wall of the cloaca but in amphibians, it develops from the ventral wall. The rhipidistians, on the other hand, present fewer obstacles. The earliest fossil amphibians show deeper similarities with *Osteolepis* of the Devonian period than any extinct and surviving dipnoans. It is universally accepted that the amphibians have originated either directly from some rhipidistians or an unknown group closely related to the rhipidistians but never from the dipnoan source. It can best be suggested that the dipnoans have diverged very early from the remote basic stock from which the amphibians originated. The rhipidistians and the dipnoans are the divergent offshoots of a common piscine ancestral stock.

Unit-III Fish Physiology

13

Fin System in Fishes

13.1 Introduction

The fish's distinguishing features are its fins. The fins constitute the major propulsive organs in fishes. These are either folds of skin or projections from the body surface. The fins lack coelom but are supported by muscles and fin rays (lepidotrichia). These supporting rays may be bony, cartilaginous, fibrous or horny. The diversity in the fin system is due to its adaptive responsiveness. A great variety of fins is observed in fishes. However, they are primarily of two types:

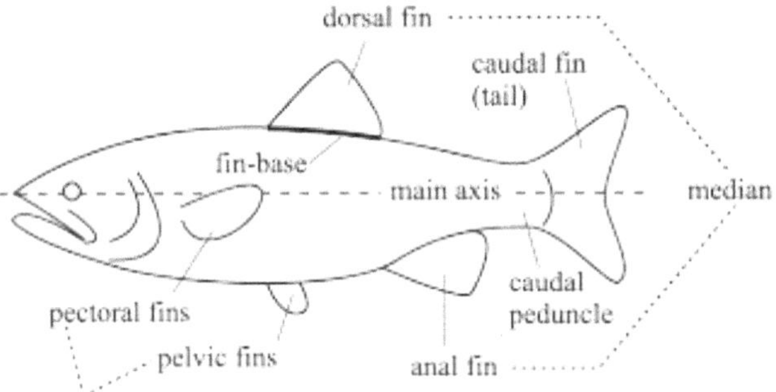

Fig. 13.1: Unpaired and paired fins in fishes (from https://www.researchgate.net)

A. Unpaired or median fins: The median fins include the dorsal, caudal and anal fins based on their positions. They are associated with the axial skeleton of fishes. The modifications of unpaired fins are related to their particular role in propulsion. In some fish, all the median fins are continuous and constitute a single fold. During swimming, dorsal fins act as stabilizers to prevent the body from rolling (Fig. 13.1).

B. Paired fins: These are either pectoral or pelvic fins. They are associated with the appendicular skeleton of the fish. These fins are supported by dermotricia (finrays) and somactids (radialia). Somactids are represented by a jointed basal piece that articulates with the girdle and has postaxial and preaxial radials. The radials are attached to the basal piece. The somactids in *Cladoselachae* are arranged parallel to one another and are referred to as orthostichous type. The rachiostichous type of fin skeleton is observed in *Ceratodus*. In elasmobranchs, the fin skeleton assumes a fan-like form and is described as the rhipidoschous type. They seemed to be homologous to the paired appendages of vertebrates. During swimming, caudal fins help drive the fish by sculling back and forth (Fig. 13.1).

13.2 Origin of Paired Fins

The median fins are thought to be the result of a continuous fold of tissue. This fold extends from the posterior region of the head and continues posteriorly around the tail and forward up to the anus. This fold is supported by a series of parallel cartilaginous rods. Each supporting rod divides during development into a lower piece, or basal, embedded in the body wall and an upper piece, or radial, lying in the fin fold. From such a continuous fin fold, the dorsal, caudal and anal fins have evolved by restriction of the radials at certain areas and the progressive degeneration of the fold between them.

The following theories are put forward to explain the origin of fins in fish:

13.2.1 Gill arch theory

The idea of the origin of paired fins from gill arches was proposed by Gegenbaur (1898) and supported by Braus, Davidoff and Fiirbringer. According to him, the paired fins are modified gill structures in which the girdles, fin rays and fin fold represent the gill arches, branchial rays and gill septa respectively. The gill arch supports the gill septum and bears the branchial rays. Such an arrangement has given rise to a biserial achipterygium as observed in *Ceratodus*. The position of the pelvic fins can be explained by the assumption that some of the posterior gill arches have been shifted posteriorly.

Objections

a) The theory lacks palaeontological, morphological and embryological supports.
b) The resemblances between the girdle and the gill arch are superficial and misleading.
c) The gill septum has only one somite but paired fins have several somites.
d) The branchial arches arise from the splanchnic wall of the coelom but girdles of the paired fins arise from somites.
e) The gills are internal while the fins are external in position.
f) If paired fins originated from the gill septum, they would be transverse to the long axis of the body and may obstruct during swimming. However, during development, paired fins appear as a dorsoventral or longitudinal fold.
g) The pectoral girdle is connected to the dermal elements of the cephalic region and not the chondral elements as in fins.

h) It does not explain the resemblance between the development of median and paired fins.

13.2.2 Fin-fold theory

The theory was postulated by Balfour, Mivert and Thacher and later supported by Wiedershiem, Parker, Goodrich and others. According to this theory, the paired fins evolved from a pair of lateral fin folds running down each side of the body behind the gills and opening up to the end of the tail. These lateral fin folds are separated anteriorly but joined posteriorly to form a median ventral fin fold. There is also a median dorsal fin fold. The condition of the lateral fin-folds in the ancestral forms were presumably like the metapleural fold of *Branchiostoma*. The existence of such lateral fin-folds is supported by the discovery of an archaic fossil vertebrate, *Jamoytius kerwoodi*, from the upper Silurian period, as well as research on some fossil cyclostomes and primitive fishes (Fig. 13.2).

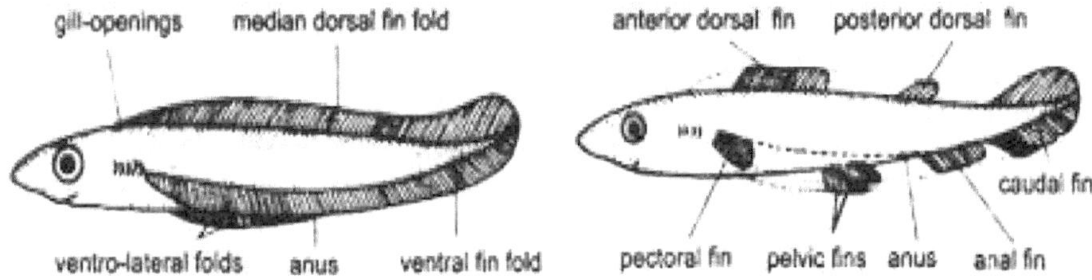

Fig. 13.2: Fin-fold theory for the origin of fins (from https://www.notesonzoology.com)

There is much embryological, anatomical and palaeontological evidence in favour of the origin of the paired fin fins from the lateral fin folds. Among the pieces of evidence are:

(a) Although the structure of the paired fins varies greatly in different fishes, the mode of development is strikingly similar. Traces of lateral fin folds may be observed in the embryos of certain elasmobranchs. Up to a certain point, the development of paired fins and median fins appears to be similar. The embryological stages of fin formation support the presence of a continuous fin fold. In paired fins, as in median fins, the skeleton is distinguished from the continuous areas of dense mesenchyme located in the fin-fold between the ventral and dorsal musculature. The muscle buds forming the radial muscles of the paired fins developed from the segments along the whole of the body length but became vestigial and disappeared beyond the fin areas.

(b) The development of paired fins may be considered evidence of the original continuity of the lateral folds. The condition of the paired fins in *Cladoselachae* added more weight to this idea. The pectoral and pelvic fins in this form are very wide and lack posterior and anterior notches from the body wall. The presence of such constrictions is a characteristic feature of other fishes. The fins are supported by a parallel arrangement of simple cartilaginous rods. The arrangement and lack of notches suggest that the fins evolved from continuous fin folds. In *Lophius*, the first pelvic bud succeeds immediately after the last pectoral one, showing the continuity of the development of the paired fins. Many fish exhibit extension and fusion of the paired fins, but these are considered secondary formations.

(c) In acanthodians (*Climatius*), a series of spines extend ventrally from behind the head to the anal fins. It also suggests the presence of a continuous fin fold. Some extinct fishes had metapleural folds that terminated in the anus rather than at the atriopore as in *Amphioxus*. The skeletal structures of the median and paired fins are almost similar indicating a common origin. According to Westoll (1958), the fossil records of ostracoderm gave a conclusive sequence to the origin of the fins.

This theory was accepted by most workers and the gill arch theory of Gegenbaur has become a theory of historical importance.

13.2.3 Fin-spine (actinotrichia) theory

Pieces of evidence are put forward against the fin-fold theory. As regards the question of metapleural folds in *Branchiostoma*, it can be suggested that this point appears baseless as *Branchiostoma* is now removed from the direct line of vertebrate evolution. Indeed, in the oldest known ostracoderm the fins were short-based. Gregory and Raven (1941) showed that the broad-based fins of the Cloadoselache were highly specialised and analogous to those of rays. Recent workers claim that so-called 'several pairs of fins' to acanthodians indicate a special kind of multiplication of the defensive spines, which may often support membranous structures. Furthermore, no true primitive fossil fish with completely continuous fins is still known (Fig. 13.3).

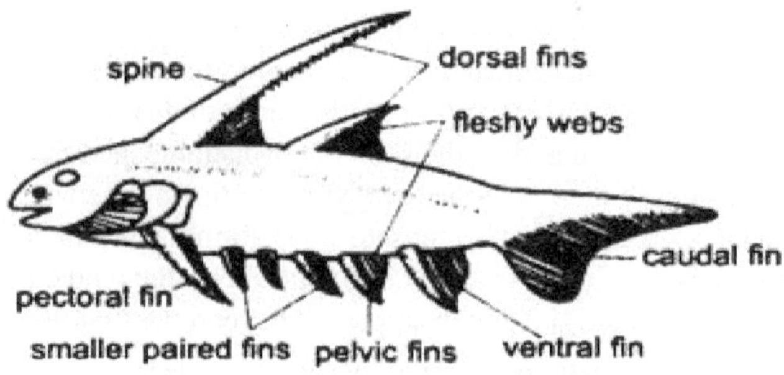

Fig. 13.3: Fin-spine theory for the origin of fins (from https://www.notesonzoology.com)

The theory suggests that the fins probably appeared in connection with some paired and unpaired median spines, as found in many ostracoderms. In the course of evolution, membranous structures appeared between these spines and the adjacent body wall.

13.2.4 External gill theory

A variation of the gill-arch theory, proposed by Kerr (1919), suggests that paired fins have been derived from external gills. The theory has many objections, such as:

a) External gills occur rarely in advanced forms among Oestichthyes and Amphibia.
b) Vascular filaments occur in the pelvic fin of *Lepidosiren* at a certain time and have a special purpose in the males of that one genus.
c) The external gill differs from paired limbs in every important detail.

13.2.5 Ostracoderm theory

Some ostracoderm possess lateral fleshy lobes, projecting from either lateral side. They may have been the precursors of pectoral fins. Other ostracoderm had ventrolateral rows of dermal spines similar to the extra fins of acanthodians. Most of these spines were lost, while some were retained in the pectoral and pelvic regions, thus, accounting for the origin of paired fins from ostracoderm ancestors.

13.3 Types of Fins

13.3.1 Dorsal Fin

The dorsal fin is located on the mid-dorsal line of the body and shows great variation in structure and disposition. The bones that support the dorsal fin are called *pterygiophores*. There are two to three of them: "proximal" (axonosts), "middle" (baseosts) and "distal". In rock-hard, spinous fins the distal pterygiophores are often fused to the middle ones, or not present at all (Fig. 13.4).

The dorsal fin may be Single, Pointed, Split, Spine triangular, Trigger and Trailing shaped.

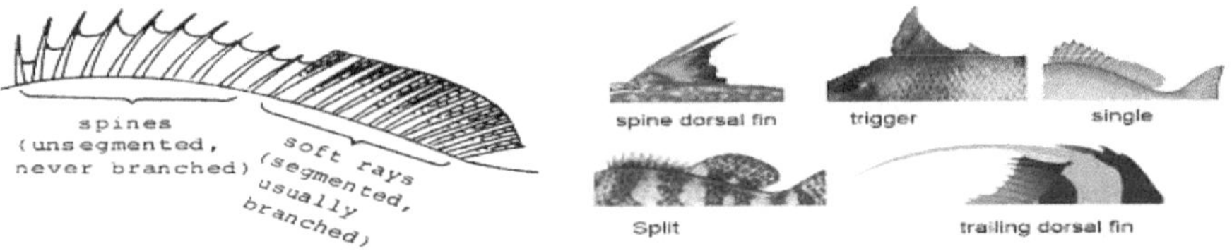

Fig. 13.4: Different types of dorsal fins in fishes (from https://www.researchgate.net and www.scuba.com)

Amongst the Chondrichthyes (*Chlamydoselachus, Heptranchias* and *Hexanchus*) possess single dorsal fins; while in the rest of the forms, the dorsal fins are two in number. The fins are supported by many horny fin rays (ceratiptricha) located beyond the cartilaginous radial. Both the ceratotricha and radials remain completely covered by the integument. But in the bony fishes, the radials lie in the muscles and the basals within the body. In higher teleosts, the radials become reduced to bony or cartilaginous nodules. The nodules are situated within the muscles of the body and are attached to the fin rays (lepidotricha). The fin rays are modified scales. Usually, the number of fin rays corresponds directly to the number of radials, but in dipnoans and sturgeons, the fin rays are the number of radials.

In many actinopterygians, the dorsal fin is preceded by many spines. The presence of a single dorsal fin is a primitive feature in these fishes. This condition is observed in the fossil forms and to some extent in *Lepisasteus* and *Acipenser*. In higher forms, there are two dorsal fins. In the crossopterygians, there are two dorsal fins, of which the posterior one is larger except *Latimeria* where the anterior one is larger. In *Polypterus*, the dorsal fin is divided into several fin lets, each having one stout spine supporting a membranous flap.

In sharks, the dorsal fins are well developed and act as a stabilizer, but in the rays, the dorsal fins are reduced because they are adapted to live at the bottom of the sea. In Sting rays and Eagle rays, the dorsal fins are absent. In bull-headed sharks, both the dorsal are preceded by a strong sharply pointed spine. The spines are defensive organs and are also associated with the poison glands as seen in Spiny dogfish (*Squalus*).

In bony fishes, the dorsal fins show extensive variations, especially in shape, size and position. It is present in all forms except the electric ray (*Electrophorus*), where it is either lost or reduced to a more filament-like structure.

In primitive bony fishes, the fins are supported by flexible rays and in the actinopterygians, the rays have become covered into stiff spines. In an eel (*Acanthenchelys*), the dorsal fin is supported by both soft rays and stiff spines. The development of the spines in the unpaired fish in many fishes added a new function of defence in addition to propulsion. The spines exhibit great variations amongst the fishes. In *Trachinus* and *Synanceia*, the spines supporting the dorsal fin are connected with the poison glands.

Extensive modification is observed in the suckerfish (*Echeneis, Remora*) where the spinous dorsal fin becomes transformed into an oval adhesive discover the head. In Angler fishes, the dorsal fin is highly modified. The first ray of this spinous fin is located on the snout and transforms into a flexible filament with a membranous appendage at its tip. In deep-sea Angler fishes inhabiting the dark regime, a luminous bulb is present on this appendage. The luminous bulb varies in shape and size and produces light to allure and attract small fishes. In *Lasiognathus*, this appendage is composed of a stout rod-like basal part and a slender filament. Besides this luminous bulb, a series of non-functional hooks are present.

In anglerfish, the anterior of the dorsal fin is modified into an illicium and esca, a biological equivalent to a fishing rod and lure. *Gymnarchus* uses only its dorsal fin for propulsion. Boxfish, pufferfish and ocean sunfish use their dorsal fin in combination with their anal fin for propulsion (tetraodontiform propulsion).

13.3.2 Adipose Fin

In some fishes, the dorsal fin comprises two parts the first and second dorsal fins. The second dorsal fin lacks any skeletal support and becomes a flabby structure and is known as an adipose fin in *Mystus* species (Fig. 13.5).

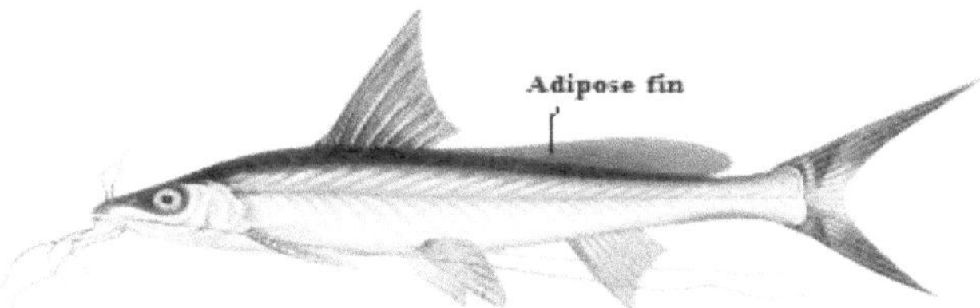

Fig. 13.5: *Mystus seengtee*

A comparative study in 2013 indicates the adipose fin can develop in two different ways. According to the salmoniform-type way, the adipose fin develops from the larval-fin fold at the same time and in the same direct manner as the other median fins. However, the characiform-type way indicates that the adipose fin develops late after the larval-fin fold has diminished and the other median fins have developed. Research published in 2014 indicates that the adipose fin has evolved repeatedly in separate lineages. It is found in Argentiniformes, Aulopiformes, Characiformes, Myctophiformes, Osmeriformes, Percopsiformes,Stomiiformes, Salmoniformes and Siluriformes. The fin may be vital for the detection of, and response to, stimuli such as touch, sound and changes in pressure.

13.3.3 Anal (Cloacal) Fin

The anal fin also shows variation in shape and size in different fishes. The bones that support the anal fin are called *pterygiophores*. There are up to two series, a proximal series (axonosts) and a distal series (baseosts). In fishes, where this fin acts as a locomotor organ, the anal fin is greatly elongated. It is usually supported by soft rays. In some forms, a few anterior rays become transformed into spines. It is generally single but in *Gadus* and some related forms, it is divided into two parts. The anal fin in the males of South American Cyprinodonts becomes modified into complicated intromittent organs. The third to fifth fin-rays are enlarged and contain either a groove or closed tube into which the reproductive ducts open.

Knifefish use their anal fins for thrust (gymnotiform propulsion). Boxfish, pufferfish and ocean sunfish use their anal fin in combination with their dorsal fin for propulsion (tetraodontiform propulsion)

13.3.4 Caudal Fin

The caudal fin plays the most important role in forward propulsion during swimming. It is highly developed in most fishes except Seahorse (*Hippocampus*) and some of the eels of Anguilliformes. In Hippocampus, the tail is prehensile. In bottom living rays, the caudal fins tend to become reduced and in the sting rays, the caudal fin is lacking. The tail in the latter forms is produced into a long whip-like tapering structure. The caudal fin in the rest of the fish exhibits a good deal of variation. The following three types of caudal fins are observed in different fishes (Fig. 13.6A to H).

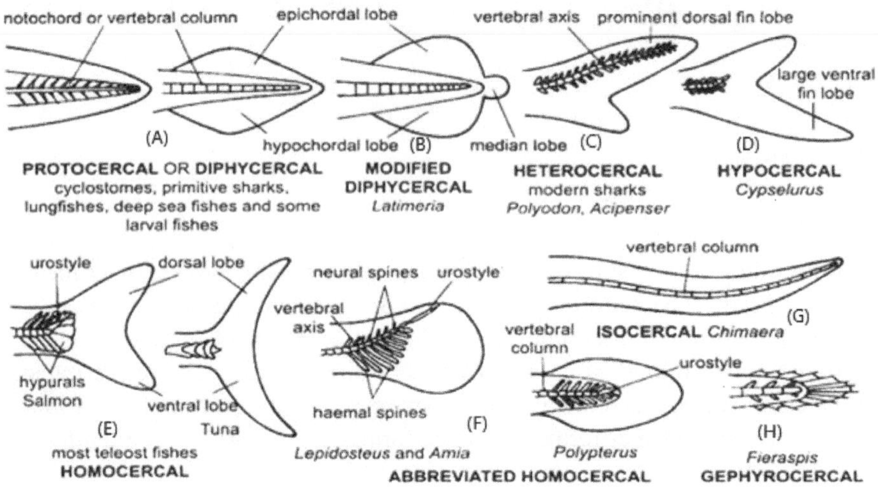

Fig. 13.6: Types of caudal fins in fishes (from https://www.notesonzoology.com)

(a) Protocercal (Diphyceral or first tail) fin (Fig. 13.6A and B): It is regarded to be the most primitive type. The vertebral column extends up to the tip of the tail and divides the caudal fin into two halves. The dorsal half is called the epichordal lobe and the ventral one is known as the hypochordal lobe. The epichordal and the hypochordal lobes of the caudal fin are equal in size and symmetry. All fishes pass through this stage during development. In *Latimeria*, the fin is slightly modified with a sharply marked median lobe.

(b) Heterocercal (Unequal) fin (Fig. 13.6C): It is regarded as the intermediate stage between protocercal and homocercal types. In this case, the vertebral column bends upwards. As a consequence, the caudal fin is divided into two unequal halves. The vertebral column is bent upwards and continues up to the tip of the fin. The epichordal lobe is greatly reduced while the hypocercal lobe is completely asymmetrical. This type of caudal fin is found in elasmobranchs, extinct crosspopterygian and primitive actinopterygians. In most elasmobranchs and early fishes, the caudal fin is usually heterocercal, but in *Chimaera* and *Chlamydoselachus*, the caudal fin is of isocercal type.

In primitive forms like *Acipensor* and *Polydon*, the caudal fin is strongly heterocercal.

From the heterocercal condition, the homocercal caudal fin is evolved by gradual reduction of the upward fleshy lobe.

(c) **Homocercal (Equal) fin** (Fig. 13.6E): It is regarded to be the most advanced type. The homocercal caudal fin is derived from the heterocercal fin and is evident from the transitional types amongst the extinct and living bony fishes. The fin is symmetrical externally but internally asymmetrical. The posterior end of the vertebral column is turned upwards and becomes greatly reduced. The tip of the vertebral column does not reach the posterior limit of the fin. There is no apparent dorsal lobe but the ventral lobe is greatly enlarged and divided into two equal superficial lobes. Most of the teleosts retain the typical homocercal condition, but in some forms, this condition is highly modified and observed in:

(1) In Anguilliformes, Blennidae, Gymnarchidae, Macururidae, Notopteridae, etc. upturned tip of the vertebral column becomes elongated and straightened out. The fin is the marginal extension above and below the elongated vertebral tip. The fin is supported by rays and is known as isocercal. Embryologically, the isoocercal (leptocercal) tail is formed by the reduction of the size of the hypo chordal lobe and elongation of the dorsal and anal fins, so that a continuous fin-fold is re-established.

(2) In Cod and Tuna, the homocercal condition is modified as the internally symmetrical tail. The upward terminal portion of the vertebral column is withdrawn.

(3) In *Flerasfer* and *Orthagoriscus*, the caudal fin completely disappears with the truncation of the vertebral column and is known as the gephyrocercal fin (Fig. 13.6H).

(4) Pseudocercal tail is found in Dipnoi. It is more difficult to interpret as it lacks embryological clues.

(5) Hypocercal (inverted heterocercal) tail (Fig. 13.6F): It is known as early Agnatha only.

In many teleosts, the caudal fin starts as diphycercal then becomes heterocercal and finally assumes the homocercal condition. This transition is very important from the phylogenetic point of view. In *Amia* and *Lepisosteus*, the caudal fin exhibits an intermediate condition between the heterocercal and homocercal types. The vertebral column is shortened and retracted at the base of the caudal fin. The fin has retained the reduced unturned fleshy epichordal lobe, but the hypo chordal lobe is greatly enlarged to form only the anteroventral lobe. The condition is known as either hemiheterocercal or abbreviated homocercal type.

Among bony fishes, the externally symmetrical tails show a great deal of variation in their shape as tabulated below:

Table 13.1: Types of Fins, Their Examples and Functions

Sl. No.	Type	Example(s)	Function(s)	Sl. No.	Type	Example(s)	Function(s)
1.	Lunate (Crescentic)	Tuna	Active swimming	5.	Rounded	Turbot, Lemon-Sole	Slow swimming
2.	Deeply Forked	Herring, Mackerel	,,	6.	Pointed	Gobby	
3.	Emarginated	Trout, Carp, Perch		7.	Double emarginated		
4.	Truncate	Flounder,					

13.3.6 Pectoral Fins

The pectoral fins in elasmobranchs are highly developed and are larger in comparison to those of bony fishes. In sharks and rays, the pectoral fins become enormously developed on the sides of the body and head. These lobe pectoral fins act as the principal locomotor organs. In bony fishes, the pectoral fins are paddle-like and small. In many bony fishes, the outer margin of the fin is provided with a stout spine. In some catfishes, the pectoral spine is associated with a basal poison gland. The spines are defensive organs and in *Doras* and *Clarias*, the pectoral spines also help in progression on land (Fig. 13.7).

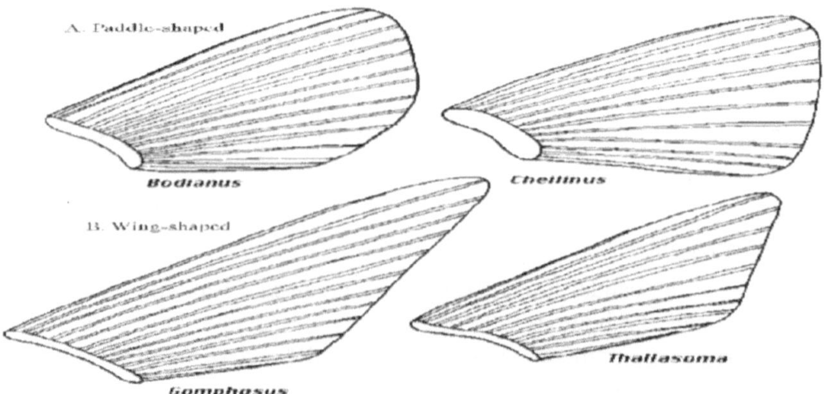

Fig. 13.7: Types of pectoral fins in fishes (from https://www.researchgate.net)

The pectoral fins are present in all fishes except some of the eels and pipe-fishes. The pectoral fins are always longer in speedy fishes but in the fishes of slow movement, these are broad and round. *Neoceratodus* possess peculiar lobate leaf-like pectorals, while in *Lepidosiren* and *Protopterus*, the pectorals become extremely elongated and filamentous. In *Exocoetus*, the pectoral fins have been extremely elongated and help to make the short aerial excursions as wings. In *Pantodon*, the pectoral is greatly elongated and beats rapidly during flight. In frog-fishes, Sea-toads and Bat-fishes, the pectoral fins assume the appearance of 'hands' and help to crawl to the bottom of the sea. The mudskipper (*Periopthalmus*) makes a short walk over marshy land and the pectoral fins are specialised for this purpose. The redfin is membranous and attached to a muscular stalk. In *Polynemus, Latris* etc., some of the fin rays supporting the pectoral fins are modified for sensory function.

These fin-rays drew out into delicate filaments containing sense organs. In sea robins and flying gurnards, certain rays of the pectoral fins may be adapted into finger-like projections. In skates and rays, the pectoral fins are used for propulsion known as rajiform propulsion.

13.3.7 Pelvic (Ventral) Fins

The position and shape of pelvic fins vary greatly in different fishes. The burrowing fishes, eels and Pipe-fishes, usually lack pelvic fins. In most of the fishes of Gadidae, the pelvic fins are reduced to filaments. The ancestral abdominal position is seen in the minnows; the thoracic position in sunfish and the jugular position, when the pelvic are anterior to the pectoral fins, as seen in the burbot (Fig. 13.8).

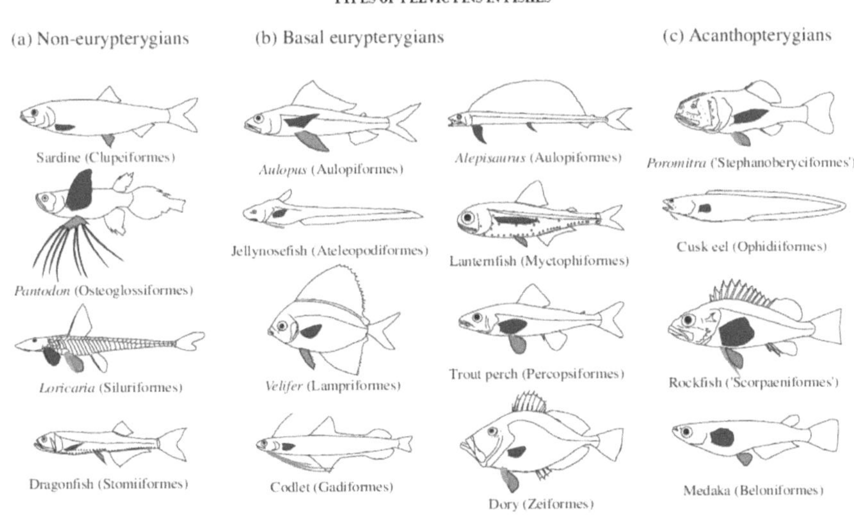

Fig. 13.8: Types of pelvic fins in fishes (from https://www.semanticscholar.org)

The specialization of the pelvic fins lies in their transformation into a sucking diet for attachment with stones or any other object. *Gastromyzon* possesses a large sucker and both paired fins participate in the formation. Another remarkable modification of the pelvic fins is observed in the skates and rays where the distal part of the pelvic fin is modified into claspers in males.

13.4 Functions of Fins

Fish use their fins for various purposes. Some important functions of fins are described below:

1. Dorsal fins act as a 'keel' for keeping the fish stable in the water and sudden direction changes. African knife fish (*Gymnarchus niloticus*) use its dorsal fin to move forward or backward by creating undulation. Lionfish and other scorpionfish have dorsal fins with hollow venomous spines which are used for self-defense. Some angelfishes (Lophiiformes) use their dorsal fin as a lure which helps to attract the prey. The modified dorsal fin of some fishes (Echeneidae) is used as a sucking disc.
2. Anal fins make stability and the anal fins of some bony fishes help in reproduction.
3. Generally, the pectoral fins help a fish for turning. Flying fish (Exocoetidae) use their long pectoral fins for gliding over the water. Mudskippers (Periophthalmidae) use pectoral fins to support themselves on land. Pectoral fins of some bottom-residence fishes such as threadfins (Polynemidae) bear touch receptors and taste buds which help to trace food. Sea Robins use their pectoral fin to glide around in the currents. Some bony fishes use their pectoral fins to help them rest on the bottom or in reef areas (e.g. Cirrhitichthys).
4. Most of the bony fish use their caudal fins for propulsion. Lunate caudal fins are characteristic features of fast swimmers such as tunas. They use it for maintaining rapid speed for a long duration.
5. Pelvic fins help the fish stabilise in the water. Pelvic fins of some fishes such as clingfishes (Gobiesocidae) are used as sucking appendages, which helps a fish hold on to stationary objects on the ocean bottom. Sea Robin fish use their pelvic fin for walking along the substrate. Some fishes such as Freshwater butterflyfish (*Pantodon buchholzi*) use their pelvic fins for gliding.

14

Integumentary System

14.1 Introduction

The integument is an envelope to separate and protect a fish from its environment; it also provides the means through which most contacts with the outer world are made (Fig. 14.1, 14.2 and 14.3). It is a large organ and is continuous with the linings of all body openings, and also covers the fins. The fish integument is dotted with various kinds of unicellular and multicellular glands along with abundant mucus-secreting glands. Poison glands of many cartilaginous fishes and some bony fishes are frequently associated with spines on the fins, tails, and gill covers. Photophores, light-emitting organs found especially in deep-sea forms, maybe modified mucous glands.

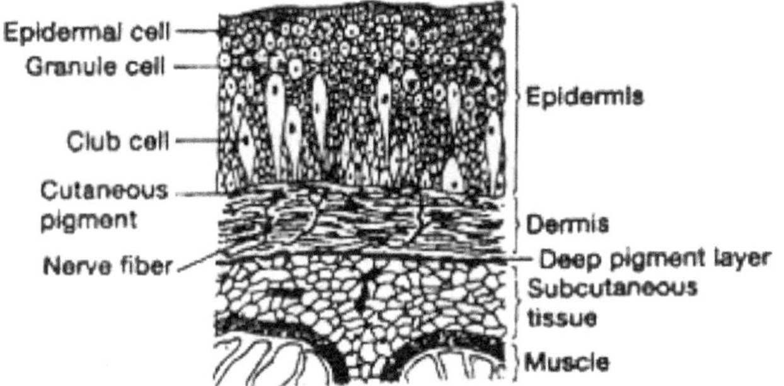

Fig. 14.1: Skin of cyclostomes (from https://www.notesonzoology.com)

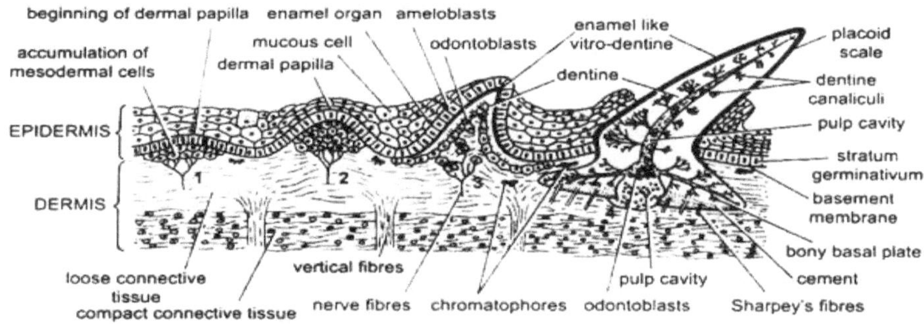

Fig. 14.2: Skin of ***Scoliodon*** (from https://www.notesonzoology.com)

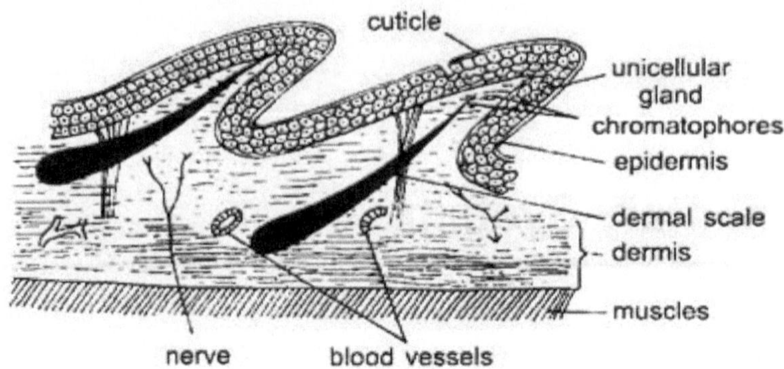

Fig. 14.3: Skin of bony fish (from https://www.notesonzoology.com)

The components of fish integument may serve important roles in protection, communication, sensory perception, locomotion, respiration, ion regulation, excretion, thermal regulation and camouflage or to permit recognition and attraction in courtship.

The integument of fishes consists of an outer epidermis and an inner dermis or corium. The two layers differ in origin, structure and function. The epidermis is essentially cellular in structure, comprised of a multilayered epithelium derived from the embryonic ectoderm. The dermis is a fibrous structure with relatively few cells and is derived from embryonic mesenchyme.

Fish epidermis varies in thickness based on species, age, region of the body and environmental conditions. In most fish species, the epidermis is thinner than the dermis. Keratinization of the epidermis is rare among fishes. Instead, the fish epidermis is generally metabolically active throughout all its layers. Because the epidermis of most fishes contains little or no pigment, it appears largely transparent by visual inspection.

The dermis contains blood vessels, nerves, scales, pigment cells, and adipose (fat) tissue, but the bulk of the typical dermal layer is comprised of fibrous connective tissue. When a fish is being skinned, the collagenous fibers of connective tissue that bind the skin to the underlying muscle and bone are very obvious. Except for scales, morphological characteristics of most dermal features are best observed microscopically.

The scales are an important component of the dermal skeleton and are the mineralised dermal structures of mesenchymal origin and variable size. Scales are usually covered completely by epidermal tissue, although portions of scales may protrude from the epidermal surface in certain species such as sharks.

The type, number and size of scales reveal much information about a fish's lifestyle. On a bony fish, scale patterns can range from a heavy coating of mail-like armor to a few large bony plates on the back, to a dense covering of thin, flexible scales, to a few localized prickles, or no scales at all. Large scales modified into bony plates function as protective armor on many slow-moving, bottom-oriented fishes such as sturgeons, many South American catfishes, poachers, pipefishes and seahorses. Conversely, the bodies of free-swimming fishes are often covered with typical scales, which provide some protection against mechanical damage without adding too much weight. Fish that are fast swimmers and those that live in fast-flowing water tend to have numerous fine scales, whereas perch and carp that live in quiet water often have rather coarse scales. Sculpins lack scales and are bottom-dwellers. Many catfish and eels frequently hide in tight places such as caves and crevices or fast-moving pelagic fish. The primitive agnathans are also scaleless. However, most tunas and anguillid eels appear scaleless but have numerous deeply embedded scales.

14.2 Epidermis

The number of cell layers in the epidermis varies from two to 10 or more. In pelagic fish species, the epidermis is frequently thickest on the dorsal side, but in benthic species, the ventral surfaces are often thicker. In salmonids, the epidermis is often thicker in the non-scaled areas of the head and the fins, than in the scaled areas of the body. Changes in ambient conditions or exposure to adverse environmental conditions, dietary deficiencies, and the presence of pathogens or other stressors can affect the structure and cellular composition of the epidermis.

14.2.1 Cell types

The fundamental structural unit of the epidermis is the epithelial cell. A great diversity of cell types exists in the various fish taxa. Some of these include holocrine mucous cells and other secretory cells, ionocytes, sensory cells and wandering cells such as leucocytes.

14.2.1.1 Epithelial cells: They are the basic cellular element of the fish epidermis and are variously known as Malpighian cells, epidermal cells, filament-containing cells, filamentous cells, polygonal cells, polyhedral cells, keratocytes, keratinocytes, principal cells and common cells. In primitive jawless fishes, homologous cells have been termed “mucous” cells, even though the term generally implies a goblet cell in other fishes.

The corresponding cell in most fishes is metabolically active throughout all layers of the epidermis. In teleost fishes, epithelial cells are capable of mitosis, although it is most common in the lower layers. Among the jawless fishes, dividing epithelial cells appear to be confined to the lower layers. Dead cells are regularly sloughed from the surface of the fish epidermis and replaced within about 4 days by living cells beneath.

Epithelial cells are small relative to many other epidermal cell types and may vary in shape depending on their position in the epidermis. The basal layer cells adjacent to the acellular basal lamina are cuboidal or columnar, whereas the superficial cells are often squamous. The presence of tonofilaments is a distinctive cytoplasmic feature of the epithelial cell. Tonofilaments are about 7–8 nm in diameter, arranged in tonofibrils or randomly distributed, and comprise an important component of the cytoskeleton of individual cells. Additionally, the attachment of tonofibrils to the desmosomal plaques that join adjacent epithelial cells enables the epidermis to respond as a whole to mechanical stress.

The exterior surface of the superficial epithelial cells of teleosts is characterized by microridges. The function of microridges is unknown, but they may provide some mechanical protection against trauma and assist in holding mucous secretions to the skin surface. Microridges also increase the absorptive surface area of epithelial cells; alter the boundary layer conditions, and maybe a factor in enabling the skin to function in gas exchange. They may also scatter light differentially, resulting in iridescent color patterns.

In fish, keratinization occurs in the surface epithelial layers. Examples are the breeding males of some species of Salmoniformes, Cypriniformes and certain teleosts, the sometimes horny projections on attachment organs on the lips and fins of ostariophysan fishes; and the polyhedral keratinized plaques on the skin of *Bagarius*, which are believed to help protect the fish from abrasion.

The mucous secretions produced by superficial epithelial cells assist in protecting fish from pathogens and other insults. These secretions form a layer called the cuticle on the skin surface, where they are mixed and modified by secretions from other cells. The cuticle layer is continually sloughed and renewed.

14.2.1.2 Goblet cells: The goblet cell is the second category of secretory cell in the fish skin, and occurs in the mucous membranes of fish. The abundance and size of goblet cells may vary in different body regions of a fish. A specific distribution of goblet cells may help to ensure an even layer of mucus over the surface of a moving fish. The goblet cells are absent in lamprey, paddlefish and mudskippers.

Goblet cells are frequently recognized in the middle to outer layers of the epidermis, although in a thin epidermis, the base of a mature goblet cell may be adjacent to the basement membrane. Immature goblet cells are rounded but become flattened laterally and increase in size as they move toward the surface of the epidermis. Upon reaching the skin surface, the goblet cell emerges.

Goblet cell abundance also differs between male and female fish of salmonid fishes. The number of goblet cells can also change seasonally and during larval metamorphosis, sexual maturation, or adaptation to seawater. Alterations in environmental conditions, such as sudden changes in temperature or exposure to UV radiation can also affect epidermal goblet cell numbers.

Exposure to toxicants can alter the size of goblet cells and can result in changes in the chemical composition of the secretion. In salmonids, the greatest volume of epidermal mucous secretions is from goblet cells. The secretions of most epidermal goblet cells are glycoproteins, but a wide array of components has been demonstrated in different fishes. A variety of roles have been attributed to goblet secretions, including lubricating, protective and possibly regulatory functions.

14.2.1.3 Club cells: Club cells are present in actinopterygians including primitive Polypteriformes, Anguilliformes and Ostariophysi. They are usually large and round to club-shaped with one or two centrally located nuclei and prominent nucleoli. Club cells are characteristic of eels that possess a secretory vacuole in the cytoplasm, whereas Ostariophysi does not. These cells are generally located in the middle epidermal layers and lack openings to the epidermal surface. The club cells of ostariophysan fishes contain an alarm pheromone that is released into the surrounding water only when the club cells are broken. Ostariophysans may detect the alarm pheromone by smell, and perform species-specific antipredator defensive reactions such as gathering into a tight school or group, diving for cover or becoming motionless.

14.2.1.4 Sacciform cells: These have been reported in rays, chimaeras, polypterids. Acanthopterygii, Paracanthopterygii, Ostariophysi and Protocanthopterygii. These cells have basally located nuclei and homogeneous to granular cytoplasm that stains readily with acid dyes. In some fishes, it can be distinguished from certain types of goblet cells. Like goblet cells, sacciform cells appear to differentiate deep in the epidermis and increase in size as they move toward the skin surface. Most mature sacciform cells open at the skin surface through an apical pore. In the Gadiformes, sacciform cells are enormous and swollen, often appearing cyst-like. Among pelagic gadoid fishes, sacciform cells appear not to open at the epidermal surface, whereas among benthic gadoids the typical surface openings are present. The functions of sacciform cell secretions may vary even within a species. Sacciform cell secretions usually include a protein, various enzymes, serotonin, cholesterol and other lipids. Sacciform cells are analogous to the granular glands of amphibian skin, producing secretions that are toxic or repellent to predators.

14.2.1.5 Thread cells: The protein released from thread cells mixes with mucous to form the thick and copious slime that gives the hagfish the nickname of "slime eel". Venomous fishes have multicellular nonducted holocrine glands associated with spines to form defensive organs. The release of venom into a wound made by the spine generally follows the rupture of the fine epidermal covering of the tip of the spine. The granular cell in the middle to upper layers of the lamprey epidermis is a large cell with a centrally located nucleus and prominent nucleolus, numerous small granules in the cytoplasm and appendages penetrating deep into the epidermis. Granular cells release their contents when lamprey skin is damaged.

14.2.1.6 Paraneurons: These cells possess a receptive site and a secretory site on their plasma membranes. These cells release signal substances and have regulating functions. The nature of substances synthesized by paraneurons and the physiological functions of these cells vary among vertebrate species. Merkel cells are paraneuronal cells that have been found in the epidermis of teleosts, lungfish and lampreys, although their functions are unknown. Additional studies in fish have demonstrated putative paraneurons among mechanoreceptor cells, electroreceptor cells, and chemosensory cells in the fish integument.

14.3 Dermis

The thickness of the dermis varies with species, position on the body, and life stage (Fig. 14.4). The dermis of most fishes is divided into two major layers. The upper (outer) layer is a loose network of collagenous connective tissue and is called the stratum spongiosum or stratum laxum, whereas the lower layer is a dense layer consisting primarily of orthogonal collagen bands and is called the stratum compactum. The relative thickness of the two layers depends on the presence or absence of scales, and the separation between dermal layers may be indistinct in nonscaled areas. In fin tissue, the dense connective tissue layer may be reduced. In lampreys and some primitive actinopterygians, the dermis consists primarily of dense collagenous tissue.

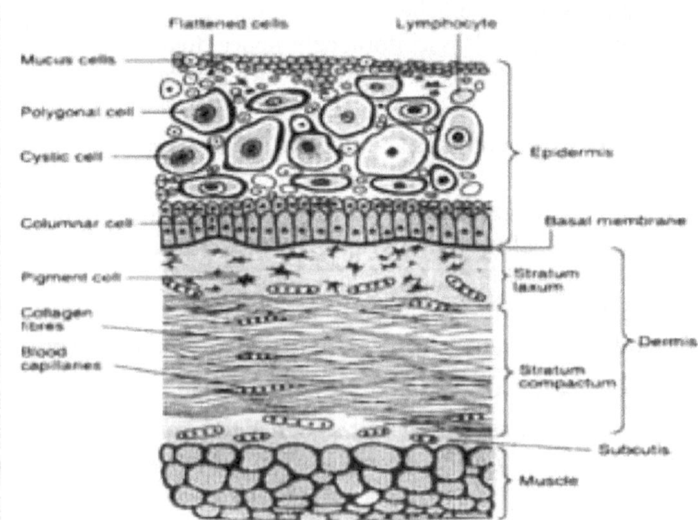

Fig. 14.4: Dermis of fish (from https://www.slideshare.net)

The dermis is separated from the epidermis by an acellular basement membrane, also called the basal lamina or adepidermal membrane. The deep face of the dermis is bounded by a single layer of cells termed the dermal endothelium. Immediately below the dermal endothelium, and separating the dermis from the underlying skeletal musculature, is a layer of well-vascularized loose connective tissue called the hypodermis or subcutis.

14.3.1 Specialized Dermal Elements

Chromatophores: Integumentary colours are primarily dependent on the presence of chromatophores in the skin. In all fishes, these cells occur in the dermis, where they may be found in the stratum spongiosum, in the hypodermis, or both.

14.3.1.1 Scales

Many types of fish inhabit in the water bodies. Some of these fishes have not changed much for millions of years. Others have evolved to use different aspects of the environment. A fish scale is a small rigid plate of the fish›s integumentary system and is produced from the mesoderm layer of the dermis. The morphology of a scale can be used to identify the species of fish it came from. Scales differ among fish types.

Scales vary enormously in size, shape, structure and extent ranging from strong and rigid armour plates in fishes such as shrimpfishes and boxfishes to microscopic or absent in fishes such as eels and anglerfishes.

The morphology of a scale can be used to identify the species of fish it came from. While different kinds of fish have developed different types of scales, each type allows the fish to function well in its environment. The more primitive scales allow for more protection than the advanced scale forms. Advanced fish use the increased speed allowed by their thinner scales to avoid harm, possibly avoiding capture by the tougher but slower primitive fish.

A good way to classify fish is by their scale type. There are four main types of fish scales. These types include placoid, ganoid, cycloid, and ctenoid.

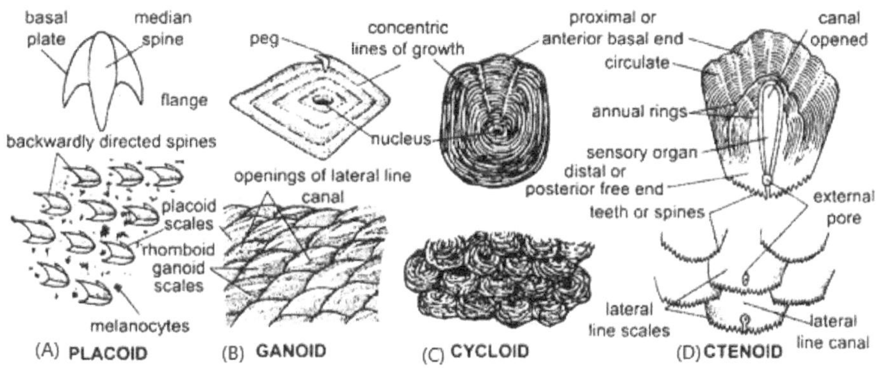

Fig. 14.5: Types of scales in fishes

14.4 Primitive Types of Scales

Primitive scales are not worse than advanced scales. This term only relates to the time frame of its evolution. Primitive fish existed long before they advanced. Some of these primitive fish have become extinct, but several still exist. Scales originated within the jawless ostracoderms, ancestors to all jawed fishes today.

14.4.1 Placoid Scales

Placoid scales or dermal denticles, the most primitive form of scales, are found in sharks and rays. These scales do not grow in size as the fish size increases. Instead, more scales are added. Placoid scales have 3 layers. The first is the pulpy vascular core. Secondly, a layer of dentine comprises the middle layer. Finally, the outside layer is made of hard vitrodentine. The placoid scales are also called dermal denticles and are structurally homologous with vertebrate teeth (Fig. 14.5A).

14.4.2 Cosmoid Sclaes

This scale is found in lungfish and some fossil fishes. Cosmoid scales are found only on ancient lobe-finned fishes, including some of the earliest lungfishes and in Crossopterygii, including the living coelacanth in a modified form as elasmoid scales. They were probably derived from a fusion of placoid-ganoid scales. Cosmoid scales increase in size through the growth of the lamellar bone layer. The upper surface is keratin. Its inner part is made of dense lamellar bone called isopedine. On top of this, lies a layer of spongy or vascular bone, followed by a complex dentine-like layer called cosmine with a superficial outer coating of vitrodentine.

14.4.3 Ganoid Scales

Ganoid scales, found in gar, are slightly less primitive than placoid scales. These scales grow in size as the fish ages. Ganoid scales are diamond-shaped. These scales are hard and thick. They are derived from cosmoid scales and often have serrated edges. They are covered with a layer of hard enamel-like dentine in the place of cosmine and a layer of inorganic bone salt called in place of vitrodentine. Ganoine covers scales, cranial bones and fin rays in some non-teleost ray finned fishes. Ganoine or ganoine-like tissues are also found in acanthodii. It has been suggested ganoine is homologous to tooth enamel or even considered a type of enamel. They are found in sturgeons, paddlefishes, gars, bowfin and bichirs (Fig. 14.5B).

14.5 Advanced Types of Scales

14.5.1 Cycloid Scales

Cycloid scales are the simplest of the advanced scale types. An example of a fish with cycloid scales is minnows. Cycloid scales increase in size as the fish ages. The growth can be seen in rings on the scale. These rings can be related to the age of the fish. Cycloid scales are thin, circular scales covered by a thin layer of epidermis and mucus. This gives the fish a slimy feel. This is a translucent layer lacking enameloid and dentine layers (Fig. 14.5C).

14.5.2 Ctenoid Scales

Ctenoid scales are the most advanced of all the scale forms. Sunfish and drum are fish with ctenoid scales. These scales originate from placodes and increase in size as the fish grows, producing rings. Its development starts near the caudal fin, along the lateral line of the fish. Ctenoid scale structure is similar to cycloid scales except tiny comb-like protrusions on the posterior of the scale. These projections allow the fish to swim faster. These scales are composed of a surface layer containing hydroxyapatite and calcium carbonate and a deeper layer of collagen. The enamel of the other scale types is reduced to superficial ridges and ctenii (Fig. 14.5D).

Types of Ctenoid scales

a. Crenate scales, where the margin of the scale bears indentations and projections.
b. Spinoid scales, where the scale bears spines that are continuous with the scale itself.
c. True ctenoid scales, where spines on the scales are distinct.

14.5.3 Other types of scales

14.5.3.1 Thecodont scales

The bony scales of thecodonts, the most abundant form of fossil fish, are well understood. The scales were formed and shed throughout the organisms' lifetimes, and quickly separated after their death.

14.5.3.2 Elasmoid scales

Elasmoid scales are thin, imbricated scales composed of a layer of dense, lamellar collagen bone called isopedine, above which is a layer of tubercles usually composed of bone. Elasmoid scales have appeared several times throughout fish evolution. They are present in lobe-finned fishes, extinct lungfishes, the coelacanths, tetrapodomorphs like *Eusthenopteron*, amiids, zebrafish and teleosts, whose cycloid and ctenoid scales represent the least mineralized elasmoid scales.

14.5.3.3 Leptoid (bony-ridge) scales

Leptoid scales are thinner and more translucent than other types of scales, and lack the hardened enamel-like or dentine layers. Further scales are added in concentric layers as the fish grows. Leptoid scales overlap in a head-to-tail configuration, like roof tiles, making them more flexible than cosmoid and ganoid scales. The scales may exhibit bands of uneven seasonal growth called annuli. These bands can be used to age the fish. Leptoid scales come in two forms: cycloid and ctenoid. These scales are found on higher-order bony fish.

Modified scales

Different groups of fish have evolved several modified scales to serve various functions:

1. Almost all fishes have a lateral line, a system of mechanoreceptors that detect water movements. In bony fishes, the scales along the lateral line have central pores that allow water to contact the sensory cells.
2. The spines of the dorsal fin of dogfish sharks and chimaeras, the stinging tail spines of stingrays, and the "saw" teeth of sawfishes and sawsharks are fused and modified placoid scales.
3. Surgeonfish have a scalpel-like blade, which is a modified scale, on either side of the caudal peduncle
4. Some herrings, anchovies, and halfbeaks have deciduous scales, which are easily shed and aid in escaping predators.
5. Male *Percina* darters have a row of enlarged caducous scales between the pelvic fins and the anus.
6. Porcupine fishes have scales modified into large external spines.
7. By contrast, pufferfish have thinner, more hidden spines than porcupine fish, which become visible only when the fish puffs up. Unlike the porcupine fish, these spines are not modified scales, but develop under the control of the same network of genes that produce feathers and hairs in other vertebrates.

Scale-less fishes

Fish without scales usually evolve alternatives to the protection scales can provide, such as tough leathery skin or bony plates:

1. Lampreys and Hagfishes have smooth skin without scales and dermal bone. Lampreys get some protection from a tough leathery skin. Hagfish exude copious quantities of slime or mucus if they are threatened. They can tie themselves in an overhand knot, scraping off the slime as they go and freeing themselves from a predator.

2. Most eels lack scales, though some species are covered with tiny smooth cycloid scales
3. Most catfish lack scales, though several families have body armour in the form of dermal plates or some sort of scute.
4. Mandarin fish lack scales and have a layer of smelly and bitter slime which blocks out disease and probably discourages predators, implying their bright colouration is aposematic.
5. Anglerfish have loose, thin skin often covered with fine forked dermal prickles or tubercles, but they do not have regular scales. They rely on camouflage to avoid the attention of predators, while their loose skin makes it difficult for predators to grab them.
6. Many bony fishes including pipefish, seahorses, boxfish, poachers and several families of sticklebacks, have developed external bony plates, structurally resembling placoid scales as protective armour against predators.
7. Seahorses lack scales but have thin skin stretched over bony plate armour arranged in rings through the length of their bodies.
8. In boxfish, the plates fuse to form a rigid shell or exoskeleton enclosing the entire body. These bony plates are not modified scales but skin that has been ossified. Because of this heavy armour boxfish are limited to slow movements, but few other fish can eat the adults.

14.6 Fin rays

Dermal rays or spines support the fins of fish. The fin supports in lampreys and hagfishes are unsegmented cartilaginous rods that provide only weak support. In both elasmobranchs and bony fishes, the dermal fin rays are known collectively as dermotrichia. These include several kinds of collagenous structures with varying types and degrees of mineralization. Dermotrichia may be segmented or unsegmented, branched or unbranched, biserial or uniserial. The fin rays (lepidotrichia) present in all Osteichthyes except dipnoans are comprised of mineralized bone and are believed to have derived from modified scales in early vertebrates.

14.7 Integumentary Extensions

Fishes have evolved a wide variety of integumentary modifications, including extensions of the skin. For example, barbels are integumentary extensions that have developed independently in many taxa as accessory-feeding structures carrying sensory organs. Barbels that differ in structure and location are present on sturgeons (Acipenseridae), goatfishes (Mullidae), marine and freshwater catfishes (Siluriformes) and some Cypriniformes such as carp (*Cyprinus carpio*) and loaches (Cobitidae).

The most frequently cited examples of the extension of skin into flaps are those of the sargassum fish (*Histrio*) and seadragon (*Phyllopteryx*). The function ascribed to the exaggerated skin flaps in these fishes is a protective resemblance to the seaweed in which they hide. Although the skin extensions in certain other fishes, such as scorpionfishes (*Scorpaena*) are less prominent, these extensions, combined with mottled coloration, also likely aid in camouflage.

Certain less permanent skin extensions assist in reproduction and may be reduced in size or shed completely after the breeding season. Keratinized breeding (nuptial) tubercles or pearl organs of Cypriniformes and certain other fishes function primarily to facilitate contact between spawning individuals. However, the antler-like breeding tubercles of breeding male stonerollers (*Campostoma anomala*) are used in defense of nests against intruding males. Outgrowths of abdominal skin or fin tissue of skin-brooding fishes enable the attachment of eggs to the skin.

15

Skeletal System in Fishes

15.1 Introduction

An endoskeleton is a skeleton that is on the inside of the body of an animal. It is an internal supporting structure composed of mineralized tissue. The endoskeleton develops within the skin or the deeper body tissues. It participates in protection, helps in movement, provides shape and support and stores minerals like calcium. Gerringer et al. (2021) suggested that changes in skeletal structure are non-linear and are driven by hydrostatic pressure, other environmental factors and evolutionary ancestry.

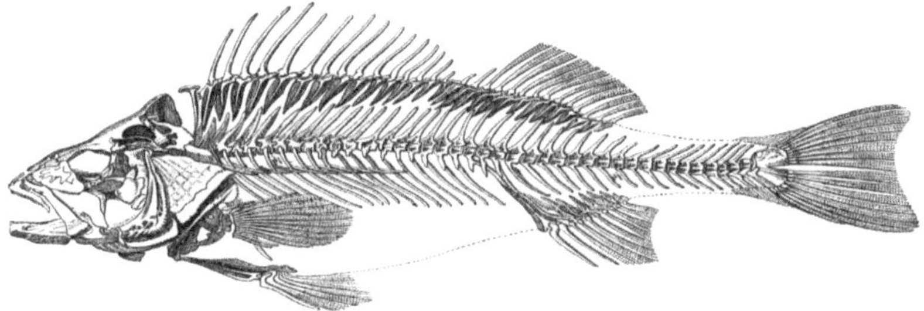

Fig. 15.1: Articulated skeletal of a fish (From https://en.wikipedia.org)

Generally, *Scoliodon* is selected as a model animal to describe the skeletons of cartilaginous fishes, while *Labeo rohia, Wallago attu* and *Mystus seenghala* are preferred to study the skeletons of bony fishes (Fig. 15.1 and 15.2).

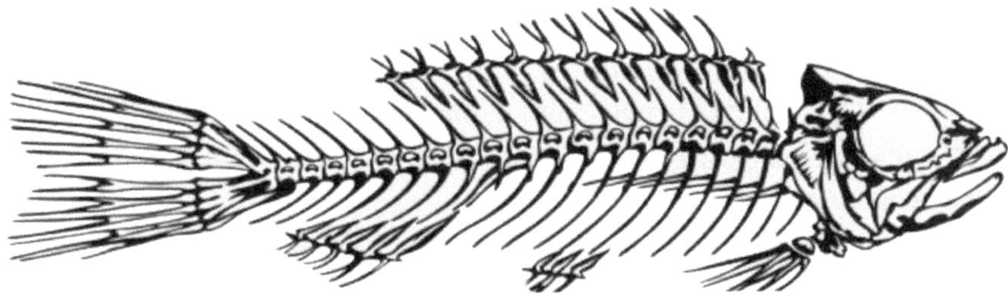

Fig. 15.2: Articulated skeleton of another fish

15.2 Composition of Fish Skeletal System

Although the skeletons of sharks and rays are more complicated, they are still mostly made of cartilage rather than bones. The skeleton of dipnoans is largely cartilaginous. The skeleton of a fish is either made of bone or cartilage. There are two different skeletal types:

A. Exoskeleton- A sturdy exterior shell of an organism
B. Endoskeleton- Inner shell and support system of an organism

The fish skeletal system consists of the vertebral column, jaw, ribs, cranium and intramuscular bones. Starting from the head, bony fish consist of the skull (=cranium). Osteichthyes have hinged jaws which aid them in feeding. But hagfish and lampreys are jawless fishes. Otoliths are distinctive features of bony fish ear plates that aid in stability. Three pairs of bones aid the gills. These pairs of bones are called gill arches and are made of bony filaments. Fins are a vital part of fish. They help with propulsion, steering, and stability. Paired fins take up the role of steering, while caudal fins and dorsal fins help in propulsion and stability respectively.

The endoskeleton of a fish may be divided into the following two parts:

The axial skeleton is made up of the skull and vertebral column, as well as the ribs and bones that support the median fins.

The appendicular skeleton is made up of girdles and fin bones.

15.3 The Axial Skeleton

Bones may recover and become stronger than the original bone since they are made of tissues that include blood vessels and nerves. The skeleton system of a fish has many small bones. Approximately 150 bones are present in the skeletal system. The bones have dense, less flexible connective tissue.

1. Pterygoid: It forms the inner cartilaginous bone of the jaw.
2. Maxillary: They are paired bones located behind or above these bones.
3. Opercula: They are flat and paired and connected to the hyomandibular from the majority of the gill cover.
4. Centrum: The centrum is the cylindrical component of each vertebra to obstructs the embryonic notochord during ossification.
5. Scapula: They are paired flat bones articulating with the coracoids anteriorly and cleithra posteriorly.
6. Coracoid: The cleithra and scapula articulate with the coracoid.
7. Radials: They connect the fin rays and articulate anteriorly with the scapula and coracoid.

15.3.1 Fish Skull

The skull (= cranium) is a two-bone structure that is encased in another bone. The skull is formed from a series of loosely connected bones (Fig. 15.3 A and B and 15.4).

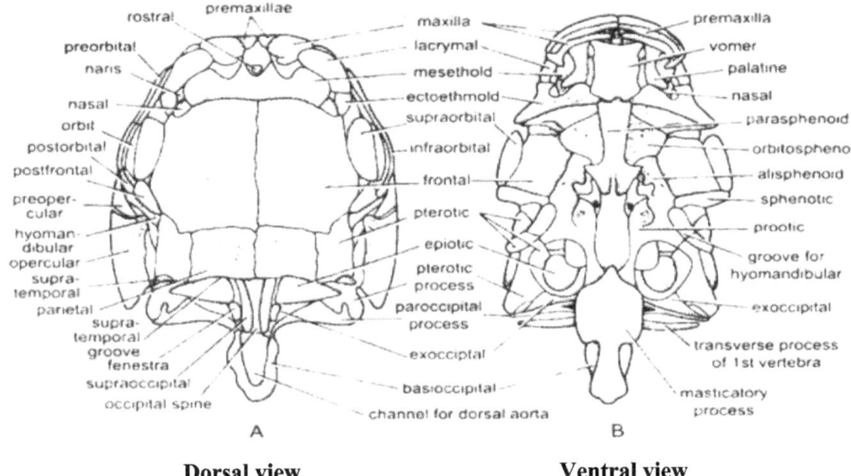

Dorsal view **Ventral view**

Fig. 15.3: Skull of Rohu (From https://www.notesonzoology.com)

Jawless fish and sharks possess cartilaginous endocranium. It is represented by cartilaginous elements that partially enclose the brain and are associated with the inner ear and nostril capsules.

In *Scoliodon*, the skull contains an intimately fused cranium, four sense capsules, olfactory organs and a visceral skeleton.

In teleosts, many investing and replacing bones participate in the formation of the skull. The typical fish skull is composed of neurocranium and branchiocranium. The neurocranium contains endocranium and dremocranium. Endocranium protects the olfactory, optic and otic capsules as well as the anterior end of the notochord. The branchiocranium includes the mandibular region, hyoid region and branchial region.

In *Catla, Cirrhinus* and *Labeo*, the fifth ceratobranchial is modified to form the inferior pharyngeal bones. But, in *Wallago* and *Mystus*, it is elongated.

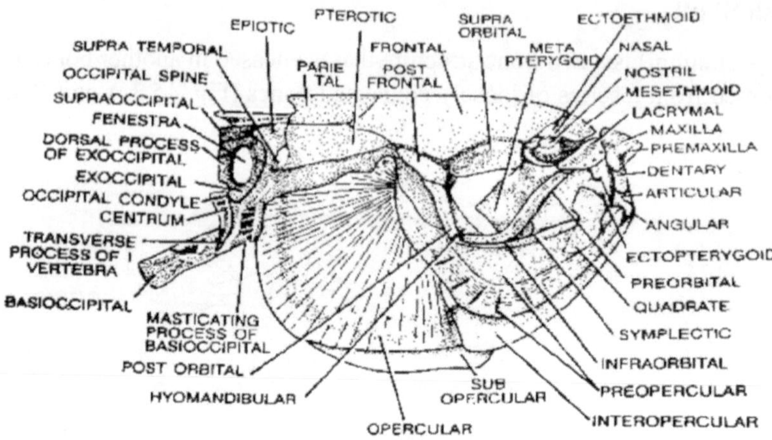

Fig. 15.4: Lateral view of Skull of Rohu (From https://www.notesonzoology.com)

In rohu, the elongated skull is of the platybasic type and contains an immovable cranium, sense capsules and visceral arches. The skull contains a posterior basal plate and anterior trabecular region. The skull is divided into the dorsal roof, occipital region, otic region, orbitotemporal region and ethmoidal (nasal) region. The dorsal roof is almost convex. The orbitotemporal contains the temporal (sphenoid) and orbital regions. The temporal region is made up of the parietal and frontal regions. The parietal region is made up of parietals, the alisphenoid and the parasphenoid. The frontal region is made up of the frontal, orbitosphenoid and parasphenoid bones. The ethmoidal (olfactory) region is made up of the nasal, ectoethmoid, lachrymal, mesethmoid, vomer and rostrum (Fig 15.5).

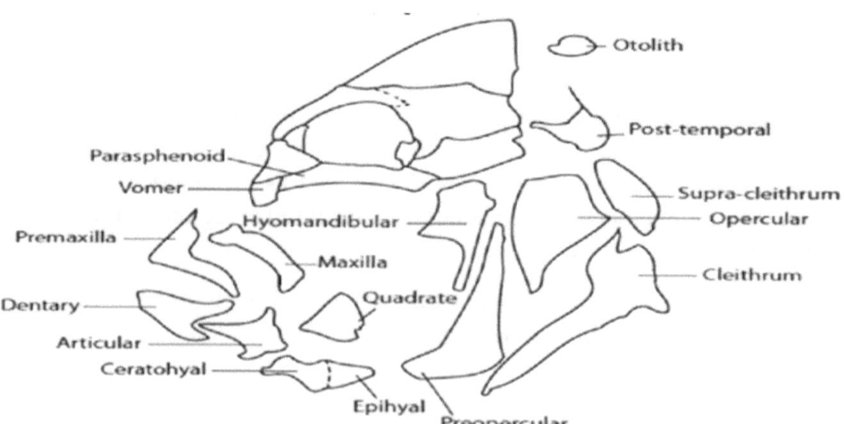

Fig. 15.5: Disarticulated skull bones of a fish (From https://www.researchgate.net)

The cranium is divided into three regions (occipital, sphenoidal and ethmoidal) in *L. rohita* but five (occipital, auditory, sphenoidal, orbital and ethmoidal) regions in *W. attu*.

1. Fontanelle: It is a large and partial opening at the top (roof) and dorsal side of the skull.
2. Parietal: It forms the roof of the skull.
3. Rostrum: It is the most anterior part of the cranium. It capsules to enclose olfactory organs.
4. The internasal septum: It separates the two olfactory capsules.
5. Foramen magnum: The occipital region forms the posterior region of the skull and contains the foramen magnum. It articulates with the first vertebra and through which the spinal cord passes. A median ridge and the occipital crest are present above it.
6. Occipital condyle: It contains replacing bones and is present on either side of the foramen magnum. The posterior portion contains a basioccipital with a single occipital condyle, a dorsal supraoccipital and two ventral basioccipitals situated on the lateral side of the foramen magnum (Fig. 15.6).

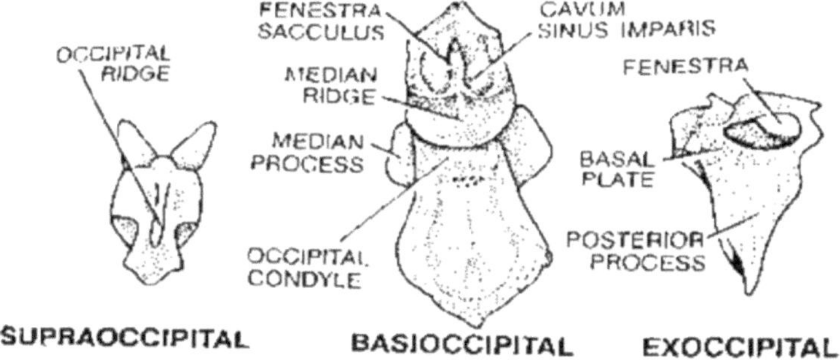

Fig. 15.6: Bones of Occipital condyle of rohu (From https://www.researchgate.net)

A. Supraoccipital: It is rectangular in *L. rohita* and *W. attu* but with an elongated occipital process in *Mystus*. It contains a lateral branch of the trigeminofacialis complex. It articulates anteriorly with the frontal, anterolaterally with the sphenotic, posteriorly with exoccipital and laterally with the pterotic and epiotic.
B. Exoccipital: It is irregular bone. A neural spine arises from it dorsally and unites with the supraoccipital. It bears a ventral foramen for the glossopharyngeal and a posterior foramen for the vagus. Its outer plate joins with the supraoccipital and epiotic. It also bears a semicircular ridge. *Mystus* bears cavum sinus impairs. It ventrally bears a perforation for the glossopharyngeal nerve, a foramen behind the vagus nerve and a small foramen for the first spinal nerve.

C. Basioccipital: It is large drain-pipe-shaped and thickest in the middle in *L. rohita* but stout and elongated in *Mystus.* It unites frontally with the parasphenoid and laterally with the exoccipital and prootic. It bears a concave occipital condyle posteriorly. It is elongated in *Mystus* and *W. attu* but drain-pipe shaped in *L. rohita.* It bears cavum sinus impairs between the horizontal plates of the exoccipitals and vertical ridges of the basioccipital. In *L. rohita* and *Mystus,* it has fenestra sacculus. It bears accessory articular process in *W. attu.* In *Mystus,* it bears two lateral depressions to lodge pars inferiors of internal ears.

In rohu, it also contains a lateral exoccipital. The exoccipital contains a basal plate, paroccipital process and dorsal process. The ventral surface of the basioccipital gives rise to the masticatory process.

7. Orbital region: It contains the four investing suborbitals and the lachrymal. The first, second, third and fourth suborbitals are splint-like, triangular, long and small bones respectively.
   A. Supraorbital ridge: It is the margin between the roof of the cranium and the orbit.
   B. Orbital: In rohu, an almost incomplete lateral wall is derived from it. It articulates posteriorly with the auditory capsule and anteriorly with the nasal capsule.
   C. Preorbital process: It is a slender cartilage for the partial encirclement of the orbit.
   D. Postorbital process: It emerges from the side of each postorbital cartilage and runs forward along the upper margin of the orbit.
   E. Post-frontal: It surrounds the orbit dorsally.
   F. Infra-orbital: it is situated at the antero-ventral border of the orbital region.
   G. Lacrymal: It is a small elliptical bone lying alongside the anterolateral border of the lateral wing of the mesethmoid right in front of the nasal opening.
8. Nasal capsule: It overlaps the ethmoid and vomer. In rohu, it is attached to the trabecular region.

Bony fishes have an additional dermal bone to form a more or less coherent skull roof in lungfish and holost fish. Dermal bone forms the outer shell of the skull. Endochondral bone makes up the inside of the skull.

1. Opercular series: It lies on the surface of the gills. It may or may not have spines. It contains opercular, preopercular, subopercular and interopercular (Fig. 15.7).

A. Preoperculum: It is present on top of the gills. *L. rohita, Mystus* and *W. attu* has a large crescent-shaped, elongated flat and elongated curved roughly triangular bone. It also may or may have spines. It articulates in front with hyomandibular and quadrate and ventrally with angular.

B. Inter-operculum: It is stout like in *Mystus* but small scute-like in *W. attu*. It lies between and below the preopercular and opercular. It's ventral border overlaps the upper half of the posterior branchiostegeal ray.

C. Operculum: It is roughly the largest, rectangular in *L. rohita* and *Mystus* but triangular in *W. attu*. It articulates with the hyomandibular through its apex.

D. Subopercular: it is elongated sabre-shaped in *L. rohita* and triangular-shaped in *Mystus*. It is the lower opercular bone. It is absent in *Wallgo attu*. It articulates anteriorly with the inter-opercular and its ventral border overlaps the hinder part of the posterior branchistegeal ray.

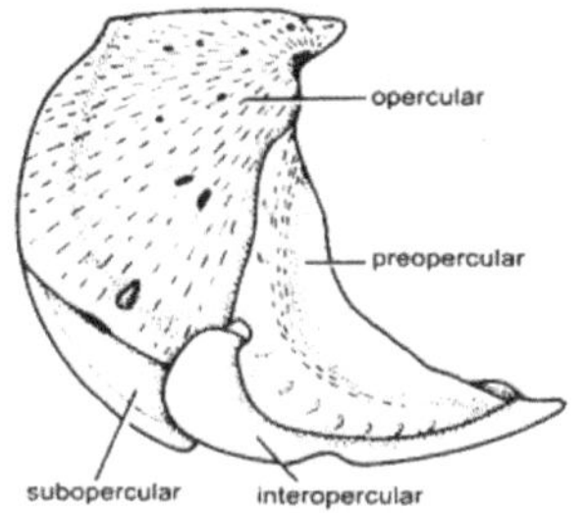

Fig. 15.7: Opercular of rohu

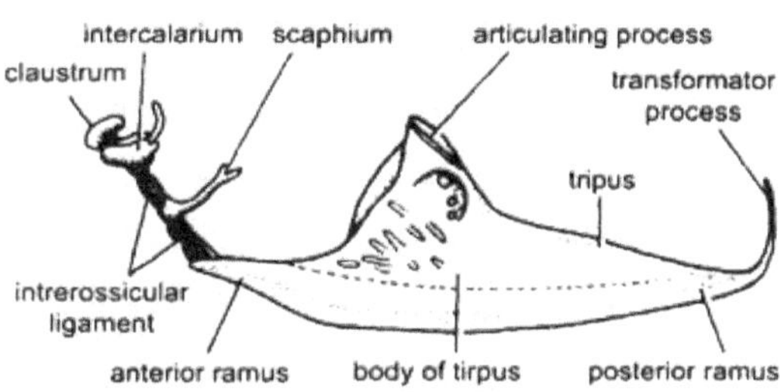

Fig. 15.8: Weberain ossicles of rohu (From https://www.notesonzoology.com)

2. Premaxilla (Upper jaw): It is very close to the mouth. In *Labeo rohita*, it is thick. *Wallago, Channa* and *Mystus* bear premaxillary teeth.
3. Maxillae: It is generally edentulous and present in the mouth after the palatine.
4. Jugal: It is present on each side of the upper jaw and attached to the posterior end of the maxilla.
5. Symplectic: It links the jaw to the rest of the cranium. It is the posterior bone of the hyoid arch.
6. Quadrate: It is present between the palatine and hyomandibular.
7. Spiracle: It is a small extra gill opening behind the eye in sharks, and some primitive bony fishes.
8. Mandible (Lower jaw): It defines a chin. Each half of the lower jaw contains articular, angular and dentary.

A. Angular: It lies between the palatine and the dentary. It articulates behind with quadrate.

B. Dentary: They are large paired dental bones united medially and create the front of the mandible. It bears 8-9 pores on its ventral side.

C. Articular: It is a moveable joint formed dorsally by a pair of paired, angular-shaped bones that are articulated and take up a portion of the posterior end of the mandible.

D. Retro Articular: They have paired bones located at each articular's lower posterior corner.

9. Mandibular arch: It is the first visceral arch. Each half of this arch divides into the upper jaw or palatopterygoquadrate and Mackel's cartilage. In teleosts, it forms the upper jaw combining six pairs of bones premaxilla, maxilla, suprapterygoid, ectopterygoid (pterygoid), metapterygoid, mesopterygoid (endopterygoid), palatine and quadrate. The suspensorium is methyostylic.

A. Premaxilla: It unites posterolaterally with the supra-orbitals and mediodorsally with the ethmoid. It terminates before the articulation of the mandible with the quadrate behind.

B. Maxilla: It is present on the opposite side of the premaxilla. It's head articulates with the palatine and the arm supports the barbel.

C. Ectopterygoid: It articulates in front with the vomer, behind with the metapterygoid and posteriorly with the hyomandibular and quadrate.

D. Palatine: It lies between the maxillary and lachrymal. It unites with the maxilla and laterally with the ethmoid.

E. Metapterygoid: It unites posteriorly with hyomandibular and behind with quadrate.

F. Quadrate: It unites anteriorly with the metapterygoid, mediolaterally with the hyomandibular and posteriorly with the preoperculum. In rohu, it is covered by ectopterygoid and endopterygoid.

10. Palatopterygoquadrate: It forms the primary upper jaw. In rohu, it becomes ossified by palatine, metapterygoid and quadrate.

11. Platoquadrate (visceral arch): In rohu, it articulates posteriorly with the auditory capsule by the otic process and anteriorly with the trabecular region by the basal process.

12. Mackel's cartilage: It forms the lower jaw. It is made up of the proximal and distal parts of the articular. The articular is covered anteriorly by angular, laterally by dermarticular, dorsally by suprangular and mesially by preaticular and sesmois articular. The distal part of the articular is replaced by coronoids and dentary.

13. Ethmoidal (olfactory) region: Its shape is formed by an olfactory capsule separated from each other by a median mesethmoid. It is most complicated and flattened in *W. attu.* Its shape is transversely elongated and median flat

in *L. rohita* and *Mystus* respectively. It contains dorsal ethmoid, lateral ethmoid, nasal, vomer and sometimes adnasal (Fig. 15.9).

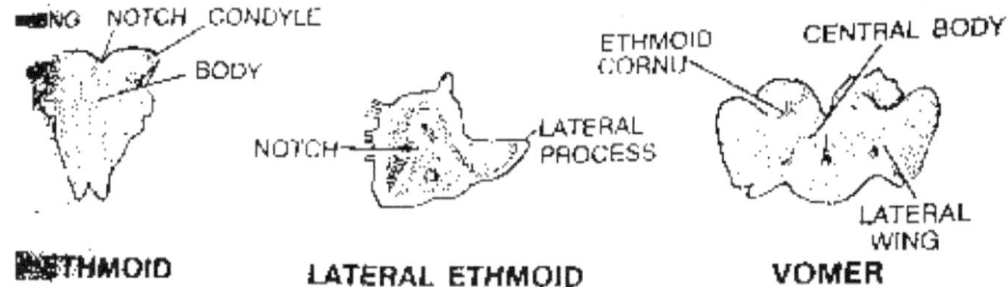

Fig. 15.9: Bones of Ethmoidal region of rohu (From https://www.researchgate.net)

A. Ethmoid (mesethmoid): It is flattened in *W. attu*, transversely most complicated dorsal and flattened in *W. attu.* Its shape is transversely elongated and median flat in *L. rohita* and *Mystus* respectively. It contains a central body. Its ventral horn articulates with the vomer posteriorly. Its dorsal horn is also known as ethmoidal cornu.

B. Lateral ethmoid (parethmoid or Ect-ethmoid): It is hammer-shaped most complicated and hammers like in *W. attu.* Its shape is roughly rectangular in *Mystus*. It is the inner part of the hammer that lies below the ethmoid and frontal. In *L. rohita*, it bears a nasal pit, nasal openings, mesial body and an outer lateral process

C. Nasal: It is the dorso-lateral splint-like tubular bone in *W. attu* but a small rectangular bone in *L. rohita.* It encloses the supra-orbital canal of the lateral line organ.

D. Vomer: Vomer is present on the ventral side of the roof of the mouth typically to support teeth. It is quadrangular, T-shaped and rhomboidal shaped in *L. rohita, Mystus* and *W. attu*. The vomer of *Mystus* and *W. attu* bear vomerine teeth. It forms the floor of the orbit. It articulates with ethmoid in front and posteriorly with parasphenoid.

14. Hyoid arch: It is the second visceral arch bearing basihyal, ceratohyal, hyomandibular and sometimes sympletic. In teleosts, the hyoid arch contains hyomandibular, interhyal, epihyal, ceratohyal, hypohyal, hyoid cornu and a median triradiate bone urohyal.

    A. Hyomandibular: It is strong elongated, flat spreaded and large flattened bone in *L. rohita, Mystus* and *W. attu*. It articulates anteriorly with the metapterygoid, posteriorly with the operculum and preoperculum and anteroventrally with the quadrate.

    B. Interhyal (stylohyal): It is located on the inner side of the preoperculum. It joins the hyoid cornu with hyomandibular.

    C. Epihyal: It is located ventrally and bears branchiostegal rays to the head along the left and right anterior edges of the gill cavity. It articulates anteriorly with ceratohyal and posteriorly with interhyal. Its posterior edge bears 3-4 branchiostegeal rays.

D. Ceratohyal: It bears a dorsally dentigerous plate and 16 branchiostegeal rays at its posterior edge.

E. Hypohyal: These are two pairs of bones, lying one behind the other.

F. Urohyal: It lies ventral to the hypohyal and is connected anteriorly to it. Its dorsal surface contains a longitudinal ridge and laterally articulates with cleitheral bone.

15. Sphenoidal (temporal) region: It has three replacing bones namely pleurosphenoid, basisphenoid, alisphenoid and median orbitosphenoid as well as two investing bones frontal and parasphenoid (Fig. 15.10).

A. Pleurosphenoid: It is a small irregular bone bounded anteriorly by the parasphenoid and posteriorly by the prootic. It articulates with the orbitosphenoid in front, prootic behind, sphenotic on the outer side and the parasphenoid on the inner side.

B. Basisphenoid: It is a small Y-shaped bone present in front of the basioccipital. Its upper concave surface forms the floor of the brain.

C. Alisphenod: It is a big bone present on each side of the basisphenoid.

D. Orbitosphenod: It is a pair of dumb-bell shaped bones in *L. rohita* and *W. attu* but boat-shaped in *Mystus*. It is also present on each side of the basisphenoid. It articulates anteriorly with lateral ethmoids and posteriorly with pleurosphenoid.

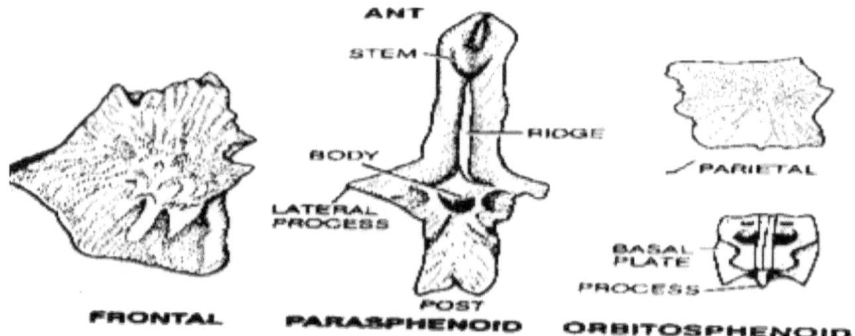

Fig. 15.10: Bones of Sphenoidal region of rohu (From https://www.researchgate.net)

E. Parasphenoid: It is cross-shaped and complicated in *L. rohita*, median long bone in *Mystus* and rhomboidal shaped in *W. attu* Its arm unites in front of the vomer. It articulates laterally with prootics and posteriorly with the basioocipital.

F. Frontal region: It contains two large frontals above, orbitosphenoids and a median long stipulate parasphenoid at the ventral side. It extends posteriorly to cover the dorsal side of the parietal along with the supra-occipital. Its inner margin articulates with pleurosphenoid and orbitosphenoid. Its shape varies from the large flattened bone in *Mystus* and *W. attu* to roughly triangular in *L. rohita*. In *W. attu,* it bears supra-orbital canal and supra-temporal groove.

G. Parietal region: It contains large rectangular parietals above and a pair of irregular pelurosphenoid below. The parietal fossa is an oval depression between the two auditory capsules.

16. Auditory (Otic) capsule: It is made up of replacing bones namely epiotic, prootic, pterotic and sphenotic. In teleosts, except rohu and *Wallago*, opisthotic is also present (Fig. 15.11).

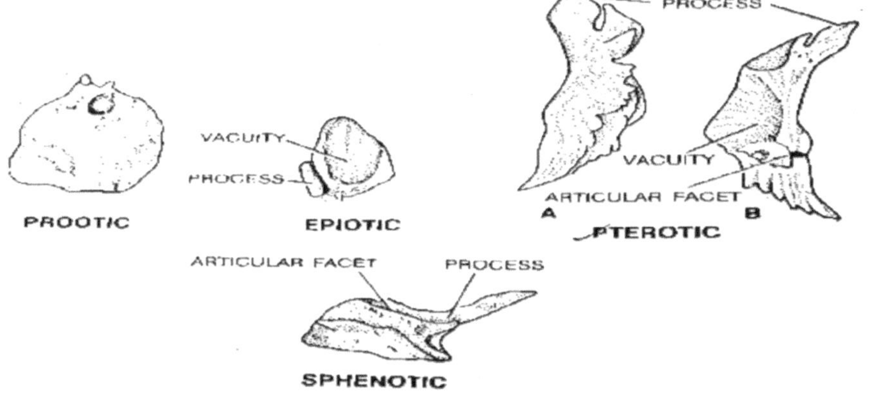

Fig. 15.11: Bones of Auditory capsule of rohu (From https://www.researchgate.net)

A. Prootic: It lies on the ventral side of the auditory region. It articulates anteriorly with parasphenoid and pleurosphenoid, laterally with sphenotic and pterotic, posteriorly with exoccipital and internally with parasphenoid. The structure of prootic is simpler in *L. rohita* compared to *Mystus* and *W. attu*. The shape of prootic is irregular in *L. rohita* and *Mystus* but square shaped in *W. attu*.

B. Epiotic: Its shape is bowl-shaped, plate-like and spatula-like in *L. rohita, Mystus* and *W. attu* respectively. It lies on the postero-lateral side of the auditory region. It articulates with exoccipital, supraoccipital and pterotic through its ventral lamella.

C. Pterotic: Its shape is irregularly triangular, roughly triangular and triangular in *L. rohita, Mystus* and *W. attu* respectively. It bears an articular facet and vacuity in *L. rohita*. It lies on the outer margin of the auditory region. It articulates anteriorly with sphenotic, posteriorly with epiotic, dorsally with supraoccipital and ventrally with prootic and exoccipital. It also bears a shallow groove for the articulation of the hyomandibular ventrally.

D. Sphenotic: Its shape is irregular triangular and elongated triangular in *L. rohita* and *W. attu* respectively. In *W. attu*, it bears supraorbital canal and infraorbital sensory canal. It articulates posteriorly with pterotic, mediolaterally with supraoccipital and anterolaterally with frontal and ventrally with prootic and pleurosphenoid.

15.3.2 Vertebral Column

The vertebrae of fish are built along the basic chordate body plan which contains a stiff rod running through the vertebral column or notochord, with a hollow tube of the spinal cord above it and the gastrointestinal tract below. The vertebral arch surrounds the spinal cord.

The vertebral column is made up of several endochondral bones known as vertebrae. It is divided into two sections: trunk and caudal. Rohu contains 37–38 (21 trunk and 16–17 caudal) vertebrae. On the other hand, *Wallago* contains 73 vertebrae (16 trunk and 57 caudal). Out of 16 trunk vertebrae, anterior five are intimately united to form complex vertebrae and the remaining are free. Fish ribs are attached to the spinal column in the trunk region. The centrum of one is the enormous spool-shaped core region.

The vertebral column consists of a centrum and vertebral arches. The vertebrae of lobe-finned fishes consist of

Pleurocentrum: It is the small plate-like bone that lies just beneath the arch to protect the upper surface of the notochord.

Intercentrum: It is a larger arch-shaped bone below the pleurocentrum to protect the lower border of the notochord.

In dipnoans, the centrum is absent. The joints between the skull and vertebral column are also lacking. They lack the premaxilla, maxilla and nasal.

(I) Complex vertebra: In *W. attu,* the first vertebra is represented by the centrum alone lying between the basioccipital and complex vertebra. It also bears a pair of ventral accessory articular processes, which fit in front of the similar processes of basioccipital and behind the similar processes of complex vertebra. Beneath the centrum, well-developed wing-like parapophyses are found, which is forked into an anterior and a posterior division. its neural crest and neural spine is crest-shaped. A radial nodule is present in the insertion of the articular process of the tripus. The centrum of the fifth vertebra is amphicoelous with well-developed parapophyses.

In *Mystus,* the centra of the first five vertebrae are fused.

The first 4-5 vertebrae fused to form a complex vertebra in most teleosts. These vertebrae lack parapophyses.

In rohu, paired prezygapophyses, postzygapophyses and a pair of short parapophyses are present. The first four (the first six in *Mystus*) trunk vertebrae are greatly altered to connect with the swim bladder and internal ear. The last 3–4 trunk vertebrae bear posteroventral processes.

(II) Typical Trunk vertebra: Various processes are projected from the arches. An arch extending from the top of the centrum is called a neural arch, while the haemal arch (chevron) is found underneath the centrum in the caudal vertebrae of fish. The centrum of a fish is usually amphicoelous, which limits the motion of the fish (Fig. 15.12A and B). Its neural spine is backwardly directed. Anterior pair of zygapophyses arise from the bases of the neural arches and the posterior pair of zygapophyses arise from the posterior edge of the centrum. Parapophyses arise from the ventrolateral grooves of the centrum and run outward and backward, bearing ribs at the distal end. From the bases of parapophyses arise a forwardly directed spine.

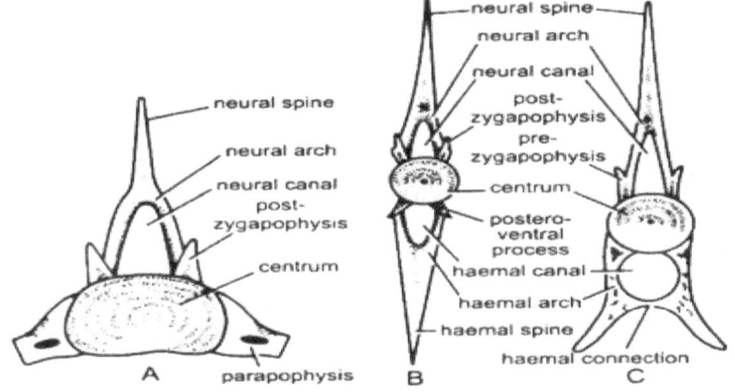

Fig. 15.12: Trunk vertebra Typical trunk vertebra Caudal vertebra
(From https://www.notesonzoology.com)

(III) Typical precaudal vertebra: It has a spool-like or deeply biconcave amphicoleous centrum. A narrow neural spine is present on the dorsal side of the neural arch. Paired transverse processes, prezygapophyses and postzygapophyses are also present. It has downwardly curved ribs articulated with the terminal end of the transverse process. A pair of short parapophyses is directed downwards in rohu (Fig. 15.13A).

(IV) Typical caudal vertebra: It bears a spool-like amphicoleous centrum along with a median dorsal, median ventral and paired dorsolateral and ventrolateral grooves. The centrum has a ventrally located haemal arch with a narrow haemal spine. A narrow and long neural spine is present on the dorsal side of the neural arch. Paired transverse processes, prezygapophyses and postzygapophyses are also present. It lacks ribs (Fig. 15.13B).

In *W. attu*, the centrum of the first and the last caudal vertebra is concave. The last caudal vertebra bears 5th-7th hypurals. Spinous urostyle arises from the posterior end of the vertebra and is directed upwards.

In rohu, the last 3 (in other teleosts, 3–5) caudal vertebrae are modified to support the caudal fin.

(V) Ribs: 17 pleural ribs are segmentally disposed slender bony rods in rohu. They are attached to the dorsal ends of the parapophyses. The first pair of ribs connects to the fifth trunk vertebra's parapophyses.

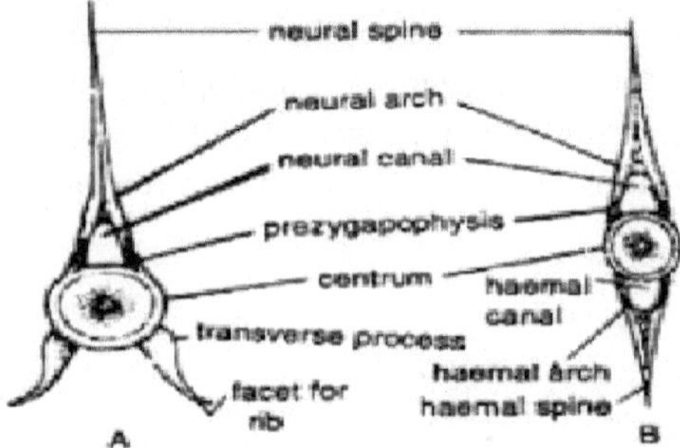

Fig. 15.13: Precaudal and caudal vertebrae

15.4 Appendicular Skeleton

The pectoral and pelvic girdles, as well as the bones of the appendages, make up the appendicular skeleton.

(I) Pectoral girdle: The pectoral fin is connected to the fish skeleton by the pectoral girdle. Each half of the pectoral girdle contains a primary and a secondary girdle. The primary girdle contains three replacing bones dorsally located scapula, ventrally located coracoid and mesocoracoid. The secondary girdle contains four investing bones cleithrum (clavicle), supracleithrum, posttemporal and postcleithrum. These four investing bones form the dermal girdle in rohu. The dorsal end of the clavicle is produced into a backwardly directed postclavicle and an upwardly directed stout supraclavicle. The supraclavicle is articulated with the post-temporal.

In rohu, it is situated behind the last branchial arch. The scapula is located in the dorsal portion of each girdle half, while the coracoid is located in the ventral portions. The junction of these two bones bears three facets: propterygium, mesopterygium and metapterygium (Fig. 15.14).

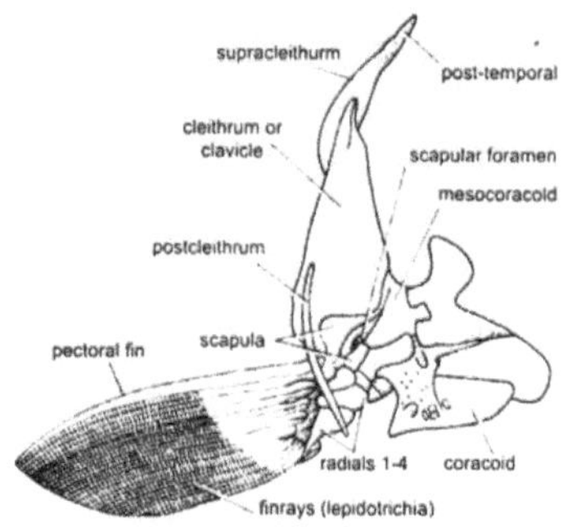

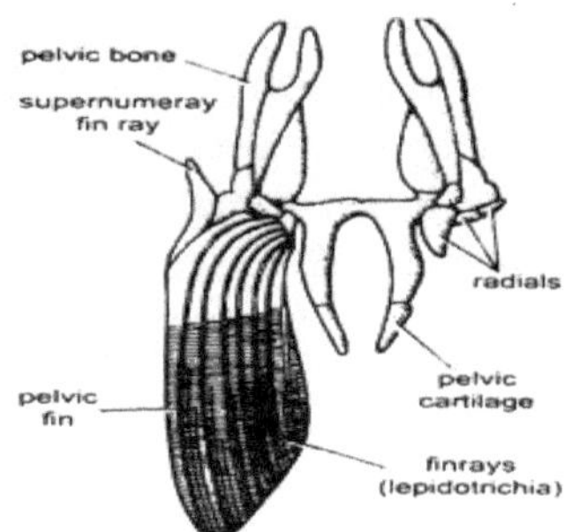

Fig. 15.14: Pectoral girdle of rohu **Fig. 15.15:** Pelvic girdle of rohu

(From https://www.notesonzoology.com)

A. Cleithrum (clavicle): It is the largest L-shaped in *W. attu* but triangular in *L. rohita*. The spine of the pectoral fin fits in its depression.

B. Postcleithrum: In rohu, post-cleithrum is a stout-curved, rod-like but absent in *W. attu.*

C. Supracleithrum: It is absent in *W. attu* but dagger-shaped in rohu.

D. Post-temporal: Its superior process articulates with pterotic while its inferior process is with basioccipital. It is deeply notched in *W. attu* to bear superior and inferior process nut not in rohu.

E. Scapula: It is a roughly triangular bone lying on the inner surface of the cleithrum. It is attached to the dorsal limb of the cleithrum, the other end is united with the coracoids and the third end is directed downward and backward and provides articulation to the pectoral fin.

F. Coracoid: It is L shaped with a short dorsal limb and a long ventral limb in *W. attu* but fenestrated in *L. rohita*. It is directed backward and provides articulation to the pectoral fin.

G. Mesocoracoid: It is curved in *W. attu* but inverted 'Y' shaped in rohu. Its one end is attached to the dorsal limb of the cleithrum where the scapula is attached.

The scapula and coracoids take part in the formation of glenoid articulation.

(II) Pelvic girdle: It is located anterior to the anal fin. It joins the fish skeleton to the pelvic fin. It contains two parts as anterior and posterior processes.

The pelvic girdle is free-floating and not connected to the spinal column or the pectoral girdle. The basipterygium is a curved piece of basal cartilage that supports the pelvic girdle's radials (Fig. 15.15).

15.5 Function of Skeleton

It provides a framework and support that surrounds and supports the body's soft organs. The heart, lungs, and brain are among the fragile organs that the skeleton protects. It aids in bodily movement, and muscles use bones as levers. The bone is a component of the skeleton and aids in the storage of minerals like calcium and phosphate. Bone marrow in the bones aids in the production of blood cells.

16

Locomotion and Swimming in Fishes

16.1 Introduction

Fish are primarily designed to swim in the water. Their fusiform body allows for efficient swimming. Swimming is the most cost-effective mode of fish locomotion because fish bodies are supported by water and thus do not require energy to counter gravity. To swim, a fish uses its fins. The caudal fin is primarily used for propulsion, while the remaining fins are used for balance control and fine maneuvering. However, some Plectognathi fish (Trunk-fish, Parrot-fish, Butterfly-fish, Porcupine-fish, and Trigger-fish, among others) have lost their ability to swim using body flexure and can only move with their other fins.

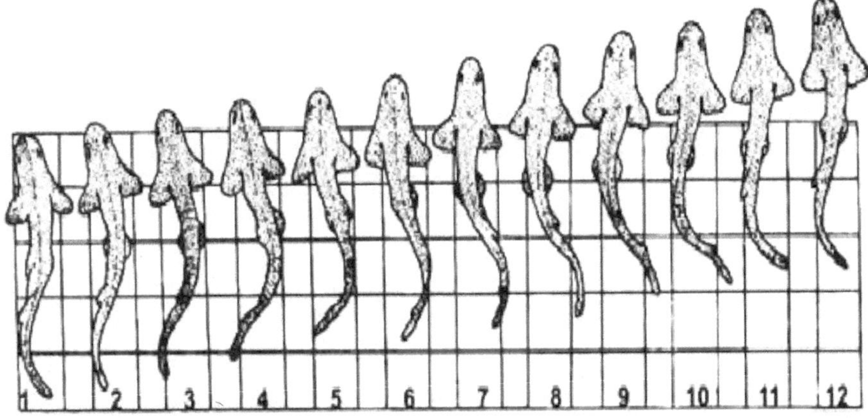

Fig. 16.1: Swimming in *Scoliodon* (from after Gray)

Sculling is possible with both the pelvic and pectoral fins. The skates and rays are the most extreme proponents of pectoral fin locomotion. Many fish, including tube-mouths (pipefish and seahorse) and pike (*Esox lucius*), rely on dorsal and anal fin undulations to move around.

The energy cost for a walking ground squirrel to travel 1 km/kg of body weight is 5.43 kcal, 1.45 kcal for a flying gull and 0.39 kcal for a swimming *Salmon*. The streamlined body of a fish provides the least resistance when moving through viscous water. Bony fish air bladders and shark fatty livers achieve nearly neutral buoyancy. Breder (1926) and Gray (1933) investigated fish locomotion.

16.2 Primary Methods of Swimming

Fish swim using one of three primary methods:

(a) By contracting and relaxing myomeres (=myosomes) alternately. Fish have a muscle power of about 0.001 HP/kg of body weight. The fish oscillates from side to side while swimming, displaying lateral undulations. This is the most widely used method. Metachordal contraction refers to the wave of curvature caused by myotome contraction that passes down each side of the body alternately from head to tail in *Scoliodon* (Fig. 16.1). During steady swimming, *Scoliodon* generates 54waves/minute.

(b) Through various fin movements.

(c) The gill aperture is opened by the sudden expulsion of water as a jet.

The tail and the caudal fin are the chief locomotory organs and are used for rapid swimming during which the tail is lashed from side to side by alternate contraction and relaxation of the myomeres on the two sides of the vertebral column. Slow movement takes place with the help of fins alone.

Aside from these three primary modes of movement, fish can also move by

(i) Climbing: The Indian climbing perch (*Anabas testudineus*) uses sharp spines on the lower parts of its gill cover to hold onto the ground while pushing itself forward with its tail, pectoral fins and operculum.

(ii) Crawling: A fish can crawl on the bottom of a body of water using their paired fins. Anglerfish (*Lophius*) crawls over the seafloor with its pectoral fins. Lizardfishes (Synodontidae) and gurnards (Trigilidae) love crawling across the ocean floor on their fins. Eels crawl on the bottom like snakes, using undulatory body movements

(iii) Flying: When a flying fish (*Exocetus*) or *Dachylopterus* is launched from the water by a powerful tail stroke, its pectoral fins expand and it floats or glides like an airplane similar to a parachute. Gliding and air travel have evolved in the families of Belonidae (Gar-fish), Dactylopteridae (Flying-gurnards), Exocoetidae (Flying-fish) and Hemirhamphidae (Half-beaks). Exocoetidae have greatly enlarged pectoral fins, which they keep folded up alongside their bodies while swimming but open up once they are out of the water. They also have asymmetrical dorsal fins, with the lower lobe being larger than the upper lobe.

(iv) In *Cypselurus*, the pectorals and pelvis are both enlarged for gliding.

(iv) Jumping: Mud skippers and gobies (Gobiidae) are excellent at jumping or hopping. Mullet (*Mugil* sp.) jumps to escape an enemy for food or simply for fun. *Salmon, Cyprinus* and *Magalops* leap from the water with great force to escape an enemy or cross a barrier. The flying fish (*Exocoetus*) leaps into the air using the powerful force of its tail, then glides for several meters in the air using enlarged pectoral fins.

(v) Skipping: Indian gobies (*Perophthalamus*) use their pectoral fins, which are bent at an angle similar to an elbow joint to hop over sandy flats left bare by the repeating tides

(vi) Tetrapod-like-walking: Lungfishes (*Protopterus* and *Neoceratodus*) move slowly and deliberately raising themselves on their unusual pectoral fins and posing for a while before moving their hard region sideways.

(vii) Walking: The lower positions of *Cephalocanthus* large pectoral fins are divided into 3–4 finger-like rays that they use for walking. The most skilled walker is the African Mud Skipper (*Periopthalmus papilio*). They have a strengthened girdle and extra muscles to help them walk by using the pectoral fins and the tail to support the end of the body. They can climb tree roots and stay out of the water for extended periods. They hunt terrestrial insects and surface-dwelling crabs with this unusual ability. Anglerfish (*Antennarius*) have well-developed and muscular pectoral fins that allow them to 'walk' on the bottom.

Clarias, Heteropneustes and *Channa* use their pectoral fins and spines to cross short distances on hand from one body of water to another to avoid drying. *Clarias* moves on land by using the tail's undulations.

The sucker fish has a sucker on the dorsal surface of its head and attaches itself to a large swimming fish or a boat allowing it to travel long distances without using its energy,

Swimming can be classified into three groups based on energy costs, according to Hoar and Randall (1978) (Fig. 16.2).

16.2.1 Sustained swimming

Sustained swimming can be maintained for long periods (>200 minutes) without muscular fatigue (Beamish 1978). Swimming speed is nearly 7 body length/sec and is maintained for long periods. Aerobic respiration provides the energy required by muscles; oxygen debt is not accrued because fatigue occurs slowly. It is used for large-scale foraging.

16.2.2 Burst swimming

It is used to avoid predators, pursue prey, or swim against water currents. High speeds of up to 20 body length/sec can be achieved, but can only be sustained for short periods. Anaerobic respiration, which uses up creatinine phosphate and muscle glycogen, generates power. Fatigue occurs quickly and can thus be sustained for short periods.

16.2.3 Prolonged swimming

In terms of speed and energy, it falls somewhere between the first and second types. Aerobic and anaerobic respiration both provide energy. Swimming sessions can last up to three hours, and longer sessions can cause fatigue. It is only used when the situation calls for it.

Fig. 16.2: Swimming in fishes

16.3 Formulation of Speed of Fish Movement

The absolute speed of a swimming fish is determined by its size and shape. Maximum speed for trout, herring or sardine-shaped fish is around 10 times body length/sec. Fish, on the other hand, can swim at a variety of speeds. In general, fish develop a speed that is proportional to their length and the frequency of their tail beats in the following manner.

$$V = \frac{1}{4} [L(3f-4)]$$

Where V = velocity (cm/sec),

L = length of the fish (cm)

f = the frequency (number) of tail beats/sec.

Table 16.1: Maximum speed of fishes

Sl.No.	Fish		Length (cm)	Maximum speed (km/hr)
	Common name	Scientific name		
1.	Sailfish	*Istiophorus*		96.0
2.	Swordfish	*Xiphias*		96.0
3.	Bonito	*Sarda*		64-80
4.	Barracuda	*Sphyraena*		43.2
5.	Sea trout	*Salmo truta*	60	22.5
6.	Sea trout	*Salmo truta*	30	10.8
7.	Sea trout	*Salmo truta*	20	8.1
8.	Pike	*Esox lucius*	16.5	7.2
9.	Dace	*Leuciscus leuciscus*	15	5.8
10.	Herring	*Clupea harengus*	20	5.4
11.	Goldfish	*Carassius auratus*	15	4.8
12.	Sea Stickleback	*Spinachia spinachia*	10	2.6
13.	Sand goby	*Pomatoschistus minutes*	6.5	0.9
14.	Blenny	*Zoarces*		0.81

According to Breder (1926), locomotion in fish can be of three types.

(I) Anguilliform (Eel-like) swimming (Fig. 16.3A): It can be found in highly flexible fish with elongated bodies. While swimming, fish create a deep sinuous wave. It is caused by the sequential contraction of the myotomes on each side of the body. Eels, lampreys, lungfish, and some sharks are examples.

(II) Ostraciform (trunk-fish-like or wig-wag) swimming: Fish that swim actively with the caudal fin but flex the fin in such a way that the body remains relatively stable. They contract all of the muscles on one side before moving on to the other. This beats the tail but does not cause the body to move in a sine wave pattern. The body is inflexible because it is enclosed in a hard boxlike protective sheath, and locomotion is caused by caudal fin undulations. Examples: *Ostracion, Tetraodon, Diodon,* Boxfish, Trunkfish etc.

(III) Carangiform (jack-like) swimming (Fig. 16.3C): This is the most common mode of fish locomotion. A large number of fish flex their bodies to an intermediate degree. Many sharks have an S-shaped body, but it is a shallow wave. Body undulations are restricted to the caudal region, as in trout, rohu, mrigal etc. However, Tuna may only flex the front half of its body (Fig. 16.3B).

Carangiform and anguilliform swimmers only contract a portion of their muscles on each side of their bodies at a time. They create the wave-like movements we see when they swim by controlling and varying the muscular activity on both sides of their body.

Balistiform movement: Some fish rarely flex their bodies and move forward by undulating their median fins. Complete waves are usually seen along the fins.

Swimming in a labriform pattern: Pectoral fins are used for locomotion by catfish, parrot fish, and surgeon fish.

Fig. 16.3: Modes of swimming A. Anguilliform, B. Subcarangiform, C. Carangiform and D. Thunniform (from twitter.com)

Powerful flight is used by the African Butterfly Fish (*Pantodon buchholizi*) and the Marbled Hatchet Fish (*Carnegiella strigata*). Their bodies are deep, with a lot of muscle support for their enlarged pectoral fins. They simply jump out of the water, gaining speed with their enlarged fins and extensive musculature while still in the water

Eels (*Anguilla* sp.), cuchia (*Amphipnous* sp.), snake-heads (Ophicephalidae), and several catfish (*Clarias* and *Saccobranchus*) have been observed crossing short land barriers between areas of water using the same actions they use when swimming

At low tide, they happily leap from one pool of water to the next in coastal, tropical environments. Sheep's-head Molly Miller (*Bathygobius soporator*) hops by curving up its tail and then straightening it suddenly while pushing it against the ground.

Dorsal and anal fins help globe fish (*Tetraodon*), porcupine fish (*Dindon*) and seahorses (*Hippcocapus*) swim. Furthermore, pectoral fins are used for slow movement

Amia, for example, has a very long dorsal fin that extends almost the entire length of the back. This fin can undulate and propel the body forward while swimming at a slow pace. Similarly, *Notopterus* and *Wallago* have very long anal fins that almost reach the tail fin and are used to propel the body forward.

17

Acoustico-Lateralis System

17.1 Introduction

Mechanical irritability is a general phenomenon exhibited by a cell. Apart from a general response to mechanical deformation, there are specialised receptors distributed internally as well as externally that act as mechanoreceptors. It includes the following:

1. Algesireceptor (Nociceptors): For pain.
2. Georeceptor: A gravity and acceleration receptor.
3. Phonoreceptor: Used for vibratory contact and sensation of hearing and also for equilibrium.
4. Rheoreceptor: Used for orientation in aquatic environments.
5. Tangoreceptor: For touch and pressure.

Mechanoreceptors may be primary or secondary. In the primary receptor, the stimulus directly acts upon the sensory nerve terminal whereas the secondary receptor communicates with the environment.

According to Northcutt (1980) and Jorgensen (1989), animals have sensory hair cells that act as mechanoreceptors.

Fish live in a wide variety of habitats with highly diverse hydrodynamic conditions, ranging from pools and lakes to slowly moving deep-sea currents and rough coastal surf regions. Fish can detect and perceive their hydrodynamic and physical environments, and use this sensory information to guide their behaviour via their mechano-sensory lateral-line system (Mogdans, 2019).

17.2 Acoustico-Lateralis (Octavo-Lateralis) System

The inner-ear region and lateral-line organs are located in the skin, forming a sensory system that conveys environmental information to the brain of a fish: the lateral-line organs respond to changes in water pressure and displacement, the inner ear responds to sound and the entire system is known as the acoustico-lateralis system.

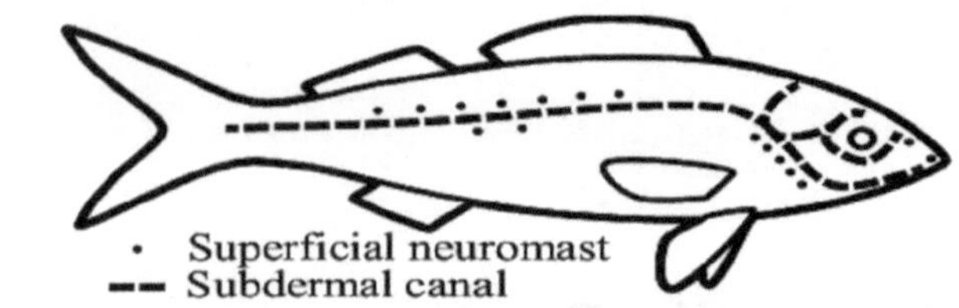

(a) Typical layout of the lateral line in a fish

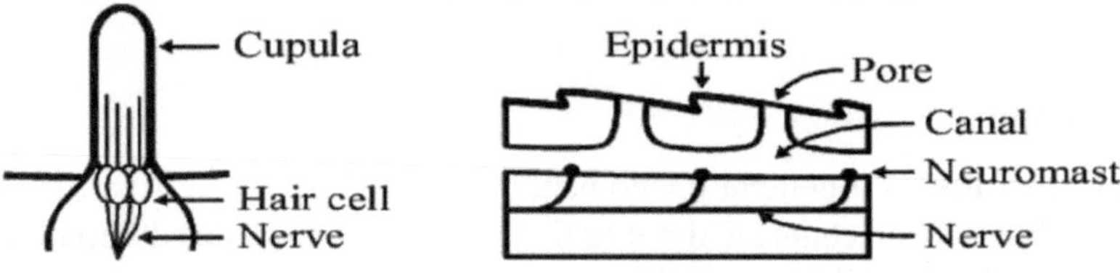

Fig. 17.1: Lateral line system of a fish (From https://www.researchgate.net)

17.3 Lateral Line System

The primitive mechano-receptor, phono-receptor or stato-receptor containing lateral line canals and neuromast organs is known as the lateral line system (mucus canal system). There are two types of lateral line systems: the head lateral system and the trunk lateral system. The neuromast's threshold for mechanical disturbances in the water varies greatly (Fig. 17.1a and 17.2).

The development of neuromast organs appears to be related to the modes of life and habitat of fishes. In free swimming and active fishes, it is well developed as the fish is exposed to water currents. In bottom living and inactive species like *Exox*, it is secondarily reduced but fairly developed in rapid stream swimmer (*Nemacheilus*). It is present on the head of blind cavefish and aids in the detection of prey, locating feed, and ingestion without sight.

Most amphibian larvae and some aquatic adult amphibians possess mechano-sensitive systems comparable to the lateral line.

Over the past 20 years, researchers have learned a great deal about the functional significance of the fish lateral line. Several studies indicate that fish do not need lateral line input for the control of locomotion (Liao et al. 2003); however, in highly turbulent waters fish use lateral line information to reduce their costs of locomotion (Liao 2006).

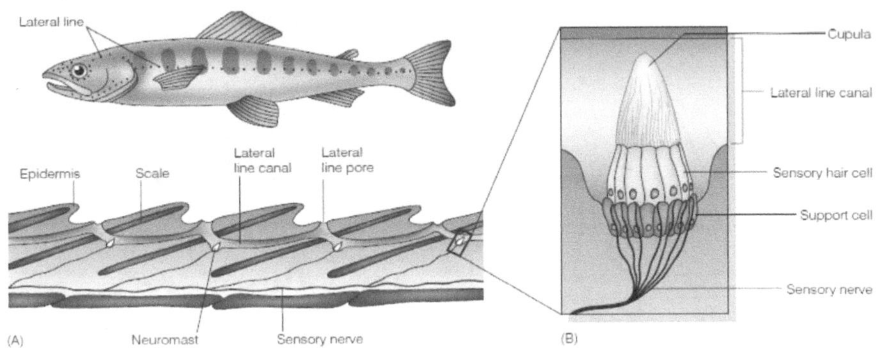

Fig. 17.2: Lateral line system of fish (From https://www.brainkart.com)

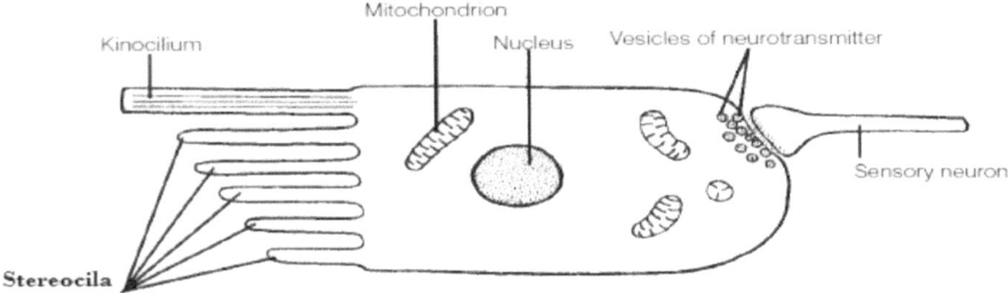

Fig. 17.3: Sensory cells of fish (From https://www.brainkart.com)

17.3.1 Neuromast organ

The neuromast organ can be found all over the skin's surface or it can sink into the grooves of the lateral line canals (Fig. 17.1b and c). Each neuromast contains a group of pear-shaped (=flask-shaped) sensory (=hair) cells along with slender and long surrounding and supporting (=sustentacular) cells. The sensory cells possess a bundle of cilia on the apical surface as hair cells (Fig. 17.3). Each hair cell consists of a globular basal body from one end which projects a series of 20-25 tiny sterocilia and a large kinocilium at one edge. The klinocilium extends into a gelatinous dome called a cupola, which is partially exposed to the external environment. The hair can be moved by surrounding water. Hair cells typically possess both glutamatergic afferent connections and cholinergic efferent connections. The receptive hair cells are modified epithelial cells and typically possess bundles of 40-50 microvilli which function as the mechanoreceptors.

The neuromast occurs freestanding on the skin (superficial neuromast) or in fluid-filled canals (canal neuromast) that are usually open to the environment through a series of pores. In different fish species, the organization of the peripheral lateral line can be quite different. For instance, some fish have fewer than 50 superficial neuromasts on each body side, whereas other fish species have up to several thousand superficial neuromasts distributed over the head, trunk and tail fin (Schmitz et al. 2008) (Fig. 17.4A and B).

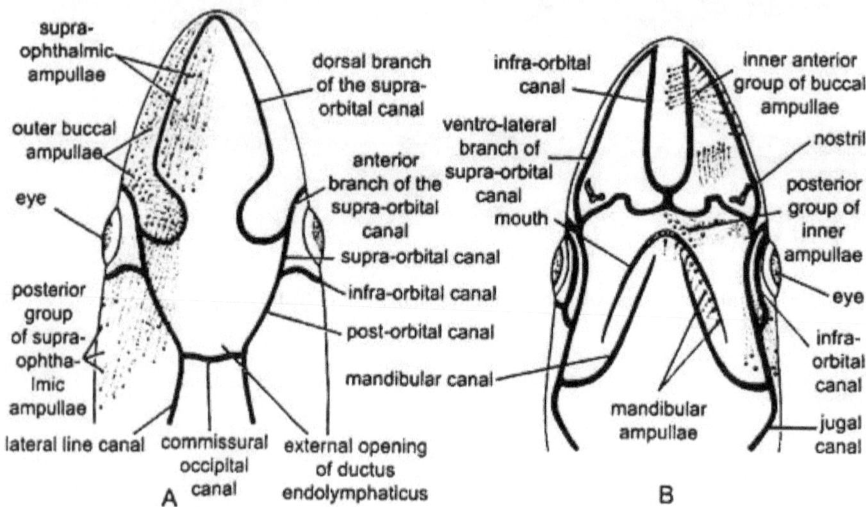

Fig. 17.4: Neuromast organs of *Scoliodon* (From https://www.notesonzoology.com)

If the neuromast is enclosed in a canal, the cupula (=gelatinous tube) is moved by mucus. Histologically, the wall of the neuromast contains simple cuboidal epithelium. A large number of mucus cells are present in the wall but absent in *Notopterus chitala*. The sensory cells are similar to the auditory organs.

The neuromast organ is regarded as the receptor unit of the lateral line system.

Functions: It performs the functions of mechanoreceptors. It helps in detecting the prey, avoiding the prey, schooling and intraspecific communication. It also acts as a hydrodynamic detector and distant touch receptors. The cupula moves freely when the water surrounding it is set in motion.

17.3.2 Lateral line canals

During development, the lateral line canal differentiates as grooves along the longitudinal axis on the dorsal, ventral and lateral sides. In most of the teleosts, the number and distribution of cephalic canals may show a loss, new development or a change in occurrence.

Lateral line canals open to the surface through regularly spaced pores and contain a fluid (=endolymph). The number of hair cells beneath a cupula varies from 6-10 in an epidermal organ. Structurally, each neuromast is similar to a hair cell from the crista.

The neuromast is found all over the fish's body and arranged in a definitive pattern on the head. On each side of the body, they exist in a line from head to tail (Fig. 17.5).

1. The canals are as open grooves in *Chlamydoselachus*.
2. The canals are opened but closed in the head in *Protopterus*.
3. In holocephalians, the grooves remain open even in the cephalic region.
4. The canals of teleosts and chondrichthyes lie deep in the skin and their path can only be traced by the distribution of pores.

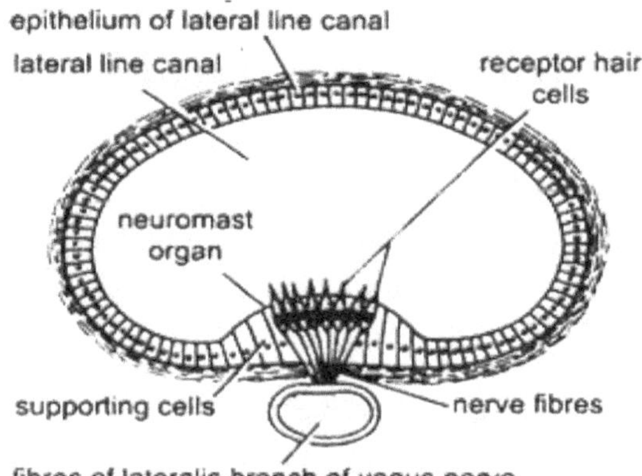

Fig. 17.5: T.S. of lateral line canal through neuromast organ
(From https://www.notesonzoology.com)

Generally, the main lateral line canal pierces very deeply along its course. The head region is divided into:

1. The supraorbital canal above the eye to reach the snout is supplied by the superior thalamic of the facial (VII) nerve.
2. The hyomandibular canal behind the eye and to the mandible is supplied by the buccal branch of the facial (VII) nerve
3. The infraorbital canal below the eye and to the snout is supplied by the hyomandibularis of the facial (VII) nerve
4. The temporal (transverse occipital) canal is supplied by the supra temporal of vagus (X) nerve
5. Mandibular canal
6. Preopercular canal

Lateral line canals are supplied by various branches of VII, IX and X nerves.

The cephalic canals are reduced in cyprinids but absent in Cobitis. In these fishes, supra-orbital and infra-orbital are not joined behind the eye. A mandibular canal may be absent but a supra-temporal cross-commiussural canal is present. Cyprinids have a large number of neuromasts to siluroids.

In *Notopterus chitala*, the main lateral line canal is not continued into cephalic canals but forms saccular dilatations. Closed cephalic canals are found in *Gymnarhcus niloticus, Lota lota* and *Neoceratodus*. The sensory canals of *Channa punctatus* have a water lumen and direct openings to the outside. In *Xenentodon*, the lateral line canals open to the exterior by pores. A complex pattern of the lateral line is present in the head region of *Hilsa ilisha*.

The lateral line system arises from skin placodes on special thickenings. In the course of development, the dorsal and ventral canals in the trunk region disappear. The body canal is extremely reduced in *Rhodeus, Mugil* has many parallel lateral lines.

17.3.3 Specialisation of lateral line canals

(a) **Sharks:** Lateral line canals run along the entire length of the body. The supraorbital canal is much developed in the rostrum and joins the infraorbital canal which may pass forward between the mouth and the nostril. The hyomandibular fails to meet postorbital and the mandibular is completely separated from the hyomandibular (Fig. 17.4).

(b) **Holocephali:** The lateral line canals remain open and their neuromast lies within the open canals.

(c) **Actinopterygii:** The occipital transverse commissure is complete across the middle line. The supraorbital and infraorbital are normally developed. The hyomandibular joins the postorbital dorsally but s does not run across the check to the infraorbital. The oral canals start from the infraorbital in the absence of jugal.

(d) **Dipnoi:** The preopercular fails to reach the postorbital but the jugal is present as usual and the oral canal remains in continuation with the infraorbital.

Functions: A simple neuromast organ is used as a distant "touch receptor". It mainly detects and locates moving animals as well as intimate mobile and immobile objects. It helps in the detection of prey, avoiding the prey, schooling, orientation and intra-specific communication. Its cupula, changes the dereliction by applying pressure in all directions. It builds up a "bow-wave" of pressure when a fish is moving. It helps in "echolocation" in darkness or a turbid medium. It can receive sound waves. It acts as a thermoreceptor and baroreceptor also. It reports vibration disturbances at 100-200 cycles/second. It detects particle velocities and accelerations with frequencies below 100 Hz.

In teleosts, specialised neuromast organs serve as an additional sensory organ system and are used to detect electric stimuli by intimate sources. It helps in information about the neighbour, the distance, sex, activity etc. It seems to detect water underflow from head to tail in the reverse direction (rheotaxis).

This behaviour is diminished when lateral line function is inhibited by cobalt chloride application. Cobalt chloride treatment disrupts ionic transport and prevents signal transduction in the lateral lines.

17.3.4 Phonoreceptor (The Internal Ear or Membranous labyrinth)

Fishes lack external and middle ears. The inner ear is a fluid-filled structure attached to three semicircular cartilaginous tubes. These tubes are set at right angles to one another and are lined with hair cells. Each semicircular tube responds only to accelerations within the plane parallel to its orientation. Thus, collectively, the three semicircular tubes grant a simultaneous sense of its movements in all three dimensions of its liquid environment. However, these tubes are not considered to be involved in sound perception (Carrier et al., 2004) (Fig. 17.6).

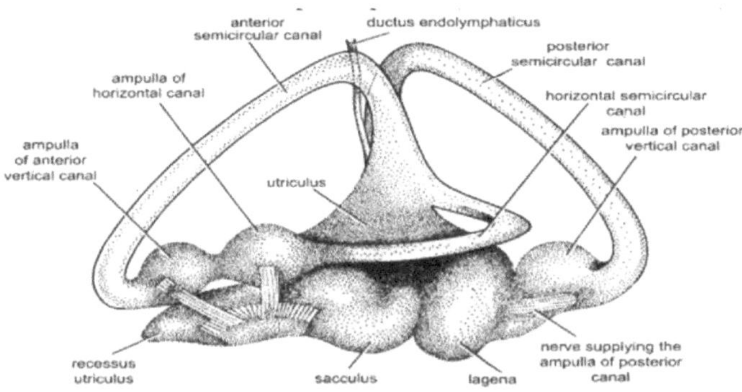

Fig. 17.6: Internal ears of *Scoliodon* (From https://www.notesonzoology.com)

The inner ear is represented by a membranous labyrinth in fishes. The membranous labyrinth contains a membranous sac in a chamber on either side of the posterior of the skull and is completely closed by otic bones. The sac is partially divided into the upper utriculus and lower sacculus.

The saccule, lagena and utricle are three sensory areas probably involved in balance and sound perception. They consist of a path of sensory hair cells on an epithelium is overlain by an otconial mass. The otoconia (Otolith) is made of calcium carbonate granules embedded in a mucopolysaccharide matrix that acts as an inertial mass (Tester et al., 1972). In other fish, the otolith is responsive to accelerations produced by a sound field, which accelerates the sensory macula relative to the otoconial mass. These otoliths, therefore, respond to gravity, providing information about their orientation in the water, be it head up, head down, on its side, right-side-up or upside down.

Membranous labyrinth has many structural peculiarities in elasmobranches. A small tube arising from the dorsal side of the sacculus opens to the surface of the head. The duct is known as an invagination canal homologous to the endolymphaticus duct of bony fish. In these fishes, seawater is filled with in place of endolymph. It may function as a vibration receptor. It is a statocyst that is responsible for detecting a particular movement and direction.

17.3.5 Macula Neglecta

Jeffrey Corwin (1981) reported that in some sharks one of these otolith-equipped parts known as the *macula neglecta*, respond particularly strongly to vibrations through the top of the skull. Based on his functional morphology studies of many shark species, he proposed that the *macula neglecta* may provide actively predatory sharks with an enhanced ability to hear sounds originating from above and in front.

17.3.6 Central Pathways into the Central Nervous System

The ear is innervated by the VIIIth cranial nerve. Studies of afferent connections and the physiology of the VIIIth nerve forms individual end organs (saccule, lagena, utricle and the macula neglecta) show projections ipsilaterally to five primary octaval nuclei: magnocellular, descending, posterior and anterior and periventricular (Corwin and Northcutt, 1982).

17.4 Relationship of Lateral Line System and Internal Ear

The internal ear originated phylogenetically as specialised sunk part of the lateral line system. They have the following relationships:

1. They develop from similar dorsoventral placodes.
2. They are histologically similar and have sensory cells.
3. They are innervated by fibres from corresponding and related centers in the brain.
4. There is a peculiar connection between the lateral line canal, swim bladder and the ear in some cases.

17.5 Modifications of Neuromast Organs

17.5.1 Ampulla of Lorenzini

It is an additional modification of the lateral line system in elasmobranches and *Plotossus anguillaris*. It forms a complicated system of jelly-filled canals over the snout and the head. The canals end in swollen bulbs containing sensory cells. It contains a long tube terminating into 8-9 ampullary sacs radically disposed around a centrum. The sacs bear sensory as well as glandular cells. The jelly does not give a positive reaction to mucus (Fig. 17.7).

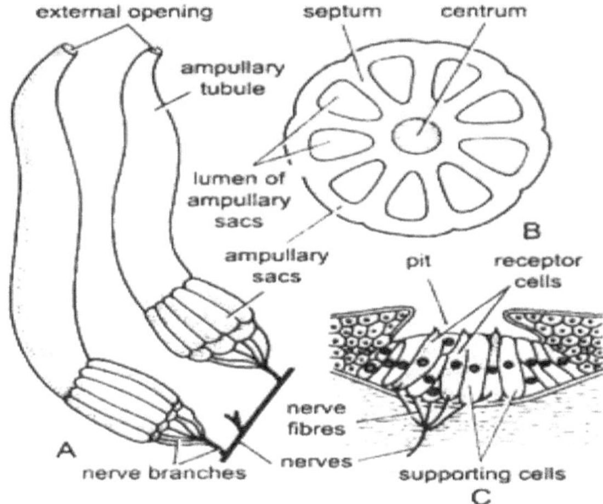

Fig. 17.7: Ampulla of Lorenzini (From https://www.notesonzoology.com)

Sand (1938) showed that it works as a thermoreceptor (range of 3-25^{0}C), hydroreceptor etc. Murray (1957) observed that it is a mechanoreceptor or better electroreceptor and baroreceptor. Kalmin (1971) demonstrated their role in detecting food buried in sand by electrolocation.

The ampulla is divided into several secondary ampullae but it is undivided in *Torpedo* and *Plottosus anguillaris*. Histologically, ampullae contain pyramidal and flask cells. These cells function as secretory and sensory cells, respectively. The spontaneous firing of the ampulla is highly sensitive to temperature changes as small as 0.05^{0}C. The sensory cells of the ampulla are supplied by axons of the facial (VIIth) nerve.

17.5.2 Pit Organs

They are ectodermal individual neuromast organs embedded into small pits in the skin. They are found on the dorsal and ventral surfaces of the head region. They are found on the dorsal surface of ray-fishes. Beneath these, nerve fibres are also present. They are frequently arranged in rows in the head and the rest of the body. Each pit organ is bud-like, round or oval and arises as a modification of the epithelium. Each pit organ is composed of two rows of cells. The outer row contains supporting cells and the inner row has sensory cells. It is supplied by the lateralis branch of the vagus (Xth) nerve. Pit organs seem to be additional organs of current receptors and distant touch.

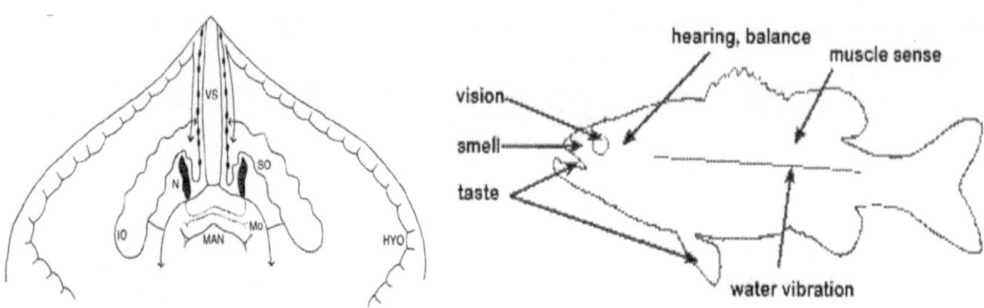

Fig. 17.8: Organ of Savi **Fig.** **17.9:** Sense organs of fish

(From https://www.researchgate.net) (From https://oceanadventures.co.za)

17.5.3 Vesicles of Savi

These are closed sacs lying on the ventral surface of *Torpedo*. It is a closed sac cut off from the rest of the epidermis. It contains gelatin-filled follicles encircling grey amorphous materials. It is supplied by branches of the Vth cranial nerve. It is responsible for the response to mechanical pressure applied to fish (Fig. 17.8).

17.5.4 Sensory Crypts

Sensory cells and supporting cells coexist in sensory crypts. They are supplied by branches of the Vth, VIIth and lateralis nerves. It is present in many elasmobranches and distributed in their head and trunk regions. They are thought to be gustatorecepetor (Fig. 17.9).

17.5.5 Organs of Fahrenholz

Dipnoi larvae have pits in their heads. They bear haired sensory cells. However, its function is still unknown.

17.5.6 Spiracular Organ

These are present in chondrichthyes, holostei and chondrostei. They show peculiar types of sense organs of unknown function. They are sensory and lateral line organs.

17.5.7 Mormyromasts

They are found in Mormyrids and gymonotid eels. It is held to be a component of the lateral-line sense organ. It is probably an electro-sensitive organ. It contains two types of electroreceptors. One of them resembling neuromast is known as mormyromast.

They assume to respond to electric impulses of greater frequencies compared to the ampullae of Lorenzini.

18

Digestive System in Fishes

18.1 Introduction

The digestive system of fish shows remarkable diversity in its morphology and function, related to both taxonomy and different feeding habits (Abdulhadi, 2005). Fish eat a wide variety of foods. Some feed on plants, others on small fish, and the remainder are omnivorous. The digestive system is drastically altered as a result of different feeding habits. In response to feeding habits, the structural plan of the buccal cavity, pharynx, and rest of the gut has changed.

In the absence of basic food, obligatory or emergency food is sometimes consumed. Food for fry and fingerlings is generally different from that for adults. Young ones with short and small intestines prefer zooplankton over algae and phytoplankton.

The anatomy of the digestive tract of fish varies greatly due to both evolutionary degrees and different types of feeding. Histologically, from the cranial end to the caudal end of the digestive tract, the wall of the alimentary canal is formed by four distinctive layers. Starting at the lumen, these layers are: mucosa, submucosa, muscularis externa and serosa (Kierszenbaum and Tres, 2012) (Fig. 18.1).

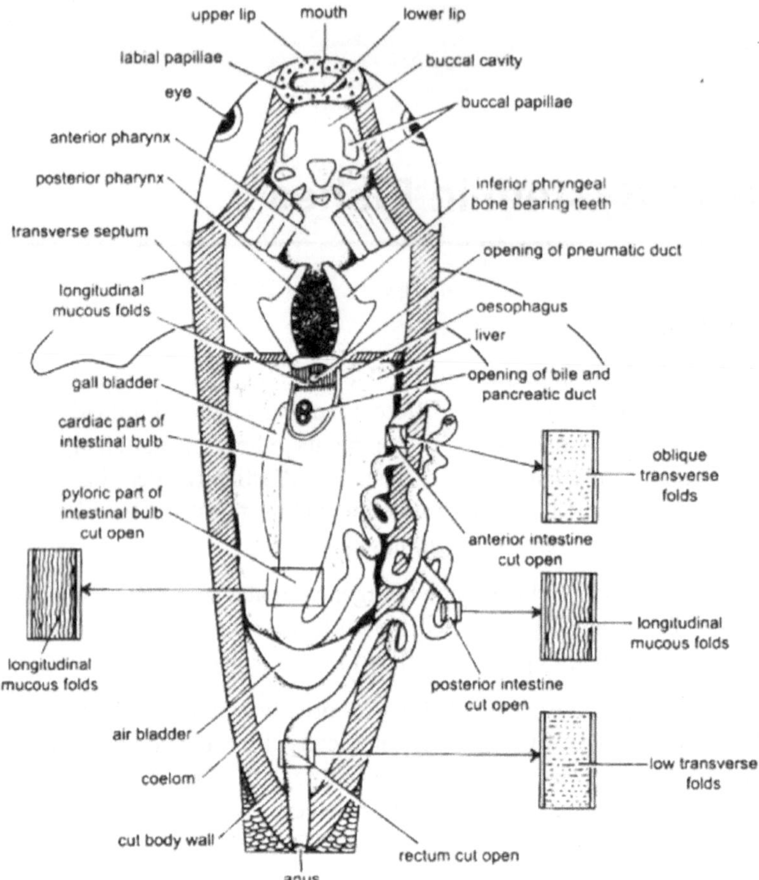

Fig. 18.1: Digestive system of rohu showing folds of mucous lining (from from https://www.notesonzoology.com)

18.2 Types of Fishes Based on Food and Feeding Habits

18.2.1 Because fish consume a variety of foods, their feeding habits differ greatly between species. Fish consume the following four types of food, according to Nikolskii (1963):

1. **Basic food:** It is found in large quantities in fish diets. The majority of fish prefer it.
2. **Obligatory food:** It is consumed if basic food is in short supply.
3. **Secondary (occasional) food:** It is consumed in small quantities by fish linked to basic food.
4. **Incidental food:** Fish rarely consume it.

18.2.2 Fishes can be classified into the following types based on their food characteristics:

1. Plankton feeders: These fish eat phytoplankton (algae, microscopic plants, decaying microvegetation) and zooplankton (rotifers, insects and small crustaceans).

Examples: *Catla catla, Cirrhinus mrigala, Gadusia chapra, Hilsa ilisha, Hypophthamicthyes molitrix, Labeo rohita, Scomber japonica, Carassius* sps.

The digestive system consists of the alimentary canal and digestive glands. It mainly incorporates food, feeding, digestion, absorption and elimination of undigested wastes.

2. Herbivorous fish: These fish eat a variety of vegetation, including leaves, stem fragments, flowers, unicellular or multicellular algae, organic debris, decaying organic matter, mud, weeds and aquatic vegetation. They can be found between the latitudes of 40^0N and 40^0S. They have a short oral cavity, a blunt snout, and many teeth that can crush, scrape, and even dig. Their relative gut length ranges between 8.0 and 20.0.

Examples: *Ctenopharyngodon idella, Cyprinus carpio, Hypophthamicthyes molitrix, Labeo rohita, Labeo gonius, Labeo bata, Labeo niloticus* (RGL: 16.9), *Tilapia mossambica, Oreochromis niloticus, Osphronemus goramy,* surgeon fishes, rabbit fishes and parrot fishes.

3. Carnivorous fish: These fish consume animal-derived materials such as mollusks, crustaceans, fish fry, fingerlings, small fishes, animal matter, detritus and so on. They have a 0.2-2.5 relative gut length.

Examples: *Clarias batrachus, Channa argus, Channa striatus, Channa marulius, Glyptothorax pectinopterum, Mysts seenghala, Notopterus chitala, Oncorhynchus mykiss, Samo salar, Wallogo attu.*

Carnivorous fish are classified into three subtypes:

(a) **Zooplankton eaters:** They can filter enough water through their digestive tract to consume zooplankton. They are species endemic to India and one of the most important in polyculture.

    Example: *Catla catla.*

(b) **Benthonic invertebrate eaters:** These include corals and cichlids from Africa's Great Lakes.

    Examples: Mobile benthos hunters (Lutjanidae, Holocentridae), crushers (Labridae) and probes capable of extracting their prey from coral-like forests (Sygnatidae).

(c) **Piscivorous:** They swallow their prey whole and have large mouths with pointed bills to keep their prey from fleeing. The digestive tract is distinguished by a true stomach and a short intestine. They play an important role in industrial aquaculture.

Examples: Trout (*Oncorhynchus mykiss*) and salmon (*Salmo salar*)

4. **Omnivorous fish:** These fish consume food that is both plant and animal in origin, such as sand and mud. They have a 6.0-8.0 relative gut length.

   Examples: *Anabas testudineus, Barbus carnaticus, Barilius bendelisis, Barilius barana, Barilius barila, Catla catla, Cirrhinus mrigala* (RGL: 8.0), *Cyprinus carpio, Heterropnesustes fossilis, Noemacheilus montanus, Puntius chilinoides, Tor tor, Tor putitora.*

5. **Detritivorous:** They eat decaying and inert organic matter at various stages of degradation at the bottom of lakes, ponds, and certain marine habitats. They have a poorly developed digestive tract with no stomach and a long intestinal tract, similar to herbivorous species.

   Example: flathead grey mullet (*Mugil cephalus*).

18.2.3 Fish can be classified into the following types based on how they capture and consume food:

1. **Grazers (Browsers):** These fish feed either singly or in a series of bites. By elongating their jaws, they can transform their mouths into long flutes.

   Examples: Blue gill, parrot fish, butterfly fish, column feeder

2. **Strainers:** Using their gill rakers, these fish filter a large amount of water.

   Examples: Basking shark, Whale shark, *Catla catla, Cirrhinus reba, Gadusia chapra, Hilsa ilisha.*

3. **Suckers:** By enlarging their buccal and opercular cavities, these fish suck food materials into their buccal cavity. By elongating their jaws, they can also transform their mouths into long flutes.

   Examples: *Anguilla anguilla, Idusidus, Tilapia mariae, Carassius auratus, Macropodus opercualris.*

4. **Parasites:** These fish get their nutrition from other fish and are then adopted in various ways.

   Examples: Lamprey, Hagfishes, Suckerfish, Anglerfish.

18.2.4 According to Nikolskii (1963), fish can be classified into the following types based on their diet:

1. **Euryphagic fishes:** These fish eat a variety of foods. Example: Mud minnows
2. **Monophagic fishes:** These fish only eat one type of food.
3. **Stenophagic fishes:** These fishes only eat certain types of food. Example: *Chaetodon trifasciatus.*

18.2.5 Based on the trophic niche, fish may be of the following types:

1. **Surface feeder:** Because of the presence of a mouth on the dorsal side, these fish feed on plankton floating on the surface of the water.

   Examples: *Catla catla, Chanda nama, Chanda ranga, Puntius ticto, Gadusia chapra, Hilsa ilisha, Hypophthalmichthyes molitrix, Oxygaster bacalia.*

2. **Column feeders:** These fish feed along the longitudinal column of an aquatic body due to the presence of a mouth on the terminal side.

   Examples: *Labeo rohita, Puntius sophore, Mystus cavassius, Mystus vittatus, Tor tor, Wallago attu*

3. **Bottom feeder:** These fish feed at the bottom of the body of water due to the presence of a mouth on the ventral side. They also consume a small amount of mud.

   Examples: *Amblypharyngodon mola, Cirrhinus mrigala, Cirrhinus reba, Channa marulius, Channa striatus, Labeo calbasu, Labeo bata, Labeo gonius, Puntius sarana.*

18.3 Alimentary Canal

The digestive system is made up of the alimentary canal and digestive glands. It primarily consists of food digestion, absorption, and waste elimination (18.2A to C).

The archenteron of a fish gives rise to the midgut, which joins to form the complete digestive tract during development. The anterior invagination of the midgut forms the foregut, while the posterior invagination forms the hindgut.

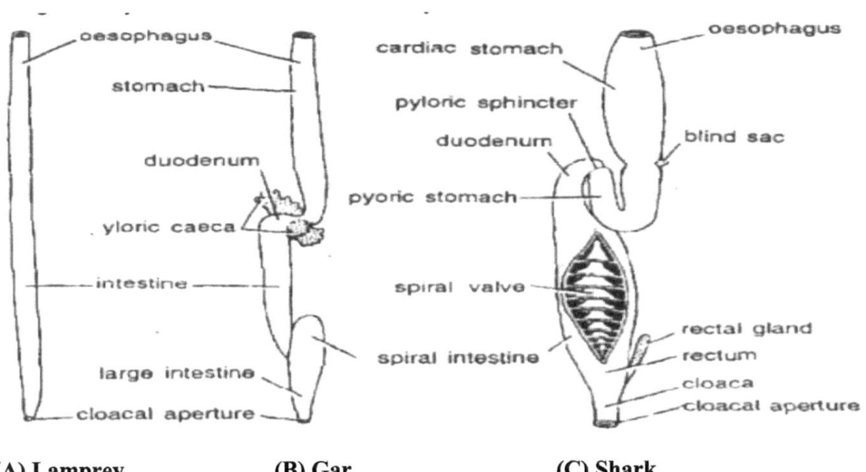

(A) Lamprey **(B) Gar** **(C) Shark**

Fig. 18.2: Alimentary canal of fishes (from https://www.researchgate.net)

The alimentary canal starts in the mouth and ends in the anus. The mouth opens into a large buccal cavity. The mucous membrane lines the buccal cavity. The alimentary canal is roughly divided into the bucopharynx, oesophagus, stomach, intestine, and rectum. It is divided into four histological layers: serosa, muscular layer, submucosa, and mucosa.

The alimentary canals of *Scoliodon* and *Clarias batrachus*/*Mystus seengala* can be considered typical examples of chondrichthyes and teleost alimentary canals. The variations are seen in the position of the mouth, structure of the buccopharynx, relative length of the gut, presence or absence of the stomach and pyloric caeca.

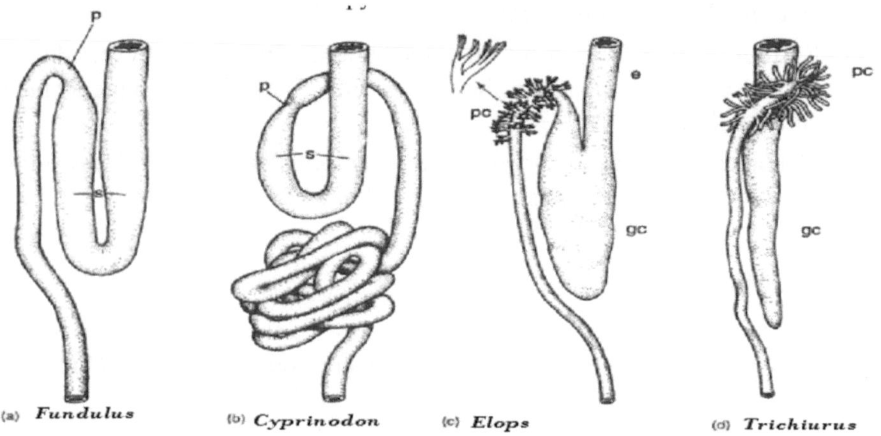

(p = Pylorus, s= stomach, o= oesophagus, gc= caecum like stomach, pc=pyloric caeca)
Fig. 18.3: Alimentary canal of fishes (from https://www.researchgate.net)

18.3.1 Mouth

In Cyclostomata, the mouth is circular and contains a suctorial disc. Predatory charcinoides have mouths on their superior positions. A long beak like mouth is present in *Fistularia villosa, Forcipiger longirostris, Syngnathus, Xenentodon cancila*. But *Hemiramphus gorakhpurensis* has a half-beaked mouth. *Anabas testudineus, Channa* species and *Nandus nandus* all have protractile mouths.

The mouth of fishes is usually guarded by the lips. Lips are suctorial in *Cirrhinus, Labeo, Puntius*, sturgeon and sucker fishes. It acts as a holdfast organ in myxine, sisroid catfishes and armoured catfishes. In various carnivorous and omnivorous fish, the lips are either thin or reduced.

18.3.2 Buccopharnyx

The buccopharynx is the first part of the alimentary canal to perform (a) food collection and transport to the oesophagus as well as (b) respiration.

The nasal cavities are quite separate from the oral cavity except for the choanichthyes. The oral cavity lacks glands except the simple mucous glands opening into it.

The tongue is just a fold that develops from the floor of the buccal cavity. The sensory receptors may be present in the tongue which lacks muscles. In rays, the tongue may be missing.

18.3.2.1 Dentition: The teeth are modified according to the purpose, they are used. The location of teeth also varies greatly. Besides their normal location in the jaws, they may be present on the tongue and the hyoid arch. Most dentition is of the polyphiodont, homodont, lyodont and usually acrodont types. Simple conical teeth bearing longitudinal grooves and ridges are present in extinct crossopterygians and *Latimeria*. It may be modified into vertically flattened triangular plates as seen in sharks. In *Chimaera* and Dipnoi, plate-like structures are formed by the fusion of several teeth to crush molluscan shells. In teleosts, the teeth are recurved to prevent the escape of slippery prey. Teeth may be large in deep-sea fishes.

In carnivorous and predatory fishes (e.g., *Wallogo attu, Mystus seenghala, Mysutus aor, Channa marulius, Channa punctatus Channa striatus, Notopterus chitala, Notopterus notopterus, Harpodon nehereus, Muraenesox telabon*), it is armed with strong teeth in its various parts and even the gill rakers are tooth-like.. It is divided into the buccal cavity and pharynx. Teeth are borne by the premaxilla, maxilla, vomer, palatine and denary. Teeth are present on the tongues of *Notopterus chitala* and *Notopterus notopterus*. Pharyngeal teeth are known as superior and inferior pharyngeal teeth. Teeth may be villiform, incisiform, canine-like, molariform or blunt knobs. They are generally sharp, pointing backward and unmasticatory. The teeth in the anterior region serve to prevent the escape of prey from the mouth. The pharyngeal teeth act as a rasping organ.

In herbivorous fishes (e.g., *Labeo rohita, Labeo gonius, Labeo calbasu* and *Cirrhinus mrigala*), is divided into the anterior respiratory and posterior masticatory parts. These lack teeth except the pharyngeal one. They are produced by the fifth cerebro-branchials and work against a dorsal, hard, and calloused pad. These teeth are used to crush prey.

The plankton feeders (*Hilsa ilisha, Gadusia chapra*) may lack teeth in the buccopharynx. However, *Atherina forskali* and *Hilsa filigera* possess minute teeth on the jaws.

Predaceous fish (*Esox, Lepidosteus* etc.) have sharply pointed teeth for grasping, puncturing and holding prey. Other such fish (*Serasalmus*) have razor-like cutting teeth.

18.3.2.2 Gill-rakers: Gill rakers are present on the inner side of each gill arch to protect gill filaments from injury. In carnivorous fishes (e.g., *Channa marulius,*

Channa striatus, Harpodon nehereus, Heteropneustes fossilis, Mystus seenghala, Notopterus chitala, Wallogo attu), gill rakers are long, hard and teeth-like rasping organs but are reduced in *Trichurus* or completely absent in *Muraenesox.*

In herbivorous fishes (*Cirrhinus mrigala, Labeo rohita, Labeo gonius*), gill rakers form a broad, sieve-like structure across gill slits to filter the water and retain the food in the bucco-pharynx.

Gill-rakers aid in the filtering of food and prevent it from escaping from the buccal cavity via the gills. The filtering mechanism is best developed in plankton feeders such as *Hilsa ilisha, Catla catla* and *Gadusia chapra* whose gill rakers are fairly long. However, gill rakers have degenerated in *Boleophthalamus* and *Syngnathus.*

Therefore, it seems that gill rakers exhibit a structural adaptation concerning the feeding habits of the fish.

In *Hilsa, Gadusia* and *Chanos*; paired epibranchial organs are present on the roof of the pharynx above the posterior branchial arches. The lumen of this organ seems to be an extension of the pharyngeal cavity and contains a spiral tube with rakers and a blind sac.

In *Tor tor, Puntius sarana* and *Puntius ticto*, gill rakers are short and stumpy.

There is a correlation between feeding habits and the structure of gill rakers. The filtering efficiency increases considerably in carnivorous and omnivorous fish, reaching its maximum in herbivorous fish.

18.3.2.3 Taste-buds and Mucus-secreting: Many carnivorous and predaceous fish like *Channa striatus, Channa gachua* and *Mystus vittatus* have sight and taste buds, but they are rare. Some fish (*Catla catla, Cirrhinus mrigala, Puntius sophore, Schizothorax plagiostomus, Tor tor*) depend more on their gustatory faculty for feeding and possess a large number of taste buds. A soft cushion pad is present on the roof of the buccal cavity in herbivorous and omnivorous fishes. This pad has many papillae with taste buds and mucus-secreting cells.

The buccopharynx is lined by mucosa, submucosa and thin muscularis with scattered longitudinal muscle fibers. The mucosa is formed by several layers of stratified epithelial cells including the deepest layer, the Malpighian layer. The mucosa contains large, club-shaped cells, flask-shaped mucous cells and a few taste buds, especially in the hindquarter region. The mucosa of *Labeo rohita, Tor tor* and *Catla catla* also bear taste buds. The submucosa contains loose connective tissue and blood capillaries. The mucosal folds are prominent in the region of the gullet.

18.3.3 Oesophagus

There is no distinct differentiation between the oesophagus and the stomach. It is meant for the reception of large pieces of food and is primarily modified to produce acid and pepsin. It is a short and narrow tube in many herbivorous and carnivorous fishes (*Cyprinus carpio, Labeo rohita, Labeo calbasu, Catla catla, Puntius sophore* and *Tor tor*). In *Labeo,* the ductus pneumaticus of the swim bladder opens into the oesophagus. Carnivorous and predatory fishes, like *Channa, Mystus seenghala* and *Wallago attu*; which feed on large prey possess a longer and distensible oesophagus (Fig. 18.3).

Oesophagus has prominent mucous folds lined by columnar epithelium. Many mucous cells are present in the mucosa. The submucosa is thinner and contains scattered longitudinal muscles arranged in bundles. The muscularis is composed predominately of circular muscle fibers. The serosa is thin.

18.3.4 Stomach

According to Barrington, the stomach is distinguished from the oesophagus only by differences in mucosal folds (thin in the oesophagus and thick in the stomach). It is generally sac-like and thick-walled in carnivorous and predaceous fishes (*Wallago attu. Mystus seenghala, Channa striatus* and *Notopterus chitala*). It is reduced in *Hilsa ilisha, Mugil carsula* and *Gadusia chapra* etc. but greatly thickened to become a gizzard-like distribution of food (Fig. 18.3).

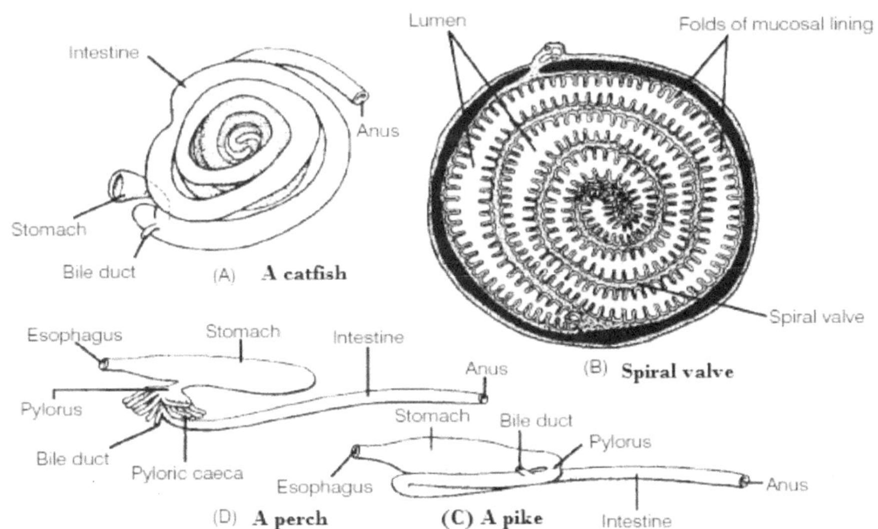

Fig. 18.4: Variation in length of intestine (from https://www.brainkart.com/article)

The stomach is absent in many fishes. The anterior part of the intestine swells to form a sac behind the oesophagus in such species. This structure serves to store food and is known as an intestinal bulb or swelling (*Barbus sarana, Catla catla, Cirrhinus mrigala, Labeo rohita. Labeo gonius, Puntius sophore, Tor tor*). The intestinal bulb stores food but lacks gastric glands. In *Scoliodon*, the pyloric stomach opens behind into the bursa entima guarded by the pyloric valve. The presence or absence of a true stomach is not related to the feeding habits of the fish. The stomach is absent in the Holocephali and Dipnoi too. According to Barrington, the absence of a stomach might have been brought about by neoteny. *Scomberesox, Xenentodon cancilla,* and the *Hippocampus* lack stomachs as well.

The significance of the division of the stomach into a descending cardiac part and an ascending pyloric part is unknown. The stomach in elasmobranchs is usually J-shaped, but in teleosts, it is V-shaped with a prolonged cardiac part as a blind pouch. Originally, it was meant for the reception of large pieces of food and secondarily, it became modified to produce acid and pepsin.

In *Protopterus*, the spleen is composed of vascular tissues and is attached to the right dorsolateral wall of the stomach.

The stomach has prominent folds in the mucous and the muscularis is very thick. The mucosa is raised into prominent folds. Numerous simple tubular gastric glands are present and open into the lumen of the stomach. The mucosa of the intestinal bulb resembles the intestine and lacks gastric glands.

18.3.5 Intestine

The length of the intestine generally varies in different fishes. It is short and nearly straight in carnivorous fishes but long, thin-walled and highly coiled in herbivorous species. Omnivorous fish have an intermediate condition. In *Clarias*, its wider proximal part is called the duodenum (Fig. 18.3 and 18.4).

Das and Munshi (1956) have shown that the ratio between gut length and body weight (=length) is fairly constant in a large number of species. However, *Pterois* has a long looped intestine. It is difficult to generalize the length of the gut and the nature of the diet due to the omnivorous habits of a large number of fish. The length of the gut depends on the average muscular area as well, and a short gut may be compensated by longer mucosal folds. A straight and short intestine is present in *Xenentodon* and *Harpadon* (piscivorous fishes).

Spiral valves are usually absent in osteichthyes except for *Polypterus*. In *Lepisosteus* and *Amia*, it may be vestigial. In *Scoliodon*, it is a 2.5 anti-clockwise longitudinal turn to increase absorptive area (Fig. 18.4b).

The fingerling outgrowths at the pylorus or anterior part of the intestine are known as the pyloric or intestinal caeca. They are present in *Notppterus, Channa,*

Mastacembelus, Hilsa and *Harpodon* etc. Their number varies from one to several hundred. It acts as an accessory food reservoir. They are not found in stomach-less fish and have no taxonomic value as they are found in large numbers. There is no relationship between the number of caeca and the length of the intestine. In Somniosus, there is a pair of pyloric caeca. *Polylpterus* has only one pyloric caeca, but has more than 200 as seen in mackerel. In *Lates*, pyloric caeca have five blind fingers-like structures, but *Channa* has only two pyloric caeca. In some cases, the pyloric caeca may be bound together by connective tissue to form a compact mass. Most likely, it increases the absorptive area of the gut.

The intestine is thin-walled. As lamina propria, the mucous membrane is folded from microvilli. The microvilli are numerous in the proximal part of the intestine where they may fuse. The occurrence of microvilli on the laminar surface of the columnar absorptive cells is a special feature of the teleost intestine. The mucosa contains columnar epithelium containing absorptive and mucous cells. Sinha (1994) reported pinocytotic vacuoles and large mitochondria in the columnar epithelium of *Labeo rohita*. The absorptive cells have a free-striated birder or brush border. The mucosa may contain lymphocytes and granulocytes. The submucosa is highly vascular. The muscularis consists of circular and longitudinal muscles. Various nonabsorptive enteroendocrine cells are reported in many teleosts. Enterochromaffin cells and argenaffin cells are present respectively in *Salmo truta* and *Mugil auratus*. The argenaffin cells probably secrete serotonin.

18.3.6 Rectum

The rectum is distinguishable by having an ileorectal valve at the junction of the intestine and rectum in many fishes (e.g. *Sciana, Tetradon* and *Muracuosox*). In elasmobranches, the rectum terminates into a cloaca. But in most teleosts, the anus opens separately to the outside. A rectal or digitiform gland opens into the dorsal side of the rectum. The rectal gland is absent in actinopterygians but seems to be present in *Latimeria* (Fig. 18.3).

The rectum has reduced muscle fibers with short and broad mucosa folds.

18.4 Digestive Glands

18.4.1 Liver

In fish, the liver is a large, bilobed or trilobed, yellow or brown coloured organ. The gall bladder is present in almost all forms except in a few sharks. The bile coming from the gall bladder contains emulsifying salts, biliverdin and bilirubin. The liver is where fats, blood sugar, and vitamins A and D are stored.

18.4.2 Pancreas

The pancreas is generally a diffused structure in the body cavity among the coils of the intestine. However, it is compact in sharks and rays. It also extends into the liver to form the hepato-pancreas. Elasmobranches have a distinct pancreas, whereas, in most teleosts, the endocrine part of the pancreas remains encapsulated and separated from the exocrine part. Such isolated areas are known as principal islands. In *Labeo rohita, Cirrhinus mrigala* and *Catla catla*, the exocrine cells are present between spleens as well.

The liver contains many hepatic cells, bile ductules and blood capillaries. The pancreas contains zymogen granules.

18.4.3 Gastric gland

Channel catfish (*Ictalurus punctatus*) HCl and pepsinogen. However, *Esox lacius* lacks gastric glands.

18.5 Physiology of Digestion

18.5.1 Mechanical digestion

Fish lack salivary glands in the oral cavity so their function is limited to capturing and crushing food. In some species, the secretary glands are replaced by mucus-producing ones, which favour the transit of food.

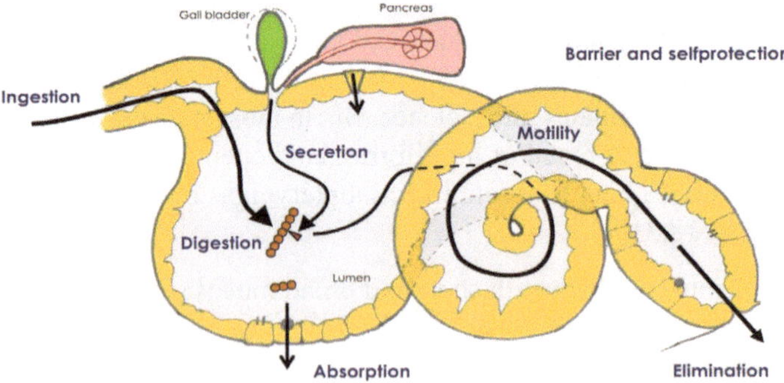

Fig. 18.5: Mechanism of digestion in a fish (from https://onlinelibrary.wiley.com)

18.5.2 Chemical digestion

The physiology of digestion is under the control of various enzymes, hormones and the nervous system. In sharks, digestion can take a long time (Fig. 18.5)

Enzymes from bile, the pancreas and the small intestine work best at a pH ranging from neutral to alkaline. Carboxypeptidase and aminopeptidase act on the C and N terminals of the peptides of proteins in food. Trypsin and chymotrypsin are involved in the alkaline digestion of proteins. Various enzymes like cellulase, cellobiase, sucrase, maltase, lactase, dextrinase and amylase digest different types of sugars present in food. Fat is digested by lipase and phospholipase secreted from the pyloric caeca and intestinal mucosa.

Diffusion, active transport, and pinocytosis are all methods of absorption. Although most digested proteins are absorbed in the intestine. Some absorption of proteins has also been reported in the stomach of the shark.

19

Aquatic Respiration in Fishes

19.1 Introduction

Fish are the most common aquatic vertebrates. Fishes' primary aquatic adaptation is the presence of gills. They use the oxygen that remains dissolved within the water via their gills. The gills use up to 80% of the oxygen dissolved in the water that passes over them. In addition to the gills, skin, the air bladder and accessory respiratory organs perform gas exchange functions in some fishes.

Fish gills are essentially the same. The gills have been specially modified to use up to 80% of the oxygen dissolved in the water that passes over them. *Cyprinus* and *Carassius* gills excrete six to ten times more nitrogenous waste than kidneys.

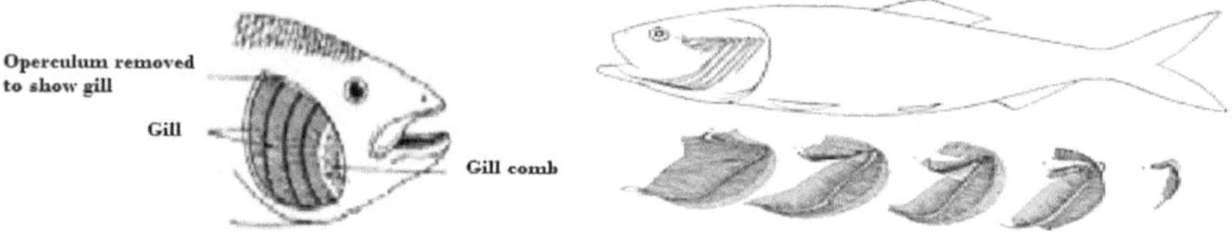

Fig. 19.1: Position of gills in a fish

Fig. 19.2: Origin of gills in a fish

The gill pouches in *Scoliodon* give birth to five-pairs of gills, and their structure varies between dogfishes. The number of gill clefts varies between fish groups, ranging from 5 to 14 pairs (Fig. 19.1).

Teleosts have four-pairs of gills, one on each lateral side of the pharynx. Each side of the pharyngeal lateral wall is perforated by five gill slits. *Hexanchus*, on the other hand, has six pairs and *Heptranchias* has seven pairs in addition to spiracles. Four gill arches on the interbranchial septa separate the gill slits.

19.2 Origin of Gills

Previously, it was thought that gill evolution occurred along two diverging lines. According to Gillis and Tidswell (2017), recent studies on the gill formation of the little skate (*Leucoraja erinacea*) have revealed evidence that all current fish species' gills have evolved from a common ancestor (Fig. 19.2).

19.3 Covering

Holocephalians are operculate. In a common branchial chamber (=gill chamber), the gills on each side are covered by an operculum. The operculum develops as an integument fold from the hyoid arch's posterior margin. The operculum extends posteriorly to completely cover the gills, leaving only a single external branchial aperture to communicate the chamber to the outside. Four opercular bones support the operculum. To form the gills, the anterior and posterior walls of each gill slit are raised in the form of vascular filamentous outgrowths.

19.4 Location

The gill pouches are separated by interbranchial septa, which serve as the pouches' walls. The opposite walls of these septa have gill filaments. Teleosts have internal pharyngeal gill slits that do not open to the outside.

19.5 Structure

In *Scoliodon*, the first gill-pouch lies between the hyoid and the first branchial arches as a hemibranch and the last one is present between the fourth and fifth branchial arches (Fig. 19.3 and 19.4).

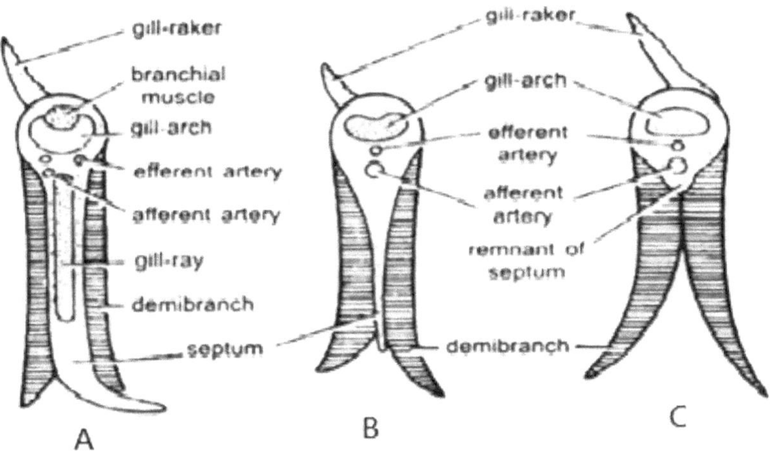

Fig. 19.3: Types of gills (A: Elasmobranchs, B: Chimaera, C: Teleost)

The gills are comb-like (holobranch) structures with an interbranchial septum held in place by bone or the gill arch. The inner face of the gill arch is concave and has teeth that resemble expanded gill rakers. The gill rakers prevent food substances from escaping from the pharyngeal cavity into the branchial chamber. It lacks spiracles but has a branchiostegitial membrane at the operculum's post edge.

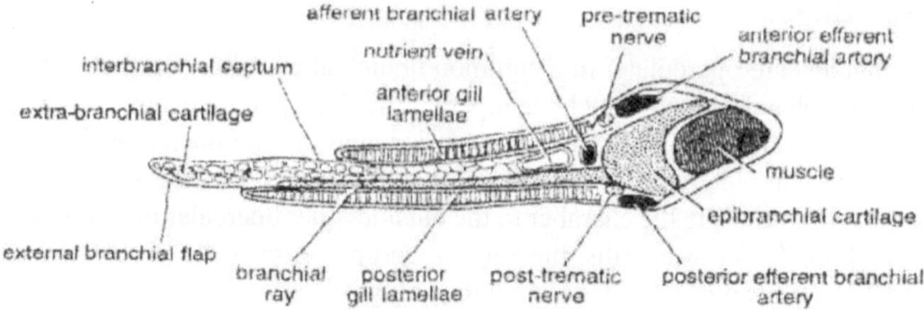

Fig. 19.4: Structure of gills in *Scoliodon* (from https://swarborno.com)

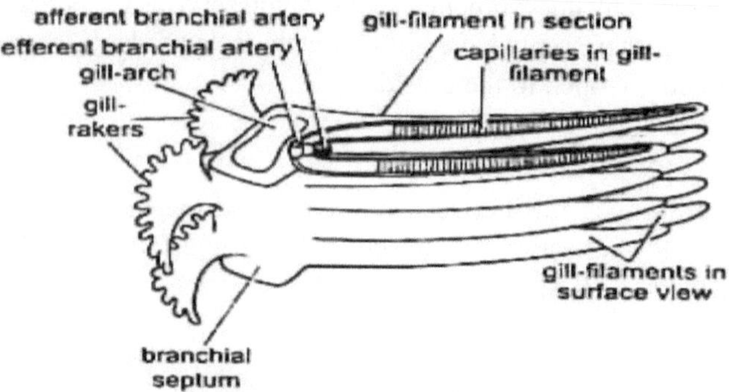

Fig. 19.5: Structure of gills in *Labeo*

Both the anterior and posterior faces of each interbranchial septum are produced into vascular and comb-like primary gill filaments. This results in the formation of two rows of primary lamellae on the outer convex border of the gill arch. Teleosts have a short interbranchial septum hence two rows of primary lamellae are freed at the distal end. The interbranchial septum extends only halfway down the length of primary lamellae in *Hilsa ilisha* and *Labeo rohita* while it is still shorter in *Rita rita* and *Channa striatus*. The gill filaments of each row are generally free of each other but in a certain case, these fuse with neighbouring filaments at the tip and base, so that narrow slits exist between them as shown in *Labeo rohita* (Fig. 19.3 and 19.5).

The gill filaments are based on gill rays of a partly bony and cartilaginous nature. Gill rays have connections with the gill arch and with each other using ligaments. The proximal end of each gill ray is bifurcated to form a passage of efferent-branchial vessels. *Rita rita*'s gill rays are entirely cartilaginous.

Both lateral sides of a primary gill filament bear numerous leaves that resemble secondary filaments and serve as the actual site of gaseous exchanges. Except for those at the distal end of primary filaments, which may be fused, these are independent. A thin layer of connective tissue and epithelium surrounds a central network of capillaries known as vascular layers to form the secondary filament.

The size and number of gill filaments determine the respiratory area of the gill. The habit of eating fish influences the development of the respiratory surface area. The respiratory surface area of the gill is greater in fast-moving fishes, while it is smaller in slow-moving forms. Tertiary lamellae are found in some air-breathing fishes (e.g., *Channa striatus*).

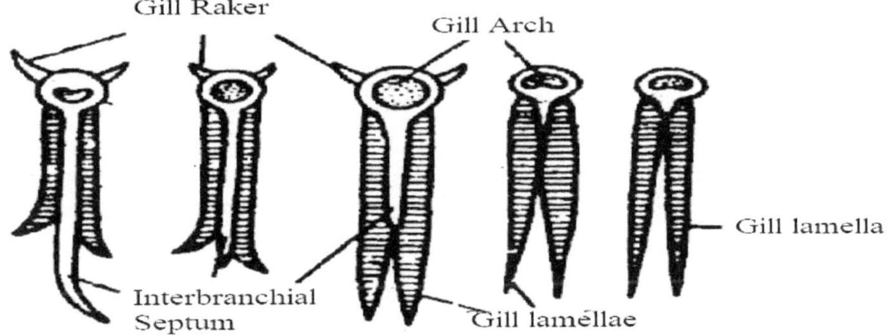

Fig. 19.6: Interbranchial septum, Gill raker, Gill arch in fishes

19.5.1 Interbranchial Septum

The interbranchial septum in primitive fishes is thick and made of tough fibrous tissue. In primitive teleosts (e.g., *Acipneser*), the septa become larger and extend up to the midway. It becomes progressively shorter in *Salmon, Rita rita, Channa striatus*. The rows of gill filaments are independent of one another but in *Labeo rohita and Hilsa ilisha*, the adjacent rows of gill filaments are fused at the tip and bear a narrow slit-like aperture (Fig. 19.6).

In selachians, the interbranchial septa are arranged in a series of independent gill slits. In these fish, the interbranchial septa are larger than rows of gill filaments. In Chimaeras, the interbranchial septa are slightly shorter than the gill filaments and the gill filaments project a little beyond the outer edge.

The condition of the interbranchial septa in sturgeons exhibits a transitional stage between the chondrichthyes and oestichthyes.

19.5.2 Spiracles

Between the mandibular and hyoidean arches, these are slit-like apertures. It is subsequently modified in various fishes. In sharks, the anterior side of the spiracular cleft bears spiracular gills. Generally, spiracular gills are represented by a *vaso-ganglion* or *rete mirabile* (network of blood vessels) or pseudobranch.

It is generally absent in bony fish, although spiracular pouches may develop. In *Amia* and *Lepidosiren,* only the spiracular pouch is present in the greatly reduced condition. Both the spiracles and the spiracular pouch are present in *Acipenser*.

Polypterus possesses a wide spiracle, a cellular ridge to separate the pouch and hyobranchial groove. Spiracles are absent in Crossopterygii. *Latimeria* has a deep spiracular pouch but a much smaller Dipnoi.

19.5.3 Pseudobranch

A pseudobranch is a series of gill filaments on the hyoidean arch. The shape, size and location of the pseudobranch vary greatly in various teleosts, but its blood supply and innervations suggest a common plan. In *Catla catla*, it presents the region anterior to the first gill filaments. It may be free or covered with a layer of the mucous membrane. It develops quite early in the embryo and may have a respiratory function but not in adults. It is absent in *Wallago attu, Mystus aor, Notopterus, Gymnarchus, Cobitis* and *Channa.* It is absent in fishes lacking choroid glands in their eyes.

Histologically, the pseudobranch may resemble gills. It contains typical chloride cells, pseudobranchial cells and mucous-secreting cells. It contains a large number of acidophilic secretory cells that replace the respiratory cells of the gills.

In sharks, rays, chondrosteans and holosteans, the pseudobranch may be derived from the posterior mandibular hemibranch. However, in teleosts, it is derived from the posterior hyoidean hermibranch.

According to Bertin (1958), pesudobranchs may be of the following types:

1. Free type: It is a row of filaments and secondary lamellae as seen in Clupeidae, Laridae, Lophidae (goosefish), Pleuronectidae (flounders) and Syngnathidae.
2. Covered type: It is a gill-like structure covered with opercular membrane and connective tissue as observed in *Catla, Glossogbius, Gadus* and *Phoxinus.*
3. Glandular type: It is deeply embedded in the thick connective tissue of the opercular cavity or the buccopharynx as in *Channa, Anabas* and *Notopterus.*

It probably produces anhydrase. It also aids in the process of filling the swim bladder and the intraocular process. It contains largely acidophilic cells. The pseudobranchial vessels join the orbital and ophthalmic vessels. It is thought to control the uptake of Na^+ and Cl^- from water. It has been associated with O_2/CO_2 tension in the blood. Srivastava and Gopesh (1981) have suggested a role in neuro-secretion for the pseudobranch.

19.5.4 Gill Arch

Each gill arch encloses afferent and efferent branchial vessels and nerves. It is covered over externally by the epithelium (having mucus glands, eosinophilic cells and taste buds). It has at least one set of abductor and adductor muscles to move gill filaments. The abductor muscles lie on the outside of the gill arch connecting it with the proximal ends of gill rays. The adductor muscles are present in the interbranchial septum and cross each other to insert opposite gill rays. In *Channa striatus* and *Rita rita*, adductor muscles arise from the base of the gill rays of one side, run obliquely and are inserted on the gill ray of the other side. In *Labeo rohita, Hilsa ilisha, Cyprinus carpio*, the muscle arises halfway from the gill ray due to a large inetrbranchial septum and is attached to the gill ray of the opposite side.

19.5.5 Hyoidean Hemibranch

These are present in the majority of fishes, but not in the most of actinopterygians. In *Acipenser, Lepisosteus* and *Polyodon*, these gills are present but *Amia* lacks the hyoidean hemibranchs. Among crossopterygians, hyoidean hemibranchs are present in all living forms except *Lepidosiren*. It is small in *Latimeria*. In Scaphirhynchus, these hemibranchs are greatly reduced. *Polypterus* lacks both the mandibular pseudobrachs and hyoidean hemibranchs.

19.5.6 Gill Rakers

The gill rakers are specially developed on the inner edge of the gill arches. These are modified dermal denticles in two rows. Its development depends on the mode of feeding. It is extremely developed in fish that consume microorganisms. It forms a sort of sieving apparatus that strains the water that bathes the gill filaments (Fig. 19.7).

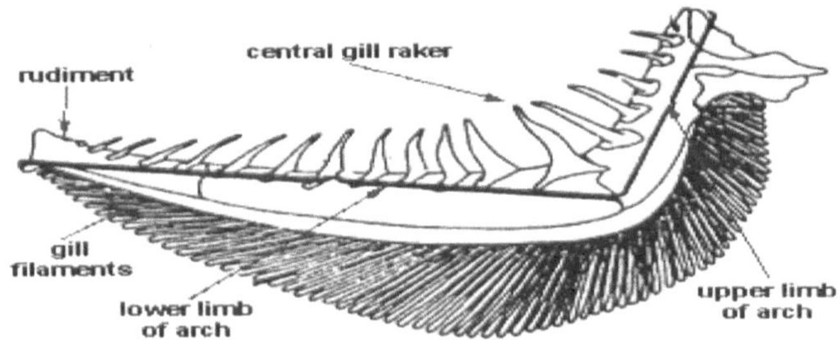

Fig. 19.7: Gill rakers in fishes

In plankton feeders (*Hilsa, Gambusia, Notopterus*), gill rakers are slender and extremely elongated to form a close-set strainer. In filter feeders, the primary gill rakers bear secondary and tertiary branches and thus appear as fine gauze. In *Esox*, the gill rakers are reduced to bony knobs to prevent the entry of larger particles.

In adult *Alosa alosa*, the lower limb of the arch bears about 80 gill rakers, but only 30 in *Aolsa fallax* The gill rakers in basking sharks (*Coelacanthus*) and whale sharks (*Rhincodon*) measure about 10-12cm in length. The gill rakers are generally absent in other sharks. It is rudimentary in crossopterygians.

19.5.7 Giil Filaments

Each gill arch has two rows of gill filaments facing the buccopharyngeal cavity. In teleosts, the interbranchial septum between two rows of lamellae is short, so that the two rows of lamellae are free at their distal ends. However, in *Laboe rohita, Hilsa ilisha* etc. the septum allows them to lead an active life and depend entirely on aquatic respiration. *Catla catla, Cirrhinus mrigala, Wallogo attu, Mystus seenghla* and other species with aerial respiration have more long filaments, but fewer and shorter filaments. Primary lamellae of two hemibranchs are alternately arranged concerning each other in *Labeo rohita* but interdigitate in *Channa striatus*.

Usually, gill filaments of each row are independent of each other, but in *Labeo rohita*, the neighbouring filaments are fused at the tips as well as at the base, leaving narrow slit-like apertures between them. In hill-stream fish, gill filaments undergo division to form additional filaments. The additional filaments are believed to increase respiratory surface area in *Garra, Gyptothorax, Botia* and *Noemacheilus*.

The gill filaments are supported by gill rays which are partly bony and partly cartilaginous and are connected with the gill arch and with each other by fibrous ligaments. Each gill ray is bifurcated at its proximal end to provide a passage for efferent branchial vessels. Completely cartilaginous rays are present in *Rita rita*. These filaments are covered by the primary epithelium and contain mucus-secreting cells, chloride cells and taste buds etc. that detect the nature of water.

19.5.8 Secondary Lamellae

Each gill filament has numerous secondary lamellae on both sides. These are flat, leaf-like structures that serve as primary gas exchange channels and vary in shape, dimension, and density per unit length of gill filament. Secondary lamellae are generally independent of one another, but they may fuse at the distal ends of gill filaments. They range from 10 to 40 on each side of the gill filament.

Each secondary lamella has a central vascular core made up of pillar cells that are surrounded by a basement membrane and an outer epithelium. Some secondary lamellae epithelium is smooth, while others are rough, with micro-ridges and micro-villi. The grooves on the surface epithelium are seen to increase the respiratory surface area.

The pillar (pilaster) cells are teleost gill characteristics that separate the epithelium on the opposite side of a lamella. Each cell has a central body with extensions at each end known as pillar cell flanges that overlap the neighbouring cell and form the boundary of blood channels in lamellae. Basement membrane columns run transversely across the secondary lamella, connecting opposite-side basement membranes and providing structural support to secondary lamellae. According to Munshi and Singh, each row's pillar cells form a complete partition between neighbouring channels in *Channa*. This configuration aids in the development of a streamlined flow and may promote the countercurrent flow of blood and water at the microcirculatory level. Thus, pillar cells appear to perform both the protection of vascular spaces and the regulation of blood flow patterns through secondary lamellae.

19.5.9 Gill area

The gill area of fish is determined by the number and size of gill lamellae. A fish's total gill area is calculated by multiplying the total length of the gill filament by the frequency of secondary lamellae and the average bilateral lamellar area. Active fishes, on average, have more gill area and a greater number of secondary lamellae per mm of gill filament than sedentary species. The total gill area is proportional to the gill sieve efficiency. It rises in proportion to the growth of the fish's body weight. The efficiency of the gills is also affected by the diffusion distance or barrier that allows gas exchange between blood and water. The lamellar epithelium, basement membrane and flange of pillar cells from the barrier that is smaller in water-breathers than in air-breathers. The diffusing capacity of gills is thus determined by the total gill area and diffusion distance of lamellae. Water breathers have a greater diffusing capacity than air breathers due to their larger gill area and shorter diffusion distance.

19.6 Vascular Supply

Generally, only one afferent and one efferent branchial vessel are present in each gill arch. The vascular central core contains capillary networks and supporting Pilaster cells. In sharks, *Labeo rohita, Clarias batrachus* and *Anabas testudineus*, there are two efferent branchial vessels. In *Anabas*, the 4th branchial arch has no secondary lamellae and the afferent and efferent arteries are joined by broad spaces. In *Channa punctatus* and *Channa striatus*, the last two branchial vessels pass directly up to the dorsal aorta. But in *Channa argus*, the afferent and efferent branchial vessels are connected by clustered loops of small blood vessels. In *Clarias* and *Heteropneustes*, the oxygenated blood passes to the dorsal aorta, but in *Anabas, Monopterus, Channa*, two circulations are parallel and there is an admixture of oxygenated blood coming from air-breathing organs with systemic blood in its passage through in the heart. Studies show that there are two types of vascular pathways in the gills (Munshi et al, 1990).

19.6.1 Respiratory (arterio-arterial) pathway

Each afferent branchial vessel transports oxygen-depleted blood into the gills. It runs the length of the gill arch and sends many afferent filamental arteries to the gill filaments. Each afferent vessel laterally divides into several secondary vessels, one for each secondary lamella. These cross gill rays before diving into

2-4 tertiary branches. These capillaries break up and form lamellar channels that are interconnected with one another in secondary lamellae, forming the vascular core of secondary lamellae. While blood circulates through these channels, gas exchange occurs. Oxygenated blood is collected by an efferent lamellar arteriole, which transports blood to a primary efferent vessel that runs along the gill filament's margin. Finally, the blood is carried to the gill arch via the main efferent branchial vessel (Fig. 19.8).

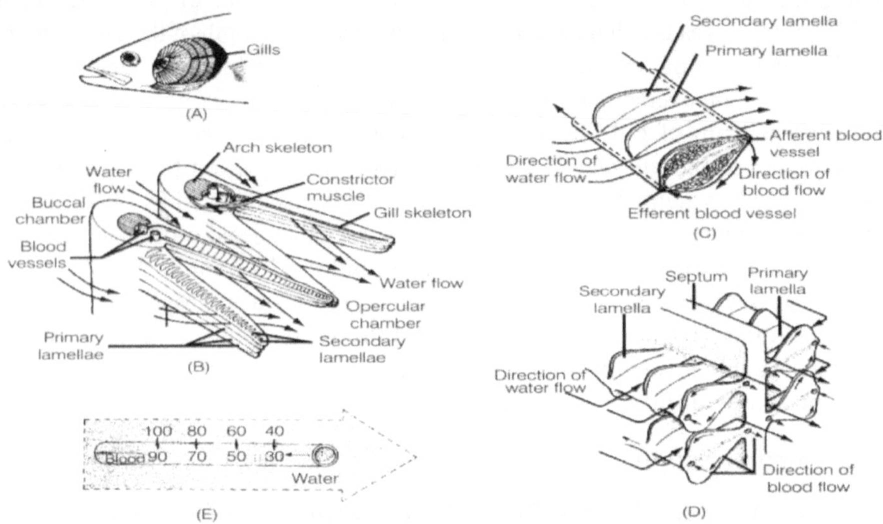

Fig. 19.8: Respiratory pathway in fishes (from https://www.brainkart.com)

The blood flows out of the afferent lamellar arterioles into a large space that is not bounded by endothelium lining except at the outer margin, giving rise to the term "sheet flow.".

19.6.2 Non-respiratory (nutritive or arterio-venous) Principle

It is made up of a complex network of sinuses and veins that carry blood directly to the heart, bypassing systemic circulation. Its function is thought to be to provide nutrition and oxygen to the filament tissue, and it may also be involved in hormone circulation. It has a nutritive blood channel, an efferent artery, a central nervous system sinus, venules and branchial veins.

19.7 Histological Aspects

(A) Gill epithelium: It is generally double-layered but EM studies of the gills of *Anabas* and *Clarias* have shown a multilayered epithelium of 5-18m thickness. It contains amoebocytes and lymphocytes. Some cells are glandular and specialised to perform various functions (Fig. 19.9).

(B) Mucous glands: These are oval and pear-shaped but typically of goblet type. It provides a thin film of mucus to prevent dehydration of respiratory epithelium. It secretes acid-neutral glycoproteins in the region of the gill arch and lamellar epithelium.

(C) Chloride cells: These are commonly found in the interlamellar epithelium of filaments and are considered to regulate the movement of chloride loss through the gill epithelium. These are found in good numbers in air-breathing fishes. It may occur in either inactive or active forms. The active cells have many mitochondria but an inactive one is found in *Arapaima gigas*. Both active and inactive chloride cells are present in *Monpterus cuchia, Clarias batrcahus* and *Channa punctatus*.

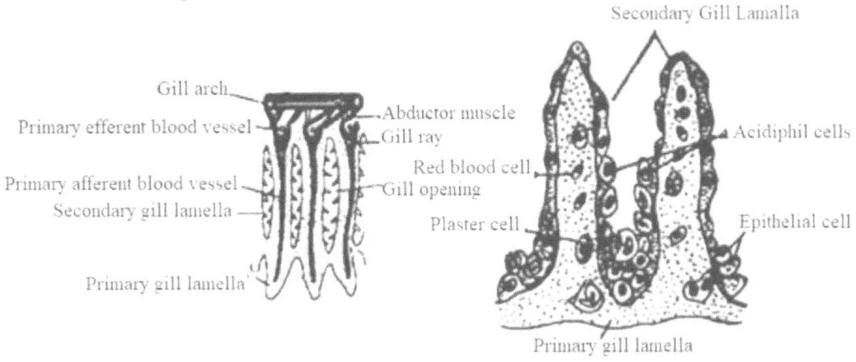

Fig. 19.9: Histology of gills in fishes

19.8 Types of Gills

19.8.1 Based on the location

(1) Internal gills: Gills lie inside the gill pouches (in sharks) or branchial chamber (in holocephalians and teleosts).

(2) External gills: The gills of many fish larval stages. It may be two types based on their development:

(a) True external gills: These are distinct from internal gills and develop as a result of integument modification.

(b) Prolongations from the internal gills: These are located outside the body. They are found in the larvae of *Protopterus* and *Lepidosiren*, as well as *Gymnarchus* and Clupisdis.

19.8.2 Based on the structure

(1) Hemibranch: A group of gill filaments on one side of the interbranchial septum. In elasmobranchs, gills are mostly hemibranchs.

(2) Pseudobranch: A modified hemibranch with no respiratory function. It is comb-like in trout. In perch, gill filaments are greatly reduced and covered by pharyngeal epithelium. The pseudobranch of *Amia* lacks a direct afferent or efferent blood connection. *Polypterus* lacks mandibular pseudobranchs.

(3) Holobranch: A holobranch is made up of two hemibranchs. It has four segments that support its bony gill arch: pharyngobranchials, epibranchials, ceratobranchials, and hypobranchials. Gills in teleosts are mostly holobranchs.

(4) Lophobranch: Gill filaments have been greatly reduced, resulting in rosette-like tufts.

Some fish still have vestiges of mandibular gills. There are four pairs of holobranhs in Latimeria, but no evidence of a mandibular pseudobracnh. The mandibular hemibranchs typically assist in closing the spiracles and obtaining oxygenated blood. The majority of sharks have a mandibular pseudobranch, a hyoidean hemibranch and four to six holobranchs on each side.

The mandibular pseudobranch is absent in holocephalians, but the hyoid arch has a posterior hemibranch. The first three gill arches contain holobranchs, while the fourth contains a hemibranch.

19.9 Ventilation of Gills

Constant water current must flow through the buccopharyngeal cavity. Most teleosts have two respiratory pumps (buccal and branchial) that work together to propel water through the gills. These two pumps are controlled by a group of muscles that cause the buccal and opercular cavities to contract or expand (Fig. 19.10). The following steps complete the process:

1. The mouth is opened, and the buccal cavity is enlarged by lateral wall expansion. To widen the angle enclosed by the gill arches and enlarge the oropharyngeal cavity, the coraco-hyoid and coraco-branchial muscles contract. Enlarging the buccal cavity reduces the water pressure in it, allowing water to be sucked in. To close the external gill slits, the gill flaps are held to the skin by exterior water pressure.

2. The mandible and gill arches relax, but the abductor muscles contract and the buccal cavity becomes a pressure pump. The mouth is closed, and operculum abduction expands the opercular cavity. However, the opercular opening remains closed, resulting in low pressure in the opercular cavity and water flowing from the buccal to the opercular cavity via the gill slits.

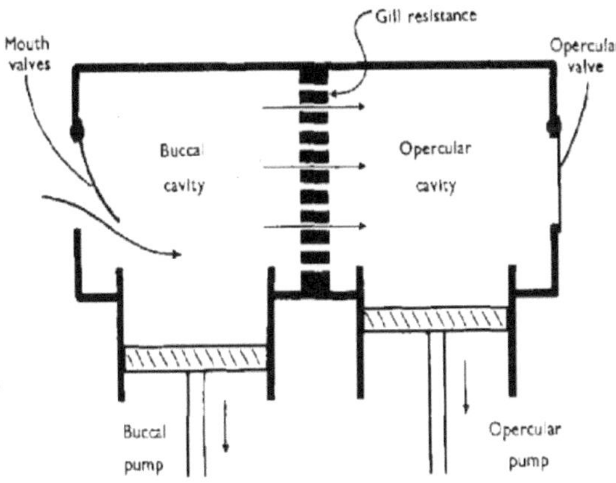

Fig. 19.10: Ventilation of gills in fishes (from https://www.researchgate.net)

3. The buccal and opercular cavities are now reduced, and pressure is applied to the water within them. Water is drawn into the gill cavity as the interarcual adductors relax. Water cannot exit the mouth due to oral valves. After reaching maximum abduction, the opercula are quickly brought back towards the body, and the water is expelled through the external branchial aperture. Water cannot enter the buccal cavity because the pressure there is greater than in the opercular cavity.
4. The next cycle soon begins.

   The mechanism of gill ventilation can change depending on the habits of the fish. Fast swimming fish leave their mouth and opercular aperture wide open, allowing a continuous current of water produced by swimming to bathe their algae. This is known as ram ventilation.

   The process is also altered in hill-stream species such as trunkfish (*Ostracion*) and globe fish (*Tetradon*).

19.10 Physiology Of Respiration

In *Scoliodon*, during respiration, the floor of the buccal cavity is lowered and the mouth is opened. Then the water rushes in to fill the greatly expanded buccal cavity. The mouth is now closed and the pharynx contracts. The water then enters the gill pouches and goes out after gaseous exchange through gill slits. The spiracles are occasionally used as accessory pathways for the entry of water for respiration instead of the mouth when it is otherwise occupied.

19.10.1 Inspiration

In teleosts, the outer opening of the gill chamber remains tightly closed to the body wall because the opercula bulge out to increase the accommodating cavities. As a consequence, water from the exterior will enter through the opened mouth and fill in the buccopharyngeal cavity (Fig. 19.2I).

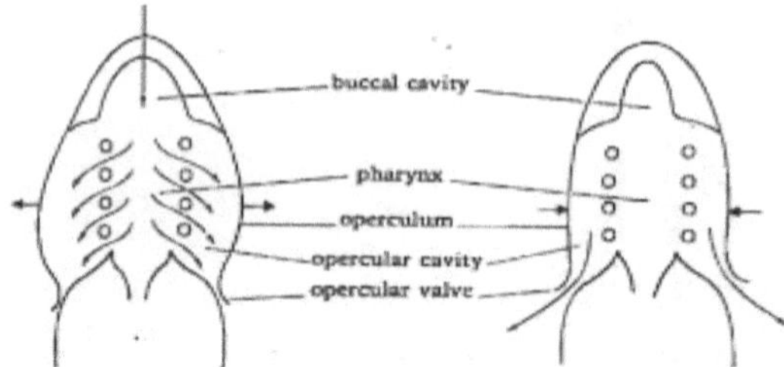

Fig. 19.11: (I) Inspiration (II) Expiration

19.10.2 Expiration

The pharyngeal and buccal cavities contract and exert pressure on the contained water as soon as water enters. As the mouth becomes closed, the contained water finds its way out through the gill slits (Fig. 19.2II).

19.11 Counter-Current Principle

Teleost gill structure is brought into close contact with secondary lamellae, allowing for gas exchange. In active fishes, the total surface area of lamellae exposed to water current can reach 1000 mm^2/g wt. The gills can use 50-80% of the oxygen in the water. Because of the significant reduction in the interbranchial septum, it is thought to be more efficient than elasmobranch gills. Furthermore, the blood and respiratory water flow in opposite directions due to the arrangement of afferent and efferent branchial vessels. It is known as the "counter-current principle," in which oxygen-containing water flows from the oral to the aboral side of the gills and blood in the lamellae flows from the aboral to the oral side, allowing for maximum exchange of respiratory gases.

20

Accessory Respiration in Fishes

20.1 Introduction

Physiological adaptations for air-breathing in bony fishes have fascinated researchers for well over a century. The metabolic rate of some fish is so high that dissolved oxygen in water becomes insufficient. Prolonged drought causes some fish to develop devices to survive in the absence of water for a short period. To overcome these challenges, fishes have developed accessory respiratory organs that can function in both aquatic and aerial environments.

As an adaptation to specific environmental conditions, accessory respiratory organs develop in addition to the gills. The development of these organs is dependent on the availability of extra oxygen. The development of an accessory respiratory organ is most common in tropical freshwater fishes. However, the development of accessory respiratory organs is essentially adaptive, allowing fish to tolerate oxygen depletion in water or to live on land for varying periods. The ability to remain outside the water is directly related to the development of the accessory respiratory organs. Air-breathing fishes include at least 450 species (Nelson, 2014) from 49 known families (Graham, 1997) of shallow stagnant freshwater in the tropical region. These structures are only found in a few marine fish species, including *Achirus, Chiloscylium indicum* and *C. griseum*. Some important air-breathing fishes are arranged in Table 20.1.

Zacconi et al. (2018) reviewed the morphology of peripheral receptors, neurotransmitter content of cells, and evolutionary implications of air-breathing organs in fishes that are involved in air-breathing control. All air-breathing fishes are bimodal breathers because they retain gills that aid in gas exchange, specifically the excretion of carbon dioxide and ammonia into the water.

Table 20.1: Name of fish and their air-breathing organs (ABOs)

Sl.No.	Air-breathing organ	Name of fish	Sl.No.	Air-breathing organ	Name of fish
1.	Air-bladder	*Erythrinus* sp. *Gymnarchus* sp. *Notopterus* sp. *Phractolaemus* sp. *Umbra* sp.	6.	Intestine	*Lepidocephalichthys guntea* *Misgurnus fossilis*
2.	Arborescent organ	*Clarias batrachus*	7.	Labyrinthine organ	*Anabas testudineus*
3.	Buccopharynx	*Channa* sp. *Electrophorus electricus* *Monopterus* sp.	8.	Operculum diverticulum	*Boleophthalmus* sp. *Periophthalmus vulgaris* *Pseudopocryptes* sp. *Symbranchus* sp.
4.	Branchial diverticulum	*Anabas testudineus* *Betta* sps. *Calrias batrachus* *Colisa fasictus* *Heteropneustes fossilis* *Macropodus* sp. *Osphronemus* sp.	9.	Skin (integument)	*Anguilla* sp. *Amphipnous cuchia* *Mastacembelus* sp. *Rita rita*
5.	Gill	*Mastacembelus* spp. *Rita rita* *Symbrachus bengalensis*	10.	Stomach	*Ancistrus* sp. *Plecostomus* sp.

20.2 Organisation of Air-Breathing Organs

Fish accessory respiratory organs can be divided into two broad categories:

A. Aerial accessory respiratory organs: Some existing organs become highly vascularized and modified to allow for aerial respiration. Such respiratory organs are found in fish from stagnant water and swamps, but they are uncommon in marine fish.

B. Aquatic accessory respiratory organs: Many fishes have modified structures to allow them to breathe in adverse conditions. Neomorphic organs are still developed in some species for this purpose.

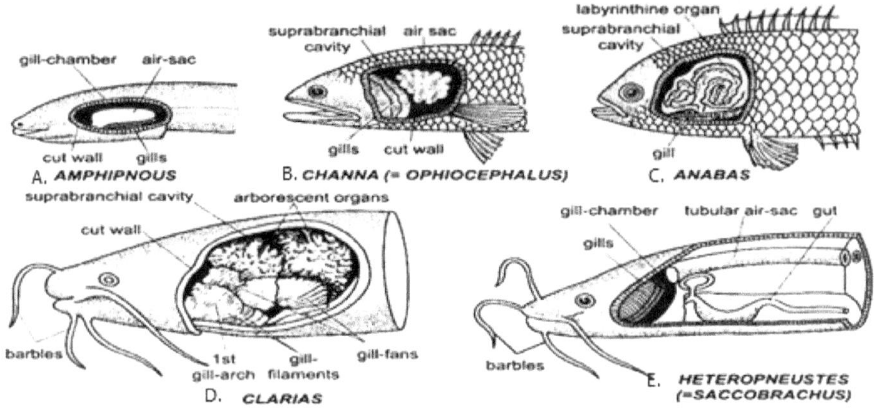

Fig. 20.1: Accessory respiratory organs in fishes

20.2.1 Skin (Integument)

The integument of many species becomes highly vascular and is kept moist by mucus to allow gaseous exchange between air and blood. Many fish embryos and larvae breathe through their skin before developing gills. Many blood vessels supply the median fin fold of many larval fishes, which aids in breathing. The respiratory structure of Sturgeon and many catfishes is the highly vascular opercular fold.

Eels travel a long distance through wet vegetation. The common eel (*Anguilla anguilla*) can breathe through its skin in both air and water. The moist skin sub serves respiration in rice eels or swamp eels (*Amphipnous* or *Monopterus cuchia*) (Fig. 20.1A), tire-track spiny eels or marbled spiny eels (*Mastacembeles armatus*), Rita (*Rita rita*), and mudskippers (*Periophthalmus* and *Boleophthalmus*).

20.2.2 Buccopharyngeal epithelium

In almost all fishes, the vascular membrane of the buccopharyngeal region aids in the absorption of oxygen from water. The highly vascularised buccopharyngeal helps mudskippers (*Periophthalmus* and *Boleophthalmus*), rice eel or swamp eel (*Amphipnous* or *Monopterus cuchia, Monopterus javanensis*) (Fig. 20.1A), and electric rays (*Electrophorus*) absorb oxygen directly from the atmosphere. These tropical fishes leave the water and spend the majority of their time skipping or walking around in damp areas, particularly around the roots of mangrove trees.

Only the second-gill arch of swamp eel (*Amphipnous* or *Monopterus cuchia*) has filaments, while the third arch has a fleshy vascular membrane. The air-breathing organs are formed by a pair of sacs located on the lateral sides of the head. Islet vascular areas are found in the epithelium as rosette or papillae. Each sac has an inhalant and exhalant aperture for air passage. When the buccopharyngeal muscles contract, the mouth closes and the air is expelled.

20.2.3 Pharyngeal diverticula

The pharynx of some fishes contains a pair of sac-like diverticula for gaseous exchange, as described below:

The accessory respiratory organs in *Ophiocephalus* (*Channa*) *marulius* and *Channa striatus* (Fig. 20.1B) are relatively simple, consisting of a pair of air chambers. These emerge from the pharynx. The air chambers are lined by thickened, vascularized epithelium. The air chambers are simple sac-like structures that serve as ling-like reservoirs. The vascular epithelium lining the chambers in *Channa striatus* folds to form some alveoli. The size of the gill filaments has been greatly reduced.

The accessory respiratory organs of the rice eel or swamp eel (*Amphipnous* or *Monopterus cuchia*) (Fig. 20.1A) are a pair of vascular sac-like diverticula from the pharynx above the gills. These diverticula open into the first gill slit anteriorly. These diverticula perform the same physiological functions as the lungs. The gills are severely reduced, with only a few rudimentary gill filaments visible on the second of three remaining gill arches. The third-gill arch contains fleshy vascular epithelium.

A pair of very small pharyngeal diverticula lined by vascular epithelium is present in mudskipper (*Periophthalmus*).

20.2.4 Stomach/Intestine

The stomach/intestine of loaches (*Misgurnus fossilis*) has been modified for aerial respiration. Air is swallowed and forced into the alimentary canal to be stored for some time. Following gaseous exchange, the air is expelled through the anus or passed out through the mouth.

Because of the significant reduction in muscle layers, the wall becomes extremely thin and transparent for this purpose. The thin wall layer is made up of a single layer of epithelium and is densely packed with blood capillaries. Mucus-secreting and glandular cells are not present. Circular and longitudinal muscle fibres are greatly reduced, forming only a thin layer.

20.2.5 Suprabranchial (dendritic) organ

The suprabranchial organ is a specialised accessory respiratory organ that *Clarias* encounters (Fig. 20.1D). The suprabranchial organs, such as the gills and rosette, are lined by thin outer epithelium layers separated by pilaster cells. Afferent and efferent blood vessels from the gill arches supply them. There are inhalants and exhalant apertures in the suprabranchial organ. These fish float to the surface and gulp air into the suprabranchial organ. As a result, the supra branchial chambers serve as the "lung". It has a complex structural organisation and is made up of the following components:

(i) A complex tree-like structure that grows from the upper ends of the second and fourth-gill arches on either side. It is made up of numerous terminal knobs, each with a cartilage core covered by a vascular membrane. Each has eight folds, implying that one such knob is formed by the union of eight-gill filaments.

(ii) Within the branchial chambers and tree-like structures are two highly vascualrised suprabranchial chambers.

(iii) The entrance to the supra branchial chamber is guarded by fan-like structures formed by the fusion of adjacent dorsal gill filaments.

20.2.6 Branchial outgrowths

There are two spacious sac-like outgrowths from the dorsal sides of the branchial chambers in climbing perch (*Anabas testudineus*) (Fig. 20.1C), *Trichogaster (Colisa) fasciatus* (Fig. 20.2). This outgrowth's epithelium lining is highly vascular and folds to increase the respiratory area. Each chamber has a distinctive rosette-like labyrinthine organ. This organ develops from the first epibranchial bone and consists of many wavy plates covered by vascular gill-like epithelium. Each branchial outgrowth freely communicates with the opercular cavity as well as the buccopharyngeal cavity. Air enters the outgrowth via the buccopharyngeal opening and exits via the external gill slits. Valve controls regulate the entrance.

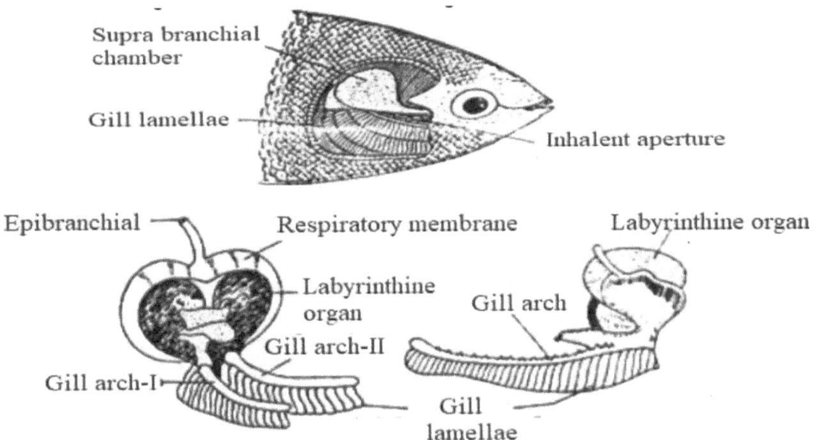

Fig. 20.2: Accessory respiratory organ of *Tricogaster fasciatus*

Anabas can be found in the branches of palm trees or other trees that have been brought there by kites or crows as these fish migrate across the land. These organs allow *Anabas* to breathe in air. These fish frequently migrate from one pond to another. Their overland movement is unusual, and it is aided by the operculum and fins. Sharp spines protrude from the free edge of each operculum. During travel, the opercula alternately spread out and attach to the ground via the spines, receiving forward propulsion from the pectoral fins and tail.

20.2.7 Pneumatic sac

The accessory respiratory organs in singhi (*Heteropneustes fossilis*) (Fig. 20.1E) are a pair of tubular pneumatic sacs, one on each side of the body, fans, and a respiratory membrane. These long tubular sacs grow from the branchial chamber and extend almost to the tail between the body musculature and the vertebral column. Each sac receives blood from its own side's fourth afferent branchial vessel via a thick branch or afferent air sac vessel. This branch gives off several capillaries that supply blood to the sac's respiratory islets. Blood is collected from the islets by efferent capillaries, which open into an efferent air sac vessel, which then transports blood to the fourth efferent branchial vessels on the same side.

In singhi, gills have four fans. The first, third, and fourth fans are formed by the fusion of the arch's gill filaments.

Although neo-epithelial cells have been found in the gills and/or lungs of *Protopterus, Polypterus* and *Amia*, the role of these receptors and their innervations in the control of breathing is unknown.

20.2.8 Swim Bladder

The swim bladder is primarily a hydrostatic organ, but in some fishes, it also serves as the "lung". The swim bladder wall is sacculated and resembles the lung in *Amia* and *Lepisosteus*. The swim bladder in *Polypterus* is more lung-like, with a pair of pulmonary arteries emerging from the last pair of epibranchial arteries. It is single in *Neoceratodus*, but bilobed in *Protopterus* and *Lepidosiren*. Spongy alveolar structures increase the 'lung's inner surface. Because the gills become useless during aestivation, the 'lung' is primarily responsible for respiratory function in these fishes. Dipnoans, like *Polypterus*, get their pulmonary arteries from the last epibranchial arteries.

The swim bladder becomes more complex in *Notopterus* and functions as a lung (Fig. 20.3). *Notopterus* invented a new function, respiration, in addition to hydrostatic, sound production, and hearing. The caudal extension of the swim bladder in *Notopterus chitala* is enlarged, and the ventral part has several fingerlike projections. The dorsal side of the gas bladder has a specialised striated muscle. The anterior part extends into an ear projection. An artery branching from the dorsal aorta forms a network of blood capillaries that spans the entire inner surface of the swim bladder's abdominal and caecal parts. The blood capillaries that cover a single epithelial layer aid in the gaseous exchange of the blood of the swim bladder and air. This habit of air-breathing is thought to be a secondary adaptation in these fishes.

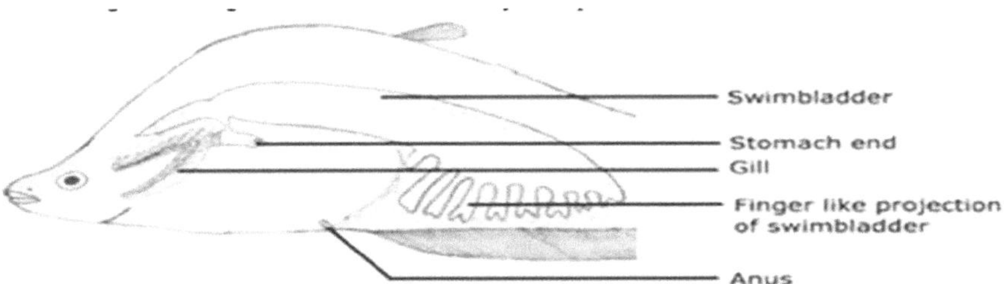

Fig. 20.3: ABO of *Notopterus notopterus* (from https://en.bdfish.org)

20.2.9 Outgrowths of pelvic fin

Delicate vascular outgrowths on the pelvic fins of *Lepidosiren* perform respiratory functions. These outgrowths appear in males during the breeding season and serve as accessory respiratory organs to compensate for aerial reparation at that time.

20.2.10 Gut epithelium

The inner epithelium of the gut plays an important role in digestion. However, in many fishes, the gut becomes modified to serve as a secondary respiratory function. The European giant loach (*Cobitis*) rises above the water and swallows a certain volume of air, which then passes back through the stomach and intestine. *Misgurus fossilis* has a bulge just behind the stomach that is lined by fine blood vessels. This bulge serves as an air reservoir and a respiratory organ. The gas is voided through the anus after gaseous exchange.

20.2.11 Rectum

The highly vascular rectum of certain fishes, *Calichthyes, Hypostomus*, and *Doras*, acts as a respiratory organ by sucking in and releasing water from the anus alternately. Because the muscular layers are reduced in these fish, the gut wall becomes thin.

20.3 Hypotheses Regarding the Origin of Accessory Respiratory Organs

There are two theories about the origin of aerial accessory respiratory organs:

(1) First Theory: Some fish have the instinct to travel to land from their primary aquatic habitat. To stay out of the water, certain devices to breathe in the air must be developed.

(2) Second Theory: When the oxygen level in the water drops significantly, the fish are forced to ascend the land. In that state, the fish take in atmospheric air from the land and pass it through the accessory respiratory structure. The fish will suffocate if they are prevented from rising to the surface by mechanical barriers. This habit of swallowing air habitats is observed in many bony fishes, particularly those living in shallow water that dries up or becomes foul due to the decomposition of aquatic vegetation. Because fish have been air-breathing for a long time, they have developed specialised accessory respiratory organs in addition to gills.

21

Air Bladder and Weberian Apparatus

21.1 Introduction

The air (=gas or swim) bladder, also known as the fish maw, is a tough sac-like structure with an overlying capillary network. It develops as an outgrowth from the oesophageal region and is found in bony fish. It demonstrates a wide range of development, structure, and function in various fish. It is usually a hydrostatic organ that controls buoyancy, allowing the body to remain at the current water depth without having to expend energy swimming. It is similar to a diver's buoyancy compensation device (BCD). It also serves as a resonating chamber for producing or receiving sound and is used as an air-breathing organ.

21.2 Origin and Development

There are two theories about the origin of the air bladder of fish:

1. According to one theory, the air bladder may have originated from a rear pair of gill pouches that had lost their external gill slits and become the storage for air or gas.
2. According to another theory, the paired lung is the forerunner of the air bladder. In other words, the presence of the lung is more primitive than the air bladder.

Fossil records indicate that placoderms and crossopterygians evolved lungs as an adaptation to the unreliable conditions of the Devonian period. Their lungs were similar to *Protopterus* and originated as a pair of ventral sac-like pouches connected to the pharynx (or oesophagus) by a single duct. It shows the following lines of evolution.

1. The air bladder of most teleosts shifted from the ventral to the dorsal side. It is supported by lateral attachment of the air bladder in *Erythrinus*. It becomes a hydrostatic organ for floatation.
2. Extensive internal folding in the air bladder leads to the lungs for respiration.
3. It is lost in elasmobranchs and a few teleosts.
4. It is evolutionarily homologous to the lungs.

In teleosts, it originates as an unpaired dorsal or dorsolateral diverticulum of the oesophagus. The diverticulum with an opening becomes divided into two halves. Of these, the left atrophies except for some primitive forms while the right one becomes well-developed and lies in the median position.

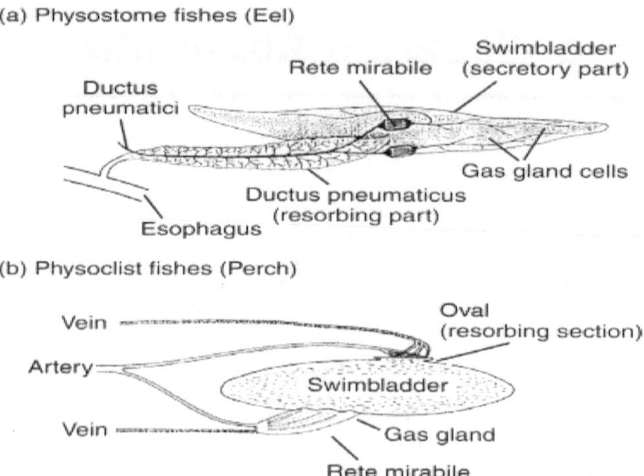

Fig. 21.1: Origin and development of swim bladder in fishes (from https://www.sciencedirect.com)

According to Spengel, the air bladder originates from the posterior pair of gill pouches but embryological evidence is lacking. Goodrich (1930) concluded that both the air bladder and lungs have developed from the posterior pair of gill pouches. However, according to Jones (1957), the air bladder may function independently of the lungs.

In embryos, the air bladder is always connected to the gut by a pneumatic duct. The duct is either lost in the course of development or retained throughout life.

21.3 Occurrence

1. In elasmobranchs, bottom-dwelling and deep-sea teleosts, the air bladder is absent in adults but a rudiment may be present.
2. In Pleuronectidae, the air bladder is present in early life and atrophied in adults.
3. Miklucjo-Maclay (1867) observed a rudimentary dorsal diverticulum from the foregut in the embryos of *Squallus, Mustelus* and *Galeus*.
4. In *Heptranchias, Scyllium, Squatina, Pristiurus, Carcharius* and many rays, small pits are recorded in the oesophageal wall.
5. Wassnezow (1932) has observed 1-6 oesophageal pits in *Pristiurus, Torpedo* and *Trygon*. These pits are lateral posterior to the 5th pouch.

21.4 Basic Structure

Between the alimentary canal and the kidneys, the air bladder is a tough sac-like structure with an overlying capillary network. The tunica externa (connective tissue and collagenous fibres), tunica interna (smooth muscle fibres and epithelial gas glands), lamina propria, and muscularis mucosa are present beneath the capillary network. They can have one, two, or more chambers. It is equipped with both oxygen-producing and oxygen-absorbing devices. The structure is vascular, with highly vascular interlacing and densely packed capillaries forming rete mirabilis. The anterior chamber evolved into an oxygen-producing red body. Retia mirabilis gas glands and blood vessels can be found in the red body. The coelic artery and portal vein supply the red gland. The activity of the red gland is controlled by the vagus nerve. The dorsal aorta and post-cardinal vein supply blood to the oxygen-absorbing region. The sympathetic nerve regulates the activities of the oxygen-absorbing region (Fig. 21.2 and 21.3).

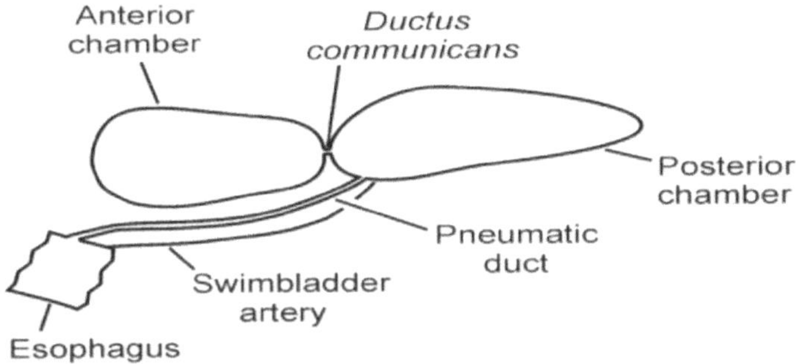

Fig. 21.2: Basic structure of swim bladder in fish

21.5 Types

Based on the presence of the ductus pneumaticus between the air bladder and oesophagus, the air bladder is divided into the following:

21.5.1 Physotomous condition (Group Physostomi) (Fig. 21.1a)

Between the air bladder and the oesophageal wall is a ductus pneumaticus (Fig. 21.4). A vessel merging from the coelicomesentric artery supplies the bladder with air, and blood is collected by the hepatic portal vein. It is found in ganoid and dipnoid fishes as well as soft-rayed teleosts. It showed the following modifications:

1. In *Polypterus*, primitive condition of the air bladder is seen. It is an unequally developed bilobed sac. The sac opens on the floor of the pharynx through the slit-like glottis. The glottis is provided with a muscular sphincter. It lacks alveolar sacculations but possesses muscular walls. It is supplied by a pair of pulmonary arteries and a hepatic vein below the sinus venosus.

2. In Dipnoi, the air bladder is called the lung and inner walls are produced into numerous alveoli. It resembles tetrapods' lungs both structurally and functionally. The air bladder lacks red bodies or red glands. It is single lobed in *Neoceratodus* but bilobed in *Protopterus* and *Lepidosiren*. The air bladder of *Neoceratodus* resembles *Lepidosiren*.
3. In *Acipensor*, the swim bladder is short and oval. Ductus pneumaticus enters the bladder ventrally and opens into the gut posterior to the pharynx. The opening is closed by simple constriction of ductus pneumaticus due to the absence of the glottis. The wall is fibrous and thick but the inner walls are smooth. Both lobes develop from the dorsal side but during the adult stage left lobe is obliterated.
4. In *Amia* and *Lepisosteus*, the unpaired air bladder extends nearly the entire length of the body. Rudiment of the left lobe appears for a short time. The ductus pneumaticus opens into oesophagus posterior to the pharynx through a dorsal slit-like glottis. The air bladder of *Lepisosteus* is more sacculated than that of *Amia*. In *Amia*, the air bladder gets arterial blood from pulmonary arteries but from the dorsal aorta in *Lepisosteus*. The blood is returned by the left ductus cuvieri in *Amia* and by the right post-cardinal in *Lepisosteus*.
5. In *Gymnarchus,* the efferent branchial arteries from the third and fourth gill arches join to form a common root for the emergence of pulmonary and coeliacomesentric arteries.
6. In *Clupea, Pellona, Caranx, Sardinella* and *Hilsa*: The ductus pnematicus opens into the fundus of the stomach and a second duct opens to the exterior near the anus. Sometimes, anterior prolongations come into close contact with the wall of space containing the internal ear.

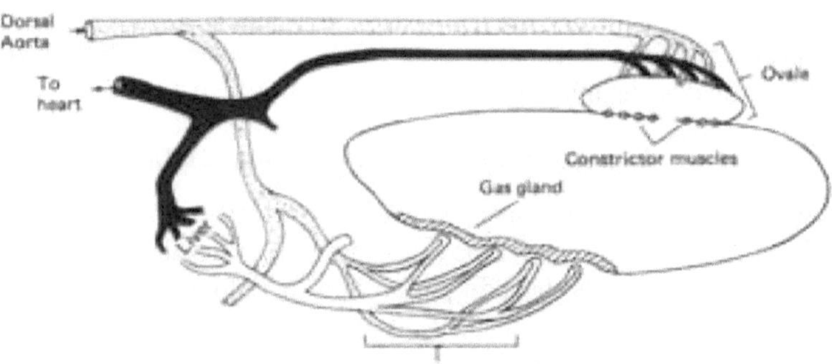

Fig. 21.3: Blood supply to swim bladder in fish

In *Clupea*, the narrow anterior end enters a canal in the basioccipital and divides into two slender branches. The anterior end of each branch dilates to form round swellings and lies in close contact with the internal ear.

7. In *Gadus*, a pair of diverticula originating from the anterior part of the bladder projects into the head region.
8. In *Otolithus*, each anterolateral end gives rise to an outgrowth that sends one anterior and a posterior horn.
9. In *Corvina loabta*, many-branched diverticula develop from the lateral wall.
10. In *Arius* and *Notopterus*, a rare longitudinal septum divides the swim bladder into two lateral chambers.

In shallow-water fish, the ratios of gases closely approximate those of the atmosphere, while deep-sea fish tend to have higher percentages of oxygen. For instance, *Synaphobranchus* has 75.1% oxygen, 20.5% nitrogen, 3.1% carbon dioxide, and 0.4% argon in its swim bladder.

21.5.2 Physoclistous condition (Group Physoclisti) (Fig. 21.1b)

There is an anteroventral secretory gas gland (which contains Retia mirabilis) and an oval-shaped posterolateral gas absorbing region. The ductus pneumaticus is closed or atrophied. The oval is caused by a degenerating ductus pneumaticus. It is supplied by the coelicomesentric artery and the dorsal aorta, but blood from the gastric gland and other parts is returned via the hepatic portal vein and the posterior cardinal vein. Spiny ray fish contain it. The ductus communicans connects its anterior and posterior compartments. Sphincters of circular and radiating muscles control the opening and closing of the aperture. It revealed the following changes:

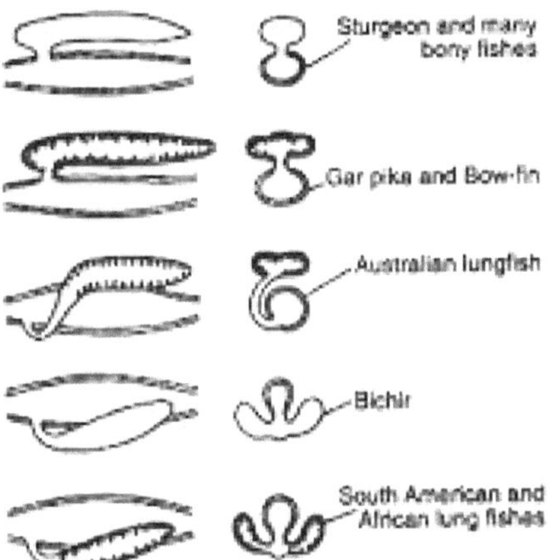

Fig. 21.4: Relationship between the swim bladder and lungs with oesophagus in some fishes

In Myctophidae, Percidae and Mugilidae: the posterior chamber with the retia mirabilia becomes flat and obliterated as an oval. The oval is a thin-walled, highly vascular area specialized for gas reabsorption. The opening of the oval is guarded by circular and radiating muscles.

21.5.3 Transitional condition (in *Anguilla*)

The ductus pneumaticus becomes enlarged to form a separate chamber containing an oval. The gas glands are also present. It is supplied by the coelicomesentric artery and the postcardinal vein.

21.6 Relationship of Air Bladder with Auditory Capsule

It evolves gradually and progressively. The connection is made by Weberian ossicles. The ossicles are not homologous to other vertebrate ear ossicles, but rather a specialization of a few anterior vertebral segments:

1. In *Gadus, Notopteru, Megalops* and Sparidae: A finger-like diverticulum from an anterior end joins the membrane containing an opening of the auditory capsule.
2. In Clupeidae (eg., *Hilsa*): A pair of tubular outgrowths centers inside the auditory capsule. Each branch again divides into two; each terminates into a swollen vesicle. These vesicles are enclosed inside the protic and pterotic. The vesicle is a process coming from the membranous labyrinth.
3. In carp and Siluroids: The connection is more advanced.

21.7 Functions

21.7.1 Hydrostasis

The swim bladder's primary function is to keep the fish's body weight equal to the volume of water it displaces. Its purpose is to re-adjust the body to its surroundings by increasing or decreasing the volume of gas. The ductus pneumaticus causes gas expulsion in the physostomous condition, but diffusion removes superfluous gas in the physoclistous condition. During this process, gas secretion reduces specific gravity, causing fish to sink.

21.7.2 Respiration

In *Amia, Chirocentrus, Lepidosteus, Megalops, Polypterus, Umbra,* ganoids and lungfishes, the air bladder serves as a lung. These fish come to the surface of the water regularly to gulp air. In physostomous fishes, it acts as an accessory organ. In physoclistous fishes (*Cyprinus, Opsanus tau, Perca fluviatilis*) the bladder stores oxygen to be utilized during deficiency. Fish living in low-oxygen water use their air bladders as a source of oxygen.

21.7.3 Sound Production

Dora, Physostoma, Malapterurus, Trigla can produce grunting, hissing or swimming sounds. The circulation of the contained air inside the swim bladder causes the vibration of an incomplete septum. The sound is produced as a consequence of the vibration of incomplete septa present on the inner wall of the swim bladder. The sound may also be produced by compression of the musculature of the swim bladder in *Polypterus* and *Lepidosiren*. In male *Cynoscion*, musclus sonorificus helps in compression.

21.7.4 Secretion and absorption of gas

The glycogen content of the gas gland epithelium of the air bladder is high, which splits to form carbon dioxide during gas secretion under the influence of carbonic anhydrase. At the same time, gas is absorbed from the air bladder if the density is reduced by attaching a float or injecting air into the bladder.

21.7.5 Buoyancy control

When *Dormitator latifrons* is in hypoxic water, their external respiratory organ emerges. At the top of the head, this accessory respiratory organ is highly vascular. After emerging from the water, the fish can inflate its gas bladder and thus stay near the surface of hypoxic water for a longer period.

21.7.6. Sensation

When a fish is subjected to pressure changes by moving into different depths of water, the enclosed air compresses or rarefies. This causes deformation of the air bladder wall, which acts as a pressure receptor similar to a manometer, barometer, or hydrophone.

21.7.7 Sound amplification

The air bladder also functions as an amplifier. Sound is produced in *Balistes* by stridulation of the pelvic girdle's basal bones.

21.8 Disorders

Swim bladder disorders can be caused by many factors including:

1. Water quality is one of the most overlooked aspects. Poor water quality can cause both acute and chronic stress in fish. Stress disrupts normal homeostasis, which can lead to negative or positive buoyancy disorders.
2. Goldfish (*Carassius auratus*) suffer from buoyancy issues.
3. Swim bladder disorders are also common in koi (*Cyprinus carpio*). Secondary changes in the swim bladder may occur in koi with spinal deformities or neurologic damage. These modifications, which may become permanent, will allow a Koi with limited mobility to survive in its natural habitat.
4. Inflation: Low water temperature and even genetic defects inflate the swim bladder of fish.
5. Bacterial and parasite infection causes inflammation of the swim bladder.

Weberian Ossicles

21.9 Introduction

The Weberian ossicles (=apparatus) form a remarkable chain of bony structures connecting the internal ear (=membranous labyrinth) with the swim bladder of several fishes. It is a series of four ossicles (tripus, intercalarium, scaphium, and claustrum) that connect the perilymphatic sac and the anterior end of the swim bladder (Fig. 21.5). It consists of the fused anterior-most vertebrae. They appear to assist in auditory, acoustic, hydrostatic and barometric functions etc.

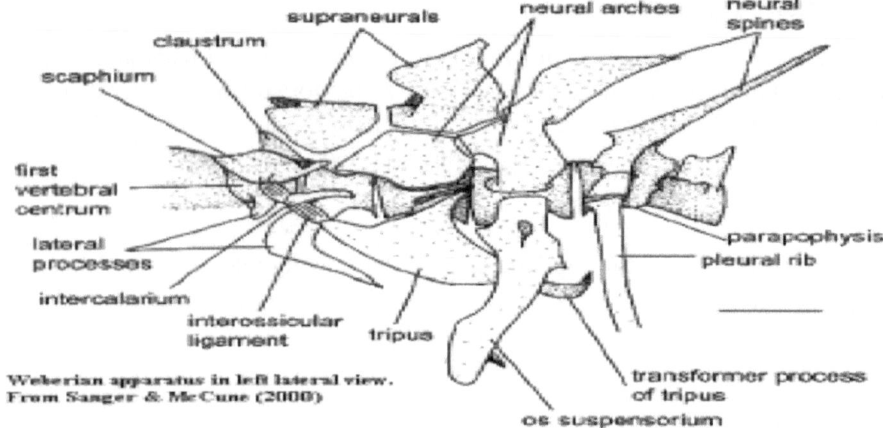

Fig. 21.5: Weberian ossicles (from https://fishionary.fisheries.org)

21.10 Historical

The Weberian ossicles were first described by Weber (1820). Bridge and Haddon (18890 found a considerable variation in the development, form and function of this organ. Frisch (1932) suggested that Weberian ossicles appear to assist in auditory function. According to Jones and Marshell (1956), Weberian ossicles permit a better and more precise adjustment of swim bladder volume. Many authors like Sorensen (18940, Mukerjee (19570 etc. have described homology on the morphology of Weberian ossicles.

21.11 Occurrence

Weberian ossicles are found in otophysine fish and comprise four orders Cypriniformes (carps and relatives), Characiformes (tetras), Gymnotiformes (South American knifefishes) and Siluriformes (catfishes) containing approximately 8000 species. It was seemingly present in a 'extinct plesiomorphic' otophysan fishes such as *Chanoides macropoma* (Patterson, 1984), *Clupavus maroccanus* (Cavin, 1999; Taverne, 1995), *Santanichthys diasii and Lusitanichthys characiformis* (Filleul & Maisey, 2004).

21.12 Location

The Weberian ossicles are placed on each side of the anterior vertebrae and extend between the membranous labyrinth and the bladder of their side.

21.13 Position

Of the four ossicles, the tripus, intercalarium and scaphium form the chain while the claustrum lies dorsal to scaphium at the anteriormost position.

21.14 Structure

Weber considered Weberian ossicles as homologous to the auditory ossicles of mammals and therefore, he called them claustrum, stapes, incus and malleus. However, Bridge and Haddon (1889) found these ossicles to be non-homologous with mammalian auditory ossicles, they designated ossicles as claustrum, scaphium, intercalarium and tripus. These terms are given by Bridge and Haddon. Thus, the Weberian ossicles include the following (Fig. 21.6):

21.14.1 Claustrum

The claustrum lies dorsal or anterior to the scaphium and in the wall of perilymphatic sac posterior prolongations. It either articulates with or forms a part of the neural arch of the first vertebrae. In *Wallago*, it is not seen as a distinct piece and appears to have fused with the neural arch of the first vertebra. It is the smallest in appearance.

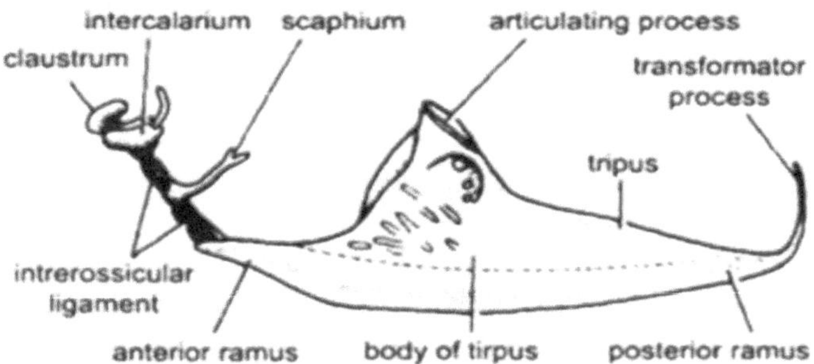

Fig. 21.6: Claustrum, stapes, incus and malleus in fishes

21.14.2. Scaphium

It is a slightly larger, broad and compressed structure. A ventral peg-like process articulates with a depression in the centrum of the first vertebra. It is connected to the intercalarium using an interossicular ligament and lies anterior to intercalarium.

21.14.3. Intercalarium

It may be a small nodule of a bone in the ligament, not connected to the vertebral column, as in siluroids (eg, *Wallago attu, Mystys seengala*), or it may be in the form of a rod like piece, reaching up to the centrum of the second vertebra, as in carps (*Labeo, Cirrhinus* and *Tor tor*). One end of the intercalarium may be bifurcated. It is connected to the tripus using of another ligament.

21.14.4 Tripus

It is the largest ossicle, lying on the anterior wall of the air bladder. it contains anterior, middle and posterior processes. The anterior process joins the interossicular ligament, while the middle process articulates between the second and third vertebrae. The posterior process is large and slightly curved and is known as the transformator. It is fenestrated in *Tor putitora*.

21.15 Homology

Despite being one of the most notable complex systems of teleost fishes and the subject of several comparative, developmental and functional studies, there is still much controversy concerning the origin, evolution and homologies of the structures forming this apparatus.

Since the Weberian ossicles were first described, many views have been put forward about their homologies. Many of the views are highly speculative as they are based largely on the morphological structure seen in adult fish. According to De Beer (1987) and Watson (1939), Weberian ossicles are mainly derived from the part of the first three vertebrae. These ossicles are not homologous to the ear bones found in mammals (Jones and Marshall 1953). Various views regarding their probable origin are summarized below:

21.15.1. Claustrum: It is believed to represent the following:

a. Interspinous ossicle (Sorensen, 1884)
b. Modified spines of the first vertebra (Wright 1884)
c. Neural arch of the first vertebra (Goodrich 1909)
d. Neural process of the first vertebra (Charnilov 1927)
e. Intercalated cartilage (Watson 1939)
f. Part of the neural arch of the first vertebra (Hora 1922, Mukerjee 1952, Rosen and Greenwood 1970)
g. Apophysis (Lagler et al, 1977)

21.15.2. Scaphium: It is believed to represent the following:

a. Neural arch of the first vertebra (Wright 1884, Sorensen 1894, Hora 1922, Charnilov 1927, Mukerjee 1952, Rosen and Greenwood 1970)

b. Neural arch of the second vertebra (Segemehl 1885)
c. Rib of the first vertebra (Goodrich 1909)
d. Partially neural arch of the first vertebra and partially mesenchyme (Watson 1939)
e. Apophysis (Lagler et al, 1977)

21.15.3. Intercalarium: It is believed to represent the following:

a. A compound bone containing neural arch and transverse process of the second vertebra (Bridge and Haddon 1893)
b. Neural arch of the second vertebra including ossified ligament or ossified ligament only (Sorensen 1894)
c. Neural arch of the second vertebra (Wright 1884, Goodrich 1909, Hora 1922, Charnilov 1929, Rosen and Greenwood 1970)
d. Partially basi-dorsal of the second vertebra and partially ossified ligament (Watson 1939)
e. Apophysis (Lagler et al, 1977)

21.15.4. Tripus: It is believed to represent the following:

a. Rib of the third vertebra, ossified ligament and ossified wall of the swim bladder (Sorensen 1894, Sorensen 1894)
b. Transverse process of the third vertebra including ossified outer wall of the swim bladder (Wright 1884)
c. Parapophysis including pleural rib of the third vertebra (Watson 1939, Charnilov 1929)
d. A compound bone formed by the transverse process of the third vertebra and the ribs of of the and fourth vertebra (Hora 1922)
e. Apophysis (Lagler et al, 1977)

21.16 Connections of Weberian Ossicles with Ear and Air-Bladder

The connection between the air bladder and the internal ear through Weberian ossicles is essentially similar. However, four types of such connections have been reported in fish. The first three types were described by Weber while the fourth by Srivastava.

21.16.1 First type connection

It is most primitive and lacks intimate anatomical association. It has low functional significance. it is found in Berycidae, Gadidae, Mormysidae, Notopteridae, Spanidae and few Serranidae.

In this type, a pro-coelomic duct or diverticulum is given out from the air bladder which simply appears at the base of the auditory capsule. In *Notopterus*, interiorly the air bladder dilates into a sub-spherical sac, which is divided into two auditory caecae to proceed interiorly towards their corresponding auditory capsule.

21.16.2 Second type connection

This connection is found in Clupidae, Chirocentridae, Myodintidae.

In this type, the anterior diverticulum of the air bladder forms a minute capillary tube or canal and extends into the head on each side. The procoelomic diverticlum of the air bladder divides into two fine capillaries canals just posterior to the dorsal aorta enclosed within the anterior air ducts. It seems to show advancement over the first one in the following respects:

a. The capillary tube being enclosed within the cartilage tube preventing it from collapsing pro-coelomic under high pressure and
b. Its penetration into the bone to establish a more intimate relationship within the internal ear

21.16.3 Third type connection

This connection is found in Cyprinidae, Characinidae, Gymnotidae, Ostariophysi, Siluroidae.

It is brought about by a series of joint articulated bones developed in connection with anterior thoracic vertebrae. All four bones of the series are connected using ligaments and the claustrum projects into the sinusimpar and communicate with the lymph contained therein. The connection attained perfection in siluroidae.

21.16.4. Fourth type connection

It has been reported in *Pama pama* of Scionidae and also in *Sciona coiter*.

It is a modification of the first type (Pandey and Shukla 2018-2019). In *S. coiter*, interiorly from the air bladder, two caecae arise and each is subdivided into three groups of clusters of small caecae, one of the clusters proceeds upwards and forwards by the sides of the corresponding auditory capsule which has on its anterior side an opening closed by a membrane. In *Pama pama*, two tubes arise from the hindermost of the bladder and proceed interiorly without dividing till the point where the air bladder ends. It should be placed between the first and second types.

21.17 Functions

21.17.1 Auditory functions

Frisch (1932) and Stetler (1938) suggested that Weberian ossicles assist in auditory functions. The ossicles transmit the vibrations of the bladder wall to the perilymph of sinus impar and then to the endolymph, to reach the saccular otolith, which are specially modified to receive these vibrations.

21.17.2 Localisation of sound

According to Evans (1926), as the vibrations received on the side of the bladder nearest the source might be stronger than on the other side. But the saccular otoliths of both sides are connected by a single transverse canal to the sinus impar to which the chain of the Weberain ossicles are attached, vibrations received by one otolith will be received equally by others and it would be impossible to distinguish right and left.

21.17.3 Sensitiveness of the sound production

Frisch (1932), Stetter (1938) reported that fishes having the connection between the air bladder and the internal ear through Weberian ossicles are sensitive to sound of higher and lower frequencies than in which such connection is not present. In this process, fluctuation in pressure makes the air bladder pulsate. The pulsation of the air bladder consequently causes ossicles to vibrate.

21.17.4 Hydrostatic function

This was attributed by Dijkgraff (1950). The pulsation of the air bladder caused by the change in its pressure is transmitted to the internal ear through the Weberian apparatus. The fish respond to changes in pressure by releasing gas from the bladder through the pneumatic duct. Weberian apparatus thus permits a better and more precise adjustment of air bladder volume (Jones and Marshall 1956).

21.17.5 Barometric function

Fishes like loaches can detect changes in atmospheric pressure and are thus forwarded to a change in weather. This is probably done by the connection of the air bladder and the internal ear through the Weberian apparatus.

22

Blood Vascular System

22.1 Introduction

The blood vascular system of fishes is modified to suit the aquatic mode of life. Fish contains the simplest single-circuit closed blood vascular system.

The structure of the heart in fish varies little. In fish, all of the chambers of the heart are essentially the same. The size and shape of the component chambers may differ. A large heart is found in extremely sluggish fishes with low oxygen requirements. The number and arrangement of valves in the conus arteriosus are the most distinguishing features in various fishes.

The heart is located ventral to the gills. It is located more anteriorly in teleosts than in elasmobranchs.

A sinus venosus, a ventricle, an auricle, and a conus arteriosus are all parts of the heart. The auricle and ventricle are considered true chambers, while the others are considered accessory chambers. The sinus venosus has a thin wall and is fed deoxygenated blood via veins. The auricle has thin walls as well and opens anteriorly into the ventricle. The ventricle is a muscular structure that gives rise to a tubular and contractile truncus/conus arteriosus. The conus arteriosus is gradually becoming extinct as evolution progresses (Fig. 22.1). The conus arteriosus continues forward as a strong tube known as the ventral aorta.

The general layout of the arteries and veins is similar. For oxygenation, the ventral aorta sends afferent branchial arteries to the gills. The number of gills is proportional to the number of afferent branchial arteries. The blood is returned to the arteries after being oxygenated in the gills. The number and location of the efferent branchial arteries differ between groups of fish (Fig. 22.1).

Blood returns to the heart via paired anterior cardinal veins and unpaired or paired posterior cardinal veins from various parts of the body. The venosus system is essentially the same in all fishes, but minor variations are common. The hepatic portal system transports blood from the alimentary canal, spleen, swim bladder, and gonads to the liver. Hepatic veins transport blood from the liver to the sinus venosus. The fishes have red blood cells that contain haemoglobin.

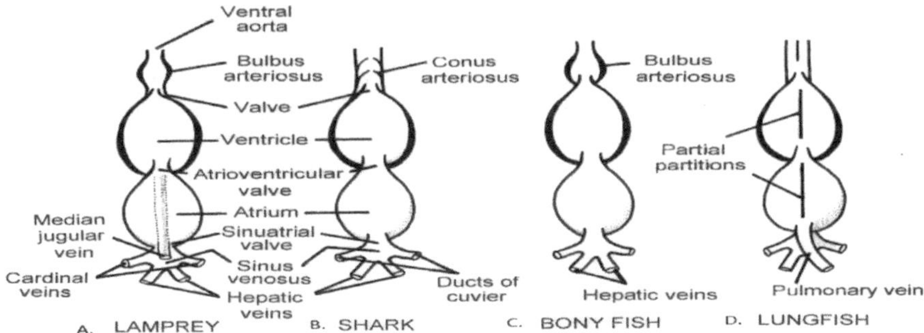

Fig. 22.1: Heart in different fishes

Six pairs of aortic arches are typically present, and their number corresponds to the number of gill pouches present in the fish. The first arch runs through the mandibular arch, the second through the hyoidean arch, and the last four through the branchial arches. They divide into a network of fine capillaries as they pass through visceral arches, supplying blood to the gills.

22.2 Cyclostomata

A sinus venosus, a ventricle, an auricle and a truncus arteriosus are all parts of the heart. Myxine's heart is similar to that of lamprey, except for the presence of common cardinal vein retention (Fig. 22.2).

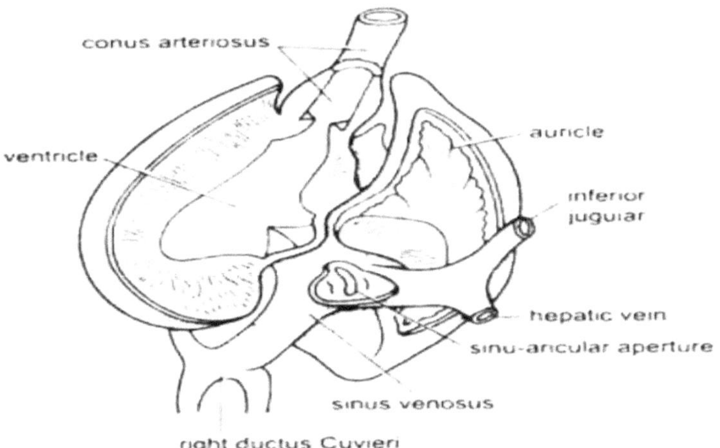

Fig. 22.2: Heart of *Petrormyzon*

Petromyzontia has eight pairs of aortic arches, which correspond to the same number of gill pouches, whereas Myxinoidea has up to fifteen pairs of aortic arches.

22.3 Elasmobranchs

In elasmobranchs, the heart is located in a large pericardial cavity and has a more primitive structure (Fig. 21.2 and 21.3). A sinus venosus, a ventricle, an auricle, and a truncus/conus arteriosus are all parts of the heart. The relative position of the auricle and ventricle, on the other hand, varies. The sinus venosus and auricle are located on the ventricle's dorsal side, as is the conus arteriosus. The valves in the conus arteriosus of elasmobranches and ganoid fishes are numerous and usually arranged in three longitudinal rows. The auricle advances over other fishes in many elasmobranchs, where it divides incompletely into left and right halves by an incomplete interauricular septum. This change is the result of the acquisition of aerial respiration.

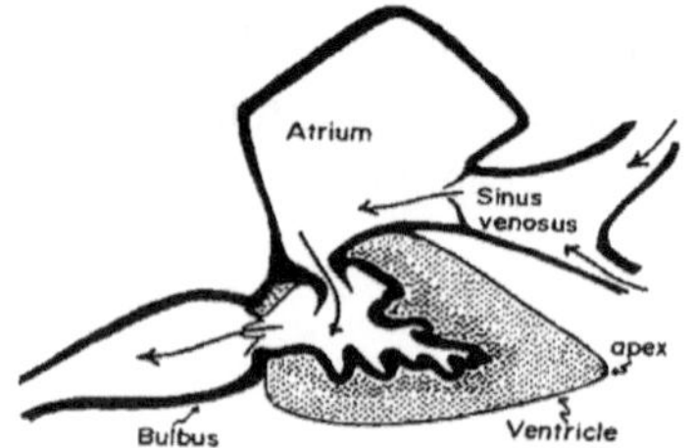

Fig. 22.3: Heart of Trout

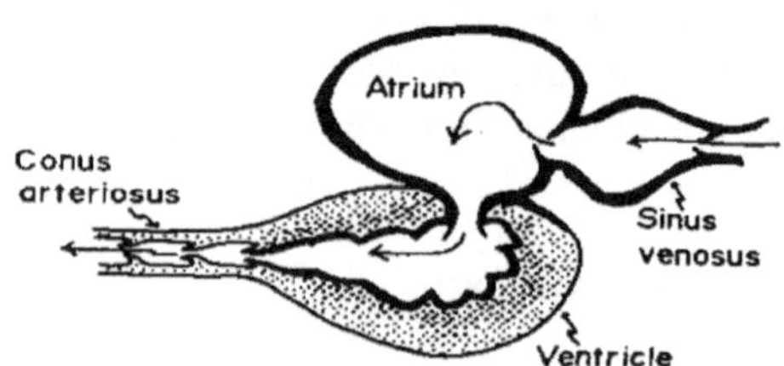

Fig. 22.4: Heart of Shark

There are six sets of valves in many elasmobranchs. However, the reduction in the number of valves in the conus arteriosus is a common trend in fish evolution (Fig.). *Squalus* has three longitudinal rows of valves, whereas *Heptanchus* has a dorsal and three proximal rows of valves.

There are five pairs of aortic arches in adult cartilaginous fishes. However, *Hexanchus* and *Heptanchus* have six and seven pairs of aortic arches, respectively.

22.4 Teleosts

The teleostean heart is small and varies greatly depending on the weight of the different fish species (Fig. 22.5 and 22.6a and b). Crossopterygii has an S-shaped tubular structure like tetrapods.

The internally undivided sinus venosus is a large chamber with a pair of lateral appendages in *Labeo, Cirrhinus* and *Catla*. A median elevation, on the other hand, divides the sinus venosus in *Amphipnous cuchia*. The sinus venosus is a thin-walled chamber in *Wallago attu, Mysutus aor* and *Clarias batrachus*, with a pair of membranous valves guarding the sinu-atrial aperture. In *Channa striatus*, the sinus venosus is small and the sinu-atrial aperture is absent. The sinu-atrial aperture in *Notopterus chitala* is protected by 8—10 nodular valves.

The ventral aorta runs forward and gives off four pairs of afferent branchial arteries in *Clarias batrachus, Mystus aor, Notopterus chitala, Rita rita, Tor putitora* and *Wallago attu*. In general, the third and fourth afferent vessels on each side originate from the ventral aorta. However, the arrangement of afferent and efferent vessels varies between teleosts. The second pair of afferent vessels in *Rita rita*, on the other hand, share a common origin from the ventral aorta.

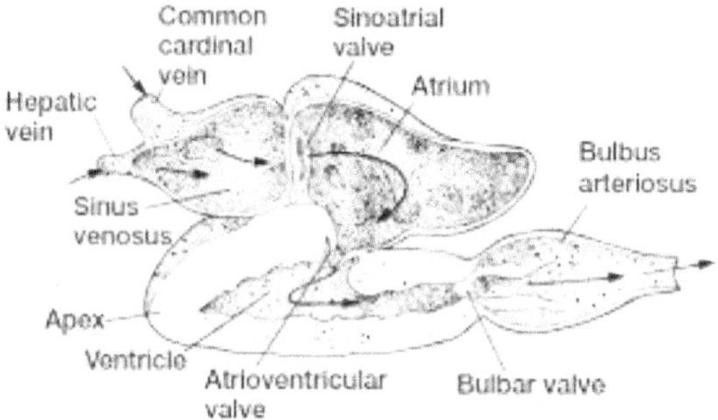

Fig. 22.5: Heart of teleost

In teleosts, each gill arch usually has only one efferent branchial vessel. *Catla catla, Notopterus chitala* and *Wallago attu*, on the other hand, have forked ventral ends to each efferent vessel. *Clarias batrachus* has one pair of efferent branchial vessels, which is considered the primitive feature in elasmobranchs. The efferent branchials may run singly or join to form epibranchials before opening directly into the lateral or dorsal aorta.

Circulus cephalicus is also found in teleosts in some form. It can be formed by the first and second pairs of efferent branchial vessels, or it can be formed by all four pairs of efferent branchial vessels.

Amia has both Conus arteriosus and Bulbus arteriosus. *Acipneser, Lepidosteus* and *Polypterus* all have muscular and contractile conus arteriosus. *Notopterus* has a muscular conus arteriosus as well. *Albula, Megalops* and *Tarpon* all have a distinct conus arteriosus with transverse rows of valves. Higher teleosts have a poorly developed conus arteriosus with one or two sets of valves. The valves in *Lepidosteus* are arranged in seven rows. Some teleosts lack a conus and instead have a bulbus arteriosus at the base of the ventral aorta. The bulbus arteriosus is a distal portion of the ventral aorta's base. It is not a part of the heart, but of the arterial system. It forms the main artery's basal trunk.

Six pairs of aortic arches are present in the embryonic stages of bony fishes, but only four pairs are seen in the adult stages. Various teleostean fishes have reported variations in the origin and arrangement of their aortic arches.

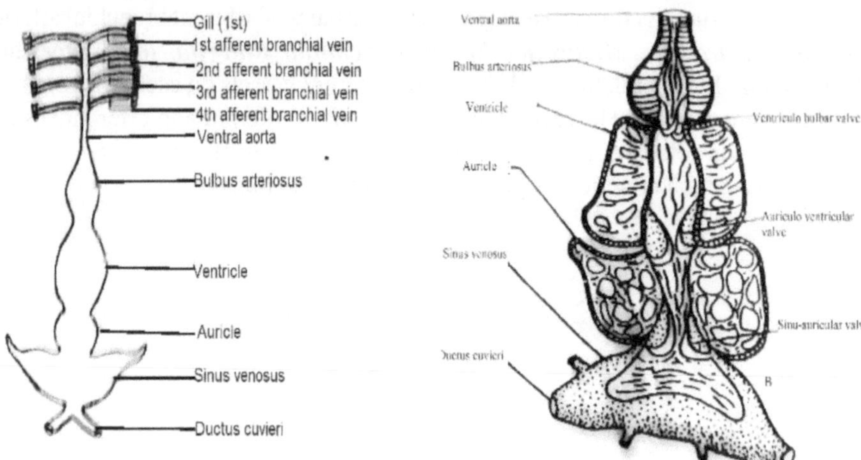

Fig. 22.6: (a) and (b) Heart of teleosts

22.5 Dipnoi

Dipnoan heart development outperforms that of other fish hearts. An incomplete septum divides the auricle into incomplete right and left halves. The swim bladder is converted into lungs, and oxygenated blood from the lungs is transported to the left auricle (Fig. 22.7a and b).

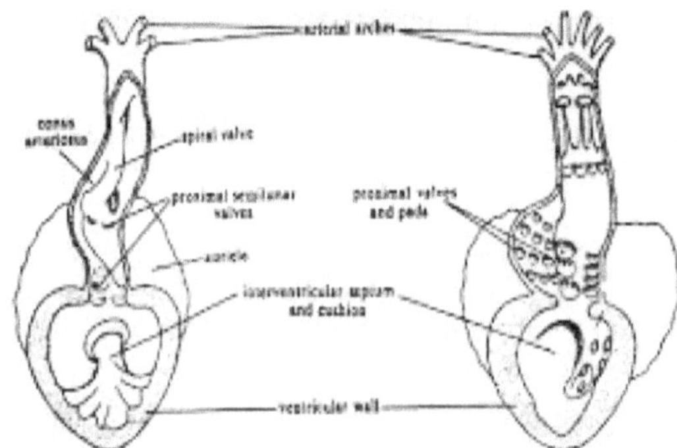

Fig. 22.7: (a) and (b) Heart of Dipnoi

Adult dipnoans have four to five pairs of gill arches, the first and second of which are lost during metamorphosis.

22.6 Blood in Fishes

Blood accounts for about 2-3% of a fish's body weight. Plasma and blood cells are found in the blood. Extremely sluggish fishes with low oxygen requirements have a large blood volume. Blood cells are found in the spleen of adult teleosts. Some marine Antarctic teleosts in the Chaenichthyidea family have colourless blood. The blood in eel leptocephalus is colourless.

The erythrocytes are oval and have haemoglobin nuclei. They range in size from 20,000 to 3000000 mm^3 of blood. An active, fast-swimming predatory fish has many more erythrocytes than a sluggish, bottom-dwelling, sedentary one.

The amount of haemoglobin in fish blood varies with the total number of cells in the blood. In lower fishes, haemoglobin is monomorphic, but in higher fishes, it is tetramorphic. Goldfish and rainbow trout have different types of haemoglobin. The level of activity of fish is linked to haemoglobin polymorphism.

Leucocytes range from 20000 to 150000 mm^3 of blood. The main types of leucocytes in fishes are large (0-3%) and small (20-38%) lymphocytes, monocytes, myelocytes. metamyelocytes. band and segmented neutrophils (7.5-37%), eosinophils (3-19%), basophils and thrombocytes (28-52.5%).

23

Excretion in Fishes

23.1 Introduction

Excretion is the elimination of nitrogenous metabolic wastes from the gut and skin of the body. Through excretion, certain substances are removed while others are conserved. Fish are either ammonotelic or ureotelic animals. Kidneys play the most important role in excretion to maintain water-salt balance.

Fish gill ammonia excretion has been investigated for 80 years, but its mechanisms are still up for debate. A new paradigm has been established by the relatively recent identification of the ammonia-transporting role of the Rhesus (Rh) proteins, a family related to the Mep/Amt family of methyl ammonia and ammonia transporters in many living organisms. These proteins also occur as glycosylated proteins in fish gills (Wright and Wood, 2009).

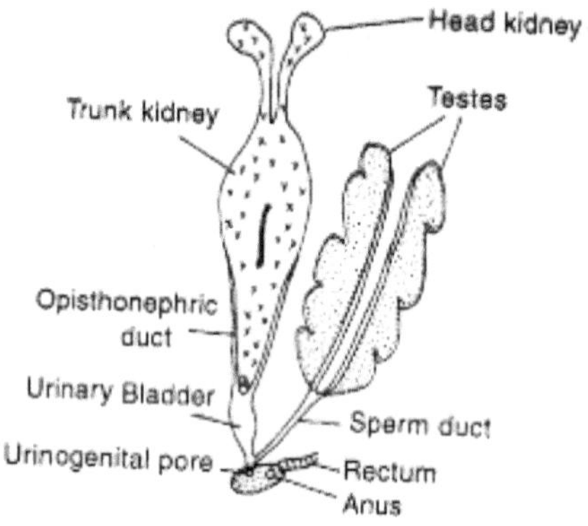

Fig. 23.1: Excretory system in a male fish

Takvam et al. (2021) stated that the monovalent ion transporters for Na^+, Cl^-, and K^+, the water transport channels aquaporins (AQP) and claudins (CLDN), and the divalent ion transporters for SO_4^{-2}, Mg^{+2}, and Ca^{+2} are all included in the renal transporters.

23.2 Excretory Organs

Although the gills are the chief respiratory organs, they are also important in excretion and osmoregulation. A pair of elongated kidneys lies above the alimentary canal close to the vertebral column. Adult fish have opisthonephric kidneys (Fig. 23.1). Although built on the same fundamental plan, the kidneys in fish show great variation in shape. Generally, kidneys are extremely elongated and extend the entire length of the body cavity. Usually, the posterior sides of the kidneys are fused and the anterior ends remain free. The anterior part of the kidney is non-renal in almost all forms.

In male elasmobranchs, the anterior part of the kidney is well formed and serves reproductive functions, but in females, this part has degenerated. In teleosts, the anterior portion of the kidney is converted into lymphatic tissue. The teleost kidney lacks connection with the testis.

The kidney of freshwater fish is often larger in body weight than that of marine fish.

23.3 Types of Kidney

Based on organogenesis, the kidneys of fish are of the following two types:

1. Head Kidney: This type of kidney originated from pronephros. It refers larval kidney which lost its excretory functions in the adult stages. It lies in the craniovertebral region of adult fishes. It compasses a mass of pseudolymphoid tissue, interregnal and chromaffin tissues and endocrine elements. Various stages of haemopoesis have been found to occur in it. It has completely degenerated in elasmobranchs and Dipnoi.
2. Trunk kidney: It originated from opisthonephros. It extends over a region, which in amniotes forms the metanephric kidney. These are paired, long and strap-shaped bodies lying retroperitoneally on either side of the dorsal aorta and posterior cardinal vein. Anteriorly, the kidney continues forward up to the hind end of the opercula and posteriorly terminates a little in front of the anus.

The two mesonephric ducts are usually present and each runs along the outer border of the corresponding kidney. The two ducts always fuse to form a common duct. The fusion may occur at the posterior end of the kidney in *Mystus* or at some point between the kidney and urinary papilla. They remain separate till the kidney opens into the urinary bladder as in *Labeo* and *Cirrhinus*. In *Discognathus lamta*, a pair of anterior and posterior mesonephric ducts is present which join to form a common duct on each side. In *Barbus* and *Mystus*, a distinct sac-like structure on one side of the common mesonephric duct is present.

A pair of abdominal pores may be present on either side of the genital aperture and are believed to be special openings of the coelom to the exterior.

The paired kidneys fuse in various ways in different fishes. Based on their fusion, Ogawa (1961a) has categorized them into the following five types:

1. Type I: There is no distinction between the head and trunk kidney. Kidneys of both sides fused throughout their length (Fig. 23.1A). For example: The kidney of Clupeidae and Salmonidae.
2. Type II: There is a clear distinction between the head and trunk kidney. Their middle and posterior portions are fused (Fig. 23.1B). For example: The kidney of marine catfish, Plotosidae, Anguillidae and Cyprinidae.
3. Type III: The head and trunk kidney are closely distinguishable. Their anterior portion is represented by two slender branches but the posterior portion is fused (Fig. 23.1C). For example: The kidney of Belonidae, Scopelidae, Mugilidae, Scombridae, Carangidae, Pleuroectidae, Cypridontidae, Cotidae and Gasterosteidae.
4. Type IV: There is no distinction between the head and trunk kidney. Only the extreme posterior portion of the kidney is fused (Fig. 23.1D). For example: The kidney of Syngnathidae and Pipefish.
5. Type V: The two kidneys are completely separated (Fig. 23.1E). For example: The kidney of Lophiidae.

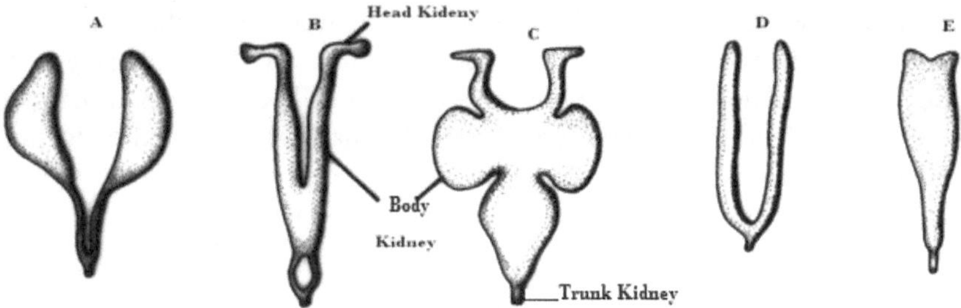

Fig. 23.2: Five types of kidneys in teleosts

In freshwater teleosts, kidneys of types 1, 2 and 3 are found and the following modifications are observed:

1. *Lepidocephalichthys thermalis* has a completely fused kidney and a slightly bifid anterior end showing double nature.
2. The trunk kidney is broad in the middle and gradually narrows in the hinder part in *Cirrhinus, Labeo* and *Barbus*.
3. In *Mystus, Amiurus, Dactylopterus* and *Arius*, the head kidney is completely separated from the remaining part.
4. *Cirrhinus mrigala, Notopterus notopterus, Eutropiichthys vacha* have head and trunk kidneys with a clear-cut distinction. The two parts are connected

by an isthmus. The head kidney of the two sides is fused in *Cirrhinus* and *Eutropiichthys* but separated in *Notopterus, Clarias* and *Heteropneustes*.

23.4 Structure of Kidney

Histologically, the trunk kidney is made up of a large number of nephrons, which are held in lymphoid tissue. A Malpighian capsule (= renal corpuscle) and tubule are found in each nephron. The structure, shape, size and number of nephrons vary greatly in different groups of fish, as described below (Fig. 23.3):

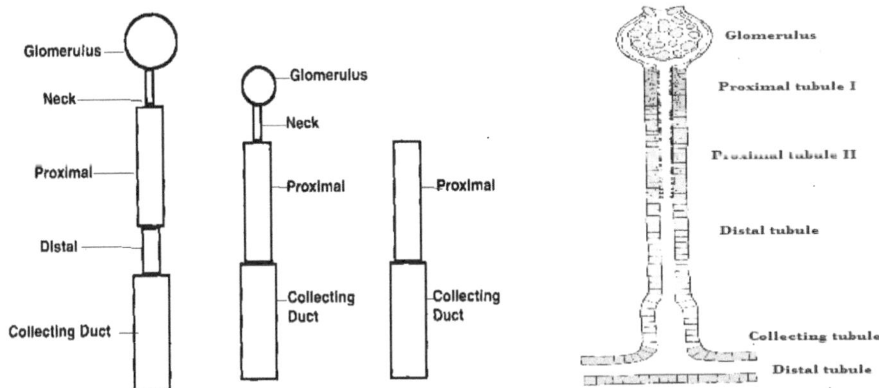

Fig. 23.3: Nephrons in fishes (A) Freshwater (B) Saltwater and (C) Aglomerular

Fig. 23.4: Details of tubules in fishes

1. In Elasmobranchs: Nephrons are very long and possess large glomeruli which exceed 10000 mm^3 of the glomerular volume of m^{-2} of the fish surface. A long signet leading off from the glomerulus and lined with cuboidal epithelium. An initial proximal segment consists of two parts (Fig. 23.4). The first part is narrower and lined with tall cuboidal cells. The second part is greater in diameter and has a higher epithelium. The presence of mitochondria with Periodic acid Schiff (PAS) positive brush border in the first part and lysosome in the second part are characteristics. A second proximal segment represents a larger part of the nephron. It is larger in diameter and lined with tall columnar cells showing striations due to the presence of numerous, closely packed mitochondria. A distal segment lined with basophilic cuboidal cells and devoid of brush border. A large collecting duct segment lined with tall columnar cells and containing mucus granules at their apical ends (Fig. 23.5).

The nephros of *Hydrolagus* (ratfish) are similar to those of elasmobranchs, but the glomeruli are less vascular and the neck and initial proximal segments are relatively shorter and simpler.

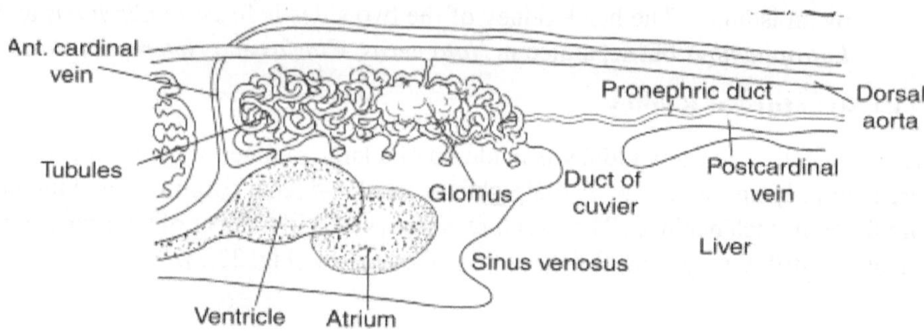

Fig. 23.5: Structure of glomerulus in fish

2. In Teleostomi: It is generally held that the bony fishes evolved in freshwater and it was only later in evolutionary history that many of them migrated to seawater.

In freshwater teleosts, the kidneys were equipped with nephrons, to act as efficient water filtration and absorption devices for hyperosmotic body fluid regulation. Their renal corpuscles contain a well-vascularised glomerulus and an inconspicuous mesangium. Their neck segment varies in length and is lined with ciliated columnar epithelium. An initial proximal segment is characterised by a brush border and lysosome. A second proximal segment with a less developed brush border contains numerous mitochondria. A narrower intermediate segment is often found between the second proximal segment and the distal segment. The distal segment is narrower in diameter and contains clear cells with elongated mitochondria. The collecting duct segment narrows further and is lined with cuboidal and goblet cells.

In marine teleosts, structural modifications in their nephrons are produced for hypo-osmotic body fluid regulation. The number, shape and size of renal corpuscles greatly vary. They are usually less numerous, poorly vascularised and grow mostly into non-functional structures in *Therapon, Arius, Myxocephalus, Paralichthys* etc. The glomerulus has a central avascular core surrounded by capillary loops, effectively reducing its functional area in *Chirocentrus, Tricanthus, Myoxocephalus* etc. In *Lophius*, the connection between the glomerulus and the tubule is lost in adults. *Opasanus tau, Porichthys notatus*, *Hippocampus, Sygnathus* and other species lack renal corpuscles. The neck segment is very short. The proximal segment comprises two or three parts. An intermediate segment is seldom present. The distal segment is inconspicuous because of its absence. The collecting tubule led off directly from the proximal segment to open into the collecting duct.

23.5 Function of Kidney

The renal tubules remove excess water, salts, waste material and foreign substances from the blood of fish. For this function, blood is filtered through the glomerulus.

During filtration, inorganic ions, glucose, urea and amino acids are filtered under the arterial blood pressure into the tubule. The filtrate flows through the tubule and during this movement, glucose and salts are reabsorbed from the filtrate. At the same time, water is absorbed so that the urine becomes hypertonic.

The gills remove 90% of the ammonia, urea and carbon dioxide in the water. Marine fish can excrete salt by using chloride cells in the gills.

The kidney in freshwater fishes removes creatine and uric acid. Their kidneys contain a large number of glomeruli, which produce copious amounts of urine (5-12% of body weight) daily.

Kidney in marine teleosts removes traces of ammonia along with Mg^{+2}, SO_4^{-2}, Na^+, Cl^-, K^+, Ca^{+2} and TMO. Their kidneys have few glomeruli and produce very little urine (3 ml/kg/day) daily.

There are several teleosts, for example: salmon, that travel between freshwater and seawater and must adjust to the reversal of osmotic gradients. They adjust their physiological processes by spending time in the intermediate brackish environment.

The main function of the kidney is water excretion.

24

Osmoregulation in Fishes

24.1 Introduction

Water is a universal biological solvent and the medium in which most of the cellular reactions of metabolism occur. Fish contain a large amount of water in their cells and extracellular fluids. It is important to a fish that its water content be nearly constant. This more or less steady state of water content in the body can only be maintained when there is a quality difference between the amount of water entering and leaving the body. The process of regulating the movement of water and its volume in the body is known as osmoregulation.

In fish, cortisol is the seawater-adapting hormone, while prolactin is the freshwater-adapting hormone. Recent evidence indicates that the growth hormone/insulin-like growth factor-I is also important in seawater adaptation in several teleosts (McCormick, 2015).

The terms and information related to osmoregulation are given below:

1. Osmolarity: It is the concentration of body fluid of an animal. It varies between 350-450mOsm/L in fish.
2. Hober (1902) coins the term osmoregulation.
3. According to Hickman and Trump (1968), elasmobranches probably do not drink seawater.
4. Evans (1980) reviewed the osmotic and ionic regulation in teleosts.
5. Isotonic fishes: The body fluids and blood of these fishes (like Hagfishes) have the same concentration as that of the surrounding water. They never face the problem of osmoregulation. There is no movement of water across the membrane, hence no osmotic pressure. A cell or an organism does not change the volume.
6. Hypotonic fishes: These fishes (like freshwater fishes) live in a medium of lower salt concentration. They face the problem of osmoregulation.
7. Hypertonic fishes: These fishes (like marine fishes) live in a medium of higher salt concentration. They also face the problem of osmoregulation.
8. Stenohaline: Animals have only a limited tolerance to change in the osmotic

concentration of the external environment or are restricted to a narrow range of salinity (Fig. 24.1). Example: *Arenicola.*

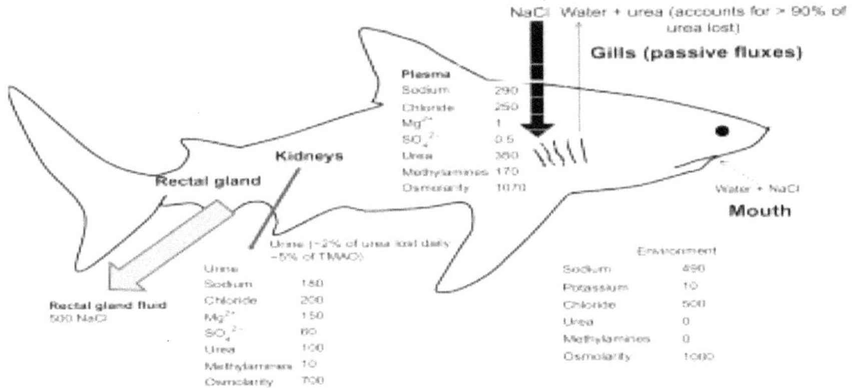

Fig 24.1: Osmoregulation in stenohaline fish (from https://www.sciencedirect.com)

9. Euryhaline: Animals that can tolerate a wider range of osmotic concentrations (Fig. 24.2). Examples: Anguilla, *Aplysia, Fundulus, Gasterosteus, Hilsa, Mystis, Phascdosoma, Salmon.*

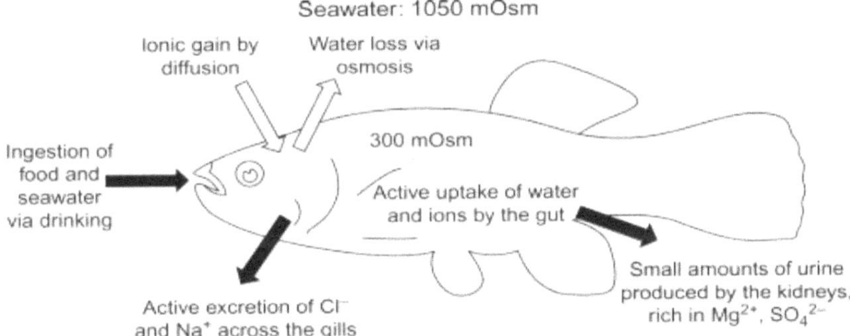

Fig. 24.2: Osmoregulation in euryhaline fish (from https://www.sciencedirect.com)

10. Osmoconformers: Animals that are osmotically labile and whose body fluid concentrations change with the medium having high tissue tolerance. Examples: *Arbacia, Golfingia, Doris, Oncidium*, Lobster, Echinoderms, Cephalopods, Ascidia and most coelenterates.

11. Osmoregulators: Animals that are osmotically stable and can maintain their osmotic concentration at a constant level. Examples: Most freshwater fishes, *Guida alba, Artemia salina, Carcinus,* mosquito larvae and chordates.

12. Volume conformers: Animals that can able to change in volume as the external osmotic concentration changes.

13. Volume regulators: Fishes that maintain a constant volume despite external osmotic changes.
14. Exosmosis: Body fluid has less solute concentration and more water concentration.
15. Endosmosis: Body fluid has more solute concentration and less water concentration.
16. Anadromous fishes: Fishes migrating from sea to freshwater.
17. Catadromous fishes: Fishes migrating from freshwater to sea.
18. Diadromous fishes: Fishes migrating from freshwater to sea and vice-versa.

24.2 Process of Osmoregulation

The nature of osmoregulatory problems is quite different in different environments and various groups of fish inhabiting them then solve them by employing a variety of mechanisms:

24.2.1 Isotonic (Isosmotic) Fishes

Inorganic ions account for the majority of the osmotic pressure of extracellular fluid in these fish. Thus, the osmotic problems are nearly absent, although ionic regulation is still essential. These produce urine relatively slowly and their composition differs significantly from that of the plasma ions other than Na^+ and Cl^-. Example: Hagfishes (Table: 24.1).

Table 24.1: Composition of extracellular fishes (From Nielsen & Mackay, 1972; unit: mOsm/L)

Sl.No.	Name of fish	Habitat	Na^+	K^+	Ca^{+2}	Mg^{+2}	Cl^-	SO_4^{-2}	HPO_4^{-2}	Urea
1.	Hagfish	Marine	554	6.8	8.8	23.4	532	1.7	2.1	3.0
2.	Lamprey	Freshwater	120	3.2	1.9	2.1	96	2.7	-	0.4
3.	Dogfish	Marine	264	4.2	3.2	1.2	256	1.0	1.1	375
4.	Shark	Freshwater	200	8.0	3.0	2.0	180	0.5	4.0	132
5.	Goldfish	Freshwater	142	2.0	6.0	3.0	107	-	-	-

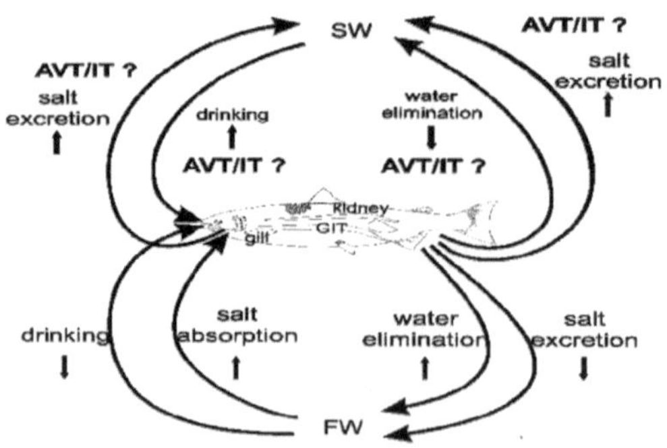

Fig. 24.3: Physiological responses of fish to different salinities (from https://www.researchgate.net)

24.2.2 Hypotonic (Hyperosmotic) Fishes

To balance the osmotic influx of water, the fishes (e.g. elasmobranches and coelacanths) produce large amounts of urine. This is made possible by a high glomerular filtration rate and minimal tubular water reabsorption (Table: 24.1 and 24.3). Some important points regarding it are given below:

1. They produce urine at the rate of 60ml/kg body weight/24 hours.
2. Urine is dilute with depression of freezing point (= -273.1^{0}K).
3. They are larger (diameter = 48-108m) and more numerous (about 10000/ kidney) glomeruli.
4. Some lack distal tubules as in aglomerular fishes and euryhaline fishes.
5. According to Evans (1980), the collecting tubule and collecting duct are major sites of urine dilution by electrolyte absorption. According to him, Cl^- is also linked to the Na^+/K^+ exchange system.
6. According to Payan (1978), gill epithelium is the site of active ion uptake.
7. These have the same range of blood concentration.
8. They take water equivalent to 30% of their body weight daily.

In these fishes, osmoregulation is brought about by

1. Renal complex: The kidney contains many more glomeruli than marine fishes. It contains the following:
   a. Neck segment: A segment containing ciliated cells.
   b. Proximal segment I: It is the site of isosmotic Na^+ and Cl^- removal. It contains a prominent brush border and numerous lysosomes.
   c. Proximal segment II: It is the site of divalent ions secretion. It contains a less-developed brush border and numerous mitochondria.
   d. Distal segment, collecting tubule, collecting duct and urinary bladder sites of reabsorption of Na+ and Cl-.
   e. A narrow ciliated intermediate segment may be absent in some species.
2. Gill epithelium and branchial glands: Electron microscopic studies of the gills of *Anabas* and *Clarias* show a multilayered epithelium covering the Lamellae vary from 5-18m in thickness. The following types of goblet cells are found in the gills of fishes:
   a. Mucous glands: They provide a thin film of mucous to prevent dehydration of respiratory epithelium.
   b. Acidophilic granular cells
   c. Basophilic mast cells
   d. Glandular cells: Either binucleated or trinucleated

e. Acidophilic chloride cells: It was discovered by Keys and Willmer (1957). These are aggregated in large numbers at the base of the lamella and are in contact with water and blood. Munshi (1964) used histochemical $AgNO_3$ techniques to localize these cells. Further studies by Munshi and Hughes (1973) reveal that it contains many mitochondria and rough endoplasmic reticulum.

3. Gut plays a minor role.

Water balance in these fishes is performed by:

a. To cope with the steady inflow of water, these produce a large amount of hypotonic urine by a large number of glomeruli.

b. Due to copious secretion of urine, these have a urinary bladder.

c. Urine contains creatine, creatinine, amino acids, urea and little ammonia.

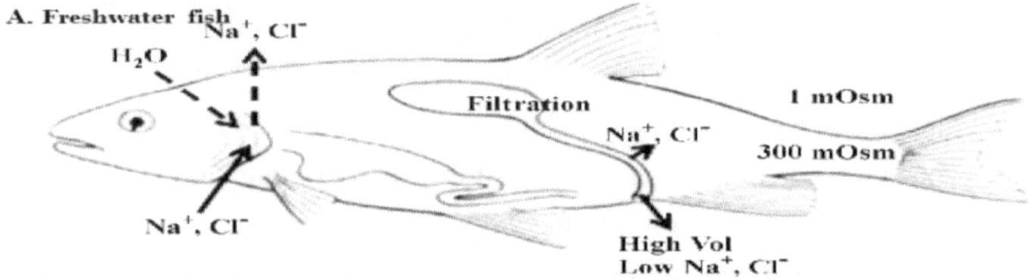

Fig. 24.4: Osmoregulation in freshwater teleost fish (from https://www.semanticscholar.org)

Salt balance in these fishes is performed by:

a. Some Cl^- is lost into the water by diffusion and some pass as faeces and urine.

b. Some ions like Li^+, Na^+, Ca^{+2}, Mg^{+2}, Cl^-, Br^-, HPO_4^{-2}, and SO_4^{-2} are absorbed by gills and oral membranes.

c. Some salts are easily replaced through the food in the alimentary canal.

d. Na^+ and Cl^- are actively absorbed by the urinary bladder from the stored urine (Renfrn, 1975).

e. A Na^+/K^+ activated ATPase system seems to mediate a Na^+/K^+ exchange system for Na^+ uptake. Cl^- is also linked to the Na^+/K^+ exchange system (Evans, 1980).

f. Payan (1978) demonstrated that Na^+ uptake by chloride cells is geared to NH_4^+ and H^+ efflux. Carbonic anhydrase is involved in it and causes the production of Na^+ and HCO_3^- from CO_2 and water. Diamox causes a distinct fall in Na^+ and Cl^- in some fishes.

24.2.3 Hypertonic (Marine Teleosts) Fishes

Marine teleosts and lamprey live in a hyperosmotic medium and face problems almost opposite to those of freshwater teleosts. The blood concentration of marine teleosts is a little higher than freshwater forms, coming to a concentration of only about half of seawater (Fig. 24.1 and 24.5). Some important points regarding these fishes are given below:

1. Smith (1930) first studied the physiological mechanisms of marine teleosts.
2. The glomeruli are small in size (27-94) and fewer (5000/kidney) even absent in at least 23 species.
3. Urine has depression of freezing point (= -273.6 to -273.9^{0}K)
4. They produce urine 3ml/kg body wt/24 hr.

In these fishes, osmoregulation is brought about by

1. Renal complex: A typical nephron of a marine glomerular teleost contains the following:
    a. Renal corpuscle: It contains a glomerulus
    b. Neck segment: It may be short

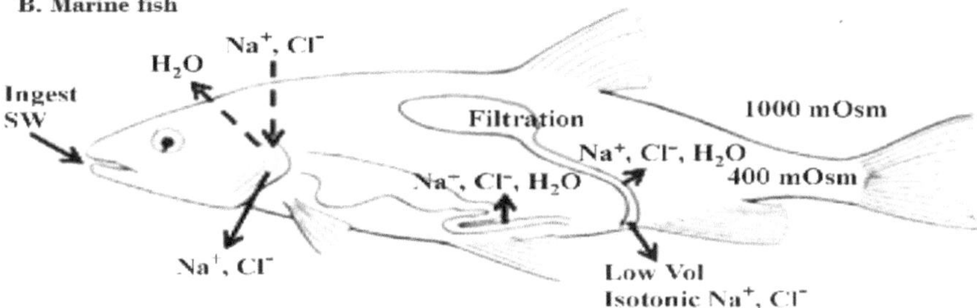

Fig. 24.5: Osmoregulation in marine teleost fish (from https://www.semanticscholar.org)

    c. Proximal segment: It is two or three
    d. Intermediate segment: It is variable.
    e. Collecting tubule:

2. Gill epithelium: Cl^- cells are quite numerous in the epithelium lining the opercular cavity. It is not prominent at the apical membrane where it should be from where extrusion normally occurs Na^+/K^+ activated ATPase has been detected in Cl^- cells also known as ionocytes.
    a. According to Evans (1980), the exact mechanism of salt extrusion by and elsewhere is not clear.
    b. It is agreed that Cl^- is expelled by active transport, but there is controversy regarding Na^+ removal.

c. Na^+ extrusion is coupled with K^+ influx through Na^+/K^+ activated ATPase.

d. Ammonia along with the excess slats (Na^+, Ca^{+2}, Mg^{+2}, Cl^-, SO_4^{-2}) is excreted by gills.

3. Gut: According to Hickman (1968), the majority of Na^+, Cl^- and water is reabsorbed by the gut and most of Mg^{+2} and SO_4^{-2} remain in the gut and are expelled with faeces. Na^+ and Cl^- are removed per the extra-renal route. Ca^{+2} is partly absorbed by the gut and a part of allowed to be removed with faeces. Reabsorbed Ca^{+2} is excreted through the renal and the extra-renal routes.

The physiology of osmoregulation in fish includes the following:

a. In aglomerular kidney, urine formation is secondary to the secretion of monovalent and divalent ions across the proximal tubule with water flowing osmotically. Urea and wastes are secreted into the kidney.

b. Na^+, Cl^- and water are absorbed isotonically in the proximal tubule and collecting tubule, but Mg^{+2}, Cl^- and SO_4^{-2} are secreted into the proximal tubule with water following passively.

c. Urinary bladder when present is more permeable to the eater. It actively absorbs water and reduces urine volume by lowering its solute concentration although a little Na^+, Cl^- may be absorbed.

d. The urine is hypotonic to its blood but rich in salt content, Na^+, Ca^{+2}, Mg^{+2}, Cl^-, SO_4^{-2}, NH_3, urea and TMBO etc.

24.2.4 Hypertonic (Marine Elasmobranch) Fishes

The blood of elasmobranches contains about the same amount of salts as that of marine teleosts and thus they are osmotically hyposmotic to the medium. They have adapted entirely different means to maintain the osmotic equilibrium of the body fluids (Table: 24.1 and 24.3). The following points are important regarding these fishes:

1. They maintain a higher internal osmotic pressure than surrounding seawater, the osmotic concentration of the blood is often 10-50 mCsm/kg more than that of seawater.
2. The urine is always slightly hyposmotic to blood (50-200 mOsm/L)
3. The structures concerned are similar to marine teleosts.

Renal complex

a. The osmotic tonicity contributed by electrolytes is less than half and the rest is made up of urea and TMAO. These ate in higher amount.

b. Na^+, K^+, Ca^{+2} and Cl^- are reabsorbed from the filtrate. Na^+ is absorbed by the active transport and Cl^- follows passively.

c. It excretes Mg^{+2} and SO_4^{-2}.

d. There is a progressive increase in GFR and urine flow occurs leading to lowered Na^+ and Cl^- levels in the blood.

e. These are capable of retaining a large concentration of urea and other dissolved salts in their body fluids which keep the body fluids either isotonic or hypertonic to seawater and thus do not lose water by osmosis.

24.3 Hormonal Control

Hormones influence the amount of urine flow by altering the filtering rate in the renal corpuscles through changes in blood pressure. Thus, the amount of urine output is altered as required. Similarly, salt balance is maintained by hormones controlling the diffusion and absorption of ions by the chloride cells in the gills. Adrenocortical hormones, thyroid hormones and produced by suprarenal are also known to influence excretion and osmoregulation in fishes. prolactin also has a role in osmoregulation (Table: 24.6).

Adrenocortical hormones play a role in the regulation of gills and kidney functions. The blood of trout contains hydrocortisone, corticosterone and other cortical compounds. The administration of corticosteroids minimises sodium elimination through the kidney and increases excretion through the gills.

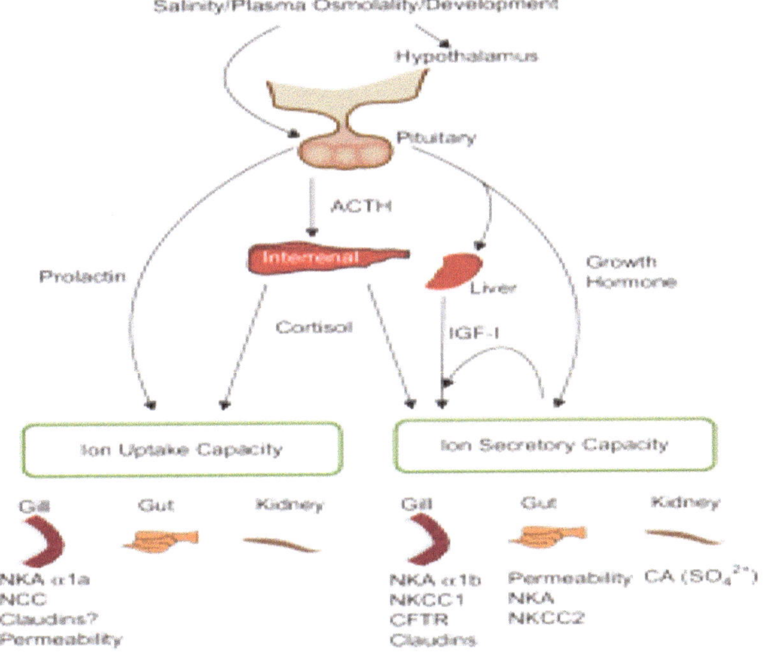

Fig. 24.6: Hormonal control of osmoregulation in fishes

The gills of *Anguilla,* when subjected to adrenaline showed a strong vasodilator effect.

The gonadal hormones may also exert a hydrating effect using sodium retention.

25

Nervous System in Fishes

25.1 Introduction

All fishes have a similar nervous system organisation. Meninx primitive is a single protective covering that covers the brain and spinal cord. The brain varies in size about the body in different species and does not occupy the entire cranial cavity (Fig. 25.1).

25.2 The Brain

Except for elasmobranchs, which have larger brains, the relative ratio of the brain has been observed to be fairly constant in all fishes. The brain-body weight ratio, also known as the cerebro-somatic index (CSI), is calculated as follows:

$$\text{CSI} = \frac{\text{Weight of brain}}{\text{Weight of body} - \text{Weight of brain}} \times 100$$

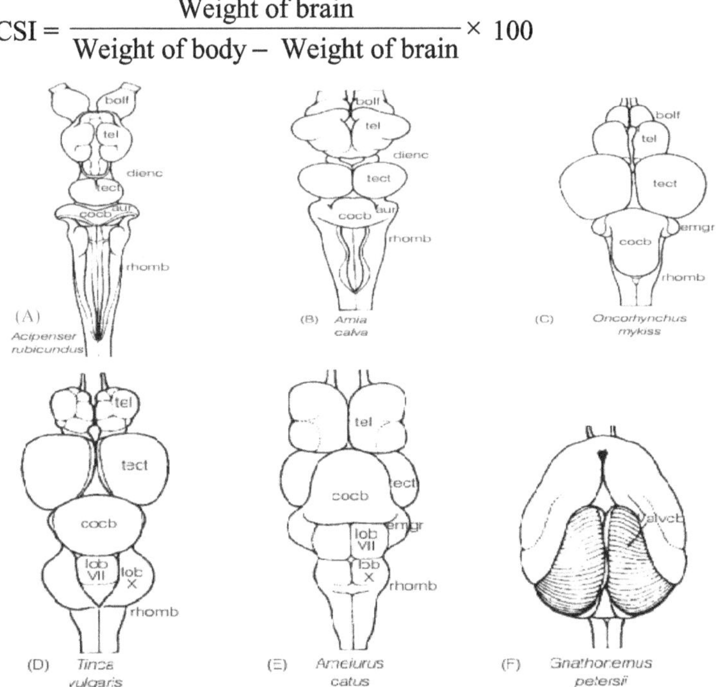

Fig. 25.1: Brain in different fishes

25.2.1 Forebrain

The earliest chondrichthyans' brains have a telencephalon made up of sessile olfactory bulbs. In general, telencephalon development is limited to the development of the olfactory centre.

The olfactory lobes are well-developed and are usually carried on long olfactory peduncles. The olfactory lobes in the rays and *Acipenser* are solid structures. *Puntius ticto* has both the olfactory lobe and the olfactory bulb, but some species may have only one or the other. In fish that hunt primarily by smell, such as hagfish, sharks, and catfish, the olfactory lobes are very large.

The earliest chondrichthyans' brains had paired cerebral hemispheres, extensive lateral ventricles, and thin walls. The cerebral hemisphere is widely separated in most elasmobranchs, but not so clearly in others. In the latter forms, there is at least a median depression, indicating the division of the anterior part of the forebrain. The cerebral hemispheres of teleosts and actinopterygians are everted. The prosencephalon is not divided into hemispheres in most teleosts, and the roof is non-nervous and thin. The cerebral hemisphere's floor is well developed. The cerebral hemispheres are quite distinct anteriorly in dipnoans.

In fish, the diencephalon is moderately developed and located behind the telencephalon. The presence of saccus vasculosus and inferior lobes, particularly in elasmobranchs, are modifications of the diencephalon's lower part. The folded and pigmented walls of the saccus vasculosus act as a pressure receptor. Deep sea fishes have a high level of development. The diencephalon forms the hypothalamus and the infundibulum on the ventral side. Its cavity forms the brain's third ventricle.

25.2.2 Midbrain

The earliest chondrichthyans have a bilobed and laminated optic tectum with the majority of cell bodies periventricularly, forming an octavolateralis complex. A dorsally located optic tectum and a ventral tegmentum are found in the midbrain. Labeo rohita lacks optic chiasma. The optic tectum is divided into two optic lobes. In all teleosts, the optic tectum projects into the optocoel and forms a pair of torus longitudinalis. Cave and blind fishes have a significantly reduced tectum.

25.2.3 Hindbrain

The unfoliated cerebellar corpus of the earliest chondrichthyans is divided into anterior and posterior lobes. The cerebellum is poorly developed in slow-moving fishes, but it is prominent in active forms. The cerebellum is elongated in sharks but shorter in rays among the elasmobranchs.

The medulla oblongata of the earliest elasmobranchs has a wide rhomboid fossa. The medulla oblongata undergoes structural changes in response to various functional responses. It has Muller's characteristic cells as well as Mauthner's giant cells. Mauthner cells are well developed in actinopterygii at nerve VIII level but absent in eels and molas. On its dorsal surface, it has one median facial lobe and two lateral vagal lobes. These lobes are well developed in many teleosts and are especially noticeable in buffalo fish (*Carpiodes tumidus*). Siluroids (*Mystus seenghala, Wallago attu*) have a more developed facial lobe because they use long barbels to search for food and explore their surroundings. Electric lobes are spherical projections of the floor of the fourth ventricle that innervate the electric organs.

Hearing-impaired species have a well-developed central acoustic lobe in the medulla oblongata. *Mystus seenghala* only has lateral acoustic lobes, which receive nerve fibres from the fish's internal ears and lateral line sense organs. *Channa striatus* has less developed paired facial and vagal lobes.

25.3 The Spinal Cord

In most fishes, the dorsal fissure is present in the spinal cord, but the ventral fissure is absent. In cyclostomes, it is flattened; in jawed fishes, it is cylindrical or quadrilateral.

25.4 The Cranial Nerves

Cranial nerves arise from the brain and are found in all fishes in ten pairs. The origin and distribution of the cranial nerves are nearly identical in all fishes. The 'special cranial nerves' (supplying a specialised head structure) are: the olfactory nerve (I), the optic nerve (II), the auditory (VIII), the glossopharyngeal (IX), and the vagus (X). These last two (IX and X) supply the lateral line system. system. The details of cranial nerves found in various fishes has been described in the chapter 8.

Although the terminalis is found in all fish except cyclostomes, its function is uncertain.

26

Endocrine System of Fishes

Endocrine glands secrete their products into the bloodstream and body tissues in conjunction with the central nervous system to control and regulate a variety of body functions (Fig. 26.1). Various endocrine glands described below have been discovered in fish that are associated with various tasks and functions.

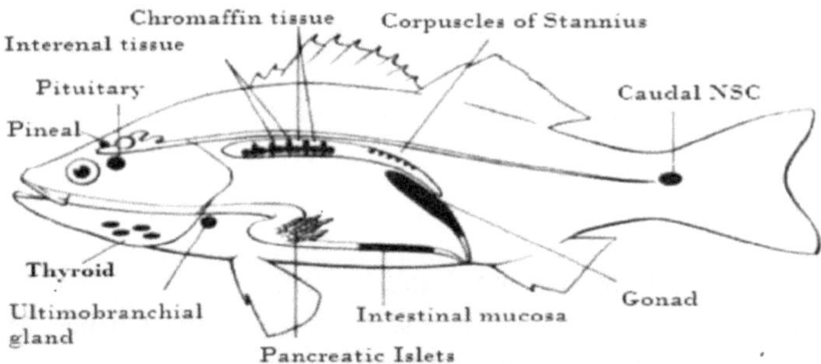

Fig. 26.1: Endocrine glands in Fishes (From https://biologyeducare.com)

26.1 Pituitary Gland

26.1.1 Introduction

The pituitary gland (=Hypophysis) is a central and discrete organ of the neuro-endocrine system in fish playing critical roles in various physiological processes such as stress response and behaviour.

The teleost hypo-thalamic-pituitary system is unique amongst vertebrates in that it has direct innervation of pars distalis by neurosecretory neurons of the hypothalamus and lacks modification of the typical vertebrate hypothalamohypophysial portal vascular system for the transport of neurohormones to pars distalis.

26.1.1.1 Occurrence

In isopondous fishes, there is a distinct communication between the mouth and a pituitary cleft and this duct opens in *Elops* and *Hilsa.* In perches and carp, there are no follicles. The anterior lobe of teleost is not homologous to mammalian pars anterior.

26.1.1.2 Location

The pituitary gland is situated beneath the diencephalon (hypothalamus), behind the optic chiasma and anterior to the saccus vasculosus. A delicate connective tissue capsule completely envelops it. When the brain is lifted in many fish, the gland detaches.

26.1.1.3 Shape

The pituitary gland has an oval body and is compressed dorsoventrally. The gland gradually tapers caudally from the rounded anterior end on the ventral side. The pituitary gland of platyfish is concave on the dorsal surface and slightly convex on the ventral surface.

26.1.1.4 Size

The size of the infundibulum varies depending on the species and sex. It is usually smaller in cyclostomes, but it grows larger in bony fishes, with prominence in the groove or depression of the parasphenoid bone receiving the gland. The anterior-posterior length, width, and depth of hypophysis of a sexually mature platyfish are 472.9μ, 178μ and 360μ respectively. There is no sella turcica like the ones found in mammals, such as *Barbus stigma* and *Xiphophorus maculatus*. Male fishes have a smaller hypophysis than female fishes. A lumen runs through the thick-walled, hollow infundibular stalk and into the third ventricle.

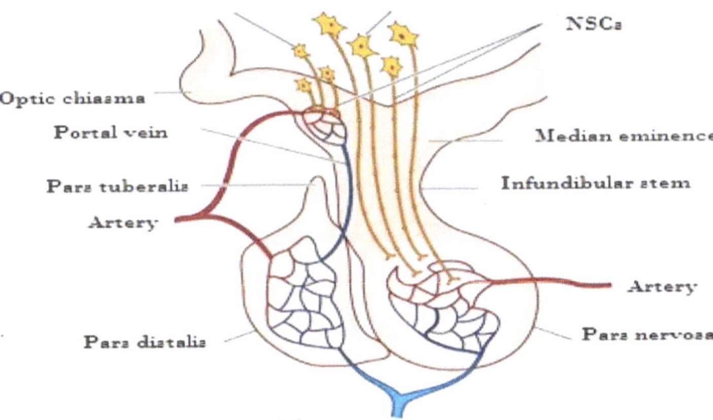

Fig. 26.2: Pituitary gland of Fishes (From https://www.britannica.com)

26.1.1.5 Origin

The pituitary gland is derived from two different components. The adenohypophysis is a glandular part that develops from the dorsal evagination of the ectodermal part of the buccal cavity called Rathke's pouch. This pouch eventually loses its connection to the buccal cavity and remains permanently attached to neurohypophysis for the rest of its life. Neurohypophysis develops from the floor of the embryonic diencephalons. The hypophysis in adult fish remains attached to it by a stalk called the infundibular stalk or neurohypophysial stalk, which occupies a position on the underside of the brain in the region of the diencephalon. Both parts are present nearby (Fig. 26.2).

26.1.1.6 Types of Pituitary gland

The pituitary gland is of the following two types based on the stalk.

1. Platybasic type: It has no stalk. The neurohypophysis consists of a flat floor of the infundibulum which sends processes into disc-shaped adenohypophysis. Examples: Glassfish (*Ambasis* and Eel).
2. Leptobasic form: It has a stalk. The neurohypophysis has a fairly well-developed infundibulum stalk and the adenohypophysis is globular or egg-shaped. Examples: Carp and catfish.

   There are many intermediates between the two.

26.1.1.7 Anatomy

The Heidenhain's azon method, Masson's poncean acid fuchsin aniline blue, the periodic acid Schiff's reaction (PAS), aldehyde fuchsin (AF) technique, Azocarmine method, Orange G method, Lead-haematoxylin (PbH) method, Chrome alum/haematoxylin and phloxine procedure, Pvronin G/methyl green procedure, Performic Acid/Alcian Blue Method, Thioglycolate– Ferricyanide Reaction, Iodination/Coupled Tetrazonium Reaction and Sakaguchi's (1925) test are used to detect the types of cells and chemicals present in the cells (Fig. 26.3).

Pickford and ATZ (1957) classified adenohypophysis as proadenohypophysis, mesoadenohypophysis and metaadenohypophysis. Gorbman (1965) classified adenohypophysis into three parts as well, but he referred to them as the rostral pars distalis, the proximal pars distalis, and the pars intermedia.

1. Rostral Pars Distalis (RPD) or Pro-Adenohypophysis: It is the smallest anteriormost part that is present in the form of a thin strip dorsal to the mesoadenohypophysis. It contains cells that only secrete PL (LH), ACTH as well as other hormones. Based on immunocytochemistry, the cells were recently classified based on hormones released by the pro-adenohypophysis. It is still thought to have a thyrotrophic function.
2. Proximal Pars Distalis (PPD) or Mesoadenohypophysis: It comprises the majority of the space between the rostral pars distalis and the pars intermedia. GTH and GH are secreted by the cells. TT cells can be found in either or both of the rostral and proximal pars distalis. Acidophil cells are rounded, oval or pyramidal.
3. Pars Intermedia or Metaadenohypophysis: It is found at the distal tapering

end of the pituitary gland. It also contains the most neurohypophysial tissue of any region. PAS is found in basophilic cells. The granular cells, on the other hand, do not show a consistent staining reaction with PAS and AF stains.

I. Adenohypophysis: Previously, staining was used to identify adenohypophysis cells. The two types of cells, chromophobes and chromophils, are classified based on their binding affinities with ribonucleoprotein. Chromophilic cells that take acidic stains are known as acidophils, while basophilic cells that take basic dye are known as basophilic and chromophobes are cells that do not take any stain.

PAS (periodic acid Schiff) and AF (aldehyde fucshin) negative cells are acidophilic cells. They are coarsely granular and appear splotched in the cytoplasm. They have round to oval nuclei on the periphery.

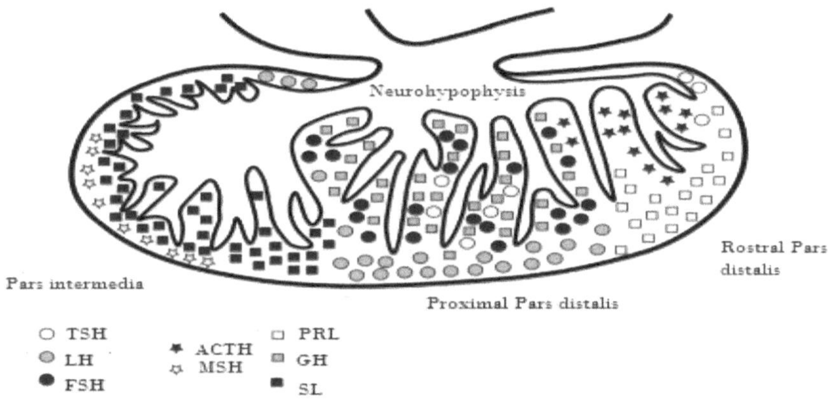

Fig. 26.3 Pituitary Regulation of Fishes (From https://www.sciencedirect.com)

Basophilic (cyanophilic) cells produce ACTH, but if they secrete TSH, they are called thyrotrophs, and if they secrete FSH, they are called gonadotrops. Basophils are spherical with large, round nuclei in the centre. Their cytoplasm is finely granular, and their structure is similar to that of pro-adenohypophysis cells. PAS and AF are both present in these cells.

II. Neurohypophysis: It takes up a significant amount of space and has many interesting and distinct features. It is made up of connective tissue, neuroglia cells, and a tangled network of nerve fibres.

The amorphous masses are known as "Herring bodies," and they have a close relationship with diencephalic neuro-secretory cells known as the nucleus preopticus via a fibre tract known as the preoptic neurohypophysial tract.

Each group of neurons in the diencephalon is known as a nucleus. The axons of the nucleus preopticus (NPO) and nucleus lateralis tuberalis (NLT) are associated with both adenohypophysis and neurohypophysis.

The NPO is located on either side of the optic recessus, just ahead of the optic chiasma. Neurosecretory stains can stain the NPO, axons, and nerve endings. Because the neurons are neurosecretory, they can be distinguished from other nuclei in the preoptic region using Gormori's chrome alum haematoxylin, aldehyde fuchsin, and alcian blue. It is divided into two sections:

I. Pars parvocellularis: It is located anteroventrally and consists of relatively small cells,

II. Pars magnocellularis: It is situated posterodorsally and comprises relatively larger cells.

26.1.1.8 Blood Supply

Pituitary vascularization has been studied in a variety of species. In brook trout (*Salvelinus fontinalis*) and Atlantic salmon (*Salmo salar*), the caudal hypothalamic artery supplies the neuro-intermediate lobe and hypophysial arteries branching off the anterior cerebral arteries supply the combined rostral proximal pars distalis. In teleosts, however, the rostral proximal pars distalis receives blood supply from extensive loopings of arterioles near the interface with the pars distalis.

These vessels, along with the interdigitations of the anterior neurohypophysis, invade the pars distalis. These anterior loops are thought to be the rudiment of the hypothalamohypophysial portal system. There are no neurovascular connections with these blood vessels, as is common in the median eminence of various vertebrates. As a portal system, this hypothesis is argued. Because pituitary cells have direct innervation by neurosecretory endings, the function of the hypothalamohypophysial portal system as a means of transporting neurohormones to the pituitary has become redundant and pure vascular.

Despite this, teleosts have been shown to have a typical hypothalamohypophysial portal system. According to Sathyanesan et al., hypothalamic artery branches form a "primary capillary plexus" in the meningeal tissue and adjacent neural tissue of the hypothalamus anterior to the pituitary stalk. This plexus forms vessels that enter the pituitary, proximal pars distalis or pars intermedia.

This portal system is thus the only and primary source of blood for the pars distalis in teleosts. Cypriniformes and Siluriformes have a smaller portal system than other teleosts.

26.1.2 Hormones and Their Functions

All hormones are necessarily polypeptides. There exists a slight difference in the pituitary hormones of the different groups of fishes.

26.1.2.1 Thyroid stimulating hormone (TSH) or Thyrotropin

TSH is secreted by pro-adenohypophysis and stimulates thyroid hormone activity. In fish, TSH is secreted under the influence of thyroid-releasing hormones (TRH) from the diencephalon. A crude extract of the hypothalamus or goldfish results in decreased radioiodine uptake by the thyroid in *Carassus auratus*, indicating the presence of TRH activity.

TSH cells in teleosts receive direct innervation from neurosecretory endings that are adjacent to cells with no synaptic contact or endings that may be separated from TSH cells by a basement membrane. TSH cells in *Tilapia mossambica* and *Carassius auratus* have direct contact with endings containing elementary neurosecretory granules and endings containing dense granules.

26.1.2.2 Gonadotrpohic hormeone (GTH) or Gonadotropin

The PPD is densely packed with gonadotropin cells, which can form a solid ventral rim of cells. These cells have irregular dilated GER cisternae with granules of varying electron density. They have glycoprotein. Cyprinoid has a similar situation. They are found throughout RPD and PPD in salmonids and eels.

GTH is regulated by the gonadotropin-releasing hormone (GTRF). Neurosecretory stimuli may pass along nerve fibres in many teleosts, piercing the laminae that separate the neuro from the adenohypophysis and penetrate the endocrine parenchyma of the pars distalis (Ball, 1981).

Nerve fibres are classified into two types: A and B. A-type fibre remains in contact with hormone-producing cells. B-type fibres make synaptic contact with granular vesicles 60-100 nm in diameter, whereas A-type synapses have granules 100-200 nm in diameter.

Teleost GNRH, like luteinizing hormone-releasing hormone (LH-RH), is found in the ventral lateral nucleus preopticus periventricularis (NPP) and posterior lateral nucleus lateral tuberis (NLT), among other places.

The presence of immunoreactive fibre tracts from NPP and NLP cells in the pituitary gland suggests that these areas are the source of the endogenous releasing hormones.

According to Peter and Crim (1978) research on *Carassius auratus*, the NLT pars posterior and anterior, which are located in the pituitary stalk, actively participate in the regulation of GTH secretion for gonadal recrudescence.

Channa punctatus, Cirrhinus reba, Clarias batrachus, Glyptothorax pectinpterus and *Heteropneustes fossilis* have all shown histological changes as their gonads mature. GTH secretion is linked to ovulation in several fish species. GTH levels in *Carassius auratus* rise on the day of ovulation. However, high levels of GTH were found during spawning in sockeye salmon (*Oncorhynchus nerka*).

Only one functional gonadotropin, known as piscine pituitary gonadotropin (PPG), is found in fishes. Its properties are similar to those of mammalian LH and FSH.

LH stimulates the appearance of secondary sexual characters in fish by promoting the release of gametes from nearly mature gonads. The *Salmon* pituitary gland secretes gonadotropins that are similar to LH. Furthermore, gonadotropins derived from the human chorion and gravid mare urine have LH-like properties that accelerate egg release in female fishes.

The presence of FSH has yet to be confirmed. Prostaglandin has recently been isolated from the testis and sperm of bluefin tuna (Thynnus thynnus) and flounder (Paralichthys olivaceus).

Piscine GTH can be of two types, according to Matty (1985):

(i) GTH rich in carbohydrates: They are involved in steroidogenesis, spermiation, ova maturation, and ovulation.

(ii) GTH deficient in carbohydrates: They are involved in vitellogenesis.

26.1.2.3 Adrenocorticotrophic Hormone (ACTH)

ACTH cells located between the rostral pars distalis and the neurohypophysis secrete it. The hypothalamus stimulates ACTH secretion via corticotrophin-releasing factor (CRF).

Carassius auratus hypothalamic and telencephalic extracts, as well as longnose suckers (*Catostomus catostomus*), stimulated ACTH secretion in vivo. Telencephalic hypothalamic CRF is similar to mammalian CRF in nature. ACTH cells in *Carassius auratus* are innervated by aminergic-like type B fibres. Neurohypophysial peptides may regulate ACTH secretion in teleosts.

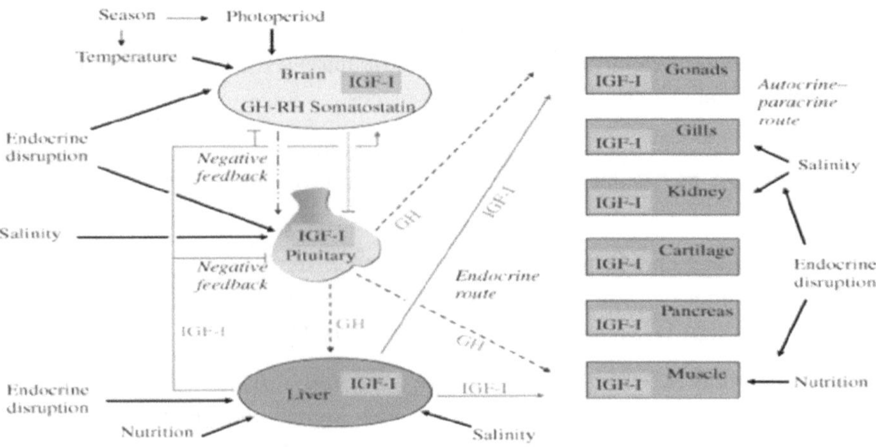

Fig. 26.4: GH–IGF-I system of fishes (From https://www.researchgate.net)

Cortisol pellets implanted in *Carassius auratus* demonstrate that corticosteroids have negative feedback effects on the brain, suppressing ACTH secretion. Cortisol added to the medium inhibits ACTH cell activity and release, implying a direct negative feedback effect of cortisol on ACTH cells.

26.1.2.4 Prolactin (PL)

It is released from the proadenohypophysis. Prolactin, in conjunction with intermedin, promotes the laying down of melanin in skin melanophores in mummichog (*Fundulus heteroclitus*).

Prolactin is one of several hormones involved in electrolytic regulation in teleosts, but its importance in maintaining homeostasis varies by species. Prolactin secretion is inhibited by a neuroendocrine control of hypothalamic origin.

26.1.2.5 Growth Hormone (GH)

Mesoadenohypophysis secretes GH, which accelerates the growth of fish body length. In teleosts, little is known about its control, mode of action on cell division, and protein synthesis. Teleost GH cells are capable of spontaneous activity and to continue to synthesise and secrete GH in vitro (Fig. 26.4).

Osmotic pressure may influence GH secretion, as GH release from cultured *Salmo gairdneri* and *Anguilla anguilla* is greater in a low sodium medium than in a high sodium medium, relative to plasma sodium levels. *Poecilia latipinna* GH release, on the other hand, does not affect osmotic pressure.

Recently, a hypothalamic zone has been identified as being responsible for GH control in *Carassius auratus*. The nucleus anterior tuberis (NAT) and sometimes the nucleus lateralis tuberis (NLT) of this fish form an area that stimulates GH secretion and may be the source of a growth hormone-releasing hormone (GRH).

Ultrastructural studies reveal hypothalamic control of GH secretion. In teleosts, GH cells of the pars distalis have direct synaptoid contact with type B endings, as in *Carassius auratus*, and *Tilapia mossambica* has direct contact but no synaptoid appearance.

Only a few species, such as *Oryzias latipes*, have synaptoid contact with type A endings on GH cells; in other teleosts, type A fibre may have direct contact with GH cells, but the endings are usually separated from the cells by a basement membrane. Thus, it is clear that neuroendocrine factors reach GH cells and that GH cells are likely regulated by a dual hormone.

26.1.2.6 Melanocyte Stimulating Hormone (MSH) or Intermedin

Intermedin can be melanophore dispersing hormone (MDH) or melanophore concentrating hormone (MCH) (MCH). MSH is secreted by the metaadenohypophysis and works in opposition to melanin hormone (MAH). It expands the pigment in the chromatophores and thus contributes to background adjustment. It also increases melanin synthesis. The teleost pars intermedia contains two types of secretory cells that can be distinguished by their staining properties.

Periodic acid Schiff-positive cells may be lead hematoxylin negative, or vice versa (Holmes and Ball, 1974). Salmonid appears to have only negative lead hematoxylin cells. These cells produce MSH, which stimulates melanin dispersion in melanocytes and skin darkening. *Salmo gairdneri*'s neuro-intermediate lobe appears to contain a melanin-concentrating factor.

The presence of MSH and/or its precursor ACTH in PbH cells of several species of teleosts has been demonstrated using immunofluorescence techniques. Thus, these findings support previous findings linking body colour or background adaptation to PbH cell activity.

The control of pars intermedia in teleosts varies by species. Neurosecretory axons in *Cymatogastes aggregate, Anguilla anguilla* and *Salmo gairdneri* do not enter the pars intermedia but instead terminate in extravascular channels bordering the pars intermedia or at the basement membrane. *Carassius auratus* and *Gillichthys mirabilis*, on the other hand, have direct innervation from neurosecretory axons.

MSH secretion in teleosts may be inhibited by a catecholaminergic mechanism. When 6OHDA is used to destroy catecholaminergic nerve terminals, it causes activation of MSH cells in *Gillichthys mirabilis*, darkening of skin or activation of MSH cells in *Anguilla anguilla*, and catecholamine directly inhibit MSH release when the former is MIH.

Catecholamine may also indirectly affect MSH secretion by promoting the release of an MIH or inhibiting the secretion of MRH nerve terminals within the pars intermedia.

Several histological studies show that pars intermedia is stimulated during reproduction. During the spawning season, *Clupea*'s pars intermedia becomes extremely active. During oogenesis and the breeding season, the number of lead hematoxylin-positive cells in *Carassius auratus* increases in both number and activity.

26.1.2.7 Oxytocin and Vasopressin

These hormones are stored in hypothalamic neurosecretory cells in fish. These endocrine substances have long been known to influence mammalian metabolism.

Vasopressin is responsible for the constriction of blood vessels in mammals, which stimulates water retention through its action in the kidney. Oxytocin stimulates mammalian uterine muscles and increases milk output in lactating mammals. They regulate osmoregulation in fish by maintaining water and salt balance.

26.2 Pineal Organ

26.2.1 Introduction

The pineal gland is a component of the endocrine system that regulates rhythmic activity in fish. It serves two purposes in fish. Melatonin (N-acetyl-5-methoxytryptamine), an indole derivative hormone, is synthesised from tryptophan. Melatonin is primarily synthesised at night because light inhibits production.

26.2.2 Location

It is located near the pituitary gland and is covered on the dorsal side by either (a) a thin translucent skin forming the pineal window, as seen in *Carassius carrasius, Cyprinus carpio* and *Salmo gairdneri* or (b) a thick skin, as seen in *Lebistes reticulatus*. Despite being a photoreceptor organ, the pineal organ has an endocrine nature and a questionable function. The removal of the pineal gland from *Lebistes* results in reduced growth rate, skeleton, pituitary, thyroid and Stannius corpuscle anomalies. The thyroid and pituitary glands have been shown to influence pineal secretion.

26.2.3 Origin

The pineal gland of fish develops from the epithalamus's posteromind mid-dorsal evagination behind the habenula and grows anteriorly through the adipose tissue.

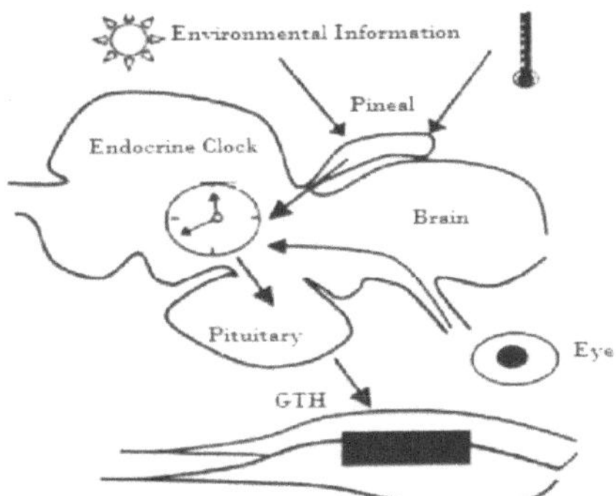

Fig. 26.5: Pineal gland of fishes (From https://www.researchgate.net)

26.2.4 Morphology

It contains a pineal sac (vesicle), pineal thalamus and pineal stalk (Srivastav, 1989) or two parts as a pineal sac (vesicle) and pineal stalk (Hafeez, 1971).

The pineal stalk varies in shape and size between fish species. It can be sessile, sessile, short and strong, or long and thin.

Nodule-like, flat and lobed, secular, tubular and compact, club-shaped, dagger-shaped, oval, slipper-shaped, or funnel-shaped pineal sacs are possible. *Wallago attu* has a conical shape, *Bagarius bagarius* has a rosette-like shape, and *Nangra punctatus* has a funnel shape. It is encased in a capsule of collagen connective tissue. Many species have hollow shells, but lanternfish have solid shells. *Glossogobius giuris'* sac wall may be uniformly thick or thin on the dorsal side and thick ventrally (Fig. 26.5).

26.2.5 Structure

The pineal sac contains the following three types of cells:

(i) Photoreceptors or sensory cells: The margin of the lumen of sensory cells varies between species. It can take the form of a gourd, a club, a fusiform or an oblong with lateral constriction. In some species, such as *Eutropiichthys vacha*, the parenchyma can protrude into the lumen as villi, which can form partitions. These cells have a two-segmented structure when examined under an electron microscope.

(ii) Supporting cells: These cells lie between sensory cells and are predominant in the basal part of the epithelium.

(iii) Sensory neurons or ganglionic cells: These cells are found in the epithelium's basal layer. They are viewed as cytoplasmic processes that protrude into the lumen.

26.2.6 Secretion

Pinealocytes secrete an apocrine substance containing glycogen, which supports its secretory nature. Srivastav et al, (1984) reported the presence of chromium, copper, nickel and zinc.

26.2.7 Hormone

In fish, for example, the pineal gland is involved in rhythmic activity. Melatonin is primarily synthesised at night because light inhibits production. The limiting factor is arylalkylamine N-acetyltransferase (AANAT), which has cyclic activity and is more active in the dark (in Pike, *Exos lucius*). As a result, melatonin is thought to play an important role in biological rhythms. The pineal gland is a photoneuroendocrine gland that communicates with the brain via a neural pathway. Several experimental studies have shown that the pineal organ can translate environmental information into rhythmic messages, and the pineal hormone melatonin is the internal chemical messenger of environmental signals or Zeitgeber and regulates a variety of functions in vertebrates. Recent research confirms the influences on gonad maturation in *Mystus vitatus*.

26.3 The Thyroid Gland

26.3.1 Introduction

In animals, the thyroid gland is an important metabolic regulator. The discrete thyroid gland in fish can be either compact/encapsulated (as in sharks, rays, and sturgeons) or diffusely arranged in the pharyngeal, heart, and kidney regions. It is thought to be the largest endocrine organ (Genten et al., 2008).

In a skate (*Raja fullonica*), Retzius (1819) described a sublingual gland. Simon (1844) identified the thyroid gland in "fishes," even though he confused it with the pseudobranch in several species. It is found in the branchial region of the head (Genten et al., 2008). Todd (1849-1852) and Owen (1866) both confirmed that the sublingual gland in fish is a compact thyroid. Baber (1881), Maurer (1886), and Gudernatsch (1891) later examined the diffuse structure of the thyroid gland in teleostean fishes (1910). Later research compared the structure/location of the gland in different fish species and uncovered the role of the thyroid as a regulator of metabolic activity and the role of the pituitary and hypothalamus in thyroid function regulation.

26.3.2 Origin

In fish, the thyroid gland emerges as a median evagination from the pharynx floor.

26.3.3 Location

The thyroid gland is located in the pharyngeal region of many teleosts, between the dorsal basibranchial cartilages and the ventral sternohyoid muscle.

Thyroid follicles in cyclostomes are dispersed around the ventral aorta and do not form a compact capsulated structure. In platyfish, follicles migrate to unusual locations such as the liver, kidney, eye, gut, spleen, gonad, and so on. It may be found under the first branchial arch on each side of bony fishes. It is found in many teleosts along afferent branchial arteries of the gills. As in *Ophiocephalus*, the thyroid surrounds the anterior and middle parts of the first, second, and sometimes third afferent branchial arteries of the ventral aorta. It runs almost the entire length of the ventral aorta and afferent arteries in *Heteropneustes*. It is concentrated around the ventral aorta, the middle ends of two pairs of afferent arteries, and the paired inferior jugular veins in *Clarias batrachus*.

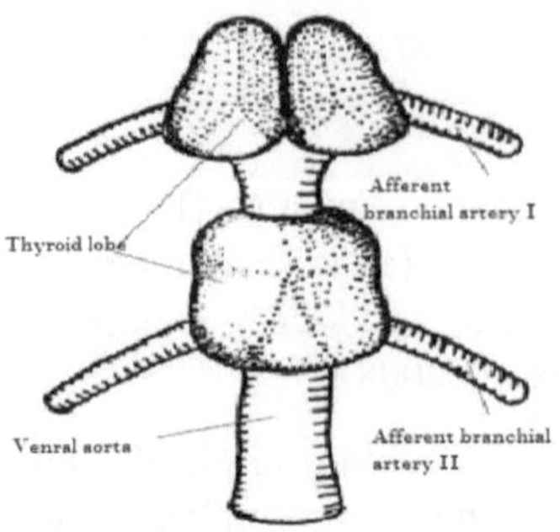

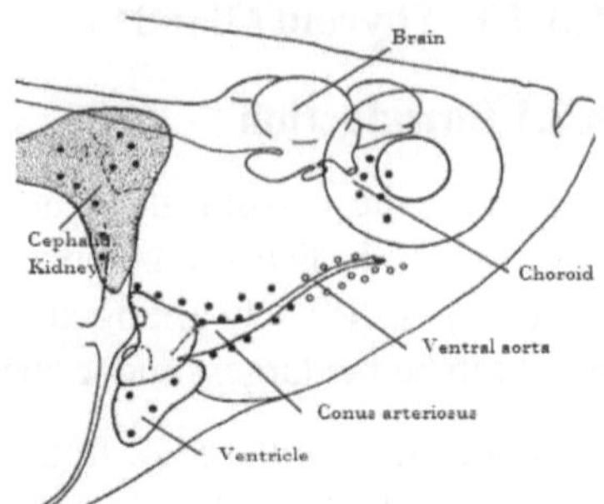

Fig. 26.6: Thyroid islets in fishes (From https://www.researchgate.net)

Fig. 26.7: Distribution of thyroid islets in fishes (From Baker et al, 1955)

26.3.4 Shape and Size

The shape of the gland varies depending on the fish group. The thyroid in cyclostomes is in the form of follicles. The thyroid is a compact and capsulated structure in elasmobranches and bony fishes (parrot fish, swordfish and *Gymnarchus niloticus*). The thyroid is encapsulated in the majority of teleosts, and thin follicles are dispersed or arranged in clusters around the base of afferent branchial arteries. The thyroid in dipnoi is made up of two interconnected lobes. Many fishes have been found to have heterotopic or ectopic thyroid follicles (Fig. 26.6).

The thyroid gland is thin-walled, sac-like, compact dark brownish and enclosed in a thin-walled capsule of connective tissue in these fishes may be correlated with the habit of air-breathing because the thyroid gland acts as thermoregulatory to adapt the fish to a semi-terrestrial environment of low thermal capacity.

It is an unpaired thin-walled brownish cylindrical in *Heteropneustes*. It is encapsulated and elongated in *Clarias batrachus*.

26.3.5 Structure

Fish account for approximately 73,500 of all vertebrate species. This diversity has resulted in significant differences within ecological niches, physiological mechanisms, and local adaptations. Within and between species, there are significant differences in thyroid morphology, physiology and regulation (Fig. 26.7).

The gland in skates and sharks is a compact, sometimes bilobed, capsulated organ (Baber, 1881), whereas, in teleosts, the gland is diffuse, encapsulated with scattered islets of follicles in the branchial region, close to the ventral aorta (Baker, 1958). Some ectopic thyroidian follicles may exist in the cephalic kidney as well as the choroid of some species (Chavin, 1956; Fournie et al., 2005). Scarids (Matthews and Smith, 1948), dipnoans (Chavin, 1976) and Comorian coelacanths have all been reported to have compact thyroids (Chavin, 1976). The angler (*Lophius piscatorius*) was studied for a peculiar thyroid gland embedded in a blood sinus (Burne, 1927)

The thyroid influences the hypothalamus, pituitary and gonads (HPG) axis differently in fish depending on gender, development, and species. The thyroid axis influences fish reproductive processes through actions at all levels of the HPG axis. The molecules that make up the HPT axis in fish are very similar to those found in mammals (Blanton and Specker, 2007). Power et al. (2001) discovered that thyroid hormones are abundant in fish eggs and are most likely of maternal origin.

The thyroid gland is made up of numerous follicles that surround the ventral aorta and afferent branchial arteries. A simple, cuboidal epithelium with an oval nucleus surrounded by basophilic cytoplasm forms each follicle. These characteristics have been studied in Epinephelus aeneus (Abbas et al, 2012), *Garra congoensis, Scyliorhinus canicula, Parachanna obscura* (Genten et al, 2008) and *Channa gachua* (Misra, 1990). The interrenal tissue was discovered within another endocrine organ in the anterior part of the kidney, containing several interrenal cells. These cells are arranged histologically as a chord among the blood vessels. Many observations suggested that these cells' function is to directly secrete corticosteroids (Genten et al, 2008). Furthermore, it was linked to the regulation of carbohydrate and protein metabolism, as well as osmoregulation in fish species, according to other studies (Jung et al, 1981, Takahashi et al, 2013).

26.3.6 Histology

Histologically, the thyroid gland in teleosts is made up of a large number of follicles, lymph sinuses, venules, and connective tissues. Follicles can be round, oval, or irregular in shape. Each follicle has a central cavity that is surrounded by a single layer of epithelial cells. The secretory activity of the epithelium influences its structure. The thin epithelium is more common in less active follicles.

There are two kinds of epithelial cells: (i) Columnar or cuboidal-shaped chief cells with oval nuclei and clear cytoplasm and (ii) Colloid cells also known as Benstay's cells. They have secretory material droplets. The follicles are held in place by connective tissue fibres that surround them. The follicle's central lumen is filled with colloids containing chromophilic and chromophobic vacuoles.

26.3.7 Blood Supply

The thyroid gland is highly vascularised and has a good blood supply. The thyroid gland is supplied with blood via a single buccal vein and two pairs of commissural vessels. A pair of veins emerges from its posterior end, merging immediately to form the posterior inferior jugular vein, which sends blood to the heart.

It also receives blood from the same vessels in *Ophiocephalus*. The buccal vein collects blood from the buccal region and opens into it after running for a short distance beneath the anterior end of the pairs of thyroid glands.

The two commissural blood vessels are highly branched and carry blood from the pharynx's floor. One pair opens at the anterior end of the gland, while the other opens in the middle of the gland on either side. The commissural vessels in *Heteropneustes* are more than two pairs.

26.3.8 Secretion of hormone

TRH action on thyrotropes appears to differ between species in teleosts. TRH treatment of pituitary cells in bighead carp (*Aristichthys nobilis*) increases TSH mRNA expression. TRH, on the other hand, does not affect TSH expression or release from the pituitary gland in common carp (*Cyprinus carpio*) or coho salmon (*Oncorhynchus kisutch*).

TRH has been shown to stimulate the secretion of GH, PL, ACTH and MSH in fish. TRH stimulates the release of proopiomelanocortin (POMC)-derived peptides (MSH and ACTH) from the anterior pituitaries of goldfish (*Carassius auratus*) as well as the synthesis and release of PL in common carp.

Fish TSH is a glycoprotein made up of a hormone-specific subunit (TSH) and a glycoprotein subunit (GSU). TSH, LH and FSH all share the subunit, whereas the subunit confers hormonal specificity. TSH mRNA is primarily expressed in teleost pituitary tissue, but ectopic expression occurs, most notably in the gonads.

TSH works by binding to TSH receptors (G protein-coupled receptors) on follicle basal membranes. Most teleost groups have two TSH receptor sequences, but only one receptor gene has been identified in the coelacanth and elephant shark genomes. TSH appears to have a stimulatory effect on TH synthesis/release and iodide uptake.

Pituitary TSH release is inhibited by DA and SS, neuropeptides and negative feedback actions by T_4 and T_3. In fish, there is evidence for feedback control of THs at the pituitary level, as THs decrease pituitary TSH expression both in vivo and in vitro.

Iodine is assimilated by diet or from the water via the gills in most fish, and the thyroid uptake of iodine requires TSH binding to follicles.

26.3.9 Hormones of the Thyroid Gland

Thyroid hormone is synthesised in colloids, from which inorganic iodine is extracted. This inorganic iodine forms a complex with tyrosine. Fish thyroid hormones, such as mono- and di-iodo-tyrosine and thyroxin, appear to be identical to those found in mammals.

These hormones are stored in the thyroid follicles and released into the bloodstream in response to metabolic demands. The release of thyroid hormone from the follicle is regulated by pituitary thyrotropic hormone (TSH), which is influenced by a genetically determined maturation process as well as environmental factors such as temperature, photoperiod and salinity.

26.3.10 Functions

Induced thyroid hyperactivity speeds up the transition into the juvenile moult stage in salmon, but a high thyroid titer slows larval growth. Thyroid hyperactivity in mud skippers (*Periophthelamus*) causes morphological and metabolic changes in response to the more terrestrial lifestyle of fish that live mostly outside the water. Salmon and stickleback thyroid glands are known to influence osmoregulation.

During the spawning migration of *Salmon*, the thyroid gland becomes hyperactive. It has been proposed that the thyroid influences goldfish growth and nitrogen metabolism, as evidenced by high ammonia excretion. Thus, thyroid action is linked to other vital processes such as fish growth and maturation and diadromous migration.

In *Salmon* and *Gastrosteus*, the thyroid hormone influences osmoregulation.

It is also known to affect goldfish growth and nitrogen metabolism.

Though inconsistent, the role of thyroid hormones in oxygen consumption in fish has been identified.

The thyroid hormone also has an effect on electrolyte movement in fish.

The thyroid hormone influences the CNS's responsiveness to constant stimuli.

Thyroid hormones are involved in carbohydrate metabolism because when the thyroid gland is active, the liver glycogen levels drop.

The thyroid, like other endocrine glands, influences fish migration.

Thyroxine also influences scale and bone formation.

Thyroxine stimulates fish maturation.

Triiodothyronine prevents the conversion of retinol to vitamin A.

26.4 Islets of Langerhans

26.4.1 Introduction

Most fish have an endocrine pancreas known as the islet of Langerhans, which is linked to the exocrine pancreas. Some species have very large islets that are visible (Brockmann bodies). The clear cell frequently described in Brockmann bodies is most likely not a real entity, but many fish species' islets most likely contain a fourth granular cell type in addition to the usual population of A, B and D-cells. During the spawning season, the size and number of islets in some fish will increase.

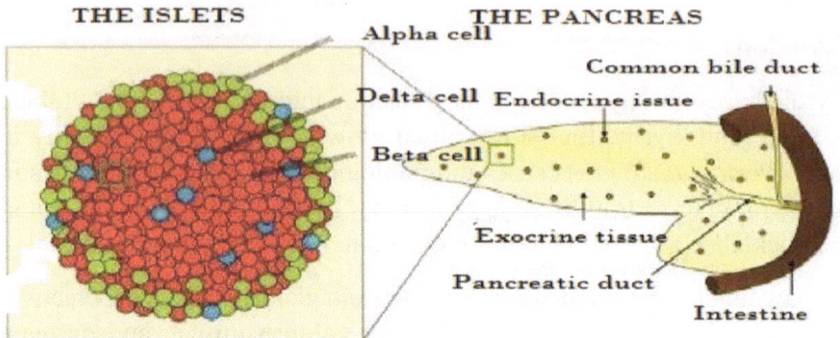

Fig. 26.8: Cells of Islets of Langerhans in fishes (From https://www.researchgate.net)

26.4.2 Morphology

Small islets separate from the pancreas and are found near the gall bladder, spleen, pyloric caeca or intestine in some fishes such as *Labeo, Cirrhinus* and *Channa.* These islets are frequently referred to as principal islets. However, similar to higher vertebrates, *Clarias batrachus* and *Heteropneustes fossilis* have a large number of large and small islets embedded in pancreatic tissues.

26.4.3 Structure

The main pancreas of adult zebrafish contains several major islets surrounded by exocrine tissue. Caudally along the intestine, a tail of single islets embedded in exocrine tissue and fat extends. In contrast, tilapia β-cells reside in Langerhans islets located along the mesentery that are not surrounded by exocrine tissue (Fig. 26.8).

The Langerhans islets are endodermal in origin and are found in most teleosts as small bodies scattered throughout the exocrine pancreas.

They contain a large number of bright eosinophilic, secretory granules and three types of cells in actively feeding fish.

26.4.4 Cells and Hormones

(i) The beta cells: It takes aldehyde fuschin. Insulin is secreted by these cells and regulates the blood sugar level in fish.

(ii) The alpha cells: It does not take aldehyde fuchsin and have two types, A1 and A2 cells, which produce glucagon-like peptides.

(iii) The delta cells: They produce Somatostatin.

26.5 The Ultimo-Branchial Gland

26.5.1 Introduction

In fish, the ultimo pharyngeal body, ultimobranchial body, post-branchial body or ultimobranchial gland contains follicles or cords of cells that exhibit endocrine secretion characteristics. Sharks, bony fish, and lungfish all have hypocalcemic activities in their ultimobranchial bodies. The amino acid sequence of salmon calcitonin was discovered and the hormone was synthesised in vitro.

Van Bemmelen (1886) described the ultimobranchial gland as supra pericardial bodies in elasmobranchs and chimeroids. Greil (1905) coined the term "ultimobranchial gland".

26.5.2 Occurrence

The ultimobranchial gland is not found in cyclostomes but is present in almost all groups of fish.

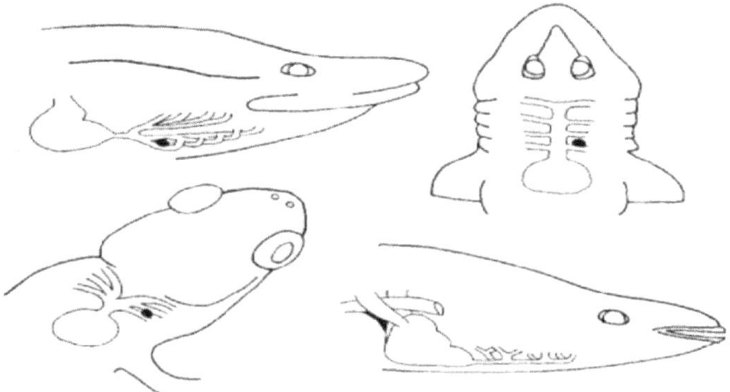

Fig. 26.9: Ultimobranchial gland in fishes (From https://bioone.org)

26.5.3 Origin

Embryonically, the gland develops from the pharyngeal epithelium near the fifth (last or ultimate) gill arch.

26.5.4 Position

It is situated in the triangle formed by the basibranchial, ceratobranchial and coracobrachialis cartilages as well as the coracobrachialis muscle (Fig. 26.9). It is only found on the left side of elasmobranchs between the pericardium and the ventral surface of the pharynx. It is small, bilateral and found in the transverse septum between the abdominal cavity and the sinus venosus just ventral to the oesophagous or near the thyroid gland in teleosts.

26.5.5 Measurement

In *Heteropneustes*, the gland measures 0.41.5 mm in diameter in an average adult of 130 to 150 mm in body length.

26.5.6 Structure

It is made up of parenchyma, which is solid and made up of cell cords and clumps of polygonal cells covered by a capillary network. The follicle has a diameter of about 125μ and is lined with columnar epithelium. It has two kinds of epithelial cells. One type of cell has a dark homogeneous nucleus and a large number of mitochondria. The other cell type has a pale nucleus and fewer mitochondria. Secretion involves both types of cells.

26.5.7 Hormone

Calcitonin is secreted by the gland and consists of 32 amino acids with a disulfide linkage at the N-terminus and an amide linkage at the C-terminus. This hormone contains no lysine or isoleucine.

Calcitonin is a hormone that regulates calcium metabolism. Calcitonin is thought to be involved in osmoregulation. In Japanese eels, eel calcitonin lowers serum osmolarity, sodium, and chloride.

26.6 Intestinal Mucosa

26.6.1 Introduction

Enterocytes, goblet cells and lamina propria leucocytes (macrophages, granulocytes, lymphocytes, and plasma cells) are found in the fish's intestinal mucosa as well as intraepithelial leucocytes (mainly Band T lymphocytes).

The fish intestinal mucosal system is involved in water pathogen disease control as well as the induction of oral tolerance to dietary antigens and microbiota.

26.6.2 Hormones

Secretin and pancreozymin are produced by the intestinal mucosa and are controlled by the nervous system to regulate pancreatic secretion.

These hormones are typically synthesised in the small intestine's anterior portion. These hormones are introduced into the stomach of carnivorous fish via an acidified homogenate of fish flesh or by injection of secretin into the gastric vein, which stimulates pancreatic secretion.

Secretin influences the flow of pancreatic enzyme-carrying liquids, whereas pancreozymin accelerates the flow of zymogens.

26.7 Urohypophysis

26.7.1 Introduction

Urohypophysis or caudal neurosecretory organ is a small oval body, present in the terminal part of the spinal cord. It is an organ deposit, which releases materials produced in the neurosecretory cells situated in the spinal cord.

26.7.2 Occurrence

The urohypophysis found only in elasmobranch and bony fishes, probably developed independently in each group but corresponds to the hypo-thalamo neurosecretory system present in vertebrates. The neurosecretory cells comprising the urohypophysis are concentrated at the hind end of the spinal cord, where they are associated with a vascular plexus to form a neurohemal organ.

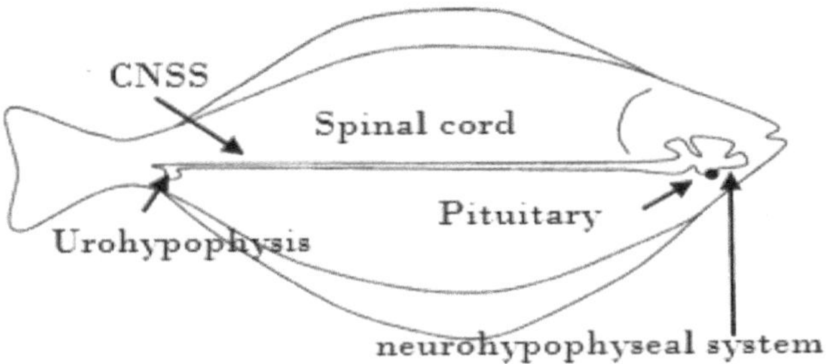

Fig. 26.10: Urophypophysis in fishes (From https://link.springer.com)

26.7.3 Shape

It may be staked as in *Fundulus* and *Gilichthys* or without a stalk as in eel. Generally, it is a median unpaired swelling, but in *Cirrhinus*, it is paired due to the development of a deep median furrow. Partial division into two has been reported in *Labeo* and *Puntius*. In *Osteobrama cotio*, it is single externally but paired internally.

26.7.4 Structure

In the caudal neurosecretory system, neurosecretory cells are diffused in the terminal part of the spinal cord. These cells are relatively large and have enlarged axons, whose cell bodies are called Dahlgren cells. Axon terminals of these cells assemble at the ventral side of the region and form Urohypophysis with blood capillaries (Fig. 26.10).

The neurosecretory cell is a large nerve cell and has a basophilic cytoplasm and a polymorphic nucleus. The cells may be acidophilic or basophilic. The cells may be unipolar, bipolar or multipolar based on the number of axons. They can be stained only with acid violet.

In Ayu (*Plecoglossus altivelis*) the Urohypophysis is extended like a bow. In carp and yellow tail, it is a conspicuous oval body. The Urohypophysis is made up of spinal cord elements like neurosecretory axon, glia, and ependymal and glia fibers and meningeal derivatives such as vascular reticulum and reticular fibres.

26.7.5 Hormones

The hormones of Urohypophysis are called "urotensins". Urotensin may be of the following versions urotensin I, II, III and IV. These are peptides and all may not be present in the same fish. However, urotensin I and II are commonly found in a fish and their release is controlled by the central nervous system. The extract of Urohypophysis produced the following effects.

1. **Urotensin I:** It increases the blood pressure of fish.
2. **Urotensin II:** It is involved in the contraction of smooth muscles such as urinary bladder, causes a marked increase in blood pressure and enhances urine flow.
3. **Urotensin III:** It induces the sodium intake across the gills (osmoregulation) of goldfish. However the effect of this component is not observed in other fishes.
4. **Urotensin IV:** It shows activity like antidiuretic hormones of the pituitary gland.

26.7.6 Functions

1. The caudal neurosecretory system is said to be related with osmoregulation. Urohypophysis extract shows the ability to contract smooth muscles of ovary and oviduct of guppy (*Poecilia reticulata*) and the sperm duct of goby (*Gillichthys mirabilis*), suggesting the possibility of involvement in reproduction and spawning.
2. They increased sodium influx in freshwater fishes and decreased its excretion in goldfish.
3. They may have kinetic influences on fishes.
4. They may participate in reproduction (Berlind, 1972).
5. They participate in caudal fin locomotion (Honma and Tamura, 1967).

26.8 Adrenal Gland

26.8.1 Introduction

The adrenal gland is an important part of the body's stress response. The adrenal gland is extremely sensitive to toxic compounds, many of which act as endocrine disruptor chemicals (EDCs). Except for *Sculpin*, no fish have a true adrenal gland. Adrenal glands in fish are represented by two distinct tissues, a steroidogenic interrenal and an aminergic chromaffin, both of which are located in the kidney region but remain distinct from one another. Two components are related to the adrenal cortex and the medulla in mammals. One of three layers of cells that line the cardinal veins in the haemapoietic head kidney.

The literature on the Cyclostomata's adrenal gland is limited and a comprehensive description of the adrenocortical and chromaffin tissue of jawless fishes is lacking. The adrenocortical tissue in the elasmobranchs organises into a single compact gland that is located on and between the kidneys and is thus referred to as the interrenal. The cytology of adrenocortical and chromaffin tissue in Teleostei has been studied relatively thoroughly (Baecker, 1928; Chavin, 1966)

26.8.2 Adrenal Cortex

26.8.2.1 Location

Inter-renal tissue or Adrenal Cortical Tissue: These cells are found throughout the body cavity close to the post-cardinal vein of Lamprey. They are found in more or less close association with posterior kidney tissue, with some species having inter-renal tissue concentrated near the left central border of the organ and others near the right central border. They are found in sharks between the kidneys. They are multilayered and located along the post-cardinal veins as they enter the head kidney in teleosts. There have been significant variations in the structure and distribution of inter-renal in fishes, which can be classified into four groups (Nandi, 1962).

1. Group I: Inter-renal cells surround the post-cardinal vein or its large branches.
2. Group II: Inter-renal cells are distributed throughout the head kidney via small and medium veins.
3. Group III: Inter-renal cells in the head kidney are associated with venous sinuses.
4. Group IV: Inter-renal tissue solidifies in a specific area of the head kidney.

26.8.2.2 Origin

It develops from the embryo's mesodermal layer. It is made up of haemopoietic or lymphoid tissue.

26.8.2.3 Anatomy

Columnar, cuboidal, round, polygonal, or spindle-shaped adrenocortical cells with large nuclei and basophilic chromatin are reported by Jones (1957). Eosin has been used to stain them. As in higher vertebrate adrenocortical cells, cytoplasmic lipoidal droplets are present (Aboim 1946).

Inter-renal cells are arranged in *Puntius ticto* as a thick glandular mass, whereas they are present in *Channa punctatus* as lobules. Each interrenal cell is columnar and eosinophilic, with a round nucleus (Fig. 26.11).

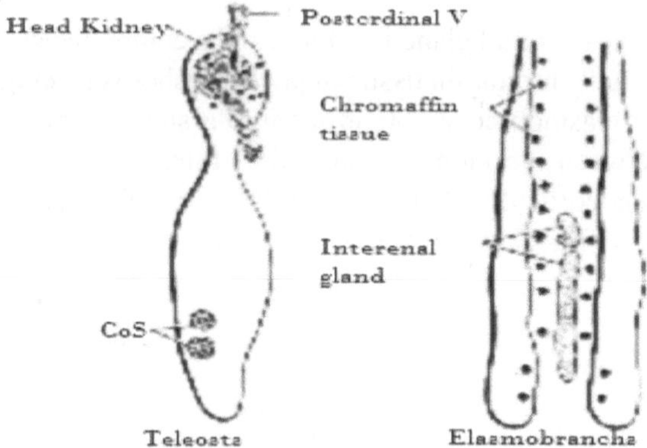

Fig. 26.11: Inter-renal glands and Corpuscles of Stannius in fishes (From https://www.sciencedirect.com)

26.8.2.4 Secretion

The adrenocorticotropic hormone regulates adrenocortical hormone secretion (ACTH). Their concentration in the blood varies depending on the species, season, and physiological condition of the fish.

26.8.2.5 Adrenal Cortical Hormone

Two hormones are secreted by the adrenal cortical tissue. They are as follows:

(i) Mineralocorticoids (e.g., Aldosterone): These are concerned with the osmoregulation of fish. *Salmo gairdneri* treated with mineralocorticoid excretes more sodium ions than normal through its gills but conserves more sodium in the kidneys and osmoregulation in the body.

(ii) Glucocorticoids (e.g., Cortisol, Cortisone and Corticosterone): They control carbohydrate metabolism, specifically blood sugar levels. They play a role in gluconeogenesis. An intramuscular injection of glucocorticoids into the oyster toadfish causes an increase in blood sugar levels, demonstrating carbohydrate metabolism control. Cortisol levels in salmon blood plasma rise during the spawning period and fall during the more sedentary stages. During the spawning phases, *Oncorhynchus* catabolizes 60% of total body protein, which correlates with a six-fold increase in plasma corticosteroids and increases in liver glycogen. Adrenaline cortical hormone administration stimulates lymphocyte release in *Astyanax* and antibody release in European perch, just as it does in higher vertebrates.

26.8.3 Chromaffin Tissue

Chromaffin Tissue or Suprarenal Bodies or Medullary Tissue: Chromaffin cells are found in lamprey in the form of strands along the dorsal aorta, as well as in the ventricle and portal vein heart. These tissues are found associated with the sympathetic chain of nerve ganglia in sharks and rays, but their distribution varies greatly in Actinopterygii.

Chromaffin cells have basophilic cytoplasm and an enlarged nucleus and are rounded, cubical, or slightly cylindrical. They were found to form a layer of one or two cells thick beneath the endothelium that wraps the anterior veins, particularly the right anterior cardinal vein. These cells were thought to be homologous to the chromaffin cells of Teleostei fish (De Smet, 1962). In some species, chromaffin cells can also be found in the posterior kidney.

Adrenaline (=epiniphrine) and noradrenaline (=nor-epiniphrine) are abundant in fish chromaffin tissue. They may be found in teleosts' heads, kidneys, or plasma. Adrenaline and noradrenaline injections cause blood pressure changes, bradycardia, branchial vasodilation, diuresis in glomerular teleosts, and hyperventilation. They could be

1. Distributed throughout the interregnal tissue and vein wall.
2. It is embedded in the vein wall.
3. It is found in teleost ovaries such as *Catla catla, Cirrhinus mrigala, Channa* spp., *Labeo rohita, Notopterus* and *Puntius* spp.
4. Related to the interrenal tissue.

26.9 The Sex Glands

26.9.1 Introduction

The fish sex organs are known as gonads (Fig. 26.12 and 26.13). They are usually paired, internal and located near the middle of the body, next to the stomach. Males have two testes that produce sperm, while females have two ovaries that produce eggs. The sperm and egg cells are the fish's sex cells.

26.9.2 Hormones

Little is known about the effect of gonadal hormones on fish reproductive behaviour. When mammalian testosterone and estrone are injected into lampreys, they develop cloacal lips and coelomic pores, which aid in the reproductive process.

Such tests for rays and sharks yielded no results, whereas ethynyl testosterone (pregnenolone), which has mild androgenic and progesterone-like effects in mammals and birds, were found to be highly androgenic in fishes.

Specialized cells in the ovaries and testes produce and secrete sex hormones. The pituitary mesoadenohypophysis regulates the release of sex hormones. In fish, these sex hormones are required for gamete maturation as well as secondary sex characteristics such as breeding tubercles, coloration and gonopodia maturation.

The blood plasma of elasmobranchs (*Raja*) and *Salmon* contains the male hormone testosterone, with a correlation between plasma level and reproductive cycle. Another gonadal steroid found in medaka (*Oryzias latipes*) and sockeye salmon is 11ketotestosterone, which is 10 times more physiologically androgenic than testosterone.

Male sex hormones are more similar to those found in vertebrates than ovarian hormones; the former has a strong influence on ovarian development in the Japanese weather-fish loach.

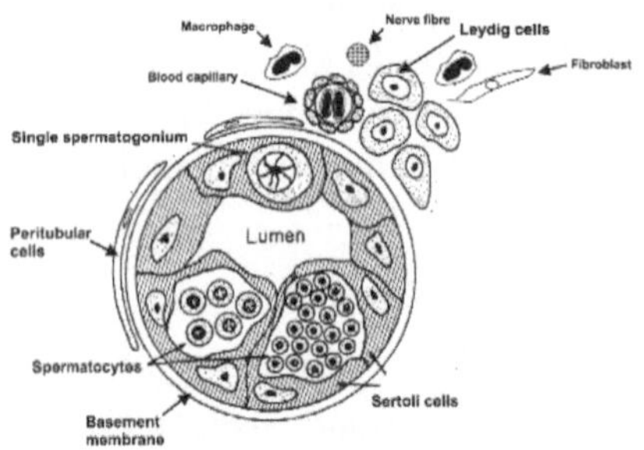

Fig. 26.12: Testis in fishes (From https://www.researchgate.net)

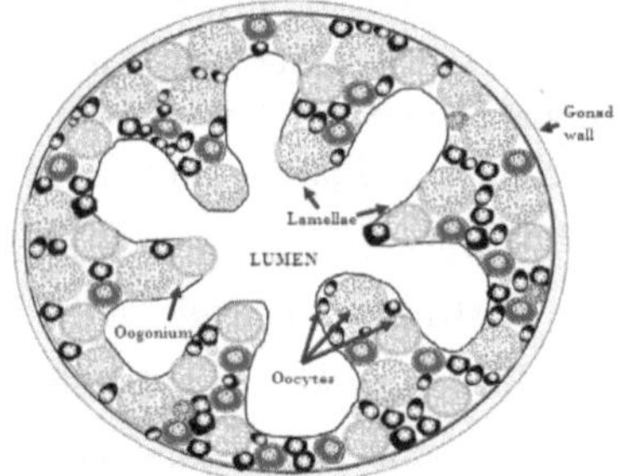

Fig. 26.13: Ovary in fishes (From https://www.researchgate.net)

In addition to estrone and estriol, the ovary secretes oestrogens, including estradiol-17B, which has been identified in many species. Progesterone is found in some fishes but has no hormonal function.

26.10 The Corpuscles of Stannius

26.10.1 Introduction

The corpuscles of Stannius are one of the characteristic endocrine glands of teleosts. Its endocrine role is still debatable. It shows osmoregulatory and renal functions. The corpuscles of Stannius were first discovered by H. Stannius (1839) as discrete glandular bodies in the kidney of a sturgeon. It secretes stanniocalcin which is the mammalian homolog of STC-1 and PTH. Although the corpuscles of Stannius are capable of limited steroid biosynthesis, they are incapable of converting cholesterol into corticoids.

26.10.2 Shape

They are generally oval or rounded. However, in goldfish, trout and salmon, these are flat structures on the peritoneal surface of the kidney. They show nodules.

26.10.3 Size

The corpuscles of Stannius range in size from 0.15 to 6.00 mm in different fish. It is largest in *Labeo rohita* and smallest in *Mystus vittatus*.

26.10.4 Position

They are partially or entirely embedded on the dorsal, dorsolateral or ventrolateral sides of bony fish kidneys. These are typically situated on the posterior side of the kidney. Their precise location is determined by the urinogenital aperture. The distance is 10-15 mm in *Channa punctatus* and 50-60mm in *Labeo rohita*. The distance is the most variable (4-30 mm) in *Clarias batrachus*. The corpuscles of Stannius are visible on the peritoneal surface of the kidney in goldfish, trout and salmon as flat, white, oval structures.

26.10.5 Number

The number of corpuscles of Stannius varies from 2-10. Their numbers are 1 pair in *Notopterus notoperus*; 2-4 in *Mystus*; 4 in *Catla catla; Cirrhinus* and *Labeo*; 8 in *Heteropneustes* respectively. The number is highly variable (2-10) in *Clarias batrachus*.

26.10.6 Origin

The pink or white corpuscle of Stannius is an outgrowth of the pronephric or mesonephric duct of the kidney. Stannius considered them to be similar to mammalian adrenal glands. They were later discovered to be anatomically different because they were derived from different embryonic tissues.

26.10.7 Nature

The steroidogenic nature of this gland has been hotly debated. The corpuscles of Stannius are embryologically distinct from steroid-producing tissues. They are outgrowths from the pronephric ducts of the kidney (Ford 1959 and Belsare 1973). This problem appears to remain unsolved until the chemical nature of corpuscular hormones is identified.

26.10.8 Location

The corpuscles of Stannius are partially or completely embedded on the dorsal, dorsolateral or ventrolateral sides of the kidney of bony fishes. Their precise position/location is determined by the urinogenital aperture.

26.10.9 Structure

The corpuscles of Stannius have an outer capsule of fibrous connective tissues that surround a mass of parenchymal columnar cells. The cells form follicles or irregular cords that are separated by connective tissue, blood pacing and blood vessels. Clusters of ganglionic cells are also found in the corpuscles of Stannius.

Gorbman and Jones (1966) discovered two types of cells in the corpuscles of Stannius in *Salmon, Gasterosteus* and *Anguilla.* These cells are AF positives and negatives. However, in Atlantic salmon (*Salmo solar*), *Labeo rohita, Cirrhinus mrigala, Catla catla, Clarias batrachus, Channa punctatu, Pseusecheneis sulcatus* and goldfish, only one type of cell is present. Endoplasmic reticulum and secretory granules are abundant in these cells. Each secretory granule is spherical and ranges in diameter from 0.5 to 1 μm.

The cells of the corpuscles of Stannius are designated as types I and type II.

Type I cells are more numerous and contain larger secretory granules. These cells produce calcium-regulating factors.

Small granules are found on type II cells. These cells synthesise potassium-regulating factors.

Some evidence suggests that type II cells are the immature or post-secretory phase of type I cells.

The granulation of the cells of the Corpuscles of Stannius has been observed to increase or decrease as the salinity of the water changes. Granules are also lost during hypophysoectomy. Removal of this gland in *Anguilla anguilla*, caused a reduction in plasma Na^+ and increases in K^+ and Ca^{+2} concentration and even resulted in death.

26.10.10 Functions

1. **Renin activity**: Extracts' of the corpuscles of Stannius of eel caused a rise in blood pressure and showed a resemblance to renin.
2. **Osmoregulation**: In eels, removing the corpuscles of Stannius (stanniectomy) resulted in a decrease in plasma Na^+ and an increase in K^+ and Ca^{+2} concentrations and even death.

   Morphological changes in the corpuscles of Stannius in response to salinity changes

   Evidence suggests that the corpuscles of Stannius have osmoregulatory effects and also increase motor activity.

   Changes in the salinity of water also result in a decrease or increase in the granulation of cells.

3. **On the kidney**: The kidney appears to be a major target organ upon which the secretions of the corpuscles of Stannius act.

   The removal of the corpuscles of Stannius did not affect the Na+ transport system in the gills of either freshwater or marine eels.

   Renal function was significantly altered, as evidenced by a decrease in urinary calcium and magnesium excretion.

4. It helps in homeostasis in fish.
5. Loss of granules has been reported after hyphophysectomy.
6. According to Narwarly and Gupta (1991), seasonal changes in the Corpuscles of Stannius with the maturation of gonads have been reported in *Notopterus*.
7. The hypocalcemic proteins (hypocalcin and teleocalcin) produced by the corpuscles of Stannius enable fish to survive both in freshwater and marine water.

27

Sense Organs in Fishes

Fishes use their sense organs to detect their bodies and their environment. Sense organs include the eyes, the lateral lines, the nostrils, the olfactory organs and the taste organs. Each of these sense organs is equipped with sensory nerve endings. Most fish possess highly developed sense organs to detect smell, sight, hearing, taste, touch, temperature, etc.

27.1 Olfaction in Fishes

27.1.1 Introduction

The majority of fish can detect odours, feeding, reproduction, migration and predator avoidance using a pair of olfactory rosettes connected to the olfactory lobes. The structure of fish olfactory organs varies greatly due to differences in their habits. Fish use their sense of smell to mediate behaviours and physiological responses related to food search and feeding, social interaction, mating, predator detection and contamination, migration and searching for spawning sites.

Some fish species have been classified as macro-osmatic or micro-osmatic based on the morphology of their olfactory systems (Atta 2013).

27.1.2 Organisation

The olfactory chamber is located in a depression in the ethmoid bone at the fore end of the head on either side. The olfactory opens to the outside through anterior and posterior nostril apertures and contains an olfactory rosette. These nostrils function as anterior and posterior apertures for water currents to enter the olfactory chamber (Fig. 27.1).

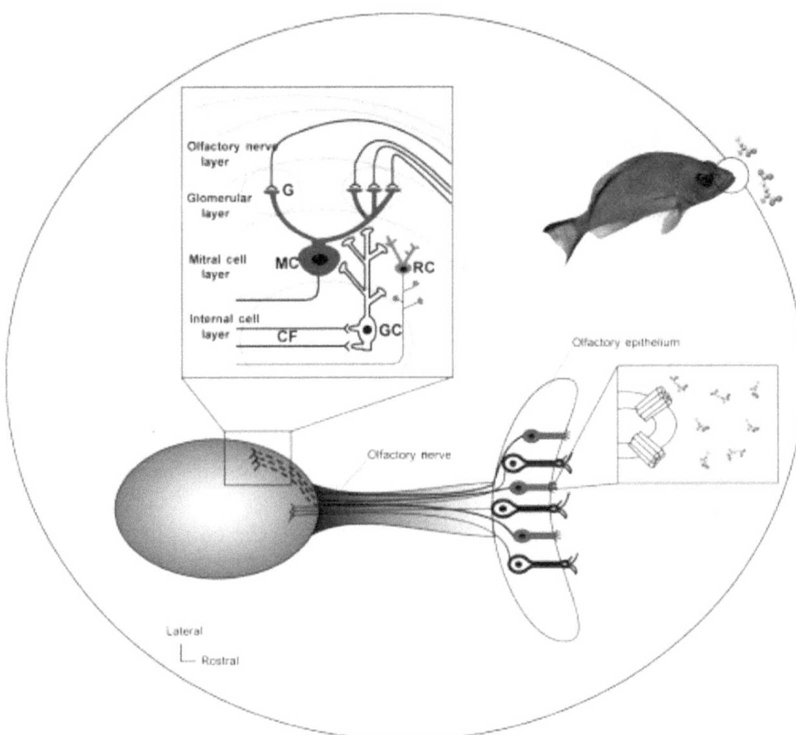

Fig. 27.1: Olfaction in fishes (From https://www.sciencedirect.com)

Carp's nostrils are close together, with a ventral bridge or nasal flap of skin separating them and extending downwards into the nasal cavity. These nostrils are separated into *Wallago attu, Channa striatus* and *Clarias batrcahus* and the anterior one is raised as a tube or funnel. These nostrils are found at a significant distance apart in eels and *Mastacembelus*, with the former opening into the olfactory chamber via small prolongations. *Cottus* and *Gastrosteus* have a single nasal aperture that alternately fills and empties the olfactory chamber.

The olfactory rosette is oval or elongated and does not fill the entire nasal cavity. It is oval in *Labeo rohita, Cirrhinus mrigala,* and *Tor tor,* but elongated in *Wallago attu* and *Mastacembelus*. The olfactory rosette is composed of numerous leaf-like structures containing a variety of functionally distinct cell populations. A macrosmatic fish's olfactory epithelium is made up of millions of olfactory neurons (Kreutzberg and Gross 1977) that are embedded between sustentacular cells and ciliate cells that support them. Fish have a diverse range of olfactory neurons, including ciliated olfactory sensory neurons and microvillous neurons, which have been described (Hansen and Zielinski 2005). Other neuronal cell types have recently been described including crypt neurons, Kappe neurons (Ahuja et al. 2014) and pear-shaped neurons (Wakisaka et al. 2017).

S-100 and Tropomyosin kinase receptor A (TrkA) expression can be used to identify crypt cells (Germanà et al. 2004). Pear-shaped neurons have been shown to detect ATP and related molecules with an adenosine moiety by expressing primarily Olfactory Marker Protein (OMP) (Wakisaka et al. 2017). Kappe neurons are distinguished by the expression of GO receptors and have a shape similar to crypt neurons, except for the distinctive "cap" structure that gives them their name. Kappe neurons' odorant specificity is still unknown (Klimenkov et al. 2020).

The olfactory rosette contains a median axis or "raphe" to which numerous leaves like lamellae are attached. The olfactory lamellae are highly vascular and their number varies in different species. The raphe and lamellae are often pigmented. The simplest type of raphe is club-shaped as seen in *Hemiramphus georgii* and *Belone vulgaris*. These species show poor development of the olfactory organ, but the eyes are more developed. Discs like olfactory organs are seen in *Channa* species but *Esox lucius* has a circular raphe. The oval-shaped raphe is found in eels. The olfactory organ is the least developed *Hemiramphus georgii* but is complicated in *Scoliodon sorrakowah. Harpodon neherius* mentions a more complicated raphe.

The number of lamellae in rosettes is said to increase with fish age and size, and new lamellae are added at the rosette's anterior extremity. The number of lamellae on the right and left sides of the fish's rosette is nearly equal. Deep-sea anglerfish have been found to have significant sexual dimorphism. Males of these fish have a large olfactory organ, whereas females have a smaller one. In *Labeo rohita*, the olfactory lamellae are crescentic but have a linguiform process along the concave margin. An accessory or nasal sac is also present in *Labeo rohita, Channa punctatus, Channa striatus, Clarias batrachus, Mastacembelus annatus* and *Wallaogo attu* but not seen in *Barilius, Puntius* and *Schizothorax.*

A simple club-shaped raphe is present in *Hemiramphus georgii, Belone vulgaris, Exoxoetus* and *Tetradon* in which the rosette is poorly developed, but the eyes are well developed.

27.1.3 Cellular Structure

Histologically, olfactory epithelium contains the following:

27.1.3.1 Olfactory or receptor cells

The olfactory epithelium is ciliated and covers both sides of the lamellae. The olfactory epithelium is absent from the middle raphe. On each side, the sensory epithelium contains 5-10 million cells. In *Barilius bendelisis*, two types of sensory and supporting cells have been identified (Singh, 1982). The supporting cells can be ciliated, columnar, or round and non-ciliated. Primary or secondary receptor cells can exist. The primary receptor is a bipolar neuron with a dark round nucleus and a cylindrical dendrite terminating at the epithelium's free surface. The dendrite's distal end forms an olfactory knob or swelling, which produces the proximal part. The olfactory nerve travels posteriorly and terminates in the olfactory bulb. 4-8 cilia radiate from the olfactory knob of each ciliated receptor cell. These cells can regenerate and are constantly renewed to compensate for environmental loss and injury.

The axonal ends of primary receptor cells connect with the dendrites of secondary neurons, which have spindle-shaped nuclei and are found in the deeper part of the epithelium. Secondary receptor cell dendritic ends synapse with primary neuron axonal ends. Their axonal ends extend up to the basement membrane and into the lamella's central core, where they aggregate to form bundles of olfactory fibres.

27.1.3.2 Supporting or sustentacular cells

Columnar epithelium extends vertically from the surface to the base of the supporting cells. In most fishes, the receptor and supporting cells are intermingled in the olfactory epithelium, forming a mosaic that makes distinguishing strata impossible. Ciliated non-sensory cells are columnar epithelium with numerous kinocilia. The beating of these cilia generates a weak current across the lamellae, assisting in the removal of water and the transport of odorous molecules in the olfactory organ.

27.1.3.3 Basal cells

Basal cells are small, undifferentiated cells with no cytoplasmic processes that reach the free surface. It is thought that these cells give rise to sensory or supporting cells. The olfactory epithelium also contains a few secretory cells of the goblet type. In adult fish, receptor cells are constantly renewed.

27.1.3.4 Secretory cells

Because of the large number of gland cells, the olfactory epithelium is highly secretory. Mucin and acid mucopolysaccharides are found in the mucus secreted by these cells, which forms a uniform layer around the olfactory epithelium. This protects the sensitive sensory hair from the osmotic effect of water.

The fish olfactory system is made up of a pair of peripheral multilamellar sensory organs known as olfactory rosettes. The olfactory nerve is made up of the axons of sensory neurons in the rosette, which form synapses in the olfactory bulbs.

27.1.4 Olfactory Bulb and Tract

The olfactory nerve emerges from receptor neurons and connects with secondary bulbar neurons to form glomeruli. The axons of the olfactory nerve are organised into several bundles. There are two main median bundles in *Cyprinus carpio*, one derived from rostral lamellae and the other from caudal lamellae. The mixed cells have a large body and dendrites that terminate in various glomeruli. Axons from sensory cells terminate in a single glomerulus, but each glomerulus receives axons from multiple receptor cells. The axons of mitral cells in the olfactory bulb form the olfactory tract, which transports information to the telencephalon. The olfactory tract is divided into two main bundles, each of which is further subdivided. Some of the fibres run directly to the hypothalamus, while others cross the anterior commissure of the brain.

27.1.5 Histochemistry of Olfactory Epithelium

Several researchers have studied the histochemistry of fish olfactory epithelium. According to the findings, receptor cell dendrites have both alkaline and acid phosphatase activity. These enzymes' activities are low in supporting cells. According to Pandey and Misra (1984), basal cells are undifferentiated and give rise to other cell types in developing lamellae while replacing degenerating cells in older ones. In *Channa punctatus*, lipids have been found in high concentrations along synaptic connections in primary neurons, the proximal limb of supporting cells, and secretory granules of goblet cells. The lipid content of basal cells, secondary neurons, and supporting cell cilia is relatively low.

The olfactory nerve fascicles make up the olfactory nerve layer, which connects peripheral sensory neurons to primary output neurons in the olfactory bulb. The glomerular layer is formed by the synapses between the olfactory neurons and the primary output neurons in the bulb, which form spherical structures composed of neuropil called glomeruli. Mitral cells, which are primary output neurons, and granule cells, which are interneurons that primarily synapse with mitral cells, make up the mitral cell and granular cell layers (Zielinski and Hara 2006). The olfactory tracts connect the olfactory bulbs to higher brain centres (Hara 2011).

27.1.6 Mechanism of Olfaction

The mechanisms underlying enzyme roles in olfaction are poorly understood. Although lipids may play an important role and olfactory hair is a site for the reception of odorant molecules, the exact nature of the chemical reaction is unknown. The water current is most likely important in olfaction. Water enters the olfactory chamber through the anterior nasal opening and exits through the posterior one via one of three routes:

1. When the fish moves forward, a strong current of water enters the olfactory chamber. Water enters through the anterior nostril and exits through the posterior. The vertically projecting flap of the skin aids in the process of forward progression. The water current is directed toward the posterior opening by the linguiform processes of the lamellae.

2. Because of cilia activity, a slow water current flows through the olfactory chamber.
3. According to several authors, the accessory sacs help to keep the water current flowing by pumping. The dilatation and compression of the accessory sac are thought to be synchronised with the respiratory movement of the fish.

27.1.7 Functions

1. Olfactory epithelium cells convey information about the chemical environment to the brain.
2. The chemical that affects the olfactory mucosa's surface stimulates sensory cells.
3. The odorant epithelium is extremely sensitive to amino acids.
4. Many fish use their olfactory sense to find food. If the lateral part of the olfactory tract is cut, the feeding response induced by food odour disappears, but cutting the medial part has no effect. The amino acid feeding response is processed through the lateral bundles of the olfactory system; this is dependent on other responses as well.
5. Fish are classified as "eye-nose" if they have an oval rosette, "eye-fishes" if they have a circular rosette and "nose-fishes" if they have an elongated rosette.
6. It plays an important role in many aspects of fish reproduction.
7. It is also useful in fish warning systems. When injured, some fishes release an "alarm" substance, while others react with a well-defined "alarm" reaction.
8. Fishes respond to fear primarily through olfaction, though sight and hearing are also involved. When a fish's skin is damaged, the alarm substance is replaced in the water.

27.2 Electroreception in Fishes

27.2.1 Introduction

Natural water, being a good conductor, is teeming with electrical current. A large number of fish have evolved to direct such an electrical field. Fish are equipped with electroreceptor organs for this purpose. These electroreceptor organs are highly sensitive lateral line organs (Fig. 27.2). The following facts about electroreception are important:

1. The active electrosensory systems can be used in electrolocation analogous to the echolocation of bats and crustaceans. They may have weak or well-developed electric organs. It can also operate passively.

2. A stimulus includes both animate (membrane potential to fish) and inanimate signals (e.g., oceanic current, magnetic field).
3. Electroreceptor includes a large number of accessory organs.
4. It accomplishes electrocommunication and social signaling etc.
5. They are capable of detecting a field of even 0.01V/cm (one flashlight per 1500km).
6. They are ampullae of Lorenzini to detect electric fields. They form a network of mucus-filled pores in the skin of cartilaginous fish and basal actinopterygians.

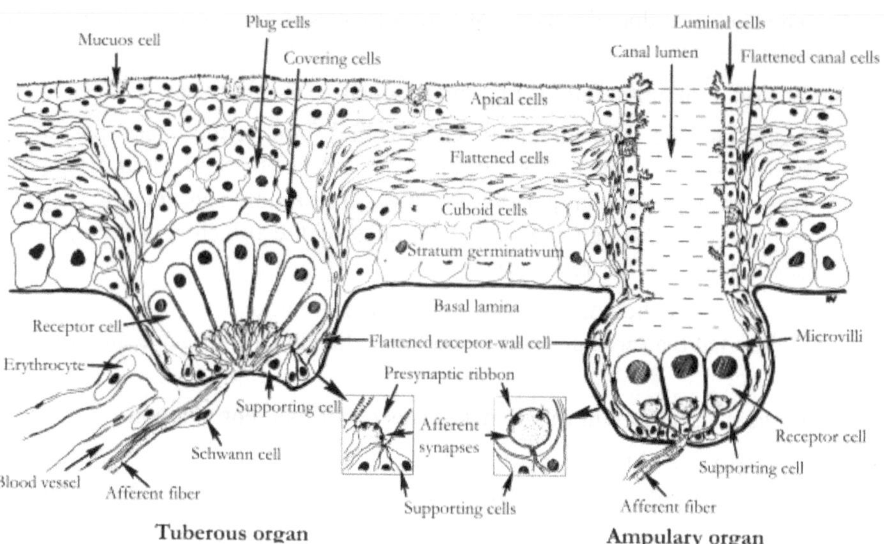

Fig. 27.2: Electroreception in fishes (From https://www.sciencedirect.com)

27.2.2 History

Electroreception is a phylogenetically widespread sensory modality that has evolved several times throughout vertebrate evolution but is most commonly found in fish, some amphibians, and a few mammals. Lorenzini (1678) discovered organs on the heads of fish that are now known as Lorenzini ampullae. Without knowing their function as electroreceptors, Franz (1921) described knollenorgans (tuberous organs) in elephantfish skin. Lissmann (1949) noticed that the African knife fish (*Gymnarchus niloticus*) was able to swim backward at the same speed and with the same dexterity around obstacles as when it swam forward, avoiding collisions. In 1950, he demonstrated that fish produce a variable electric field and that the fish reacts to changes in the electric field around it. Murray (1960) discovered that these organs have an electroreceptive function.

27.2.3 Grouping of Fishes

Ampullary electroreceptors are found in sharks, skates, rays and chimaeras; bichirs and reedfishes (Polypteriformes), sturgeons and paddlefishes (Acipenseriformes), lungfishes, coelacanths and caecilians; urodeles and teleosts (Siluriformes, Gymnotiformes and some Osteoglossiformes) generally occupying freshwater habitats (Fig. 27.3). These are derived forms of sensory hair cells similar to those in the mechanosensory neuromast organs of the lateral line (Gillis et al, 2012).

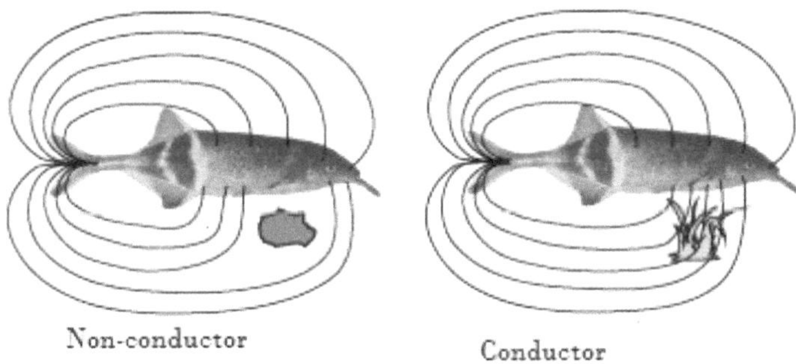

Fig. 27.3: Electric fishes (From https://www.sciencedirect.com)

I. Based on the presence /absence of electroreceptor and electric organs

A. Fishes lacking electroreceptor organs

B. Fishes have electroreceptor organs

(a) Fishes lacking electric organs

1. Marine fish including sharks, catfishes and rays.
2. Freshwater fishes: e.g., Catfishes, Dipnoi, Eels, Rays, Branchiopterygii

(b) Fishes having electric organs

1. Gymnarhcids, *Sternopygus, Stenarchus, Eigenmennia* and other wave fishes.
2. *Electrophorus, Gymnotus, Hypopomus* and Mormyrids are examples of pulse fishes.

II. Based on signals

A. Active fish: The signals of these fish are accompanied by their activity from the aforementioned sources. Fish generate a weak electric field and sense the different distortions of that field created by objects that conduct or resist electricity.

B. Passive fishes: These fish detect signals from an external source of either animate or inanimate origin. They are capable of detecting changes in the earth's current. e.g., Sharks, Flatfishes etc. Fish use it to supplement or replace their other senses when detecting prey and predators.

III. Based on the nature of the cell

A. Tonic receptors: These are fairly and rhythmically active and give long-lasting responses to low frequency. Its channel leads from the receptor cavity to the exterior.

B. Phasic receptors: These are sensitive to high frequency. These are too insensitive to keep electric stimuli going. It lacks a channel to the exterior.

IV. Based on the taxonomic approach

Table 27.1: Nature and response of electroreceptors in fishes

Sl.No.	Groups	Morphological nature	Physiological nature	Name of response
1.	Elasmobranchs	Ampullary (definitive)	Tonic	Ampulla of Lorenzini
2.	Catfishes	" Tubular receptor	" "	Small pit organs Ampullary organ
3.	Gymnotidae (Electric eel)	Ampullary Tuberous "	" Phasic Tonic	Ampullary type I Ampullary type II Ampullary type III
4.	Mormyridae (Elephantfish)	Ampullary Tuberous " "	Tonic Phasic " Tonic	Small neuromast type I (A) Medium neuromast type II (B) Large neuromast type III (C) Large neuromast type IV (D)
5.	Gymnarchidae	Ampullary Tuberous " "	Tonic Phasic " Tonic	Type A Type B Type C Type D

Table 27.2: Comparison of electroreceptors of various fishes:

Sl.No.	Name of fish	Habitat	Length of the ampullary organ (mm) (=a)	Disc width (cm) (=b)	= x	Reference
1.	*Potamotrygon circularis*	Freshwater	0.3–0.5	15–23	0.02–0.022	Szabo et al. (1972)
2.	*Dasyatis garouaensis*	Freshwater	0.7–2.1	23–32	0.03–0.065	Raschi and Mackanos (1989)
3.	*Himantura signifer*	Freshwater, brackish	1.0–2.3	21.5	0.046–0.107	Raschi et al. (1997)

4.	*Aptychotrema rostrata*	Marine, brackish	4.7–55.5	24	0.196–2.315	Wueringer and Tibbits (2008)
5.	*Glaucostegus typus*	Marine, freshwater, brackish	6.7–53.4	15.1	0.444–3.536	Wueringer and Tibbits (2008)

27.2.4 Morphology of Electroreceptor

1. Mormyridae: Three types of receptors are found over the entire head and along the dorsal and ventral surface, but they are absent on the sides of the body and caudal peduncle. The medium (phasic) receptors are similarly distributed as smaller (tonic) but are about twice as many as a tonic. Large (phasic) receptors are about the same in number but are somewhat uniformly distributed. These receptors are innervated by either anterior or posterior lateral line nerves. e.g., *Gnathamemus petteroii.*
2. Gymnotidae: The phasic receptors are much more numerous than the tonic receptors. The density of receptors is greatest in the head region and falls gradually towards the posterior. Electroreceptors over the entire body are innervated by the anterior lateral line nerves.
3. Catfishes: The electroreceptors are distributed over the entire body surface, but tend to be concentrated in the head region. Receptors are even found on fins. Innervations of the receptors are done by the lateral line nerves. e.g,, *Mlapterurus.*
4. Elasmobrnachs: Ampullae of Lorenzini are found in most elasmobranchs. However, the distribution varies by species. In freshwater electric fish, electroreceptors are present on the surface of the skin. Their receptors are unpigmented.

Anatomy of electroreceptors

The functional units of the chondrichthyan electro-sensory system are a series of Ampullae of Lorenzini is connected to a network of canals that radiate away from the ampullae and terminate at pores in the skin (Fig. 27.4). The canal pore extends from a somatic pore, widening proximally to an ampullary bulb. The ampulla is formed by several alveoli arranged in a grape-like formation where the epithelium of adjacent alveoli and the canal are separated by the medial zone. A sensory nerve fibre extends from the proximal end of the ampulla.

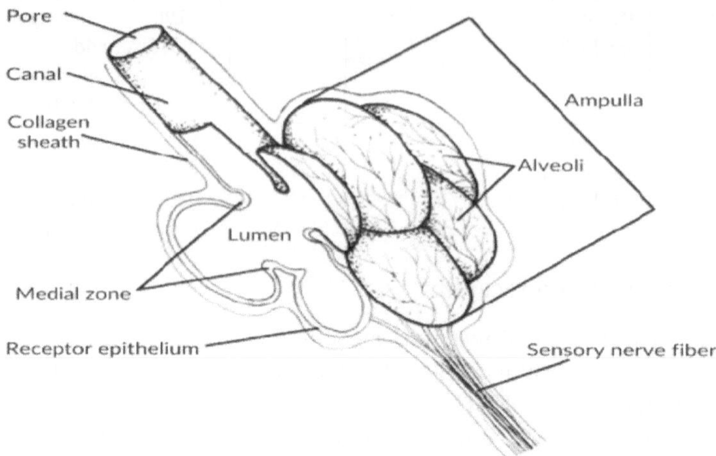

Fig. 27.4: A single ampulla of Lorenzini of a rhinobatid (*Aptychotrema rostrata*). (from Wueringer and Tibbetts, 2008).

27.2.5 Histology of tonic receptors

1. Gymnotidae and Mormyridae: These possess a canal that opens to the exterior and usually enlarges at its base to form an ampulla. Many receptor cells are embedded in the wall of the ampulla. Only a small part of their circumference is present in the lumen, although the surface in this region may be increased by microvilli. These receptor cells synapse with the single afferent fibre on their inner surface and are largely surrounded by supporting cells.

All over the body, the epidermis contains a specialised zone containing many flattened cell layers. Probably, this is responsible for the very superficial resistive barrier in the skin. This layer is closely opposed to cells of the wall of the receptor canal. The zonular tight junction between the cells of the canal walls and supporting cells around the receptor cells, prevents leakage of current through intercellular clefts and tends to channel current through the receptor cells. Receptor cells are sensitive to the potential across the skin and are insensitive to gradients tangential to the skin.

2. Catfish: These receptors are similar to tonic receptors of freshwater electric fish, except that, generally the ampullary canal is shorter. Its skin lacks multiple layers of flattened epithelial cells. In *Kryptopterus*, a cluster of receptor cells is located in a small beehive-shaped cavity that has an opening to the exterior on one side.

The receptor cells have a small internally negative potential and most of a long-lasting voltage pulse applied at the receptor opening is developed across their inner rather than their outer faces.

3. Elasmobranchs: Electro-receptor lies deep beneath the surface, but they are connected to the exterior by canals. The canal wall is of very high resistance. The receptor cells are embedded in the wall of the ampullary cavity with only a small portion of their surface exposed to the lumen.

In Skate, there may be a signal cilium extending into the lumen. At their bases, the cells form typical synapses with innervating nerve fibres. Several nerve fibres innervate the receptor cells in a single

27.2.6 Operation of tonic receptor

The potential across the receptor cell is largely developed across its nerve face, which is of relatively higher resistance. This face is pre-synaptic to the innervating nerve fibres. A transmitter substance is secreted from the inner face in the absence of stimulation. This transmitter depolarizes the nerve fibre and causes its tonic discharge. Stimuli that depolarize the inner face of the receptor cell cytoplasmic surface are more positive than hyperpolarising the inner face decreasing the rate of transmitter secretion and thereby decreasing the discharge frequency in the nerve.

27.2.7 Histology of phasic receptors

1. Gymnotidae: It contains 10 or more receptor cells in a receptor cavity. All are innervated by a single nerve fibre. These receptors protrude into the receptor cavity and only their bases are attached. The largest part of the surface is the cavity and this portion is measured by microvilli.

Another face of the receptor cells behaves as a series capacity while the inner face has an electrically excitable as well as a secretary membrane. The inner face may generate graded responses or even spikes.

2. Mormyridae: It contains

(a) Large receptors: 1-8 receptors.

(b) Medium receptors: These are physiologically complex. There is an outer and an inner receptor cavity connected by a small channel. The outer cavity is covered by loose epithelial tissue. The outer epithelial layer is not covered by fattened epidermal cells which are closely opposed to the wall of the outer receptor cavity. Receptor cells occur in both cavities. In the outer cavity, the cells are embedded in the wall of the inner cavity. The cells protrude into the cavity and are covered by microvilli.

27.2.8 Function

Electroreception participates in prey detection, predator avoidance, mate detection, geo-navigation and even communication (Tricas and Sisneros, 2004). In general, terrestrial animals have little use for electroreception, because the high resistance of air limits the flow of electric current.

27.3 Photoreception in Fishes

27.3.1 Introduction

Living organisms are provided with specialised structures that in conjunction with the nervous system detect what goes on outside or inside the body. Such specialised organs are known as receptors. Usually, a receptor is sensitive to only one kind of stimulus like light, temperature, sound etc. All receptors are transducers as they convert various forms of energy from the stimulus into electrical energy for the nerve impulse.

Light has been defined as the radiant energy capable of radiating and stimulating the photoreceptors and causing the sensation of vision.

The photoreceptor in fish is the eye (=*survey indriyanaam nayanam pradhanam*). Fish have monocular vision because both of their eyes cannot be focused on the same object at the same time.

27.3.2 Structure of Eye

The anatomy of a typical piscine is similar to that of other vertebrates. In the elasmobranchs, the eyes are very large and held in position by an optic pedicel (Fig. 27.5). In some sharks and rays, the pedicle is slender and elastic and helps in the protrusion of the eyeball. The deep-sea Holocephali has the biggest eye. An enormous eye in deep-sea teleosts has a large pupil, a rounded cornea, and an exceptionally large lens.

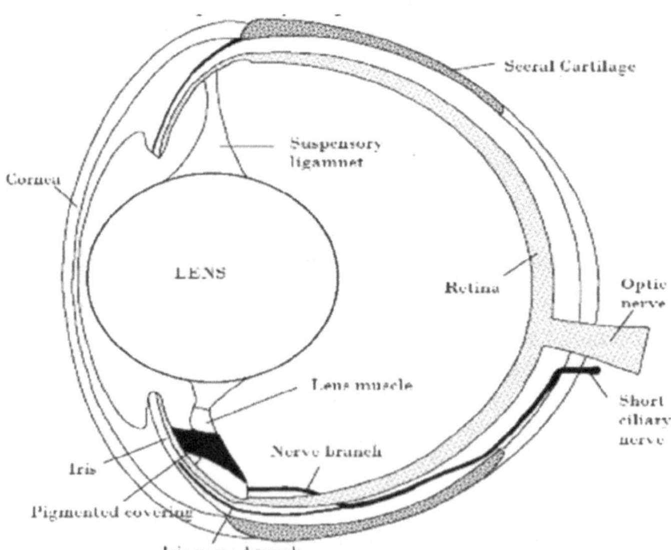

Fig. 27.5: VS of the eye of fish (from https://www.researchgate.net)

In some fish, the eye is adapted to both water and air. Even in some other deep-sea fish, the eyes are adapted for vision in dim light. In four-eye fish (*Anableps*) and *Dialommus*, each eye has two pupils. *Gigantura* possesses two retinas and a tubular eye. In catfishes, eyes are reduced and even absent in cave fishes. Flatfish bear both eyes on one side only. The large eyes of *Periopthalmus*, move in all directions. In *Benthobates moreshyi*, the eyes are degenerated but disappear in *Ipnops murrayi.*

The eye is somewhat flattened and has normal components. The eyeball is held in place by a series of ligaments in the bony orbit and is moved by the coordinated action of six ocular muscles.

The fourth rectus muscle arises from the orbit and diverges to insert around the equator of the eyeball, while the two oblique muscles arise from the ethmoidal region. In elasmobranchs, eyes are held in position inside the orbit by an optic pedicel. Sabha et al. (1979) reveal that ocular muscles are composite like the red and white muscle fibres. The specific muscles responsible for the movement of the lens within the eye differ in cyclostomes, chondrichthyes and teleosts. Fish with well-developed lentis muscles may have acute vision.

The eyeball has the following three layers:

1. **Sclerotic:** The transparent cornea is covered over by a tough, flexible, opaque, white envelope known as the sclerotic. It may be further strengthened by cartilaginous or even bony elements. In elasmobranchs, cartilage is present in the sclerotic.
2. **Choroid:** It is connected to the iris and lies between the retina and the sclerotic. The innermost part lying just behind the retina is modified into a horseshoe-shaped, highly vascular structure, the choroid gland. The choroid gland is responsible for meeting the retina's high oxygen demand. The melanin pigmentation of the choroid serves to prevent internal reflection. Teleosts have an argentum (argentia) between the sclerotic and choroid layers and lack a choroid plexus. A vascular fold from the choroid known as the falciform process pierces the retina near the optic nerve and continues up to the lens, ending in a small muscular knob or Campanula Halleri (the retractor lentis) (Fig. 27.6).

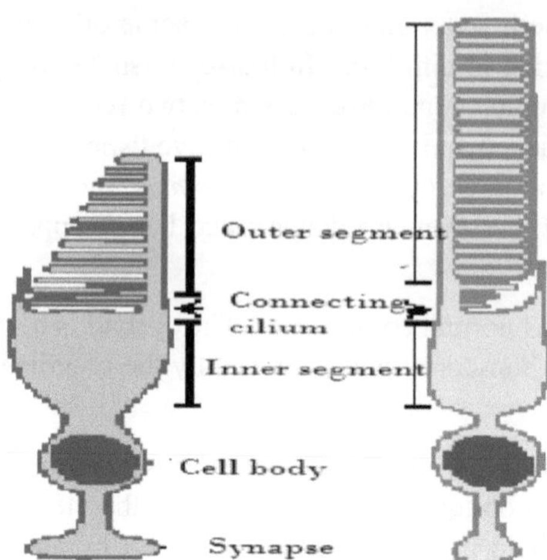

Fig. 27.6: Cellular structure of the eye in fishes (From https://www.mdpi.com)

3. **Retina:** The retina is the innermost layer and is made up of a complex series of cells arranged in different layers. In elasmobranchs, intrinsic muscles are absent in the ciliary body. In elasmobranchs except in *Mustilus* and *Myobatis*, the retinal layer lacks cones. Most species of the retina have both rod cells (for scotopic vision) and cone cells (for photopic vision), and most have colour vision. Some fish can see ultraviolet light, while others are sensitive to polarised light. These are the layers from sclerotic to vitri:
    1. Epithelial layer: It covers the outermost circumference of the cup-shaped light receptor and is immediately adjacent to the choroid in a pigmented light receptor epithelial layer, which enhances the light-absorbing qualities of the choroid and is capable of migration.
    2. Visual cell layer: Three types of visual cells occur in the teleost retina, rods and single and twin cones. Pure rods and retinas occur in deep-sea teleosts.
    3. Bipolar cell layer: It constitutes an intermediate connection between visual cells proper and ganglion cells.
    4. Ganglionic cell layer: On the one hand, the dendrites from the tripolar cells are connected with the cones or with groups of rods collectively and on the other hand, they are interlocked with dendrites from the ganglionic cells.

Many variations are observed in the number and arrangement of rods and cones in teleosts. They may have twin cones or triple cones. The rods are closely set and connected to a single nerve fibre. On the other hand, cones are widely spread and each is connected to a single nerve fibre. The ratio of rods to cones is determined by the ecology of the concerned fish species; for example, those that are primarily active during the day in clear waters will have more cones than those that live in low-light environments. In deep-sea teleosts, the rods are well developed but cones are lacking.

The eye contains the following components:

1. **Cornea:** The cornea develops as the anterior, nearly flat and transparent part of the sclerotic. It is the outermost part that is streamlined in active fishes but irregular or concentrically ridged in others. The cornea is a flat and non-muscular structure that obtains nourishment from the intraocular fluid. The cornea is very round, allowing the fish to receive images of its surroundings from nearly a full hemisphere around the eye or 360^0.
2. **Iris:** The iris forms the pupil and controls the amount of light reaching the retina. Most bony fish except flatfishes and eels have a fixed pupil and iris devoid of muscles. Most elasmobranchs exhibit extensive but slow contraction of the iris.
3. **Lens:** The lens is nearly spherical or globular and sits very close to the cornea. The space between the cornea and the lens is filled with a watery fluid, the aqueous humour. The space between the lens and the retina is filled with a transparent jelly-like material, the vitreous humour. The lens is formed of concentric layers made up of a transparent colony of cells devoid of nerve or blood supply. In elasmobranchs, a light-reflecting layer, known as the tapetum lucidum is present. However, in *Cetorhinus* and *Laemargus*, the tapetum is wanting.

27.3.3 Function

Fishes have monocular vision and the two eyes cannot be focused on the same object. However, certain deep-sea fishes have a telescopic vision.

The amount of light available to a fish eye is greatly reduced from that in air, particularly so with increasing depth of turbidity. Therefore, the task of the fish eye is to collect sufficient light from as many directions as possible.

A mechanism that stabilises images during rapid head movements is required. The vestibulo-ocular reflex, a reflex eye movement that stabilises images on the retina by producing eye movements in the opposite direction of head movements, maintains the image in the centre of the visual field.

(1) **Formation of image:** The optical density of cornea, aqueous humour and vitreous humour, approximate water (refractive index -1.33), consequently the role of image formation falls entirely to the lens. The firmness of the lens does not permit flexing the sphere to alter the surface relations for focusing; instead, the retinal lens distance is altered. In teleosts, the fluid nature of the aqueous humor permits backward displacement within the anterior chamber

through the movement of the lens through the contraction of the retractor lentis. The lens, when relaxed, is almost in contact with the cornea, which itself is bulging.

Accommodation results from changing the distance between the lens and the retina. Accommodation is caused by the change in the position of the lens. This is accomplished by the falciform process in conjunction with campanula Halleri in teleosts. But in elasmobranchs, accommodation is brought about by the activities of smooth muscle fibres. In deep-sea teleosts, the pupil and lens are exceptionally large.

(2) **Process of Photoreception:** Stimulation of the sensory part of the retina by the light of the focused image causes a chemical deposition of photo-liable substances near the tips of rods and cones. The products of decomposition stimulate the receptors to initiate a series of nerve impulses that culminate in sight.

Two kinds of light-sensitive pigments rhodopsin and porphyropsin are produced by the retina of fish. Marine fishes produce purple-coloured rhodopsin, while in freshwater fishes; pink-coloured porphyropsin is produced. In migratory fishes (e.g., eels, *Hilsa*), both rhodopsin and porphyropsin are produced.

(3) **Bimodal vision:** In *Anableps anableps*, the retina is differentiated into upper and lower parts. The upper part is adapted for aquatic vision and the lower part is for aerial vision. The retinal unit concerned with aquatic vision contains more rods (Munshi and Singh, 1978). The lens is elongated along the dorsolateral axis. A high percentage of cones reaching the front, half of the retina may be associated with the greatly increased brightness of aerial conditions; in contrast, a high percentage of rods in the upper half of the retina may be associated with the greatly decreased brightness of aquatic vision.

(4) **UV vision:** Four visual pigments that absorb different wavelengths of light mediate fish vision. Each pigment is made up of a chromophore and a transmembrane protein called opsin. Opsin mutations have enabled visual diversity, including variation in wavelength absorption. Because of a mutation in the opsin on the SWS-1 pigment, some vertebrates can absorb UV light (360 nm), allow them to see objects that reflect UV light. This visual trait has been developed and maintained by a diverse range of fish species throughout evolution, implying that it is advantageous. UV vision could be associated with foraging, communication, and mate selection.

(5) Fish eyes keep track of their lifetime mercury exposure.

(6) Ancient fish and humans share a key feature of vision.

27.4 Sound Production in Fishes (Phonoreceptors)

27.4.1 Introduction

Sound is probably the most effective channel for long-range communication underwater. Several fish are known to produce characteristic sounds with the help of either stridulating organs or the air bladder. Sound production has many folds of significance such as:

A. Discriminate between calls them with that of others.
B. Characteristic sound helps the individuals to congregate as a shoal.
C. Helps during the breeding season and increases the chances of fertilization of ova.
D. Helps in recognition of themselves with those having reduced eyes or bottom dwellers.
E. Helps in the detection of the direction and distance of the sound source.
F. Helps in the location of prey or predator (fishes are insensitive to sounds above 2-3 kHz.).
G. Highly excited minnows (=Cyprinidae), loaches (Cobitidae) and eels produce high-pitched sounds in different ways. The globe fishes (=Tetradontidae) gravel very much like dogs.
H. Sound production may have certain adaptive and non-adaptive significance in *Amphipnous (Monopterus) cuchia.*

Rice et al, (2022) mapped the most comprehensive dataset of sound production in fishes onto a family-level phylogeny of Actinopterygii. Family-level analyses allowed investigation of sound production based on acoustic recordings and morphological specializations (82%) indicative of qualitative descriptions (18%) and a conservative estimate of the distribution and ancestry of a character.

27.4.2 Sound Frequencies

The sound is deep and loud and can be heard even at a distance of 30.5 m. The sounds produced by fish are less than 1000 Hz. The majority of sounds are produced with frequencies ranging from about 45 – 60 Hz (goliath grouper and black drum). Higher frequency harmonics are produced by drumming above 1000 Hz (e.g., silver perch). The following table shows the frequencies of sounds produced by some fish:

Table 27.3: Type of sound and their range produced by fishes

Sl.No.	Name of fish	Type of sound	Range (Hz)	Reporter
1.	Codfish	Lower frequency	40-50	Hawkins and Chapman (1966)
2.	Drumfish	Low frequency	75-100	Kellogg (1955)
3.	Gulf-top sail catfish	High frequency	250	Tavolga (1962)
4.	Toadfish and silver perch	Higher frequency	300-350	Winn (1964)

27.4.3 Occurrence and Type

At present, hundreds of species of fish have been reported to be sound producers. The type of sound production is different in different fishes. Aristotle and Pliny gave several examples. They can hiss, growl, grunt and crack as reported in catfish, globefish, *Carnax hippos* and drums respectively. Some fishes produce acute sharp and repeating sounds at long intervals as in gurnards. Sorenson (1894) reported sound production in certain catfishes. Dufosse (1874) described sound production in several marine fishes. Pandey and Agarwal (1982) reported sound production in freshwater mud eel (*Amphipnous cuchia*) under stress.

27.4.4 Organs and Mechanisms of Sound Production

A little relationship occurs between the morphology of sound production during mechanisms and phylogenetic position in fishes (Fig. 27.7). Many deep-sea forms appear to possess the means for sound production but have not yet been recorded. Three general types of sonic mechanisms are present in the fish:

A. Stridulatory sound
B. Hydrodynamic (quickly changing speed and direction) sound and
C. Swim bladder (drumming) sound

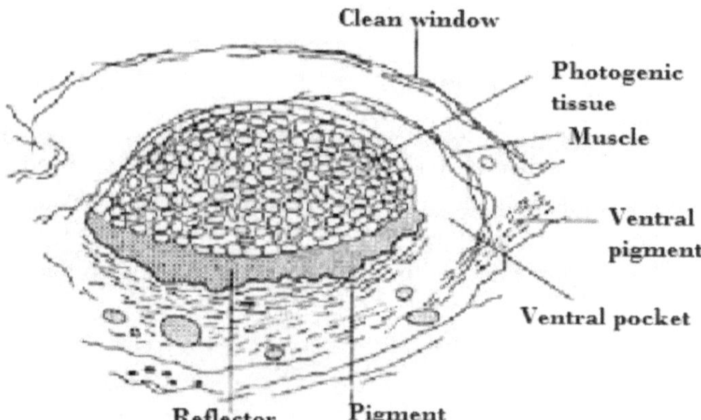

Fig. 27.7: Sound Production in fish

A. Stridulatory Sound

The component frequencies of stridulatory sounds range from 100 to 8000 Hz, while predominant frequencies are generally between 1000 and 4000 Hz.

They are produced by the friction of different types of teeth, fins, spines or bones.

The majority of teleosts possess opposing patches of denticles in their pharynx and is at least capable of sound production during feeding. The grunt (*Haemulon*) stridulates pharyngeal denticles in connection with other activities such as alarm and territoriality.

Almost all predatory fishes produce sounds when feeding and even some forms that browse on sessile plants and animals can emit sounds when crushing rocks and corals.

Incisor teeth are capable of biting through the exoskeleton and thus produce strong metallic sounds. If the food of moderate hardness is there, the teeth will produce sound. Sometimes fish will gnash their teeth without the direct presence of food.

Fish (1954) listed some sharks and rays that produce sound when feeding, although elasmobranchs are not known to be sound produces in any specialised sense (Fig. 27.8). Moulton (1958) listed some fishes that produce sounds using modified molar teeth. For example: Puffers, Filefish, Parrot fish etc.

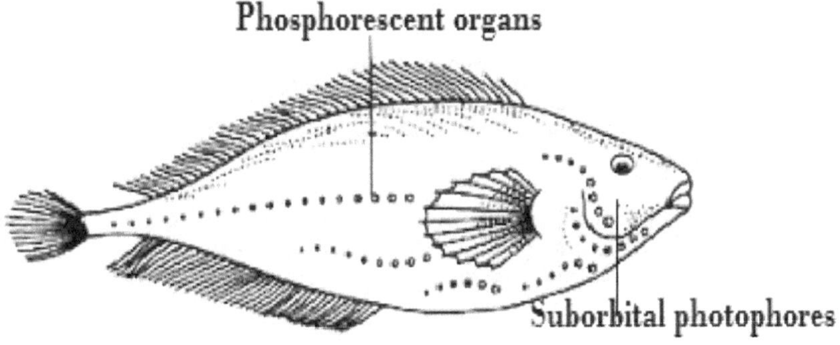

Fig. 27.8: Sound Producing organs in fishes

The sea catfish (*Galeichthyes felis*) and the gaff-topsail catfish (*Bagra marinus*) produce high-pitched squeaks when the enlarged pectoral fin spines are moved. The erection of specialised dorsal fin spines in triggerfish produces stridulation and low-intensity sound may be produced by stickleback (*Gastresteus aculeatus*) in dorsal fin spines take place.

The clown fish (*Amphiprion*) and the sea horse (*Hippocampus*) produce sound by the friction of adjacent bones which occurs in stridulation of the posterior margin of the skull against some vertebral element. Sound production by similar means has also been reported in pipe fish (*Syngnathus louisiance*).

The Indian loach (*Botia hymenophysa*) is attributed to the stridulation of Weberian ossicles to produce sound. The triggerfish produced sound by rubbing the pectoral fins against the body sides where skin thinly covered lateral evaginations of the swim bladder.

A characid fish (*Glandulocanda inequalis*) produces sound by gulping air at the water surface (Nelson, 1965).

In *Callomystax*, the neural spines of the 3rd, 4th and 5th vertebrae are fused to form a laterally compressed bony plate. The hind portion of the bony plate is divided into two parts to receive the first interspinous bone of the dorsal fin between them. The interspinous bone is ridged on both its surfaces like a file, while the bony plates on either side of it are traversed by parallel, closely set vertical ridges. The vertical movements of the 6th and succeeding vertebrae cause the interspinous bone to rub against the ridges, to produce a characteristic grating sound.

Pro-operculum of *Cottus scorpius* and hyomandibular of *Dcatylopterus* are modified for stridulation. In triggerfish (*Balistes*) stridulation takes place between the post clavicle and a longitudinally grooved area on the inner surface of the corresponding cleithrum.

Sound is produced by friction between superior and inferior pharyngeal teeth in *Scomber* and the sunfish (*Orthagoriscus mola*).

B. Hydrodynamic Sound

Swimming movements especially during rapid changes of direction or velocity result in the production of such sound.

C. Swim Bladder Sound

The swim bladder is either modified specially to produce sound or it produces sound through adjacent structures (Fig. 27.9). It is variously used by a variety of fish drums (Sciaenidae), gurnards (Trigtidae) or grenadiers (Macrouridae).

1. Swim bladder as a drum: Triggerfish (*Balistes*) produce sound by beating their pectoral fins against the areas of the body wall that cover the swim bladder (Moulton, 1958).
2. Swim bladder as resonator: In triggerfish (*Balistes*), the resonator produces sound by stridulations of cleithrum against the post-clavicles and the granting of pharyngeal teeth. In carangids and haemulids, the sound is produced by pharyngeal teeth alone.
3. *Epinephelus striatus* produces sound through a pair of sonic muscles (Winn, 1962). The sonic muscles connect the skull with the swim bladder. In some

fishes, sonic muscles are completely attached to the wall of the swim bladder. In such cases, the bladder can be dissected out of the body cavity and can function as a sound-producing mechanism just by stimulation of nerves leading to sonic muscles.

4. Various species of triggerfish produce sound by rubbing the pectoral fins against the body's sides when skin thinly covered lateral evaginations of the swim bladder.

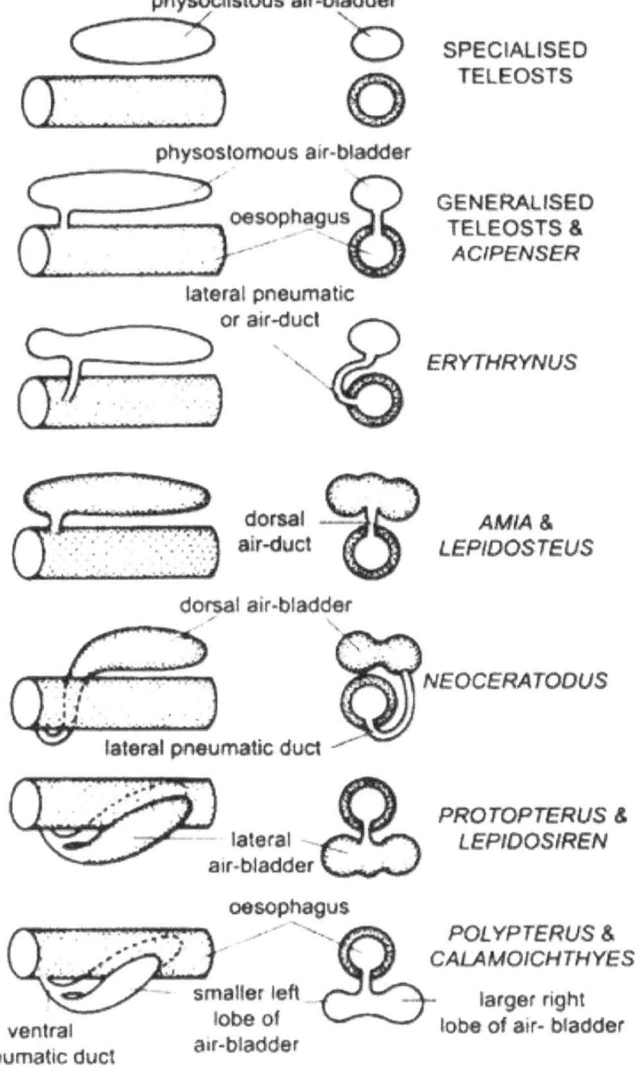

Fig. 27.9: Air-bladder as sound producing-organs in fishes

5. Expulsion of gases: Liberation of gas bubbles from the swim bladder through the pneumatic duct is often accompanied by sound as recorded in European eel (*Anguilla anguilla*), minnow (*Phoxinus phoxinus*) and American eel (*Anguilla rostrata*). Highly excited minnows (Cyprinidae), loaches (Cobitidae) and eel (Anguillidae) produce high-pitched sounds by the release of air. In *Misgurnus fossilis* and Herrings, the gas expelled from the gas bladder through a duct that ends at the anus produces a strange sound.
6. In a catfish: The vibration of the bladder wall: Muscles attached to the bladder wall is responsible for the vibrating mechanism of the gas bladder. The muscle causes rapid changes in bladder volume through fast contraction and expansion; as a result, a typical grunting noise is resonated.

In toad-fish (*Ospanus tau*) and sea robin (*Trigla*), striated muscles originate from the dorsal body wall are inserted in the gas bladder. In toadfish, striated muscles produce volume changes, which in turn produce audible signals. The swim bladder of toadfish is so self-contained that it can be taken out and made to produce sound by elastic stimulation.

In drums (*Micropogon undulatus*), sonorofic muscles are attached to the gas bladder. These muscles in some cases run on each side of the abdomen to a central tendon located above the bladder. Striated muscles originate on the dorsal body wall and are inserted in the swim bladder to make the bladder wall vibrate.

In fish (*Malapterurus*), the swim bladder is vibrated by a flexible and highly elastic spring mechanism. These springs are modified parts of the 4th vertebra. The expanded backwardly downwards extremities of transverse processes are attached to an embedded front part of the bladder. Springs are connected through two strong muscles with the occipital region of the skull. A humming sound is produced by vibrations of springs and consequently of bladder walls. In these species, the gas bladder is usually divided internally into chambers using septa and the intensity of sound is increased by vibratory movements of the free edges of the septa.

In a few teleosts, the muscles are directly attached to the wall of the air bladder. In some species, the gas bladder is related to mucus that is connected to the pectoral girdle or the muscles of the wall of the bladder may not be connected to any part of the skeleton.

Pandey and Agarwal (1982) observed that when *Puntius* move near the snout of *Amphipnous cuchia*, catch them with a quick spurt and engulf them. While doing so, the sound is produced followed by a kind of whistling sound.

The sound produced by the elastic springs of *Malapterurus electricus* is comparable to the hissing of a cat. Hippocampus produces a characteristic sound that is more intense during the breeding season. The sound produced by *Sciana aquila* has been compared to bellowing, purring, buzzing and whistling.

27.4.5 Significance of Sound Production in Fishes

Fishes produce different types of sounds using different mechanisms and for different reasons:

A. It helps individuals to form shoals and is advantageous in breeding behaviour.
B. It attracts the opposite sex in the breeding season.
C. It increases the chances of fertilization of the ova.
D. It is useful for fish with reduced eyes to recognise the members of the species.
E. It is valuable as a warning signal against the enemy or aggressor.
F. It may be intentionally used as signals to predators or competitors, to attract mates, or as a fright response. It is unintentionally produced as a by-product of feeding or swimming.

27.5 Electric Organs in Fishes

27.5.1 Introduction

Fish have a variety of adaptive structures that have evolved in response to the unique challenges they face. The electric organs, phosphorescent organs, poison glands and sound-producing organs are the most important of these structures or organs.

Fish are unique in that they can generate an electric current. Electrical organs are complex and specialised structures that allow the carrier to generate, store, and discharge electricity. Man has known about the ability of some fish to discharge electricity as an effector action since ancient times, and this may have been the first bioelectric potential observed by him.

These fish use electric discharges for navigation, communication, mating, defence and prey immobilisation.

27.5.2 History

The electric organs of torpedo, ray and electric eel were studied by Walsh et al, (1770). Hunter discovered what is now called Hunter's organ. Darwin (1859) used them as an instance of convergent evolution. Galvani and Volta are considered the founders of electrophysiology and electrochemistry respectively. Lissmann (1951) pioneered a paper on *Gymnarchus*. *Torpedo californica* electrocytes were used in the first sequencing of the acetylcholine receptor by Noda et al, (1982), while *Electrophorus* electrocytes were used in the first sequencing of the voltage-gated sodium channel by Noda et al, (1984).

LaPotin et al, (2022) reconstructed the evolution of electric organs by tracing the fish's genomic history. The development of electric organs in fish is related to the underlying genetic causes of certain diseases in humans (Vingum, 2022).

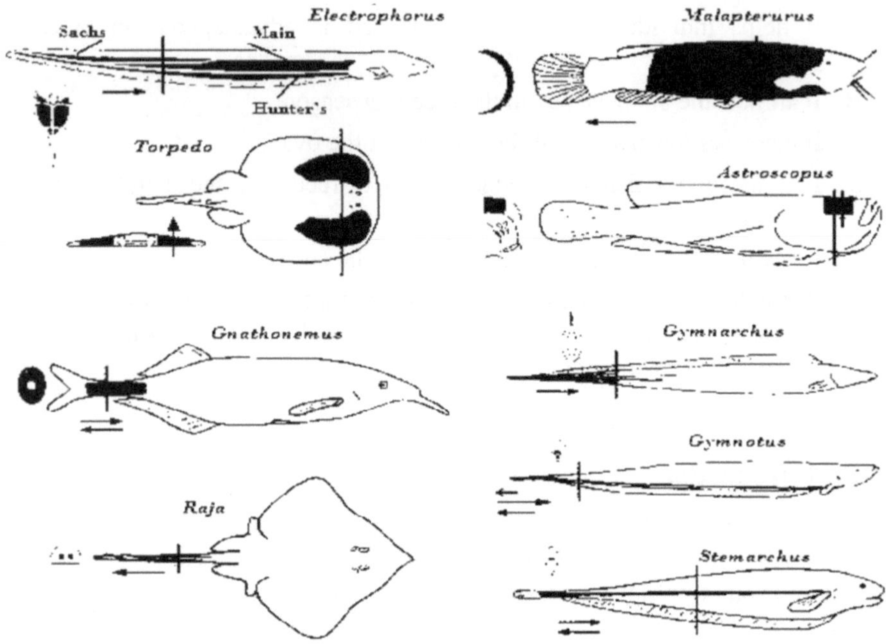

Fig. 27.10: Electric organs in fishes

27.5.3 Milestones

The following points are important regarding electric organs (Fig. 27.10):

1. About 250 species of fish possess electric organs. Most of these are restricted to the tropical freshwater of South America and Africa.
2. The electrogenic organs of fishes are structures specialised for the production of an electric field outside the body.
3. The shape and the position of the electric organs differ greatly in different species of fish.
4. All such fishes possess similar electrogenic microscopic structures.
5. These are specially modified skeletal or branchial muscle cells (as in *Torpedo*) or modified ocular muscles (*Astroscopus*).
6. Some fishes have been known to give an electric shock since ancient times (*Torpedo, Malapterurus, Momyrus* etc.).
7. Many fishes possess small electric organs that give a weak discharge (skates, rays, eels etc.).

8. Due to the transparent cytoplasm, it looks like muscles or gelatinous mass.
9. The wave duration or spike varies from 1 (Mormyridae) to 200ms (Rajidae)
10. The voltage of electric current varies from 0.5 (Mormyridae) to 600V (*Electrophorus*)

27.5.4 ORIGIN

The electric organ's development studies revealed that:

1. Elctrocyte (Bennet, 1970) is a transformed muscle fibre. It is variously known as electroplaques, electroplates or electroplaxes.
2. In rays (*Torpedo*), it is derived from some of the branchial muscles that have lost their function.
3. In skates (*Raja*) and mormyrids (*Gymnarchus, Mormyrus*), it is derived from the modified lateral muscle of the tail.
4. In catfish (*Malapterurus*), it develops from the dermis rather than the muscular tissue.
5. In star-gazers (*Astroscopus*), it develops from eye muscles.
6. *Narcine timlei, Narcine brunnea, Narke dipterygia* and *Uranoscopus* are also found in Bangladesh.

27.5.5 Examples

Electric organs are derived from modified muscle or in some cases nerve tissue and have evolved at least six times among the elasmobranchs and teleosts (Table 27.4).

Class – Elasmobranchi

1. Electric rays: *Torpedo, Narcine, Electrophorus.*
2. Rajidae: *Raja*

Class – Teleostomi

A. Amrine fishes
    1. Uranoscopidae: *Astroscopus* (star gazer)
B. Freshwater fishes
    2. Mormyridae: *Gymnarchus, Mormyrus* (elephant nose fish)
    3. Gymnotidae: *Elctrophorus electricus* (knife fish)
    4. Siluridae: *Malapterurus* (catfish)

Table 27.4: Habitat and nature of electric discharge of fishes

Sl.No.	Name of the fish	Habitat	No.	Electric discharge	Potential of discharge
	Elsasmobranchs				
1.	*Narcine*	Marine	2 pairs	37-220 volt	
2.	*Raja (Raia)*	,,	,,	4-14 volt	Small, weak
3.	*Torpedo*	,,	1 pair	30-60 volt	Large, weak/ strong
	Teleostei				
1.	*Uranoscopus*	Marine	2 pairs	50 volt	
2.	*Electrophorus*	Rivers of South	3 pairs	250-550 volt	Large, weak/
3.	*Gnathonemus*	America	2 pairs	50-300 volt	strong
		Rivers of South			
4.	*Gymnarchus*	America and	4 pairs	50-300 volt	
5.	Gymnotids	Africa	4 pairs	50-300 volt	Constant, weak
		Rivers of Africa			Small, weak
6.	*Malapterurus*	Rivers of South	1 pair	350-450 volt	
7.	*Mormyrus*	America and	-	-	Large
		Africa			Variable, weak
		Rivers of Africa			
		Rivers of Africa			

27.5.6 Architecture

Each electric organ is made up of a large number of electroplates or electrocytes, which are thin, disc-like flattened cells. The electrocytes are embedded in a jelly-like extracellular material and are joined by connective tissue to form an elongated tube or compartment. It only has one face, which is supplied by nerve fibres and jelly recessive blood capillaries. The jelly receives blood supply, and one end of electroplate is connected to a nerve fibre. Elctrocytes are multinucleate cells with nearly clear cytoplasm.

It may have distinctive folding and convolutions on one or both of its surfaces. It has a smooth innervated surface and a non-innervated surface that is highly folded and bears papillae. The letter represents a structural adaptation involving the discharge capacity of an elctrocyte by lowering internal resistance. Connective tissue separates it from the other. *Torpedo* has studied the evolution of the electric organ.

27.5.7 Structure of Electric Organ

The shape and the location of electric organs differ in different groups of fish or species of fish but they are histologically similar.

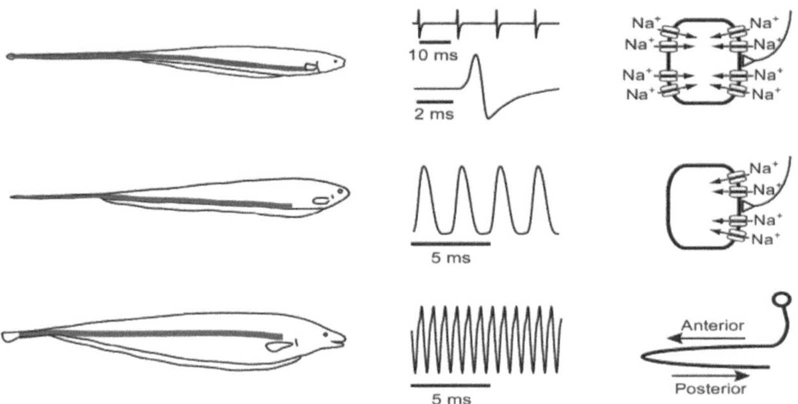

Fig. 27.11: Electric organ discharges and electrocytes from gymnotids (from https://www.researchgate.net)

1. ***Torpedo*** (power 30-50 volt) and ***Narcine*:** It has evolved into two large and small organs on each side of the hand, with the smaller one enclosed within the larger. Each organ is supplied by thick branches of the VII, IX, and X cranial nerves originating from a special lobe of the medulla oblongata and forming a large, flat kidney-shaped mass on either side of the middle line. The organ is made up of 46 vertical hexagonal columns, each with 400 elctrocytes. The innervated side of the elctrocytes is electrically negative to the other side, and they all work in tandem. This species' current flows from the fish's dorsal positive side to its ventral negative side. However, *Torpedo occidentalis* delivers an electric discharge of about 1000 watts.

   Torpedo has developed an electric organ amongst elasmobranchs.

2. ***Electrophorus electricus*** (power 270-600 volt)**:** Each side has two electric organs, one of which is much larger than the other. "Hunter's organ" refers to the smaller organ on the ventral side of the body. It accounts for nearly half of the body's mass and contains approximately 70 compartments, each of which contains a series of 600 elctrocytes. The electric organs run the entire length of the body, and the nerves that supply them emerge from the spinal cord. Its current is strong enough to cause a severe shock to man, and the organ's polarity is reversed from head to rail. Peak power may exceed 100 watts.

3. ***Malapterurus electricus*** (power 350-450 volt): The electric organ is embedded in the skin, which forms a thick, semi-transparent layer over the body. Many connective tissue septa divide the electric organs into compartments that fill into one another. Each compartment contains a large number of elctrocytes, the neural ends of which are oriented toward the tail. The processes of a single nerve cell in the grey matter of the spinal cord are connected to each half of the organ. The direction of polarity is from tail to head.

4. ***Uranoscopus* (*Astroscopus*)**: The electric organs are of a complicated structure and take the form of oval patches behind the eye. Although small in size, these organs can give a shock of quite powerful intensity. The electric organs are innervated by oculomotor nerves
5. ***Steatogenys elegans***: It has a probable electric organ in a groove below the dermis, from the anterior margin of the mandible up to the base of the pectoral fin.
6. ***Gymnarchus niloticus*:** It has a large electric organ and well-developed sensory organs (Fig. 27.12).

27.5.8 Conduction of Impulse

Elctrocytes have an 84mV resting transmembrane potential. As a result, each column of electropaque contains alternating membranes with opposite polarities.

In most cases, one surface of an electroplaque is directly innervated. Innervations on stalks extending from the caudal electroplaque surface are uncommon. The nerve impulse causes a post-synaptic potential in the stalk, which causes a spike to propagate into the electroplaque. The stalked caudal protein is the first to depolarize, followed by the neon stalked rostral surface. As a result, there is a diphasic action potential. These spikes have an extremely short duration of 0.3 milliseconds and are specially designed to function as part of the electrolocation system.

27.5.9 Mechanism of Discharge

The posterior innervated surface reverses its polarity by 6.7 millivolts during discharge, but the potential of the non-innervated end remains uncharged. As a result, there is a total potential difference of 150 millivolts between the cell's two surfaces. The voltages of successive electrocytes are added together in series and large current flows.

It has been suggested that the source of energy for electric current is the movement of first Na^+ and then K^+ from string to weak solutions (Fig. 27.11). The Na^+ first inter into the cell due to an increase in Na^+ permeability of the membrane, later the K^+ passes into the cell during a delayed increase in the K^+ permeability of the membrane. Thus, the immediate cause of the change in potential across the nervous face of the cell is the passage of the ions across the cell boundaries. The electric organ in most of the fish appears to be under the control of the central nervous system.

27.5.10 Functions

1. Defense and offense: Electric eels, rays, and skates produce a powerful electric discharge that can be used defensively against predators or offensively to stun and capture prey.

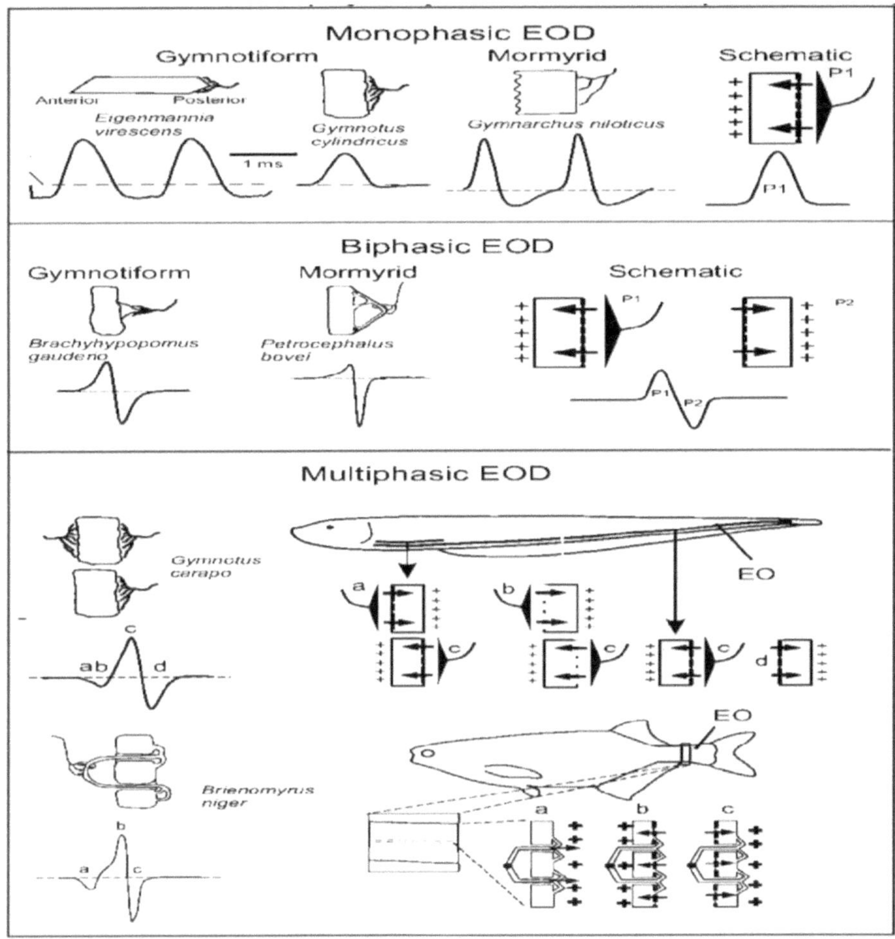

Fig. 27.12: Mechanism of Electric organ discharges and electrocytes in gymnotids and mormyrids (from https://www.researchgate.net)

2. Electro-location: This phenomenon occurs when a fish uses successive discharges to maintain stable configurations of an electric field around its body. Such pulses are emitted at rates ranging from 65 to 2000 times/s and have a constant voltage of 0.03mV. Their amplitude, on the other hand, varies from species to species. The stable configuration varies as a conductivity-varying object approaches the field. The altered configuration provides the animal with precise information about changes in the environment.
3. Individual fishes use the electric discharge to maintain their territoriality.
4. It aids in species or even sex identification.

28

Bioluminescence in Fishes

28.1 Introduction

The use of biological light in active transport, nerve transmission, and muscle contraction is a feature shared by all living organisms. Surprisingly, a large number of living organisms have evolved the ability to convert chemical energy into light, which is manifested as bioluminescence.

Bioluminescence is a fascinating biological phenomenon found in both terrestrial and aquatic living organisms. The light emitted by living organisms as a result of chemical reactions in their bodies is known as bioluminescence. Cold light is produced by this type of chemiluminescence. Although the majority of bioluminescent organisms are found in the ocean, almost none are found in freshwater. The production of light may be autogenic or bacteriogenic

The arrangement of the luciferin determines the colour of bioluminescence, which is bluish-green in dinoflagellates, yellow in fireflies and land snails (*Quantula straita*), and greenish in lanternfish. In the dark ocean, bioluminescent algae can be seen as pink or green spots. Red luminescence is used by dragonfish (*Malacosteus niger*).

Damp wood emits a glow, according to both Aristotle and Pliny the Elder. EN Harvey (1920) was the first to publish research on bioluminescence. According to Davies et al. (2016), bioluminescence appeared in ray-finned fishes in 27 distinct evolutionary events.

28.2 Occurrence

Bioluminescence is widely distributed among various animal groups. No terrestrial vertebrate, however, is bioluminescent. This phenomenon is only observed in a few aquatic bacteria and mollusks. Harvey et al, (1952) published a monograph on the distribution of luminescent organisms. Harvey (1940) discovered luminescence in 666 genera from 13 phyla and 28 classes out of 33 phyla and 80 classes. Fish have the most diverse and complex bioluminescent adaptations in the world.

The following table lists bioluminescent organisms from the various phyla of the kingdom Animalia (Table 28.1):

Table 28.1: Enlistment of animals having bioluminescent organs

Sl.No.	Phylum	Bioluminescent organisms
1.	Protozoa	*Ceratium, Collozoum, Gonyaulax, Noctiluca, Purodinium, Speherozoum, Thalassicolla*
2.	Coelenterata	*Cavernularia, Cestus, Diphyes, Liriope, Peliagia, Pennatula, Pleurobranchia*
3.	Nemertea	*Emplectonema*
4.	Annelida	*Chaetopterus, Odontosyllis, Polynae, Tomopteris, Polycircus*
5.	Arthropoda	*Acanthophyra, Cypridina, Ceratoplanus, Elaterids, Euphausia, Lamphorids, Lampris, Metridia, Odontosyllis, Photinus, Photoporus, Phrixothrix, Pontella, Pyrophorus, Sergestes*
6.	Mollusca	*Heteroteuhthis, Lycoteuthis, Loligo, Pholas dactylus, Spirula, Triopa*
7.	Echinodermata	*Amphiura, Ophiothrix, Ophiacantha*
8.	Protochordata	*Balanoglossus, Doliolum, Oikopleura, Salpa, Pyrosoma, Ptychodera*
9.	Vertebrata	*Anomalops, Bentholoatis, Centroscyllium, Harpodon, Leiognathus, Monocentris, Malacocephalus, Photobelpharon, Physiculus, Photostomiasis, Spinax, Stomias,*

In fish alone, there are about 1500 species belonging to 46 families that are known to be luminescent (Fig. 28.1 and Table 28.2). The important families of bioluminescent fishes are Batrachoididae, Brotulidae, Ceratoidae, Halosauridae, Lophidae, Myctophidae, Stomiatoidae, Sternoptychidae and Zoarcidae. For example: small deep-sea lantern shark (*Etmopterus spinax*), tiny cookie-cutter shark (*Isistius brasiliensis*), dragon fishes (*Malacosteus, Pachystomias, Aristostomias*), *Spinax, Etmopterus, Centrisoyllium, Bentho bates* etc.

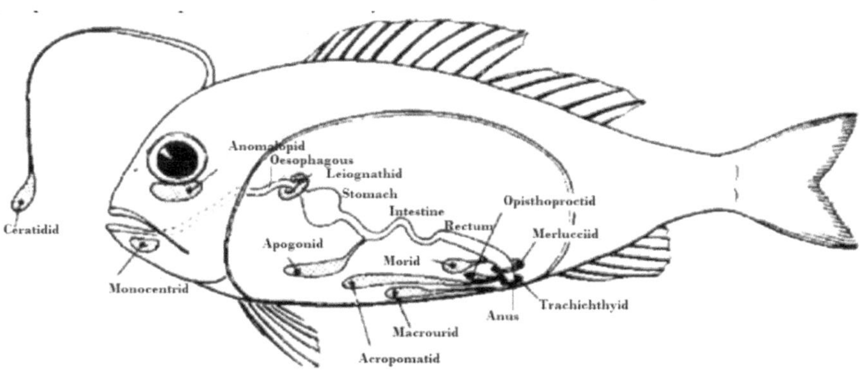

Fig. 28.1: Diagrammatic fish to indicate the locations and openings of the light organs of several different families of luminous fishes that culture symbiotic luminous bacteria (From https://www.semanticscholar.org)

Table 28.2: Orders and families of bioluminescent fishes

Sl.No.	Order	Families
1.	Squaliformes	Dalatiidae, Etmopteridae, Somniosidae
2.	Anguilliformes	Congridae
3.	Aulopiformes	Chlorophthalmidae, Evermannellidae, Paralepididae, Scopelarchidae
4.	Batrachoidiformes	Batrachoididae
5.	Beryciformes	Anomalopidae, Monocentridae, Trachichthyidae
6.	Clupeiformes	Engraulidae
7.	Gadiformes	Macrouridae, Merlucciidae, Moridae
8.	Lophiiformes	Centrophrynidae, Ceratiidae, Diceratiidae, Gigantactinidae, Himantolophidae, Linophrynidae, Melanocetidae, Oneirodidae, Thaumatichthyidae, Ogcocephalidae
9.	Myctophiformes	Myctophidae, Neoscopelidae
10.	Osmeriformes	Alepocephalidae, Microstomidae, Opisthoproctidae, Platytroctidae
11.	Perciformes	Acropomatidae, Apogonidae, Chiasmodontidae, Epigonidae, Howellidae, Leiognathidae, Pempheridae, Scianidae
12.	Saccopharyngiformes	Eurypharyngidae, Saccopharyngidae
13.	Stomiiformes	Gonostomatidae, Phosichthyidae, Sternoptychidae, Stomiidae

The following are the characteristics of bioluminescent fishes:

1. Mostly deep sea (bathypelagic) fishes possess luminescent organs as photophores and show their specific pattern of distribution on the body of a male and a female. They show sexual dimorphism. They live at a medium depth of 500-2500m in low light and low temperature and high pressure. Two-thirds of the mesopelagic fish fauna is luminescent.
2. Photophores are probably glandular epithelium (Fig. 28.2).
3. The intensity of the photophore is <0.05 of the candle power.
4. Generally, photophores occur along the lateral and ventral sides of the body and head.
5. Photophores were described by Risso (1820).
6. One or two suborbital organs are present in *Opostomias micripnus, Scopelus benoitii* and *Pachystomias microdon*.
7. Few elasmobranchs inhabiting the surface zone bear luminous organs.
8. In *Pachystomias*, a pair of large complicated luminescent organs is present under the eyes.
9. *Porichthys notatus* contains 840 photophores to emit light.

10. Variation in the number and disposition of luminous organs is observed in different fishes.
11. The occurrence of the bioluminescent organs is quite sporadic and follows the kinetic rate theory.
12. It appears to be a vestigial system in organic evolution.
13. It was an incidental concomitant of chemical reactions.
14. It has evolved with the primitive electron transport system.
15. It is dependent on oxygen, pH, temperature, salt etc.
16. From peculiar ecological conditions, it is of great adaptive significance.
17. Fishes produce either blue (*Stomias*), green (*Coelorhynchus*), blue-green (*Anomalops*), yellow (*Echiostoma barbatum*), red or white colour or more then one colour at once.
18. Luciferin is species specific.
19. Luciferin and luciferase are secreted as granules.

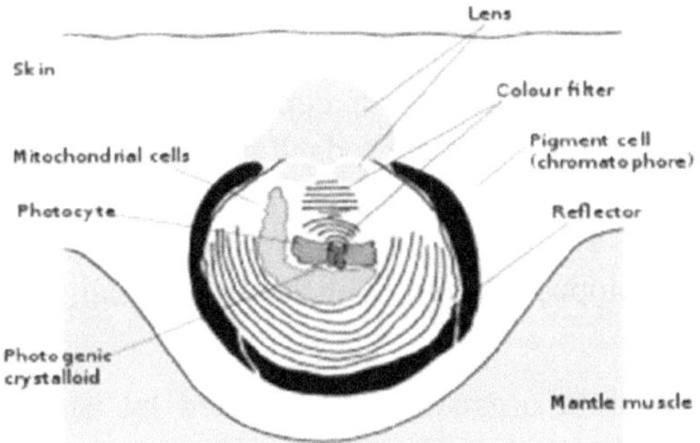

Fig. 28.2: Photophore in fishes (from https://en.wikipedia.org)

Table 28.3: Distribution of photophores in fishes

Sl.No.	Fish	Distribution	Sl.No.	Fish/group	Distribution
1.	*Apogon*	within the abdomen	5.	*Scopelus* and *Halosauropsis*	In one or two rows from head to tail
2.	*Aeropoma*	In the ventral musculature	6.	*Opostomias*	Transverse bands on the body
3.	*Etmopterus*	over the entire body	7.	*Porichthyes*	Along the lateral line
4.	*Lampanyctas leucopsanes*	Some parts of the body	8.	Ceratoidae	Elongated first fin rays of pectoral and dorsal fins.

28.3 Production of Light

Light is produced by fish photophores in two ways:

1. **Self-luminous photophores:** Most fish are self-luminous, which means that the light is emitted by glandular cells that are supported by pigmented reflectors. It has undergone significant structural evolution. These are found in the elasmobranch families Squalidae and Torponidae as well as the teleost families Stomiatidae and Myctophidae (=Scopelidae). Their photophores are represented in simple cases by radially arranged glandular tubules. The peripheral nerves supply these tubules.

There are two kinds of self-luminous photophores:

(a) Extracellular luminescence: Extracellular luminescence is common in marine invertebrates. e.g., *Phyllirrhoe bucephala, Pholas dactylus, Chaetopterus, Balanoglossus, Odontosyllis.*

(b) Intracellular luminescence: It occurs in advanced fishes, terrestrial arthropods and cephalopods. e.g., *Noctiluca*, Ctenophora, *Polynoe, Ophiothrix, Stomias, Photostomias* etc.

The dioptic parts of a complex photophore are well-developed. Suborbital organs of *Pachystomias microdon* are quite complex, with a cup-like structure and many concentric layers. Numerous glandular cells are lodged in a cup, which is surrounded by pigmented layers. A lens-like body is present in the cup's mouth, and an overlying integument functions as an iris. The reflecting layer, which is made up of spicules, is thick and developed. A branch of the Vth cranial nerve innervates these organs.

The firefly (*Photinus pyralis*) has a light organ that is activated by nervous influences and the luminescence is intracellular. The organ occupies the entire ventral surface of the 6th and 7th segments in the male, but only about 2/3 of these segments are in the female. The reflecting layer cells contain urate crystals, whereas the photogenic layer is made up of large cells filled with yellow granules. The trachea, which penetrates layers and branches in the photogenic layer, ensures an adequate supply of oxygen. Each branch terminates in a tracheal end-cell, which produces tracheoles that enter photogenic cells. Numerous mitochondria can be found in the outer cytoplasmic zones of photogenic cells.

2. **Photogenic symbiotic bacteria**: These are found specialised areas of *Anomalops, Leiopgnathus, Malacocephalus laevis, Monocentris japonicas, Photoblepharon* and anglerfishes. These are capable of producing light for a considerable period.

Structurally, these photophores contain a large number of glandular tubules that secrete luminous bacteria. In *Macrocephalus* and *Coelorhynchus*, the gland opens by a duct on the ventral surface of a fish, a little in front of the anus. The gland is richly supplied by blood capillaries. The photophores of *Opisthoproctus* are present near the anal aperture. *Photoblepharon* and *Anomalops* possess an elongated luminous organ below each eye having a long parallel glandular tube with a rich blood supply. These organs have pores opening to the exterior at the anterior end and a reflector layer at the hind end.

The provision for off-and-on light production is developed in many fishes.

The simple photophores contain a series of radially arranged glandular tubules that receive branches from the adjoining central or spinal nerves.

28.4 Structure of Luminescent Organs in Fishes

1. Elasmobranchs: Self-luminous organs restrict to *Etmopterus, Centriscyllum, Spinax* and electric rays. But rat tails and searssids produce luminosity by producing extracellular luminous slime.
2. Myctophidae: Both nocturnal and abyssal fishes have highly developed photophores. Their photophores are small pearl-like beads located on each side of the body in a series along the dorsal and ventral sides. The photogenic parts of photophores were depicted as a small circular disc positioned between two overlapping scales. To represent the lens, the outer scale is flattened and transparent. The lower scale is concave and covered by a silvery reflecting layer that is backed up by a pigment layer. The photogenic disc is located between the two surfaces of the aforementioned scales.
3. *Opisthproctus*: The luminous organ is cup-shaped and has an aperture that faces outwards. It is located in the rectal region. A pigmented layer, a refractory layer, glandular tissue, and a lens are all present. The innermost pigmented layer prevents light from leaving the chamber. The reflector is thick and shaped to reflect light toward the photophores aperture. The photogenic glandular photophores are highly vascular and form the lining of the cup's cavity. The phortophores apertures are outfitted with a homogeneous mass of relatively smaller but highly refractive lenses.
4. *Sternoptyx*: More complicated photophores are found on the ventral side of the abdomen, as well as a lateral series in front of the gill opening, a row on each side, and three above and three behind the base of the pectoral fins. Each photophore has a dorsal chamber that is covered by a pigmented layer on the outside and a layer of glandular cells on the inside. Light from the dorsal chamber is emitted through small holes outfitted with fibres into tube-like projections of photophores. The tubes are also lined with reflective layers made of guanine crystals. The light that emerges from the dorsal chamber is distributed in a wider arch by the curved surface of the reflecting layer so that its final emergence out of the photophores follows a predictable pattern despite variations in intensity and angles. It is held that such photophores function for ventral camouflage.

28.5 Mechanism of Bioluminescence

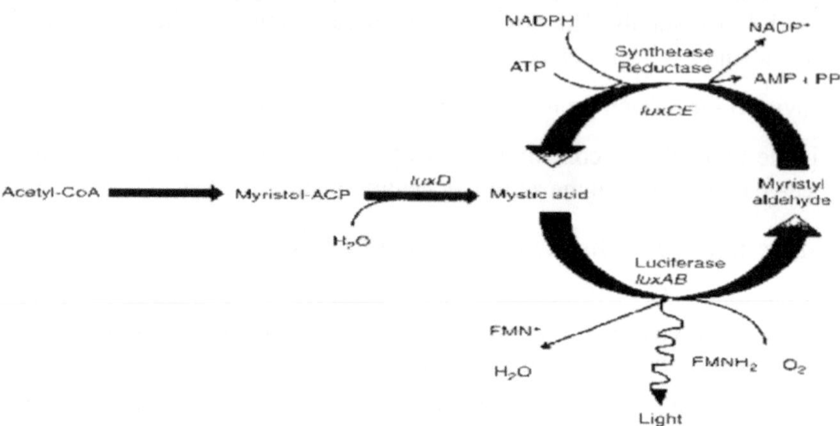

Fig. 28.3: Mechanism of bioluminescence in fishes (From https://www.sciencedirect.com)

Biochemical analysis reveals that photophores contain luciferin and luciferase. Luciferin can be extracting photophores in hot water while luciferase is obtained by extracting the organ in cold water. Dubois (1887) postulated the oxidation of luciferin (Fig. 28.3).

The luciferin of *Apogon* and *Parapriacanthus* are identical and closely resembled *Cypridina*. Luciferin is an indole derivative containing tryptamine, arginine and isoleucine. Luciferin is oxidized to oxyluciferin when mixed with luciferase.

28.6 Luciferin-Luciferase System

Most bioluminescent reactions involve luciferin and luciferase but in some cases, photoprotein and calcium are needed instead of luciferase. The photoprotein in crystal jelly known as "green fluorescent protein", is a reporter gene. According to Krönström et al, (2005), nitric oxide was suggested to modulate adrenaline-stimulated light emission in hatchet fishes (*Argyropelecus*). In the case of cartilaginous fishes, luminescence is controlled by the application of melatonin as reported for the pygmy shark (*Squaliolus aliae*) (Claes et al, 2012) (Fig. 28.4).

The dissolution of luciferin and luciferase in water permits oxidative reaction. Many catalytic agents (e.g. saponin) increase luminescence. Dubois showed that the lanterns of the fireflies continued materials that could be resolved into "thermolabile luciferase" and "thermostatic-luciferin". McElory et al, (1955) showed that luciferin contains luciferin proper, ATP, Mg^{+2} or Mn^{+2}. Luciferin is thermostable, heterocyclic phenol derivative, synthesised in D or L form. Only the natural D form participates in luminescence.

Luciferase is euglobulin having mw = 1010^4. It is thermolabile and catalyses the adenylation of luciferin to produce luciferyl adenylate.

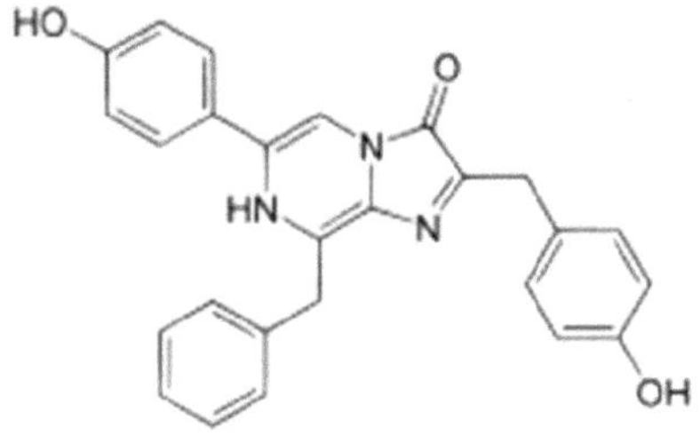

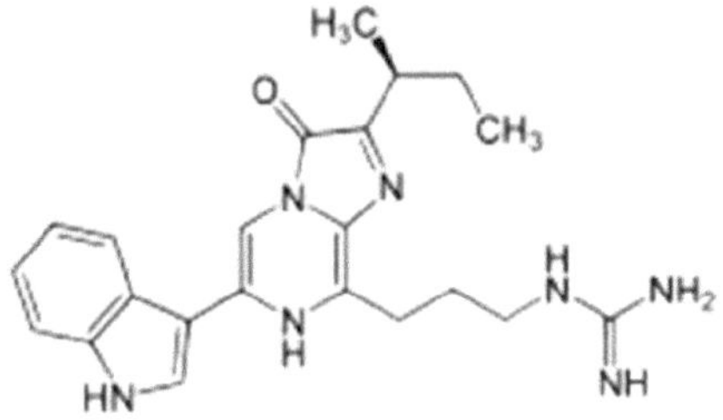

Coelenterazine

Vargulin (Cypridinluciferin)

Fig. 28.4: Luciferin in different animals

McElory et al, (1955) clarified the reactions in cases of fireflies. Firefly luminescence works only in the presence of ATP and Na^+, Mg^{+2}, Mn^{+2} or Ca^{+2}.

The reactions are enzyme substrate type as shown below:

Luciferin $\longrightarrow$ Luciferyl adenylate $\longrightarrow$ Oxyluciferyl adenylate

28.7 Efficiency of Light

A. Based on comparing the energy of visible light emitted with total radiant energy: There is very little emission from the emitted light that is visible from a bioluminescent organ. Then efficiency of firefly = 90%, cypridina = 20% and bacteria <1%. These values are more than the efficiency calculated on a similar basis for artificial systems (e.g., carbon lamp filament lamp 5%).

B. Based on the caloric equivalent of visible light emitted with the caloric equivalent of oxygen which goes into the photoemissive reaction. It may be expressed in terms of the number of molecules of oxygen required to give 1 quantum of light.

Bacteria = 33 and 193

Cypridina = 50

Luminol oxidized with H_2O_2 = 350

28.8 Control Of Bioluminescence

Photophores are directly controlled by the nervous and endocrine systems. The photophores are innervated, and the luminescence is controlled by the nervous system. Efferent fibres activate photocytes, but some endocrine glands may also activate them. They are extremely sensitive to either adrenalin or non-adrenalin. Exposure to electric shock stimulates the production of light. The nervous system directly controls the activities of photogenic bacteria. The sympathetic nerve controls the production of light in a self-luminous photophore.

Certain fishes use mechanical methods for concealing light organs. *Photostomias* can rotate their luminescent organs downward by contraction of a muscle, thus causing the bright surface to be concealed. *Indiacanthus* can pull down a pigmented sheath in front of the light-producing surface. In *Anomalops*, the light organ is provided with a hinge at the anterior edge, so that it is rotated downwards till it comes into contact with a black-pigmented tissue. In *Photoblepharon*, a membranous lack curtain lies along the ventral edge of the organ and is pulled up to conceal the light.

28.9 Significance of Bioluminescence

The significance of bioluminescence in fishes is quite diverse and may be described as:

1. Sexual attraction: In deep-sea fishes, there exists a specific pattern in the distribution of photophores on the body of the male and the female. It exhibits a clear-cut sexual dimorphism. Thus finding out an appropriate mate and mating may depend on the recognition of these luminescent signals in deep water.
2. Protection: *Malacocephalus laevis* has a glandular organ, which is controlled by muscles near the anus.
3. Attraction of prey: In deep-sea forms, the light produced by luminous organs helps to search the prey in a dark environment. *Anomalops* uses photophores as torches to illuminate the area. Some fishes make use of their luminescent organs to attract prey near the mouth. The luminescent bulbs occurring on the tentacles perform this function.
4. Defense: A sudden flash of light from a large photophore may dazzle and confuse the enemy for a moment, enabling the fish to escape.
5. In *Stomias*, the photophores are sensory in function and help to detect the where about of wandering prey.
6. Intraspecific communication: Fishes use light signals for intraspecific recognition, schooling and mating. Some fishes possess species-specific structures that purely on account of their placement and position may also assist specific recognition.
7. Interspecific communication: Some fishes use luminous organs located in or on the head to illuminate their surroundings, in search of prey and to detect predators.

Unit-IV Reproduction, Development, Genomics and Behaviour of Fish

29

Reproduction in Fishes

Fishes are generally unisexual and reproduce by several methods. Some families (Sparidae and Serranidae) are hermaphrodite also. Alz (1964) reported that in certain species, hermaphrodism becomes a normal way of life. In hermaphroditic fishes, self-fertilisation is a common process. Parthenogenesis could have been produced experimentally (Austin and Walton 1960). There maybe a few secondary organs that increase reproductive fitness. The genital papilla is a small, fleshy tube behind the anus in some fishes, from which the sperm are released; the sex of a fish can often be determined by the shape of its papilla.

Based on reproductive strategies adopted by different fishes, they are categorised into three groups:

1. **Nonguarders:** They do not guard their egg and larvae and are of two types:
   i) **Open substrate spawners or pegalophils** (lay eggs in open places): many schooling/shoaling fish like sardines, mackerels, tunas, *Coregonus, Alosa, Osmerus, Leuciscus, Abramis, Lota*)
   ii) **Brood Hiders** (lay eggs in hidden places): Mostly benthic spawners. For example *Salmo, Thymallus, Rhodeus*. Among salmon and trout the female digs a nest (redd) with her tail in gravel.)
2. **Guarders:** Male or Female or both the parents guard the eggs and larvae after laying and may be of the following two types:
   i) Substrate choosers: Example: *Polypterus, Silurus*
   ii) Nest spawners: Examples *Gobius, Gasterosteus, Pungitius, Betta, Micropterus, Cottus*
3. **Bearers:** They may be of two types:
   i) External Bearers: Examples: *Sarotherodon, Symphysodon, Scleropagus*
   ii) Internal Bearers: Examples *Jenynsia, Gambusia, Lebistes*

29.1 Male Reproductive System

29.1.1 Introduction

The male reproductive system contains paired testes, seminal vesicles, vas deferens, urinogenital sinus and genital apertures.

Cartilaginous fishes: testes > Leydig's gland > seminal vesicle > cloaca > claspers.

Most Bony Fishes: testes > vas deferens > urogenital pore.

1. Testes: Most male fish have two testes of similar size. In sharks, the testes on the right side are usually larger. The jawless fish have only a single testis, located in the midline of the body, although even this forms from the fusion of paired structures in the embryo. In dipnoans, the testes are enveloped by lymphoid tissue (Fig. 29.1).

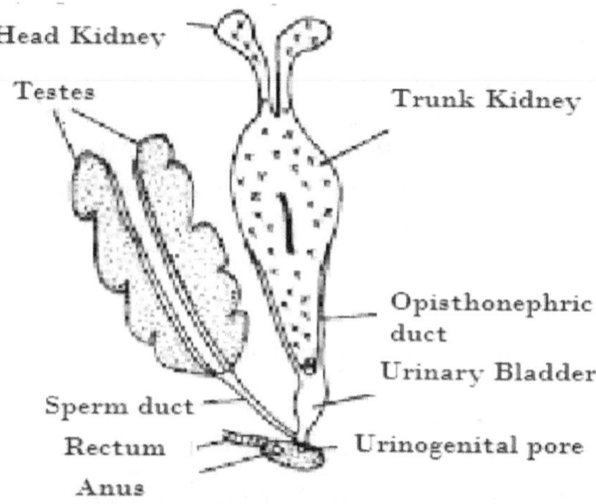

Fig. 29.1: Testis in a fish

Testes are elongated and flattened structures, situated on either side, ventral to the kidney and attached to the body wall and air bladder using mesorchia. They may be equal in size. Two sperm ducts join posteriorly to open into the urinogenital papilla. The testes may show indentation along with their margin which becomes prominent during the breeding season. In *Rita* and *Glyptothorax*, the testes present a peculiar appearance and bear a large number of lobules which become finger-likeduring breeding season. In *Mystus seenghala* and *Barbus* tor, the anterior 3/4th part of each testis is only functional. In some species, the entire testis is functional and serves to produce sperm. In *Amphipnous* and *Notopterus*, the testes are unpaired apparently because of narrower and bilaterally compressed body cavities. The shape of the testes varies with the maturity of the fish. They may be oval, elongated or thread-like. The immature testis is transparent and upon maturity becomes creamy white to pink during the breeding season.

In most elasmobranchs, the connection of the efferent ductules arising from the testis and the archinephric duct occurs in the anterior region of the opisthonephros. In *Polypterus*, the kidney tubules at the posterior part of the kidney usually degenerate and sub-serve reproductive function. In *Protopterus* and many teleosts, the separation of the testis duct from the archinephric ducts represents specialised function.

Under tunica albuginea, the testis contains seminiferous tubules. The tubules are lined with germ cells that from puberty into old age develop into spermatozoa. The developing sperm travel through the seminiferous tubules to the rete testis, to the efferent ducts and then to the epididymis where newly created sperm cells mature. The sperm move into the vas deferens and are eventually expelled through the urethra and out of the urethral orifice through muscular contractions. The manner of transportation of the spermatozoa from the testis differs in many fishes.

Most fish do not possess seminiferous tubules. Instead, the sperm are produced in sperm ampullae. These are seasonal structures, releasing their contents during the breeding season, and then being reabsorbed by the body. Before the next breeding season, new sperm ampullae begin to form and ripen. The ampullae are otherwise essentially identical to the seminiferous tubules in higher vertebrates, including the same range of cell types.

Fish can present cystic or semi-cystic spermatogenesis during the release phase of germ cells in cysts to the seminiferous tubules lumen.

2. Seminal vesicles: These are paired glandular structures present as outgrowths of the hindered ends of vasa deferentia in *Clarias batrachus, Clarias lazera* and *Heteropneustes fossilis*. The seminal vesicles are secretory and show periodical changes in the correlation with the testicular cycle. But in *Rita rita* and *Mystus vittatus*, the posterior part of the testis is glandular. The fluid secreted by seminal vesicles probably keeps the sperms in an active but viable condition or it may help in nourishing the sperms. Sundarraj and Goswami (1965) have shown that the seminal vesicles of *Heterpneustes fossilis* respond to the administration of testosterone propionate.

3. Vas deferens: Among the teleosts, the main sperm duct arises from the posterior mesodermal surface of testes and leads to the urinogenital papilla. In most of the teleosts, the paired testes fused posteriorly and the vas deferens is combined into a single sperm duct. The principal sperm duct gives rise to smaller ducts which penetrate ventrally and laterally to form a drainage system of variable complexity.

29.1.2 Histology of Male Reproductive System

The testicular structures throughout variable in teleosts, yet based on the differentiation of germinal tissues, the following types of testes are recognized by Billard et al. (1982).

1. Lobular type: Most teleosts consist of a large number of seminiferous lobules which are closely bound together by a thin layer of connective tissue (Fig. 29.2). The lobules are of various sizes and are highly convoluted structures, separated from each other by a thin connective tissue stroma. The walls of lobules are not lined by a permanent germinal epithelium. the lobules open into a spermatic duct and are not bound together into a compact mass. The lobule may be surrounded by lobule boundary cells which resemble connective tissue. The spaces between the lobules are filled with connective tissue, blood capillaries and interstitial cells.

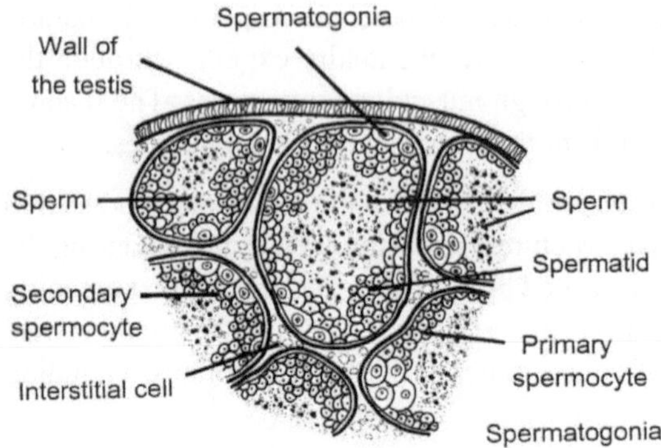

Fig. 29.2: Spermatogenesis in fishes (From https://www.researchgate.net)

2. Tubular type: It is found only in Antheriniformes (eg, *Poecilia reticulata*). Here, the tubules are arranged regularly between the external tunica propria and the central cavity. There is no structure comparable to tubular lumen in this type.

29.1.3 Different Phases of Testicular Cycle

Several morpho-histological changes are seen in the testes of teleosts and the cycle can be divided into different phases as mentioned below:

1. **Resting phase or early immature phase**: It is a quiescent period between August and September soon after spawning. The testes are thin, slender, translucent and pale in colour. Histologically, seminiferous tubules are small in size and full of spermatogonia (Fig. 29.3).
2. **Preparatory phase or late immature phase**: Normally, it is a period from October to December. The morphological appearance is almost similar to the resting phase, except that there is a slight increase in the volume and weight of the testes. Histologically, slow mitotic activity is seen and the spermatogonia starts dividing.

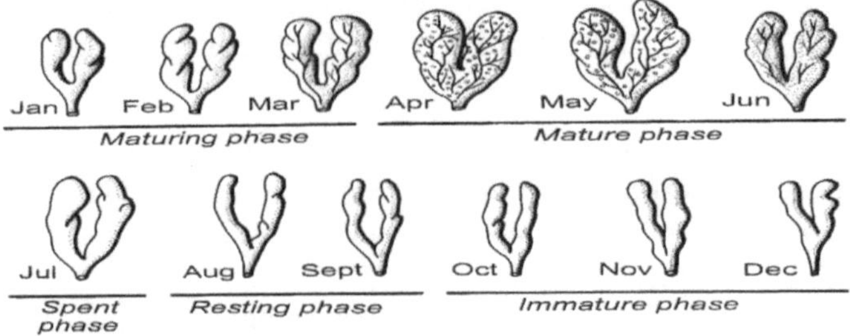

Fig. 29.3: Different stages of testes in fishes (From https://www.researchgate.net)

3. **Maturing phase or Revival phase**: It expands from January to March. There is a gradual increase in the volume and weight of testes. Testes look more vascular and opaque. Intense spermtogenesis takes place. Spermatogonia reduces in number. Interstitial Leydig cells become very prominent.
4. **Mature phase or spawning phase**: It is a period from April to June. During this phase, the testes show a well-marked increase in their volume and eight. They are turgid and pink in colour. Milk oozes out on pressing the abdomen. Histologically, seminiferous tubules acquire larger diameters and are full of sperms.
5. **Spent phase or post-spawning phase**: It ranges from mid-July to early August. The testes become flaccid due to excessive discharge of sperm. The volume and weight are considerably reduced. Histologically, empty and collapsing seminiferous tubules are seen, some of which contain residual or unexcelled sperms.

29.1.4 Other Cells

The interstitial cells are present in several species of teleosts. Craig and Bennet (1931) has correlated changes in these cells. Typical interstitial cells have been identified in *Gastroesteus acculeatus, Tilapia, Lebistes* etc. These cells are present singly or in small groups between the tubules. Seasonal changes in the morphology of the interstitial cells have been reported in some teleosts by Guraya (1976).

The lobule boundary cells lie within the wall of lobules and stain for lipids and cholesterol. Hence, Marshall and Lofts (1957) considered the lobule boundary cells as homologous to Leyding cells of mammals.

Hardisty et al, (1967) observed Leydig's cells in elasmobranchs and Marshall (1960) in Latemeria and many teleosts.

Sertoli cells are prominent in all fishes (Weibe 1968 and Pandey 1969). The spermatids and sperms are associated with the Sertoli cells and possibly they draw their nourishment from them during transformation. According to Lagois (1965), Sertoli cells are related to phagocytosis of the residual sperm.

29.2 Female Reproductive System

29.2.1 Introduction

Female bony fish gonads begin with the urogenital opening, also known as a genital papilla. The ovaries of most female fish are located in the abdominal cavity, ventral to the kidneys. The ovaries are elongated, sac-like structures that are attached to the body wall via the mesovarium. The two ovaries can be equal or unequal, and they are connected posteriorly by an oviduct. Only the right ovary fully develops in some elasmobranchs. There is only one ovary in primitive jawless fish and some teleosts are formed by the fusion of the paired organs in the embryo. There are three types of fish ovaries:

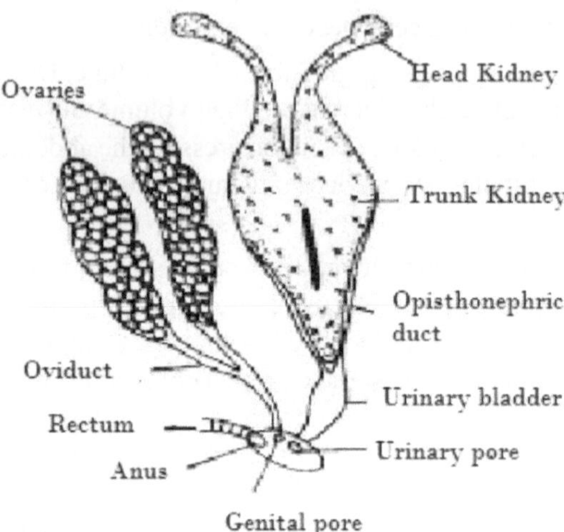

Fig. 29.4: Ovaries in fish

A. Gymnovarian: Oocytes are released directly into the coelomic cavity, then into the ostium, and finally out of the oviduct. Lungfish, sturgeon, and bowfin all have these ovaries.

B. Secondary gymnovarian: These ovaries shed ova into the coelom, where they enter the oviduct directly. Salmonids and a few teleosts have secondary gymnovaries.

C. Cystovarian: The oviducts transport the oocytes to the outside. Most teleosts have this type of ovary lumen continuity with the oviduct.

Cartilaginous fishes: ovary > ostium tubae > oviduct > shell gland > uterus > cloaca.

Most Bony Fishes: ovary > oviduct > urogenital pore

29.2.2 Histology of Ovary

The ovarian wall contains the following three layers:

(A) Peritoneum: the outermost thin layer.

(B) Tunica albuginea: the middle layer composed of connective tissue, muscle fibres, and blood capillaries and

(C) Germinal epithelium: the innermost layer that projects lamellae into the ovocoel during the spawning period. The ovocoel is the site of oocyte maturation, which can be seen at various stages of development. Oogonia from the germinal epithelium and can be found in small groups or nests in the lamellae. An oogonium has a nucleus that is surrounded by a thin layer of chromatophobic ooplasm and goes through several stages before producing a ripe ovum. During maturation, an oogonium grows in size due to ooplasm accumulation, and several cytological changes occur.

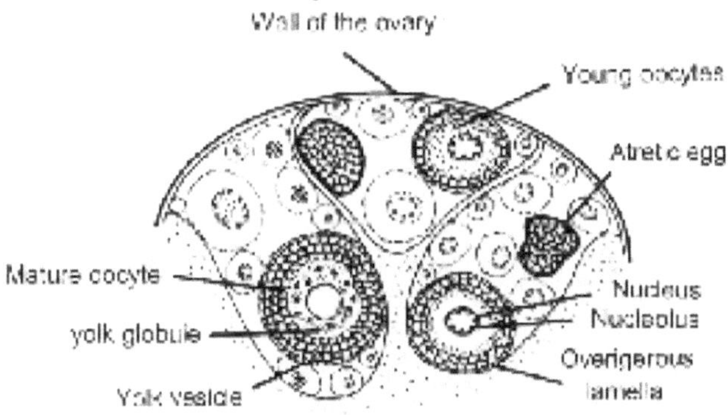

Fig. 29.5: Oogenesis in fishes (From https://www.researchgate.net)

29.2.3 Stages of Oocytes

1. Oocyte I: The youngest oocyte, which is slightly larger than the oogonium. It has a spherical nucleus with two or three nucleoli and basophilic cytoplasm.
2. Oocyte II: The oocyte has grown in size, and several small nucleoli can be seen along the nuclear membrane periphery. In some oocytes, a yolk nucleus can be seen close to the nuclear membrane. Later, the yolk nucleolus migrates to the oocyte periphery.

   Oocytes I and II represent the immature or resting phase of the ovary. They also have arteric and discharged follicles in common.
3. Oocyte III: The oocyte continues to grow. Around the cytoplasm, a thin layer of follicular cells appears. A few nucleoli can be seen in the oocyte cytoplasm, appearing to have passed through the nuclear membrane.

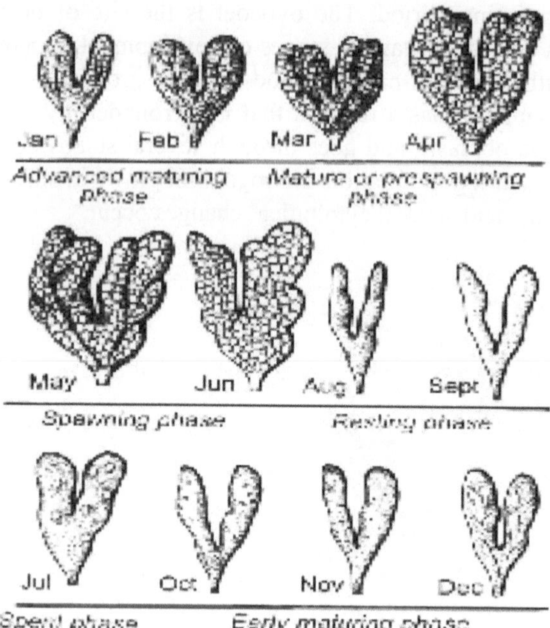

Fig. 29.6: Different stages of ovaries in fishes (From https://www.researchgate.net)

4. Oocyte IV: The oocyte grows in size and can be identified by the presence of many small, clear vacuoles or yolk vesicles along with the ooplasm periphery. The yolk vesicle appears empty and does not take stain in the early stages. The nuclear membrane becomes undulated in many oocytes of this stage, and the nuclei enter the pockets of the nuclear membrane and are thought to pass out into the ooplasm.

   Oocyte stages III and IV represent the ovary's early maturing phase.

5. Oocyte V: The oocyte grows in size and the number of yolk vesicles increases, eventually filling the entire ooplasm. Between the ooplasm and the follicular layer of zona granulosa is a vitelline membrane composed of zona radiata. During this stage, nucleoli are thought to be extruding.

6. Oocyte VI: Mature yolk granules appear in the ooplasm and accumulate in the peripheral region. The yolk granule gradually appears throughout the ooplasm, and the smaller ones fuse to form larger granules. The oocyte grows significantly in size, and a thin layer of fibroblasts known as the theca can be seen outside of the follicular layer.

   Oocyte stages IV, V, and VI represent the ovary's advanced maturing.

7. Oocyte VII: The yolk vesicles fuse and grow in size, and the yolk globules grow in size due to the deposition of more yolk. Some yolk vesicles are pushed outward to form a layer of cortical alveoli. The nucleus is now moving to the periphery.

Oocyte VII represents the mature or pre-spawning phase of the ovary.

The fully mature egg is yellowish, translucent, and filled with yolk globules. The yolk vesicles are dispersed throughout the egg, and the nucleus is usually invisible. Several ripe eggs are present in the ovary at the same time during the spawning period. They might also produce ovarian hormones.

Spawning phase: The turgid and take ovaries are up the entire body cavity. The ovarian wall is thin and transparent, allowing many translucent ova to be seen. A few ova can be seen in the oviduct and extruded by gently pressing the abdomen. The ovary is said to be in the process of running.

29.2.4 Maturation and Spawning

The maturation of oocytes in teleosts varies. Some species spawn only once, others twice, and still others several times per year. They are classified into three categories:

1. Asynchronisation: Several stages of oocytes are seen in the ovary at the same time, indicating a long spawning period and the fish spawns several times during the season. Example: the majority of freshwater teleosts.
2. Partial synchronisation: Two groups of oocytes are seen in the ovary. There is a short spawning period and spawning occurs once a year. Example: *Clarias barachus*.
3. Total synchronisation: All the oocytes mature at the same time. Example: *Onchorhynchus*.

Many Indian marine fishes appear to be continuous breeders, with a spawning period that can last up to 7-9 months.

Maturation and spawning times differ between species and even between areas of the same species. It is determined by a variety of ecological and physiological factors. Temperature and photoperiod are important factors in controlling gonad maturation.

The ovary shrinks, flaccidifies and turns a dull colour after spawning. The ovary contains a few unspawned ova as well as many small ova.

29.2.5 Gonadosomatic Index

The Gonadosomatic index predicts a species' spawning season by studying the relative occurrence of mature fish in a population and measuring the diameter of ova of a species can be calculated using the following formula:

$$GSI = \frac{\text{Weight of the gonads}}{\text{weight of the fish}} \text{ X } 100$$

A species gonadosomatic index has been widely used to indicate the maturity and periodicity of fish spawning. The GSI increases with fish maturation and shows a peak during the period of maturity. It drops dramatically after spawning.

29.2.6 Fecundity

Fecundity is a measure of a female fish's reproductive capacity and is defined as "The number of ova that are likely to be laid by a fish during the spawning season." It differs between species, and different individuals within the same species may exhibit variation depending on size, age, nutritional status, environmental factors, and genetic composition. Species fecundity can be determined by any of the following methods:

(i) **Volumetric Method**: The mature ovaries are removed from the abdomen and the total volume is calculated. Small pieces of the ovary are now taken at random from the anterior, middle, and posterior parts. The volume of each sample is calculated, and the number of ova in each is counted using a microscope. After that, the total number of ova in the total volume of the ovaries is calculated.

(ii) **Gravimetric Method**: The mature ovaries are fixed in 10% formalin to determine fecundity. The ovaries' weight is calculated. Three 100 mg samples are taken at random from the anterior, middle, and posterior parts. A binocular microscope is used to count the number of ova in each sample. The ova's total volume is calculated as follows:

$$\text{Fecundity} = \frac{S \times W}{100}$$

S = Average number of ova from three samples of 100 mg each, W = Total weight of the ovary.

30

Sexual Dimorphism in Fishes

30.1 Introduction

Sexual dimorphism is a condition in which sexes of the same species exhibit morphological differences that are not directly related to reproduction. It is a common and noticeable feature of the animal world. It is extremely important in biodiversity assessments, as well as in biometry, breeding biology, induced breeding, breeding, pheromone biology and other related fields.

Male-male reproductive competition has resulted in the evolution of a wide range of sexually dimorphic traits. In aggressive interactions between rivals, aggressive utility traits such as "battle" teeth and blunt heads reinforced as battering rams are used as weapons. Passive displays like ornamental feathering and song-calling evolved primarily through sexual selection. These differences can be subtle or extreme, and they can be subject to sexual and natural selection.

Sexual dimorphism refers to differences in appearance between males and females of the same species, such as differences in color, shape, size, and structure, markings or behavioural or cognitive traits caused by the genetic inheritance of one or the other sexual pattern.

Sexual selection has been blamed for sexual dimorphism. Sexual dimorphism predicts that differences in reproductive roles between sexes may influence selection patterns and lead to sex differences in morphological attributes such as body shape (Casselman and Schulte-Hostedde, 2004). Alternatively, sexual dimorphism can evolve through differential sex selection, which favours both dimorphic niches and, as a result, dimorphic trophic structures (Hedrick and Temeles. 1989; Herler, et al, 2010).

Many fishes exhibit sexual dimorphism. Females are typically larger than males their age. Males are larger than females in some species, such as the gudgeon, *Gobio gobio* (Mann, 1980) and the filefish, *Brachaluteres ulvarum* (Akagawa et al., 1995).

Lampreys turn a brilliant orange with black blotches during migration. A urinogenital papilla or a narrow penis-like tube formed by the united lips of the cloaca develops in the male. In males, the base of the dorsal fin thickens as well.

Females' cloacal lips become elongated and red. Furthermore, the female develops an anal fin.

While in some fishes secondary sexual characters are discernible during the breeding season and external morphological differences pertain to the following features:

1. Size of fish
2. Length/shape/texture of fins
3. Coloration
4. Genital papilla
5. Ovipositor
6. Shape of mouth

30.2 Classes of Sexual Dimorphism

1. Monomorphic fishes: Fish that have no differences in shape or colour between sexes. Example: most of pelagic fishes like sardines, seer fish, carangids, etc.
2. Temporary dimorphic fishes: Fish have different colours during the breeding season, courtship and spawning. Example: *Cyprinus carpio*
3. Permanent dimorphic fishes: Fishes' colours and shapes are always different. They have some sporadic organs, such as claspers in elasmobranchs and holocephalians and gonopodium in guppies (*Gambusia*).

Fig. 30.1: Male and female *Gambusia*

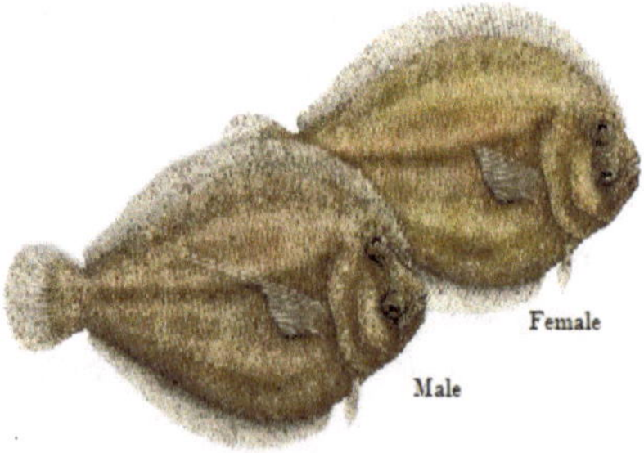

Fig. 30.2: Male and female *Bothus*

A clasper is a pair of elongated rod-like structures with a cartilaginous skeleton and a groove running the length of them. The claspers are introduced into the females during copulation, and their grooves form a canal for the spermatic fluid. The male *Chimaera* has a club-shaped frontal clasper on his head.

In *Gambusia*, the vas deferens extends as a tube up to the end of the anterior part of the anal fin, where it is modified to form the gonopodium, an elongated tubular or grooved copulatory organ (Fig. 30.1). The pectoral fins are modified as gonopodium in *Xenodexia*.

The male of *Mystus seenghala* has a conical genital papilla (Fig. 30.3). Some marine Ophididae have a penis-like intermittent organ. Male white sucker (*Catostomus*) anal fins are enlarged to transfer milt into the female body. A special tube for copulation is present in the male four-eyed fish (*Anableps*).

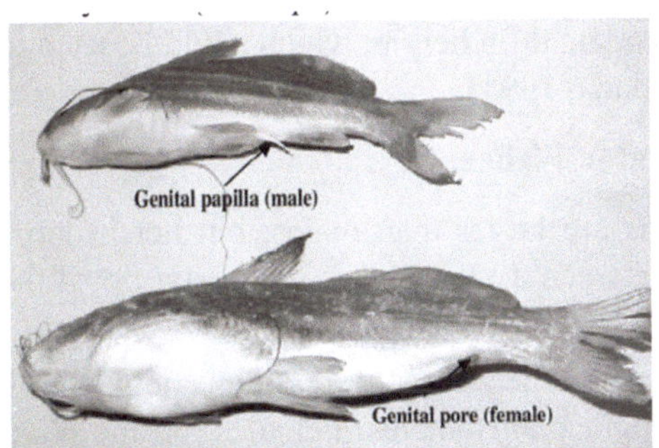

Fig. 30.3: Male and female *Mystus bleekeri*

Fig. 30.4: Male and female Swordtail fish

The fin rays of the pectoral fins of male flatfish (*Bothus*) are elongated to form feelers, whereas females are coloured on the side (Fig. 30.1). In *Mollenesia*, the dorsal fin is enlarged to form a brilliantly coloured slit-like structure. In the male swordfish (*Xiphophorus*), the lower lobe of the caudal fin is elongated to form a blade-like structure (Fig. 30.4).

Some cyprinidids (*Garra, Barilius*) have horny tubercles on the males' heads, which become more prominent during the breeding season.

Male *Amia* has a black spot at the base of the caudal fin.

Callionymus males are yellow or orange with deep blue stripes down the side and a row of blue or green spots. Their female is a dull yellow-brown with green spots.

30.3 Sexual Dimorphism in Tropical Freshwater Fishes

A. Indian major carps: Examples: *Catla catla, Labeo rohita, Cirrhinus mrigala*. Milt oozes through the genital aperture of their males, who have a rough pectoral fin. Females have a smooth pectoral fin, a reddish and swollen genital opening, a bulging abdomen and eggs that ooze out.

B. Peninsular carp: *Puntius pulchellus*. Males have a pink snout and a darker body colour. Females have a smooth snout and a white belly.

C. Magur (*Clarias batrachus*): The male anal papilla is long and pointed whereas the female anal papilla is round.

D. Singhi (*Heteropneutes fossilis*): Mature females are larger than males.

E. Tilapia (*Oreochromis mossambicus, Oreochromis niloticus*): Males in *Oreochromis* are larger than females and have prominent breeding colours

(Trewavas, 1982). Males can be distinguished from females in *Oreochromis aureus* and *Oreochromis galilaeus* by the number of openings in their genital papilla (two for males, three for females) and the more pointed shape of the anal fin in males, which is rounded in females (Brzeski, and Doyle, 1988). *Oreochromis mossambicus* has two sets of traits that tend to accelerate in males: jaw structure and dorsal and anal fin height, which could be used to sex individuals (Oliveira and Almada, 1995).

30.4 Sexual Dimorphism in Cold-water Fishes

A. Trouts: Females in small streams are larger than males, but not in large streams. There is a distinction between a kype (hook-like distal tip of the lower jaw) and belly coloration.

B. Mahsser (*Tor tor*): Males have greater operculum head width, snout length, predorsal length, and preanal length than females. Females, on the other hand, had greater dorsal fin insertion body depth, pectoral fin length, anal fin height, and ventral fin height than males.

Male *Tor putitora* have fleshy lips and the lower one is projected backwards into fleshy appendages, whereas females have normal lips and a pointed snout.

Male *Tor musullah* has more prominent tubercles on the snout than females.

30.5 Sexual Dimorphism in Ornamental Freshwater Fishes

A. Guppy, molly, swordtail, and platy (*Poecilia reticulata, Xephophorus halleri*): Males are smaller, brightly colored, and more attractive than females with longer dorsal and caudal fins.

B. Goldfish (*Carassius auratus*) (Fig. 30.5): There is no sexual dimorphism in growth, though females have been observed to grow faster than males on occasion.

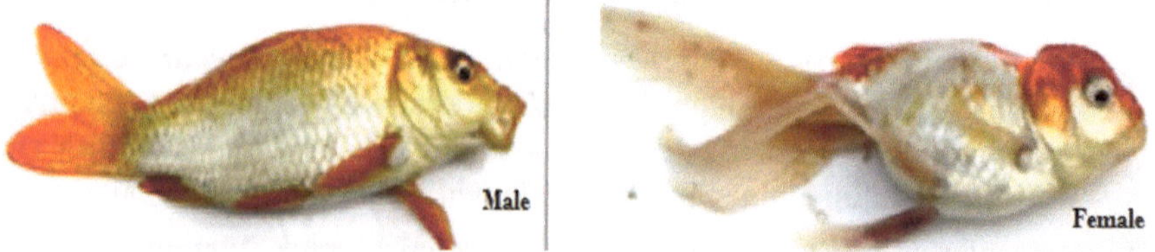

Fig. 30.5: Male and female goldfish (*Carassius auratus*)

C. Ticto barb (*Puntius ticto*): The male fish have a slight black blotch on their dorsal and ventral fins, which none of the female fish have. On both sides, there was a distinct dark pinkish colour in the lower portion of the male. Females have a slightly pinkish and dark yellow colour on both sides. The dorsal, ventral, and anal fins of male fishes are dark pinkish and slightly orange, while female fishes have slightly pinkish and dark orange fins.

D. Rosy barb (*Puntius conconius*): *Adult males are noticeably smaller, slimmer, and brighter than females.*

E. Siamese fighting fish (*Betta spelendes*): Females are smaller and lack the vibrant colours of males. Males have brighter colors, longer fins, and are more aggressive than females. Males are well-known for their magnificent bubble nests.

F. Gourami

G. Tetras

30.6 Sexual dimorphism in Brackish/marine water fishes

A. Sea bass (*Lates calcalifer*) (Fig. 30.6),

B. Grouper (*Epinephelus tauvina*) and

C. Grey mullets (*Mugil cephalus*) (Fig. 30.7).

Sexual dimorphism in deep-sea fishes is particularly fascinating in ceratoid Angler fishes, where females have luminescent illicium and males are smaller, parasitic on females and even lack luminescence.

Fig. 30.6: Male and female Sea bass

Fig. 30.7: Male and female Grey mullets

Female black dragon fishes (*Idiacanthus*) are black and serpentine with a chin barbell and grow to be at least 30 cm long. Males are yellowish with no teeth on pelvic fins or barbells and grow to a length of about 3 cm. During the breeding season, flashes from males' powerful cheek lights aid in sexual attraction. Astronesthes, Malacosteidae (*Aristostomiasis, Malacosteus, Photostomius*) and Melanostomias have cheek-light organs.

In Myctophidae, The caudal peduncle contains a light organ. Males have one or more prominent luminous glands on the upper side of the peduncle, whereas females have the organ on the lower side or do not have it at all.

Ray-finned fish are an ancient and diverse group with the most sexual dimorphism. Males are typically larger than females in species with male-male combat or male paternal care, ranging from dwarf males to males more than 12 times heavier than females.

In *Lamprologus callipterus*, males are described as being up to 60 times larger than females. Males are thought to benefit from their larger size because they collect and defend empty snail shells which females breed. Males are stronger and can collect larger shells. The female's body size remains small because she lays her eggs inside the empty shells to reproduce. Another example is the dragonet, where males are significantly larger than females and have longer fins.

Gobiusculus flavescens, also known as two-spotted gobies, showed sexual selection for female ornamentation. Selection for ornamentation within this species, on the other hand, suggests that showy female traits can be selected through female-female competition or male mate choice. Because carotenoid-based ornamentation indicates mate quality, female two-spotted guppies with colourful orange bellies during the breeding season are thought to be attractive to males. Males make significant investments in their offspring during incubation, which results in a sexual preference for colourful females due to higher egg quality.

Sequential hermaphroditic fish exhibit sexual dimorphism as well. Size is important in male reproductive success in protogynous mating systems where males dominate mating with many females. Males are larger than females of comparable age, but it is unclear whether this is due to a growth spurt during the sexual transition or to a history of faster growth in sex-changing individuals. Larger males can stifle female growth and control environmental resources.

In other cases, males will experience noticeable changes in body size, while females will experience morphological changes that can only be seen inside the body. Male sockeye salmon, for example, grow larger at maturity, including an increase in body depth, hump height, and snout length. Females have minor changes in snout length, but the most noticeable difference is a significant increase in gonad size, which accounts for approximately 25% of body mass.

31

Courtship and Mating in Fishes

31.1 Introduction

Courtship is the behavioural interaction that occurs between males and females before, during and just after mating. It encompasses all expressive movements associated with pair formation and mating. "Courtship is the heterosexual communication system that leads to the consummatory sexual act," says DeMorris. Arthropods, mollusks, fish, reptiles, and mammals are the most affected. It consists of various activities and serves a specific purpose. Many courtship activities ensure that people only mate with people of their species. The distinction between species' courtship behaviour allows females to identify males of their species. It is concerned with reproductive fitness and provides opportunities for survival (Fig. 31.1).

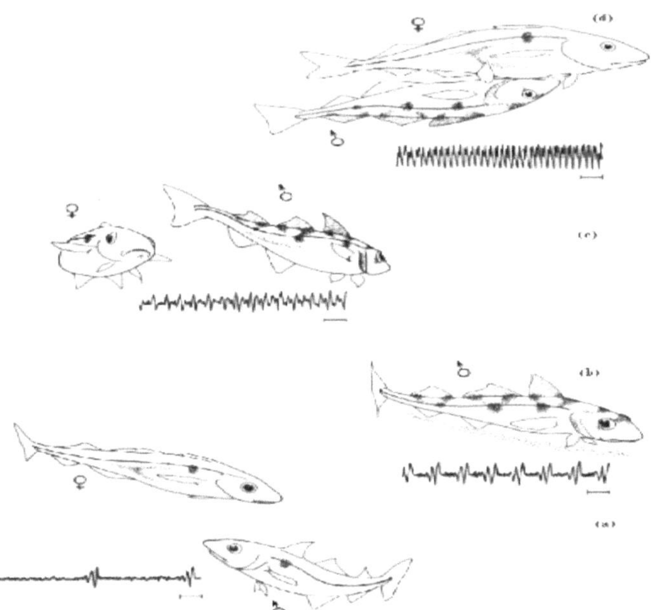

Fig. 31.1: Different stages of the courtship behaviour of haddock (from https://www.researchgate.net)

It can be brief and brief, or it can last a long time with vigorous and elaborate displays. It may temporarily decline shortly after coitus or after a lengthy unsuccessful courtship. If an enemy appears, it may drop dramatically. If the animal becomes ill, it may fall gradually.

Animals that are courting "put themselves on display." Courtship songs, flight, and dance are rituals that show the courting animals' personalities.

31.2 Functions

The courtship fulfills the following four functions:

1. Mate finding: This is a highly organised process that involves one or more senses such as light, smell, sound, touch, and even taste. Rubbing, pushing, twisting, licking, clasping, coiling, necking, and other behaviours are common in courting animals.
2. Persuasion: After recognising a potential mate, the male approaches the female. Males play a more active role than females.
3. Synchronisation: The occurrence of the same behaviour in different people.
4. Reproductive isolation: A species' inability to successfully breed with related species due to geographical, behavioral, physiological, or genetic barriers or differences.

Courtship is a diverse traditional behaviour in which animals' clues to mating are displayed, and many animals have male-selection courtship customs. It refers to all male and female interactive connections that move forward and point to the reproduction of the egg via sperm. The size of the male influences the selection of a female mate.

Many male fishes adopted techniques to impress their mate by the tip and turn dance towards the female, zig-zag dance, swimming back and forth in rapid activities all the performing females keep an eye on the male to nest or the male point to the entrance of the nest if the female enters in the nest the male shivers on her tail to stimulate her to spawn. When two males, one dull-coloured large and the other bright-coloured orange, are offered to two virgin females, the females prefer the bright orange-coloured large male.

Increases or decreases in the occurrence of displays and courtship duration, or the presentation of male-type behaviour by masculinized females, affect the partitioning of parental care between the sexes.

Uncertainty these concentration activities include interspecific communication and interaction such as mate selection and courtship, then participating in these two tactics can expose perceptions of why and how receivers and signalers co-adapted (Sargent et al, 1998).

The presence of morphological differences between the genders often indicates the presence of courtship in the species. These can take the form of special adaptations in the male for sperm transfer to the female.

Secondary sexual characteristics may be present all year, or they may appear at the start of the breeding season and then disappear. This is especially true of the bright, vivid colours displayed by some males.

During the breeding season, in species where there is a definite pairing, the female will signal her readiness to the male in some way, such as quivering and/or adopting a certain ritualistic posture or position in the water.

A complex courtship does not indicate post-fertilization care in the Common Dragonet (*Callionymus lyra*). Once released, the eggs are free to float in the sea.

Normally, small groups of fish will suddenly separate from the large school pairs, often rising to the surface to spawn quickly and then returning to the school. As the male guides the female away from the school, a brief courtship dance can be observed.

During the breeding season, the bitterling (*Rhodeus amorus*) is aggressive. The species reproduces by depositing eggs into the gills of the Swan Mussel, where they are safe from predators until they hatch and leave the safety of the mussel's shell. Regardless of the male's beauty, she will not produce the long ovipositing tube she uses to lay her eggs into the Swan Mussel's gills unless there is a living Swan Mussel present. Furthermore, when there are no males present, she will produce the ovipositor (and even lay her eggs).

31.3 Nest Building

The African Aba (*Gymnarchus niloticus*) builds a 100 cm wide nest. When the young emerges, the sides rise above the water level once more, and they are protected by one of the adults.

The male Paradise Fish (*Macropodus opercularis*) constructs a bubble nest and then attracts the female. He collects the eggs after spawning and inserts them one by one into the bubbles of the nest. The female can lay up to 500 eggs. Then he stays to protect the nest. At least four *Microctenapoma* species construct bubble nests.

The courting male of three spinned sticklebacks (*Gasterosteus aculeatus*) first builds a nest, then courts and lures his female to the nest to induce her to lay eggs. During the breeding season, both males and females appear greyish green as the illuminating neon lights and living in schools. Tinbergen et al. have described this fish's courtship. (1954). This fish's courtship includes jerking, leaping, jumping, circling, leading, zig-zag dance, thrusting and pushing, quiering, and spawning.

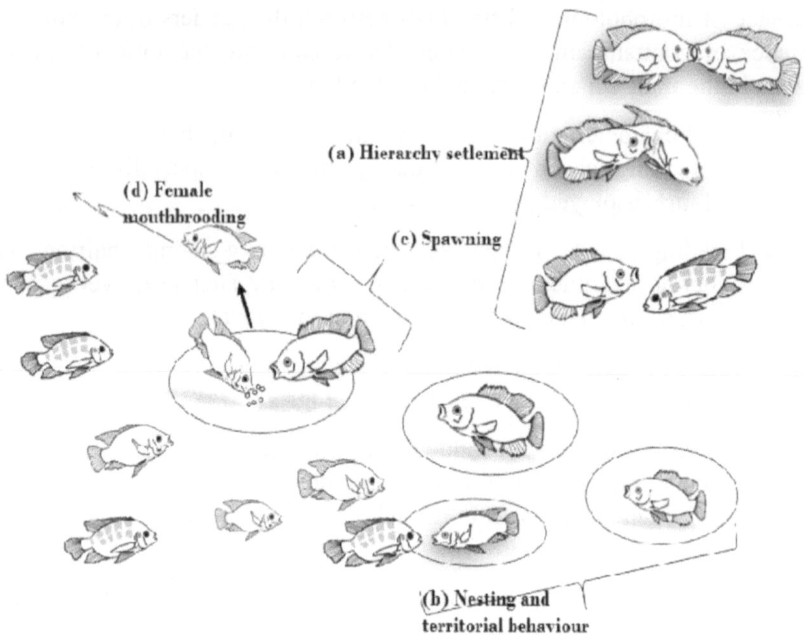

Fig. 31.2: Nesting and territorial behavior of Nile Tilapia (from https://www.mdpi.com)

31.4 Territorial and Aggressive Fish

The individuals first show off their fins, but if neither is willing to back down, physical interaction is unavoidable. The opponents line up face to face and open their mouths as wide as they can. Then they push against each other, as if kissing, until one of them turns sideways and looks away, signalling defeat.

During the breeding season, the jewelfish (*Hemichromis bimaculatus*) becomes solitary and territorial. During the breeding season, females develop nuptial colours. In this case, either sex could imitate courtship. Their two males will fight only if they are near equals and after an elaborate display. Bernards and Van Room (1950) described this fish's courtship. This fish courtship behaviour is similar to that of three spinned sticklebacks. Jewelfish completed courtship by jerking, quiering, and trembling; nipping, skimming and rubbing; spawning, guarding, egg deposition, and parental care.

There is no actual fighting in *Herichthys*. The ritualised intimidation displays, on the other hand, may well continue until one of the contestants succumbs to exhaustion (Fig. 31.3).

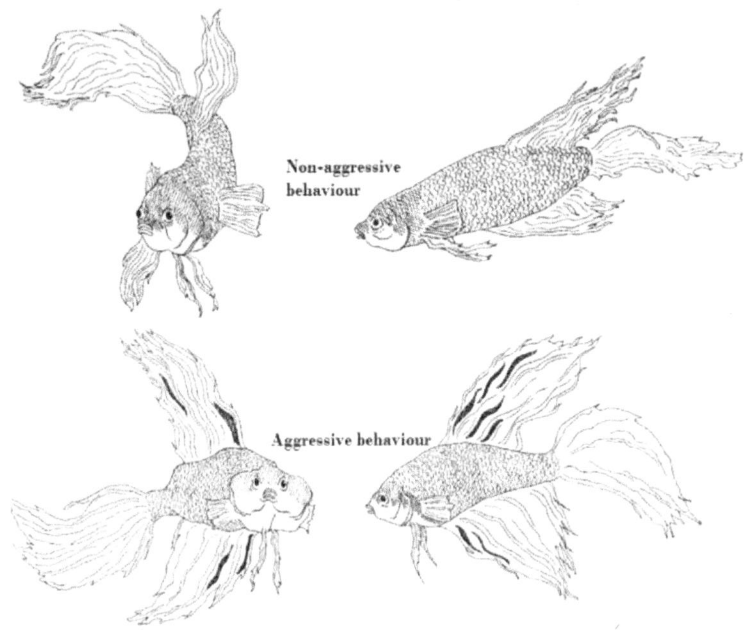

Fig. 31.3: non-aggressive and aggressive behaviour Male Siamese fighting fish (from https://www.researchgate.net)

A male *Tilapia natalensis* defends his territory. Before the female appears, he digs a spawning pit (Fig. 31.2). This fish's courtship consists of six steps: inviting, leading, tail wagging, cleaning, circling and spawning. Bernards (1952) described the courtship of this fish as similar to that of the jewel fish.

Guppy fish (*Lebistes reticulatis*) live in large schools and are extremely social. During the breeding season, male guppies are extremely aggressive. Courtship is completed in six steps in this fish: leading, inviting, luring, swimming jerkily, S-shaped curving, starling jump and checking (Fig. 31.4).

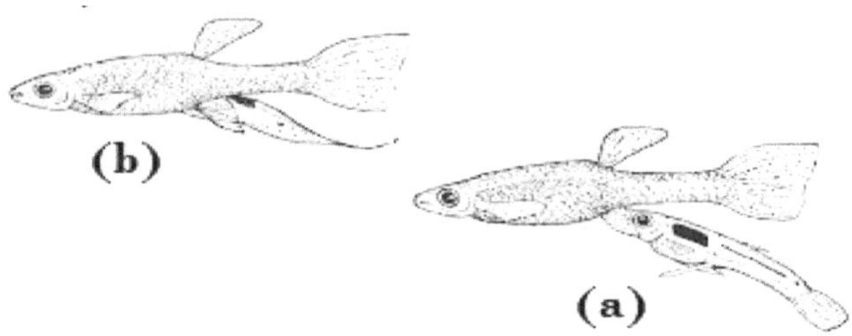

Fig. 31.4: Territorial behaviour in Guppy fish

If the female of the Siamese fighting fish (*Betta splendens*) is slow to respond, the male may become aggressive. Males will inflict significant damage on each other in the artificially well-lit and specially limited conditions of a fish tank.

31.5 Mating in Fishes

Mating is the act of two people of opposite sexes meeting for sexual reproduction. The union of opposite sex organs occurs during mating. Scientists have been able to identify different types of mating systems, such as monogamy and polygyny by classifying social interactions (Fig. 31.5).

The teleosts have a wide range of mating systems. Many species appear to mate indifferently, while others appear to require a lengthy period of courtship before selecting a partner. In some species, a male and female may stay together for years; in others, males may have multiple mates; and females frequently mate with multiple males. Some species may self-fertilize; others are almost certainly simultaneous hermaphrodites, producing eggs and sperm simultaneously and releasing either when spawning. Males can change into females (protandrous hermaphrodites) and females can change into males in many species. (protogynous hermaphrodites). Some species have no males at all.

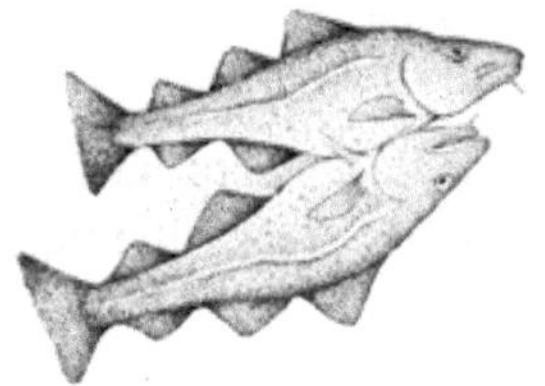

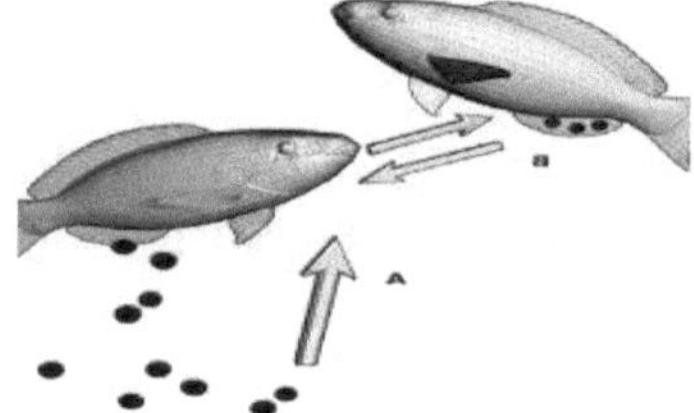

Fig. 31.5: Mating in fishes

Most teleost species are oviparous, meaning they fertilise externally by releasing both eggs and sperm into the water. Internal fertilisation occurs in 500 to 600 teleost species, but is more common in Chondrichthyes.

Teloests breed seasonally, with a few exceptions breeding continuously. There is considerable variation in the time of year when seasonal breeders breed. Freshwater temperate zone fish spawn in the spring and early summer, while salmonids spawn in the autumn. Carp spawn between April and August, depending on the climate and conditions in which they live. Water oxygen levels, food availability, fish size, age, number of times the fish has spawned before, and water temperature are all known factors that influence when and how many eggs each carp will spawn at any given time.

Because larger individuals are better able to defend their young, the requirement for each sex to have the largest possible partner may lead to assortative mating, with males and females of roughly the same size pairing off. Where there is a division of labour between the sexes, the need to select a larger partner will be felt most strongly by the sex that is not required to perform the majority of brood defence. If this occurs, a typical male female size ratio of pairs may form (Barlow and Green 1970).

The male inserts the organ into the female's sex opening, which has hook-like adaptations that allow the fish to grip the female and ensure impregnation. She is fertilised if a female remains stationary and her partner contacts her vent with his gonopodium. The sperm is kept in the female oviduct.

31.5.1 Goldfish

They are egg layers. They usually begin breeding after a significant temperature change, which occurs in the spring. Males chase females, bumping and nudging them to get them to release their eggs. While the female goldfish spawns her eggs, the male goldfish fertilises them. Their eggs are sticky and stick to aquatic vegetation. Within 48 to 72 hours, the eggs hatch. The fry takes on its final shape within a week or so, though it may take a year for them to develop a mature goldfish colour; until then, they are a metallic brown like its wild ancestors. The fry grows quickly in the first few weeks of life.

31.5.2 Siamese Fighting Fish

Male Siamese fighting fish construct bubble nests of varying sizes on the water's surface. When a male becomes interested in a female, his gills flare, body twists and fins spread. The female's skin darkens and she curves her body back and forth. The act of spawning occurs in a "nuptial embrace" in which the male wraps his body around the female, with each embrace resulting in the release of 10-40 eggs until the female is egg-exhausted. The male releases milt into the water from his side, and fertilisation occurs externally. The male uses his mouth to retrieve sinking eggs and deposit them in the bubble nest during and after spawning. When the female has laid all of her eggs, she is chased away from the male's territory because she is likely to eat the eggs out of hunger. The eggs are then cared for by the male. He keeps them in the bubble nest, ensuring that none fall to the bottom and repairing it as needed. The newly hatched larvae remain in the nest for the next 2-3 days, until their yolk sacs are fully absorbed. Following that, the fry leaves the nest and the free-swimming stage begins.

32

Development in Fishes

32.1 Introduction

Development or embryogenesis refers to the events that occur after fertilisation. Embryogenesis in fish includes the formation of stages up to fingerlings. The process is similar in all fishes, but there are some differences between teleosts and elasmobranchs. The development of telolecithal eggs in fish begins soon after sperm fertilisation.

Billard et al. (1986) recorded the length of embryogenesis in the following fish species:

Table 32.1: Name of the fish and their period of embryogenesis

Sl.No.	Name of fish	Duration	Sl.No.	Name of fish	Duration
1.	Grass carp	24-30 days	5.	Tench	60-70 days
2.	Silver carp	24-30 days	6.	Rays	60-120 days
3.	Big head carp	30-50 days	7.	Skates	150-180 days
4.	Indigenous major carps	60-70 days	8.	Sharks and Chimaera	270-660 days

The following are the various events that contribute to the development of a teleostean fish:

32.2.1. Unfertilized egg release

Female teleosts spawn thousands of eggs at a time. The eggs are small and round, ranging in size from 1.2 to 1.5mm. Each egg is surrounded by a tough elastic and semipermeable vitelline membrane and contains a large amount of yolk surrounded by a thin layer of cytoplasm on the periphery.

32.2.2 Fertilisation

External fertilisation is common in freshwater teleosts. Fertilization usually occurs while the egg is still within the follicle in marine viviparous teleosts. It occurs in the presence of a small concentration of Ca^{+2} and Mg^{+2} in monospermic teleosts. Only one sperm fuses with the egg nucleus in polyspermic fishes, with the rest reabsorbed and used as food (Fig. 32.1).

Fig. 32.1: External fertilization in fishes

As in *Oryzias latipes*, the egg becomes transparent or translucent within 10 minutes of fertilisation. It swells and expands to 1.8mm in size. The vitelline membrane separates from the cytoplasm's peripheral layer, forming a previtelline space. The cytoplasm of the egg migrates towards the animal pole to form a distinct blastodermic cap, while the yolk in the rest of the egg condenses and presses together.

Inception lasts on average for 45 minutes with a cell 0.7 mm in diameter.

32.2.3. Cleavage (Segmentation)

The zygote has meroblastic properties. It usually begins 30 minutes after fertilisation, but in *Oryzias latipes* it begins 1 hour later. The average number of cleavages is six. It is meroblastic (incomplete) in nature and is only found in the blastodermic cap of the cytoplasm. The cells will divide at 15-minute intervals as they cleave. This cycle will last 45 minutes to 2 hours, depending on the complexity of the fish (Fig. 32.2).

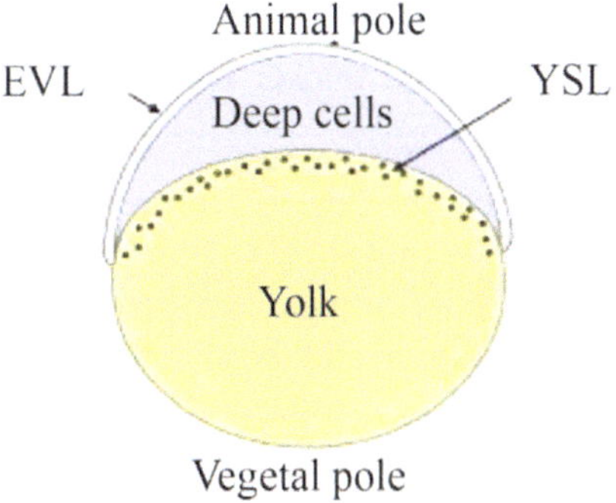

Fig. 32.2: Cleavage in fishes

The first cleavage is always meridional and runs from the animal pole to the vegetal pole. The nuclei in the first cleavage become visible, resulting in two blastomeres.

The second cleavage is at an angle to the first and vertical. It produces four blastomeres of equal size.

The third cleavage follows in the footsteps of the first. The blastodisc is rectangular.

The fourth cleavage follows in the footsteps of the second. The blastodiscs are shrinking in size.

Cleavage planes are all vertical in the early stages, so all blastomeres lie on one plane only. Later on, it becomes horizontal as well, so that the blastomeres are arranged in more than one row. Morula is produced by cleavage about 4 hours after fertilisation.

The last cleavage has become asynchronous.

MORULA: After several cleavages, the blastodermic cap transforms into a ball of cells. These protrude slightly above the surface of the yolk but occupy a nearly identical area to the original blastodermic cap. Morula refers to the embryo at this stage.

The blastomeres divide and redivide the world. A yolk syncytial layer (YSL) is produced between the yolk and the convex floor of the blastoderms during this process. This layer is known as periblast. It has many nuclei that are dispersed within a thin layer of cytoplasm. The blastoderm is represented by the cellular mass remaining above this layer.

32.2.4. Blastula

Between the blastoderm and the periblast, a blastocoel is formed at the end of the morula. This embryo represents the blastula stage. Balstocoel formation is followed by blastoderm enlargement and flattening, resulting in a radially symmetrical embryo with bilateral symmetry. The blastoderm gives rise to the embryo, while periblast cells digest and supply the yolk to the developing embryo. The embryonic material is destined for the central and thicker blastoderm. Its distal end forms the posterior end (Fig. 33.3).

FATE MAP: The fate map is a diagram that depicts the fate of each cell in the embryo. It depicts the position of various prospective (potential or future) organs in the early embryo. The cells in the blastoderm's anterior region are prospective ectodermal cells; those in the middle form the mesoderm; and those in the caudal region are destined to form the embryo's endoderm. The prospective notochordal cells are located in the middle (Fig. 32.5). The prospective mesodermal area is present all around the blastoderm margin in *Salmo*. The mesodermal area extends primarily along the blastoderm margin and is not continuous at the anterior end of *Fundulus*.

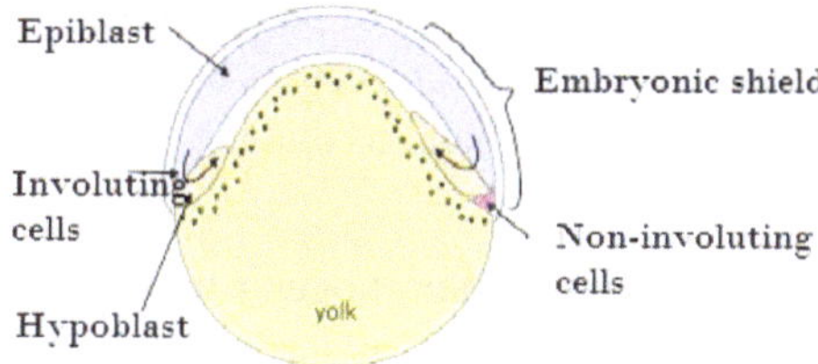

Fig. 32.3: Blastulation in fishes

32.2.5. Gastrulation

It is the rearrangement and reorganization of blastoderm to transform the disposition of the embryo into the body plan of an adult. In fishes, it begins when the blastodisc has grown over about 1/3rd of the egg by a combination of cpiboly and emboly. The gastrulation period will last anywhere from 5 to 10 hours (Fig. 32.4).

EPIBOLY: It is the accumulation of extra embryonic material on the surface of the yolk. It is linked to periblast growth and the formation of the germ ring. Throughout this procedure:

a. Blastodermal cells form a layer over the yolk.
b. Prospective ectodermal cells grow over and cover the yolk mass, forming an epiblast layer.
c. At the same time, the periblast develops and forms an inner yolk covering.
d. The periblast and epiblast form a yolk sac around the yolk.
e. Epiblast pulling down may also occur.

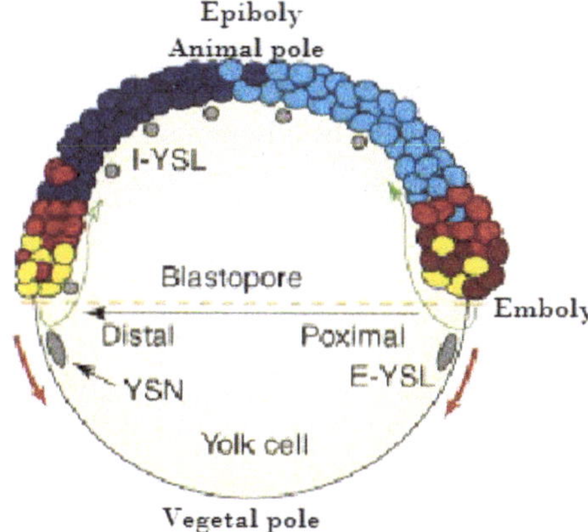

Fig. 32.4: Gastrulation in fishes

There are various explanations for the growth and contraction of periblast. Lewis (1949) proposed that a yolk gel layer forms, which contracts and causes blastoderm to pull down over the surface of the yolk. Trinkaus (1951) claims that the gel layer isolates as the periblast grows over time.

EMBOLY: It is the invagination of the blastoderm rim. Its morphogenic movement is accompanied by a marked convergence of invaginating and superficial material to the future embryo's dorsal midline. It starts with blastodermic cells invading the cytoplasm and the yolk to form a prominent germ ring. The germ ring begins to shrink gradually towards the vegetal pole. This is followed by blastoderm growth and spread, which gradually engulfs an increasing area of the yolk surface. This is the first step in the formation of an embryonic shield.

As the germ ring descends and cells are added from all sides, the triangular area narrowly elongates into the embryonic shield, which quickly differentiates into an inner embryonic and outer extra-embryonic area.

The germ ring reaches a little below the equator of the egg surface after about four hours of fertilisation. After six hours of fertilisation, the entochordo-mesoderm differentiates from the embryo's epiblast. The notochord, neural keel, and somite become differentiated after about eleven hours of fertilisation. The neural keel is covered by an epiblast and is supported by an entodermal layer. The ectoderm in the anterior trunk region is made up of neural keel cells that spread over the prospective notochord.

The secondary hypoblast differentiated after the primary hypoblast had moved forward. The notochordal and prechordal plate cells move caudally to involute over the endodermal cells. Prospective notochord cells occupy a median position, and the neural ectoderm forms a strip-like neural plate over them. Prospective mesodermal cells converge on the blastodisc's dorsolateral edges and involute to occupy a position between the entoderm and the epiblast.

The rudiments of various larval organs begin to appear after the formation of three germinal layers. The fate of germinal layers is discussed briefly below:

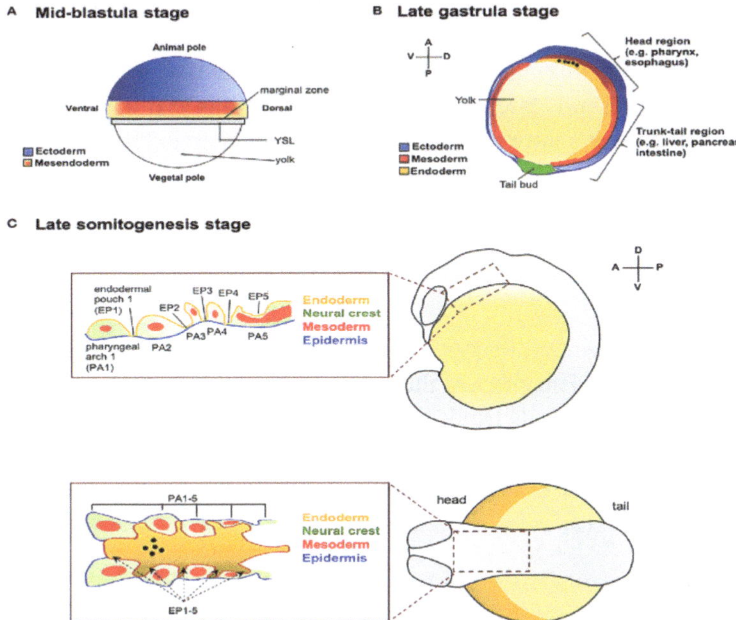

Fig. 32.5: Fate map of zebrafish (From https://www.researchgate.net)

Ectoderm: Epidermis and its derivatives, nervous system, olfactory epithelium, retina, lens, membranous labyrinth, neural crest, pituitary.
By invagination: stomodaeum, proctodaeum.
By evagination: Rathke's pouch

Endoderm: Form gut rudiment:	Pharyngeal pouches, Thymus, Parathyroid gland.
From ventral evaginations:	Thyroid, lungs, air bladder, liver, gall bladder, caeca.
Mesoderm: Epimere as dermatome:	Mesenchyme, connective tissue, blood vessels, dermis and dermal skeleton.
Epimere as myotome:	Skeletal muscles, myotomal muscle, appendicular skeleton.
Epimere as sclerotome:	Part of the skull and vertebral column.
Hypomere as somatoderm:	Blood vessels, the skeleton of the body wall and limbs, dermis, Parietal peritoneum
Hypomere as splanchnoderm:	Smooth muscle, gut, heart, visceral peritoneum, adrenal gland, germinal epithelium
Mesenchyme:	Skeleton and muscles of the head, dentine, blood, part of the dermis, endothelium of blood vessels and heart.
Mesomere as nephrotomes:	Kidney tubule, duct system of the kidney and gonads.

A dorsal-ventral axis must form, and the developing embryo's major proteins involved are BMP and Wnts. BMP2B induces cells, whereas chordin can prevent BMPs from dorsalizing the tissue. Wnt8 stimulates embryonic tissue development in the ventral, lateral, and posterior regions. To aid in proper development, fish have a centre called the Nieuwkoop centre. The interaction of FGFs, Wnt, and retinoic acid appears to be the cause of anterior and posterior axis formation. FGFs, retinoic acid and Wnts are required for posterior gene activation.

32.3 Postembryonic Development

A small cylindrical and bilaterally symmetrical embryo is formed as a result of embryonic development. The body of the embryo separates from the yolk sac. Its head protrudes internally, its trunk lies over the yolk, and its tail protrudes behind. The embryo grows in size, while the yolk sac contracts. Finally, the embryo hatches and becomes a free-swimming larva. Prolarva is another name for a newly hatched fish (Alevin in *Salmon*).

32.4 Larval Development (Hatching)

Young ones from the time of hatching are known as larvae or fry. A fry having a yolk sac is known as 'sac-fry' and with the absorption of the yolk sac; it becomes a post-larva or advanced fry'. During these 48 to 72 hours, the external indicators of the fish develop.

In some species of *Salmon*, trout and many catfishes, the development is direct and the advanced fry resembles the adult.

In several species of Clupeidae, Cyprinidae, Lophidae, Anguillidae, Scorpaenidae and other families, development is indirect and metamorphosis occurs during which larval characteristics are lost and adult characteristics develop. Fingerlings are fry that have lost their larval characteristics. The fingerlings are active feeders, resulting in rapid body growth and gonad maturation to give rise to the adult. Females are typically larger than males. Males, on the other hand, reach sexual maturity earlier than females.

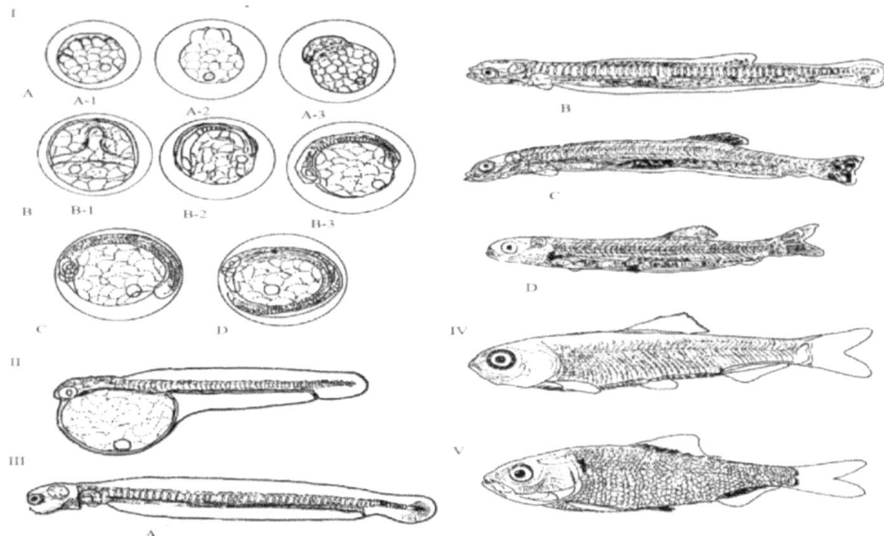

Fig. 32.6: Development of fish (From https://link.springer.com)

The pharyngula period occurs, during which the fish begins to show fully formed organs. This can last anywhere between 24 and 48 hours. This is the time when the upper and lower vertebrate embryonic development can be distinguished.

33

Fish Genomics

33.1 Introduction

World fisheries face the threats of overfishing and climate change that have resulted in the extinction of genetically unique stocks and loss of genetic diversity. Genomics has brought new tools that can help address fundamental questions in fisheries management such as stock identification, population structure, and adaptive response to environmental change (Bernatchez et al, 2017).

Genomics allows identifying genetic basis associated with detecting genetic variations caused by environmental changes. Therefore it has many applications in fisheries and aquaculture including identification of fish stocks for capture fisheries management, conservation of fish genetic resources, genome selection for genetic enhancement, disease resistance, and sex determination and control.

A genome is a collection of all functional and nonfunctional DNA sequences in an organism and provides insights to make a species unique in aquaculture and/or fisheries. Genomics originated with the advance of molecular genetics and the development and application of various biotechnologies and became a discipline around 1987.

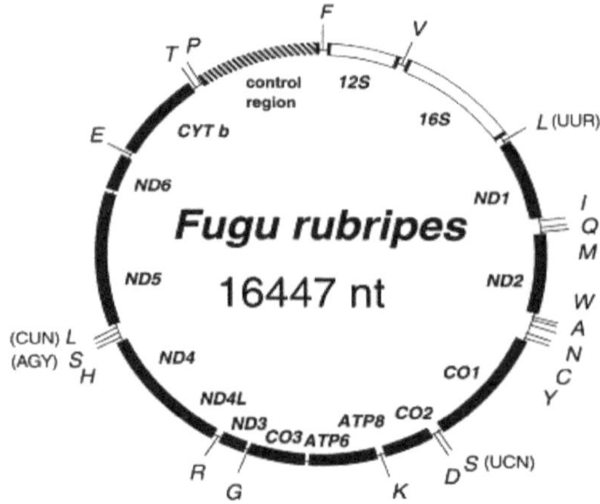

Fig. 33.1: The mitochondrial genome of *Fugu rubripes* (from https://www.sciencedirect.com)

33.2 Generation of sequencing technologies

1. The First-generation sequencing technology: It is Sanger's DNA sequencing technology invented in the 1970s (Sanger et al, 1977). The subsequent improvements in this technology led to the development of ABI DNA sequencing systems (MacKenzie & Jentoft, 2016). It was used for sequencing the human genome and the mitochondrial genomes of model fish organisms, including Japanese pufferfish (*Fugu rubripes,* Aparicio et al, 2002) and zebrafish (*Danio rerio,* Howe et al, 2013) (Fig. 33.1).

2. The Second-generation sequencing technology: It includes pyro-sequencing (454 Life Sciences) (Margulies et al, 2005), bridge amplification-based sequencing (Ruparel et al, 2005) and sequencing by oligonucleotide ligation and detection (SOLiD) (Heather & Chain, 2016). These have resulted in the "genomics revolution".

3. The third-generation sequencing technology: It was represented by Oxford Nanopore Technologies (ONT) or SMRT Pacific Biosciences (PacBio) sequencing platforms. It can produce long reads and require a relatively short running time (Eid et al, 2009).

   The combined use of second and third-generation sequencing technologies appears to be a practical approach for *de novo* sequencing of non-model organisms (Wang et al, 2019) (Fig. 33.2).

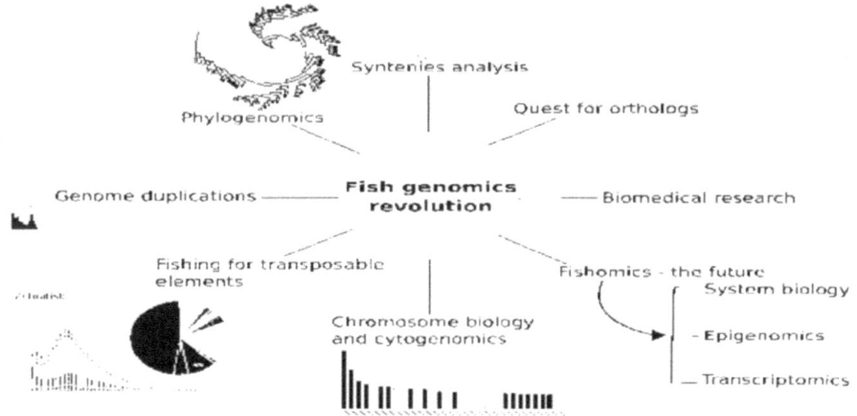

Fig. 33.2: Fish genomics and its effect research (from https://link.springer.com)

4. The fourth-generation sequencing technology: It includes Hi-C technology such as *in situ* sequencing. It combines traditional imaging analysis techniques and NGS technologies to sequence nucleic acids directly in cells and tissue (Ke et al, 2016).

The combination of Illumina, PacBio, and Hi-C technologies has been applied to construct the chromosomal-level genome assembly of yellow catfish, an economically important freshwater aquaculture fish species in Asia (Gong et al, 2018).

33.3 Fish genome sequencing

It was conducted *de novo* using next-generation sequencing technologies such as 454 and Illumina and PacBio. Most sequencing projects were based on linkage mapping and quantitative traits linkage (QTL), which enabled chromosome-level genome assemblies (Conte et al, 2017). The estimated genome sizes ranged from 544 Mb in turbot (Figueras et al, 2016) to 2.97 Gb in Atlantic salmon (Lien et al, 2016). Among the 14 major species, brown trout possessed the second-largest genome of 2.18 Gb (Berthelot et al, 2014). Common carp had a genome of 1.83 Gb (Xu et al, 2014). The flatfish species and northern snakehead had relatively small genomes (Xu et al, 2017). The genome size of grass carp was approximately 1.07 Gb for a female and 0.9 Gb for a male (Wang et al, 2015).

The scaffold N50, an indicator of genome completeness and ranged from 137 kb in Pacific bluefin tuna to 7.7 Mb in channel catfish (Chen et al, 2016). The grass carp genome had the second-largest N50 of 6.5 Mb, comprising 114 scaffolds (Wang et al, 2015). Only four species had the scaffold N50 of genomic sequences <1 Mb. It is worth noting that multiple factors, such as the coverage and assembly software, may affect the estimated scaffold N50 values.

The gene assembly of *Clarias batrachus* contained 10,041 scaffolds, with a scaffold N50 of 361.2 kb. The assembly covered a total of 821 Mb, similar to the genome size of 854 Mb estimated from ALLPATHS-LG but smaller than the estimated 1.17 Gb based on the bulk fluorometric assay method (Hinegardner & Rosen, 1972). The completeness of the genome assembly was assessed by mapping the 248 core eukaryotic genes (CEGs) from CEGMA v2.5 to the genome sequence (Parra et al, 2007).

33.4 Genome annotation

Genome annotation is mainly concerned with the identification of protein-coding genes. The predicted protein-coding genes in the genome of these 14 fish species ranged from 19,877 in Northern snakehead to 52,610 in common carp. The number of predicted protein-coding genes almost doubled in common carp compared to zebrafish (*Danio rerio*), demonstrating that the tetraploid genome of common carp likely retained a large portion of duplicated genes since the latest whole-genome duplication (Xu et al, 2019). The Atlantic salmon genome appeared to have experienced a crucial post-Ss4R (salmonid-specific fourth vertebrate whole-genome duplication) rediploidization that resulted in large genomic reorganizations and was demonstrated by the bursts of transposon-mediated repeat expansions (Lien et al, 2016). In rainbow trout, two ancestral sub-genomes remained extremely collinear after 100 million years. Interestingly, only 22,184 protein-coding genes predicted in the rainbow trout genome, attributed to the loss of half of the duplicated genes through pseudogenization (Berthelot et al, 2014).

A total of 22,914 genes were annotated from the *Clarias batrachus* genome sequence, of which 19,834 genes (86.6%) were supported by RNA-Seq evidence from the gill and the air-breathing organ (Li et al, 2018).

33.5 Genomic composition

The base composition is an important genomic feature related to genomic function (Smarda et al, 2014). In teleost, the genomic composition content was found to be higher in seawater than in freshwater fish and higher in migratory than non-migratory species (Tarallo et al, 2016). Among the 14 species, genomic composition content ranged from 31.5% in channel catfish to 45.8% in Atlantic cod. In European seabass, a lower percentage of genomic composition was observed in noncoding regions (39.6%) compared with protein-coding regions (52.6%), and the genomic composition content in the third codon position was 60.7%. An inverse relationship between the genome size and genomic composition contents reported in previous studies was found insignificant based on the genomic data of these 14 species, indicating more genomic data are needed for the validation of this assertion (Tine et al, 2014).

The *Clarias batrachus* genome had a genomic composition content of 39.2%, similar to those of other fish species (Li et al, 2018).

33.6 Genetic structure

Repetitive elements account for a significant portion of the genome. A genomic analysis of 52 fish species found the proportion of repetitive elements was positively correlated with genome sizes and repetitive element categories were enriched in marine versus freshwater species (Yuan et al, 2018). Pleuronectiformes flatfish possessed 6.0–9.0% repeats, which is comparable to takifugu (7.1%) (Aparicio et al, 2002) and stickleback (13.48%) (Jones et al, 2012).

Repetitive elements account for 58–60% of the salmon genome and are associated with its genome duplication (Lien et al, 2016). Repetitive elements in common carp and rainbow trout were found to be relatively high, which could also be explained by genome duplication (Berthelot et al, 2014). The high content of repetitive elements in the genome may increase the birth of new genes; however, it may also result in genomic instability due to possible abnormal recombination or splicing (Yuan et al, 2018). Repetitive elements of *Clarias batrachus* comprised 30.3% of the genome (Li et al, 2018).

33.7 Information on Fish Genomics

During the past decade, the remarkable increase in fish genome sequencing has revolutionized comparative and evolutionary genomics. Fish genomics has been transformed rapidly, with the availability of high-quality chromosome-level genome assemblies and large collections of sequencing datasets, which are roadmaps for striking discoveries.

Recent genome projects have demonstrated that comparative genomic analysis combined with transcriptomic analysis allows the elucidation of the genomic basis for adaptation to mudskippers (*Bolelphthalmus pectinirostris, Scartelaos histophorus, Periophthalmodon schlosseri* and *Periophthalmus magnuspinnatus*) (You et al, 2014).

594 fish genomes have been sequenced in the past two decades. All these studies have used data obtained by traditional cytometric methods and also have largely disregarded other genome attributes namely genetic composition, number of chromosomes, number of genes, and protein count.

The availability of fish genomes and the advances in genomics have allowed for addressing many challenging issues such as overfishing and germplasm degradation faced by fisheries and aquaculture. Although much progress has been achieved in genomic application to fisheries and aquaculture, the full potential remains to be explored and reaped.

The most current data on genome attributes of fishes as generated by the whole genome sequencing projects to understand the trends and effect of taxonomy on the genome attributes (genome size, number of chromosomes, number of genes, and protein count) and the interrelation of genome attributes. The trends state that a maximum number of fish genomes were sequenced in 2020, Cichliformes represents the highest number of published genomes, Illumina is the most used technology for sequencing fish genomes etc.

Comparative genomics with channel catfish (*Ictalurus punctatus*) revealed specific adaptations of *Clarias batrachus* in DNA repair, enzyme activator activity, and small GTPase regulator activity. Comparative analysis with 11 non-air-breathing fish species suggested adaptive evolution in gene expression and nitrogenous waste metabolic processes.

34

Parental Care in Fishes

34.1 Introduction

Fishes are the first aquatic vertebrates whose life cycle is largely affected by the aquatic habitat. They exhibit much variation in their social life. Some species show no attraction between males and females, conversely, some provide high social attraction and parental care. Therefore, parental care is developed for their social life and habitat.

Parental care may be defined as an association between the parents and the offspring, to increase the chances of the survival of their young ones. It is a form of altruism. In fishes, parental care includes all the post-spawning care of the offspring by the parents. Male fish are more actively involved in this process compared to female individuals.

Parental care seems to be a very important factor for the survival of species and the reproduction of the young. In the species, where internal fertilisation occurs, a parental acre is usually provided by the female. It is simply the hiding of eggs.

34.2 Estimation of Parental Care

A. 77 percent of fish show no parental care
B. 17 percent of fishes show care of eggs only and,
C. 6 percent of fishes care for their eggs and are newly hatched young.
D. According to Reynolds et al, (2002), the ratio of fish genera that exhibit male-only: biparental: female-only care is 9:3:1.

There is the evolution of parental care in fishes; a review of the teleost fishes reveals 21 transitions. All of these agree with the hypothesized transitions and in some cases, the direction of evolution is inferred by simple pedigree analysis (Gittleman, 1981).

34.3 Grouping of Parental Care

On the ethological and evolutionary bases, parental care in fish may be grouped as

A. Deposition of eggs in suitable places
B. Carrying of eggs and larvae by parents

C. Preparation of Nest
D. Formation of integumentary cup
E. Ovoviviparity
F. Viviparity
G. Special devices
H. Care for young ones

34.3.1 Deposition Of Eggs in Suitable Places

Fishes developed various ways to deposit their eggs in suitable forms for protection such as

a. Salmonids exhibit the most primitive type of parental care. The female makes a furrow in the gravely bottom of a running stream for the eggs, which are then covered by a layer of fine gravel.

b. **Masses of definite forms**: Yellow perch (*Perca flavescens*) deposits eggs in a single mass (Fig. 34.1C). The eggs are held together by a membrane and form long floating bands so that eggs are not scattered away.

c. **Transparent mass**: The eggs of angler fish (*Lophius*) are invested by a gelatinous outer coat and joined together to form a transparent mass of enormous size (Fig. 34.1B).

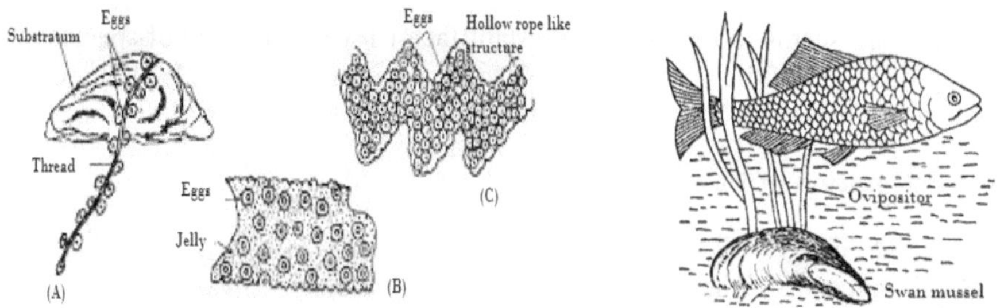

Fig. 34.1: Eggs deposited on (A) sticky thread secreted by kidney of flying fish. (B) Gelatinous sticky outer coat by Angler fish and (C) rope like structure by yellow perch (D) on dead shells of mussel.

d. **Cylindrical mass**: The adhesive floating eggs of *Fierasfer* forms a mass of cylindrical appearance of 2-3 inches.

e. **Cervices of the rocks:** The lumpsucker (*Cyclopterus*) deposits its spawn in the cervices of the rocks above the water surface at the spring tides. The spawn may contain 80000-136000 eggs. In this case, fish presses their head often into spawn to allow the water to penetrate there ensuring proper aeration of the eggs.

f. **The dead shells of mussels and oysters**: The eggs of bitterlings (*Rhodeus*) are deposited in the mantle cavity of the female whose oviduct is drawn out as an ovipositor (Fig. 34.1D). After oviposition, the male bitterling fertilizes the eggs and after the fertilisation, the male guards the eggs. According to Cunningham, the male *Rhodeus* is sexually excited by the sight of the shell of a mussel in which the eggs have been deposited. Similar parental care has also been reported in Gobies, Bleeis and Bullheads etc.

g. **Development of sticky threads**: The eggs of skippers, garpikes and flying fishes have sticky threads (Fig. 34.1A). The end of the thread helps them to adhere to any aquatic substratum or become entangled with other eggs of the same species.

34.3.2 Carrying of Eggs and Larvae By Parents

a. **Mouth cavity as shelter**: Female cichlids (*Tilapia*), male marine catfish (*Arius*) and cardinal fishes protect their eggs in this manner (Fig. 34.2A).

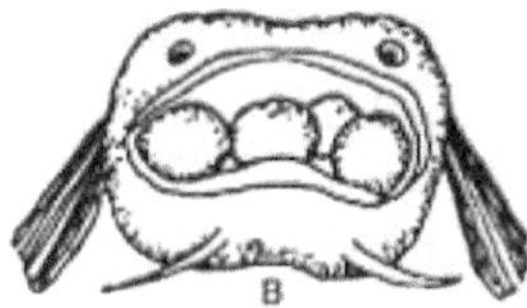

Fig. 34.2: (A) Female *Tilapia* carries eggs in her mouth. (B) Male Brazilian catfish carrying eggs in his pouch.

b. **The labial pouch as shelter**: A male Brazilian catfish (*Loricaria typus*) develops an enlarged lower lip forming a sort of pouch in which incubation takes place (Fig. 34.2B).

   These fishes swim upward to give the egg mucous which sticks them to the mouth. Even after hatching, the young fries do not leave the shelter of his/her mouth. When still more developed, the fry may be seen swimming about in a small group accompanied by the mother. On finding danger, they may return to shelter.

c. **Formation of egg ball:** In butterfish (*Pholis*), the eggs are rolled into a round ball and generally males guard the egg ball till the hatching of young ones (Fig. 34.3).

d. **Bony hook:** In Australian *Kurtus indicus*, the male of New Guinea fish has a bony hook projecting from the forehead and supported by a bony process from the skull (Fig. 34.4). The eggs are grouped into two bunches. The male carries one bunch on each side of the head.

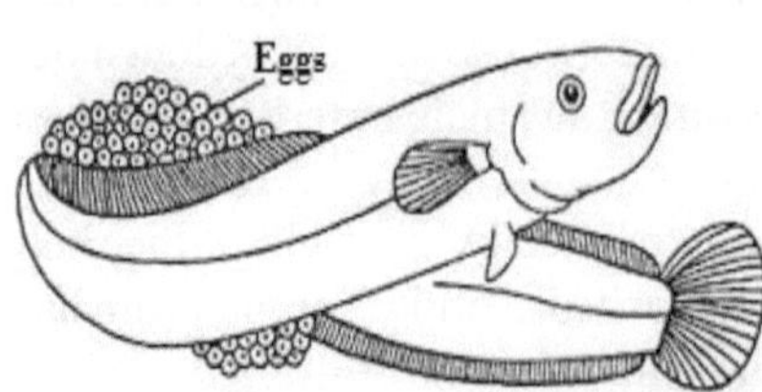

Fig. 34.3: Formation of egg ball by *Pholis*

Fig. 34.4: Formation of bony hook by *Kurtus*

34.3.3 Preparation of Nest

The nest is a structure where eggs are laid and provide a suitable and safe place for the development of their young. The extent and variation in the nest are largely restricted. Fishes use algae, stones, aquatic vegetation, secretion of the body etc. for the formation of a nest. The nest may be of the following types:

a. **Basin-like nest:** 1. Male Darter (*Eteostoma congregate*) selects a suitable place to a make nest (Fig. 34.5). Any female coming to the territory is allowed to remain there. Now, the female makes a basin-like depression to lay eggs. She sinks into the nest to deposit the eggs. The eggs are fertilized by males and covered by a sticky secretion of their kidneys. The eggs remain attached to the stone till hatching.

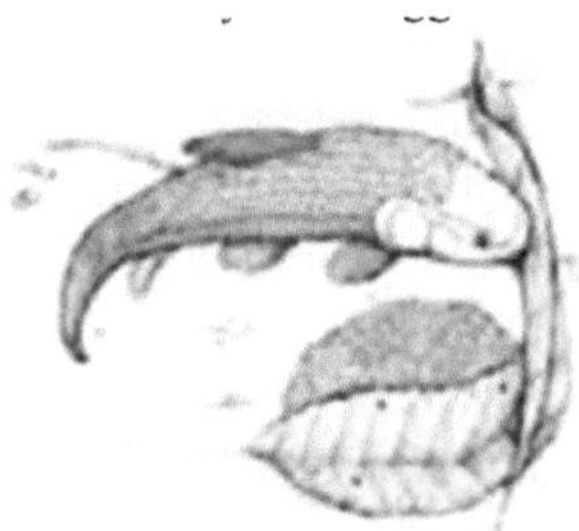

Fig. 34.5: Basin-like nest by Male darter

Fig. 34.6: Circular nest of *Amia calva*

2. Sunfish (*Masturus lanceolatus*) scoops out a shallow nest at the bottom of the water body. Pebbles are removed leaving behind the large stones. A layer of sand or gravel remains attached to the eggs. The male guards the eggs the hatching of young ones.

3. Cichlids (*Haplochromis burtoni*) build their nest. Both parents are engaged to guard the eggs. Typically, the female shows carrying of eggs but the male is an active defender of the brood against the predator.

   Female *Pterophyllum scalare* is actively involved in fanning more bont of eggs for more duration on the first day of post-spawning.

Female *Tilapia mariae* participates in more aeration of egg bonts than the males. The male did more call the young's by repeated opening and shut of dorsal, pelvic, anal and caudal fins.

Both the individuals of *Etrolus maculates* signal their young's by pelvic fins flickering.

Females of *Cichlosoma nigrofasciatum* participated more actively in fin-digging but both individuals are equally aggressive in defense of their broods.

4. American cyprinids (*Chub* and Shoiners) make a nest containing large stones but left the eggs uncared.
5. African osteoglossids (*Heterotis*) make a nest clean space among vegetation and lay eggs there.
6. Both the individuals of Amiuridae (American catfish) prepare a crude nest in mud to lay eggs.

b. Circular nest: 1. Male bowfish (*Amia calva*) constructs a crude circular nest at the margin of the lake with aquatic vegetation (Fig. 34.6). One or more females attend the nest for egg laying. The male fertilizes and guards the eggs till they hatch.

2. Both the individuals of American catfish (*Auris*) built a crude circular excavation of about 50cm diameter at the river bed. Fertilised eggs are covered there with large stones and left uncared.

c. Hole (Burrow) nest or oval pits: 1. The male African lungfish (*Protopterus*) scoops out a hole in the swamp. Their nest is surrounded by elongated aquatic weeds. The male fish provides constant aeration for developing as many as 5000 eggs.

2. The male American lungfish (*Lepidosiren*) constructs a burrow nest in the swamp bed. The nest measures 1-2m in diameter with a vertical and much larger horizontal partition in which eggs are deposited. Males take care of eggs. During the care of eggs, the male develops a long red filament from the pelvic fins performing the functions of aeration without coming on the surface to gulp the air.

d. Barrel-shaped eggs: The male three-spined stickleback fish (*Gastrosteus aculatus*) of American ponds and lakes, constructs a barrel-shaped more elaborated nest (comparable with the nest of a sparrow) by the initiation of courtship (Fig. 34.7). For this, the male selects a territory among the stem where water is shallow and flows regularly but not swiftly. He makes a shallow pit in the sand and collects algae, roots, stems and weeds to glue them together by secretion of kidneys. When the nest assumes a considerable size, the male passes through the nest forcibly in search of a mate to perform a zigzag dance around her displaying his red breast underside, often, the male shows a muzzling of the female at base of the tail. If the female is ready to lay eggs, she responds by covering her head and tail upwards. The signal of the female stimulates the male to swim toward the nest which is followed by the female. On reaching the nest, the male stretches to open the entrance with his snout. The female proceeds to deposit 2-3 eggs within the nest. She swims off through the opposite side of the nest. The male enters and fertilises the eggs. He comes out through the back door. The next day, he seeks out another female and repeats the process until sufficient eggs are deposited. He keeps the eggs aerated by fanning water with his tail. Then the male guards them persistently until they hatch. The two doorways allow a constant current of fresh cold water to bathe the eggs. The male pulls down the upper part of the nest, leaving the foundations as a kind of cradle for the fry. He sucks them into his mouth and splits them back into the nest. He relaxes his attention gradually.

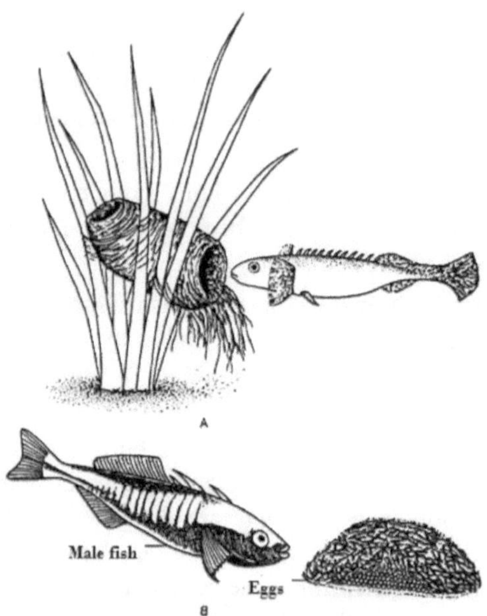

Fig. 34.7: Barrel shaped by Stickleback fish

e. **Cup-shaped nest:** *Apletes quadracus* makes the most elaborate nest attached to rooted plants near the bottom. After egg laying, the male extends the nest up several times by which several clutches of eggs are stalked vertically.

f. **Pear-shaped nest:** A male of fifteen-spined stickleback (*Spinachia*) makes a pear-shaped nest from the frond (stem-like leaf) bounded by the sticky secretion of the kidney. The secretion forms the thread-like structure that passes around the fronds till the formation of the nest. Now a female is invited there to lay eggs. The male fertilises the eggs and takes care of them till they hatch.

g. **Floating nest:** The mormyrids (*Gymnarchus*) and Anabantidae (Labyrinth fishes) construct a large floating nest from aquatic vegetation. The wall of the nest remains projected a few centimeters above the surface of the water.

h. **Foamy nest:** 1. Male fighting fish (*Betta*) prepare a foamy unusual nest by flowing air bubbles with their sticky mucus (Fig. 34.8). The mucus adheres together to form a floating mass of dome-like foam. After this, the male collects the fertilized eggs in his mouth, gives them a coating of mucus and finally sticks them to the undersurface of the nest.

   2. Male paradise fish (*Macropodus*) also prepare identical foamy nests where the eggs are up to the level of the water surface. The nest floats without being carried by the male.

i. ***Petromyzon, Acipnsor, Salmon, Oncorhynchus*** etc. make a nest of well-defined structure in the gravelly bottom but destroy the nest after spawning.

j. **Deep-hole nest:** Male African lungfish (*Protopterus*) makes a simple nest as a deep hole in the mud along the river banks. Females lay eggs but the eggs are guarded by the male.

Fig. 34.8: Foamy nest by Fighting fish

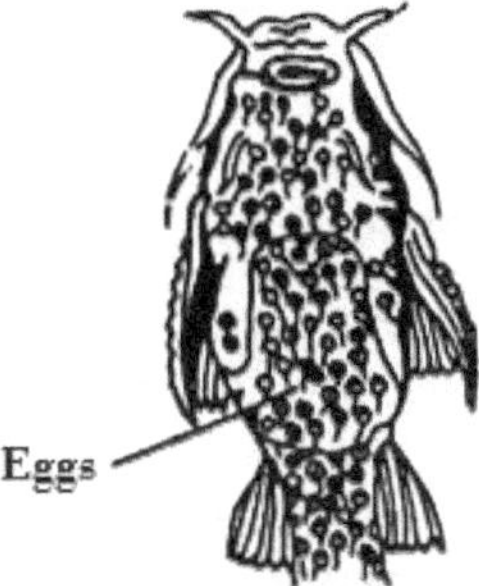

Fig. 34.9: Integumentary cup by Brazilian catfish

34.3.4 Formation of Integumentary Cups/Pouch

a. The ventral surface of Brazilian catfish (*Platystacus*) becomes soft and spongy during the breeding season (Fig. 34.9). No sooner the eggs are fertilized, the female presses her body against the eggs in such a manner that each egg becomes attached to the spongy skin till the hatching by a small stalked cup.

b. In the Brazilian catfish (*Loricaria typus*), the lower lip of the male is enlarged to form a pouch for the incubation of the eggs.

c. Male pipe fish (*Syngnathus*) and male seahorses (*Hippocampus*) are responsible for taking care of eggs till hatching (Fig. 34.10 A and B).

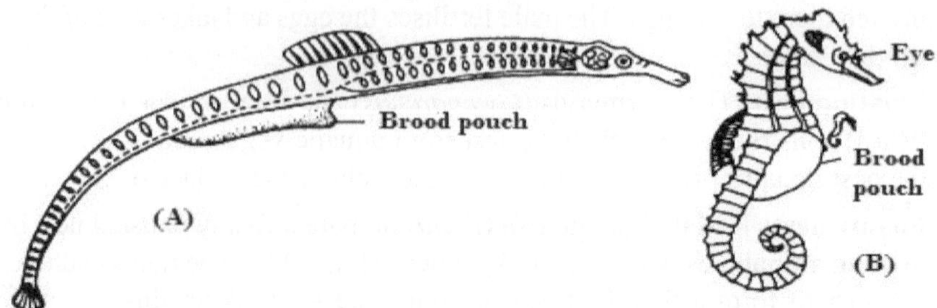

Fig. 34.10: Formation of the integumentary cup by Pipe fish and seahorse

In these fishes, eggs and fries are always undertaken by males in brood pouches present on the lower surface of their abdomen. Females transfer the eggs into the pouch after courtship. During incubation, the pouch remains closed by the flaps of skin.

In solenostomids, the pelvic fins of females form the pouch. Inside, the pouch numerous long filamentous structures are present which keep the eggs in position.

34.4 Ovoviviparity

Females of *Mystus vulgaris* assimilate nutrients (mucoprotein, fat, urea and monosaccharide) from their uterine walls. The embryos take up these nutrients from their proper development.

34.5 Viviparity

To provide the highest degree of parental care, some fish (*Scoliodon*, surf pecrh, *Cymatogaster*, scuplines etc.) have evolved internal incubation (Fig. 34.11 and 34.13). They directly give birth to their young ones. In some species, uterine milk or embryotrophe is secreted by the maternal epithelium for the young ones through the yolk sac placenta.

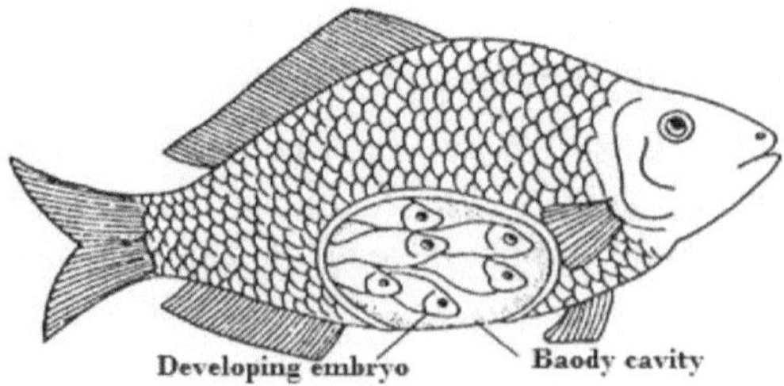

Fig. 34.11: Viviparity *in Cymatogaster*

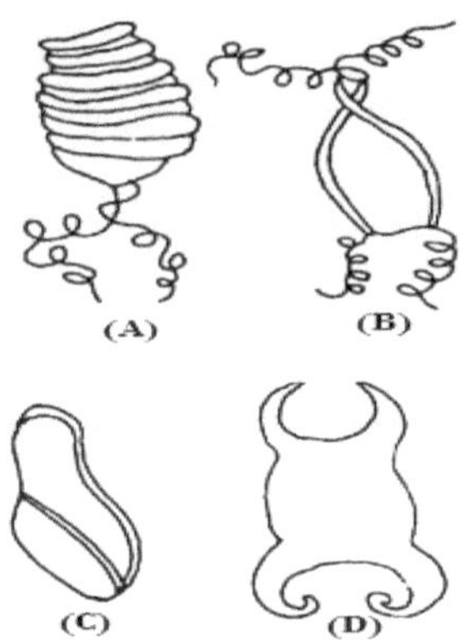

Fig. 34.12: Different types of mermaid purse

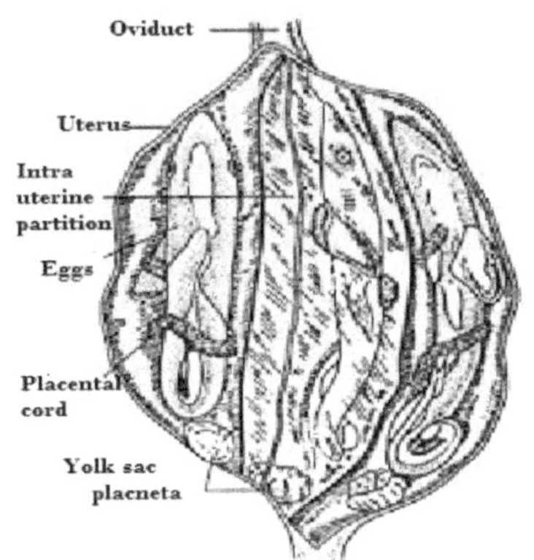

Fig. 34.13: Yolk sac placenta in fishes

More than a dozen families of sharks showed the development of eggs in their uterus. Some sharks have developed a structure for nutritive and respiratory exchange.

In *Gambusia* and *Lebistes*, a pseudo-placenta is formed to nourish young ones.

In Embioticidae, the embryo leaves the follicle and develops in the lumen of the ovary.

34.6 Special Devices

Some sharks and rays produce a special type of protective and rearing cases of various appearances known as a Mermaids purse (Fig. 34.12). This shell is secreted by the shell gland of the oviduct. The corner of the shell bears long twisted elastic filaments that serve to attach the eggs to seaweeds.

34.7 Other Cares

Gasteroesteidae, Centrachidae and Ictaluridae have been actively found to defend their young ones. These fishes place their young ones in a safe place. The hatchlings of Tilapia and pipe fishes are protected by their parents.

Goldberg et al, (2020) suggested that the relatively high prevalence of male-only care in fish may be in part explained by sexual selection through a female preference for caring males.

35

Age and Growth of Fishes

35.1 Introduction

Age and growth studies are vital aspects of modern fisheries because this information is pivotal for fish population and management policies.

Age and growth are important factors in the study of the natural history of fish. Age determination is an ancient practice that aids in systematic studies of fish and fisheries. Age denotes the activities of the organism about its size, whereas growth can be defined as an increase in any dimension of an organism from time or age until maturation. A fish ages as it grows, and vice versa. Fish age determination is important to marine resource management.

Petersen's method and fish tagging provide relevant information on the growth rate, age at maturity, number of spawning periods per life spawn, age at capture, an abundance of year classes, longevity, and mortality rates.

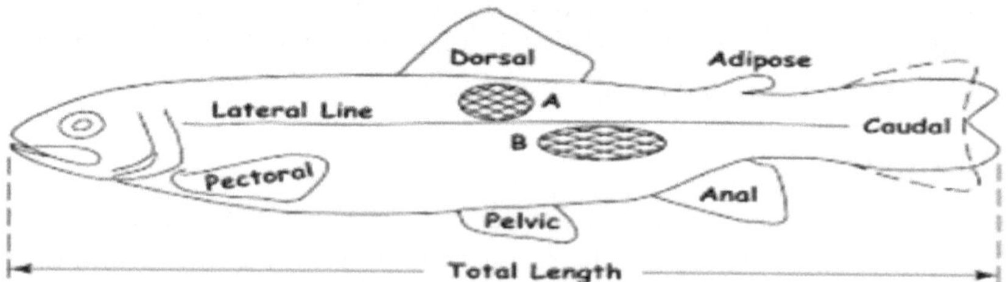

Fig. 35.1: Area for taking scales from a soft-rayed fish (from www. basu.org.in).

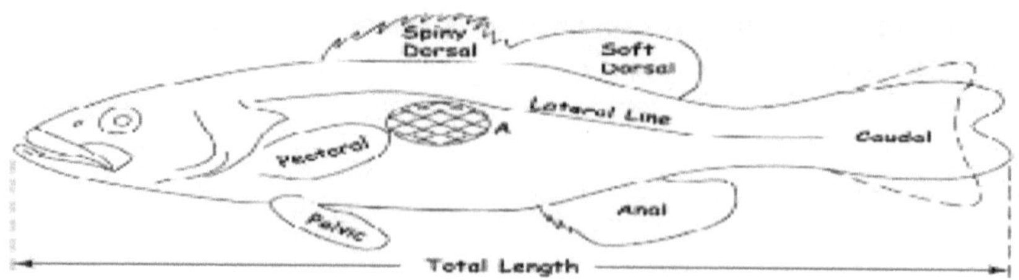

Fig. 35.2: Area for taking scales from a spiny-rayed fish (from www. basu.org.in).

35.2 Methods of Age Determination of Fishes

Methods of Age Determination		
Length frequency method	Known age method	Layers laid down by hard parts of fishes such as Scales, Vertebrae, Otolith, Finrays, Opercula

A. Widely accepted methods:

1. Scale method
2. Otolith method and
3. Bone method

B. Other methods

4. Fin-ray method
5. RNA-DNA ratio method and
6. C^{14} uptake method

The age determined by one method may be checked with other methods such as direct observations of fish reared by random sampling of the population (Fig. 35.3).

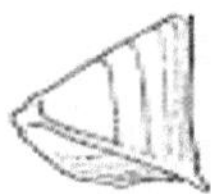

Fig. 35.3: Different parts of body of the fish used in Age Determination

35.2.1 Scale method

Scales are the most complex dermal derivatives (Fig. 35.1 and 35.2). The bony ridge scales are arranged in an overlapping pattern to show the following characteristics:

(a) Focus: The central zone or nucleus that forms first. It could be far from the centre.

(b) Circuli: A series of concentric bony ridges surrounding the focus that act as growth rings or lines of growth.

(c) Redii: These are the grooves radiating from the centre to the edge.

(d) Annuli: These are annual growth marks or growth periods in a year. The number of annuli on the scale represents the fish's age (Fig. 35.4). The distance between the focus and the first annulus is always the greatest, indicating the greatest growth during the first year. The distance between the respective annuli decreases as their number increases.

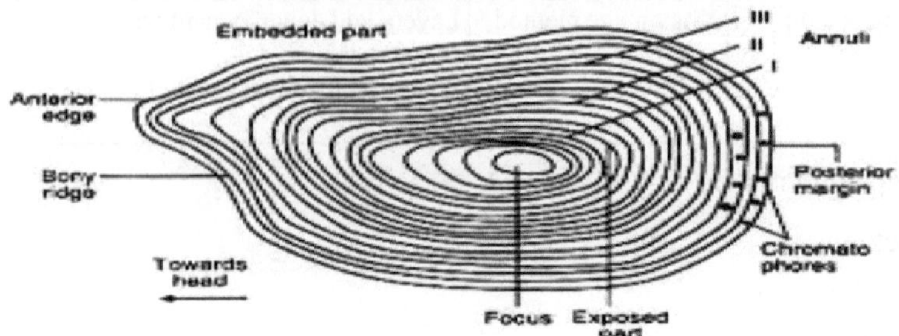

Fig. 35.4: Counting of annuli for Age Determination

The above characteristics represent winter growth in most fish and are systematically laid down once per year as the fish grows. The annuli, in particular, serve as year marks and provide the best rate at the front of the scale. Annuli are classified into two types:

(i) True (=real) annuli: They have a closely spaced circuli zone followed by a widely spaced circuli zone. They are actual sites of heavy ichthylepidin deposition, and their height is determined by the amount of calcification.

Two complete circuli surround the trough. On the antero-lateral axis of scale, it is roughly twice as wide as on the posterior side. The circuli that fall in the wide part of the trough is incomplete, ending on the scale's side and not growing completely around it. Their incompleteness grows in proportion to the concentricity of the focus. That trough narrows at the antero-lateral corners. The outer "cuts across" the incomplete circuli in the troughs of the anterior part.

Annuli formation can be disrupted by abnormal growth patterns; thus, it is essential to compare the observations to the normal standard growth pattern of the particular fish species.

(ii) Falls annuli: They move closer to the true annulus of the preceding years than to the normal annulus of the following year. It may appear as a result of resorption or a cessation or slowing of growth caused by food, starvation, disease, injury, parasitism, spawning, or a temporary temperature fluctuation (Fig. 35.4).

This method has been used around the world by Moghe (1946), Menon (1950), Jhingran (1957), Tandon and Johal (1996) and Bhatia and Dua (2005). Rao (1935), Chacko et al. (1948), Pillay (1951), Das (1959), Quasim and Bhatt (1964), Luther (1973), Dan (1980), Johal and Tandon (1992), Seshappa (1999), and Singh (2009) all worked on scale-based age determination in India. Based on their nature and shape, scales may be of the following types:

a) Ctenoid (posterior part comb-like)
b) Cycloid (smooth and circular) (Fig. 35.5)
c) Placoid (dermal dentacles)
d) Rhomboid (diamond shaped)

Further, based on the chemical composition, scales may be of the following types:

a) Cosmoid (having cosmine)
b) Ganoid (having ganoine and vasodentine)
c) Vitrodentoid (having vitrodentine)

For such purposes, reading the annulus is the most reliable for temperate fish species. Annuli formation can be disrupted by abnormal growth patterns, so it is critical to check and compare the results with the normal standard growth pattern.

According to Nikolaskii (1963), a person's age group is more or less universally accepted.

The presence of additional or supplementary rings complicates age determination. Age estimation is likely to be incorrect if the scales used for age determination are regenerated because regenerated scales represent fewer annuli than normal scales. The correct age of fish can be determined by counting the annual rings rather than the seasonal rings.

Procedure: Typically, 10-12 scales are removed from the region just below the origin of the dorsal fin, which is above the lateral line in soft-rayed fins but below the line in spiny-rayed fins. The scales are shocked in a weak NaOH solution for 10-15 minutes. After that, they are washed and mounted. Temporary mounts of clean scales are created, either wet or dry. Scales are mounted in glycerine during the wet process. Permanent mounts are created using one of two methods:

(I) Impression technique: Strong pressure is used to imprint dry skins on untreated cellulose acetate. Sometimes only light pressure is required after the surface softening of plastic with acetone. The imprints aid in counting the annuli.

(II) Common mounting media include water glass (=sodium silicate), glycerine water glass mixture, euperol gum, gum arabic, polyvinyl alcohol, glycerine gelatin mixture, and others. The glycerine gelatin mixture is the most commonly used of these because it is simple to prepare, less expensive, and relatively permanent.

Scales are sometimes stained with borax carmine, alizarine, or alizarine sulphate.

Table 35.1: Relationship between number of annuli, estimated age and age group.

Sl.No.	Number of annuli on the scale	Estimated age	Designation of age group
1.	None	Less than one year old	0^+
2.	One	One year old	1
3.	One	More than one year but less than two years old	1^+
4.	Two	Two years old	2
5.	Two	More than two years but less than three years old	2^+
6.	Three	Three years old	3

The + indicates the commencement of the growth of the following year.

Much temperate fish, such as carp, salmon, hearings, cods, and others, can be scaled. It is the most widely used method. It also provides historical information for other variables. It cannot be used on fish that live in areas with more or less uniform temperature conditions.

Furthermore, temperature variation, gonad maturity, starvation, low feeding intensity, and food scarcity govern the formation of an annulus on the scales. The abundance of food available during the summer and autumn seasons results in a rapid and widely spaced annulus on the scales.

35.2.2 Otoliths (ear stone or ear bone) method

Lapillus, sagitta, and asteriscus are all found in otoliths. Sagitta is the largest and most oval bone used to determine the age. Otoliths have annulations caused by calcium carbonate deposition. The age of a fish can be determined by observing the annulations.

Procedure: In gobiidae, otoliths are larger massive structures, whereas in cyprinidae, they are smaller. Otoliths are extracted from fresh specimens. Their preservation is hardly necessary. Grinding is usually done with a grinder. It can also be successfully untaken by placing the otolith on the tip of a finger and grinding on the ground or a glass plate with carbarandm powder. Otoliths are sectioned by cutting them in a transverse plate and then examined along the sectional surface. Polishing is usually done with a liquid with a high refractive index, such as immersion oil or creosote. The reflected light is used to examine the otoliths.

The method can be used to determine the age of scaleless fish. The disadvantage of this method is that studying annulations in the otolith necessitates the death of a fish. As a result, in small populations, it is a disadvantageous method.

Otolith images were recently used to train a convolutional neural network for automatically predicting fish age, paving the way for deep learning to require less human effort and expertise.

35.2.3 Bone method

Because of differential calcium carbonate deposition, some fish bones, such as the operculum, vertebral centra, supra-occipital, scapula, dorsal and pectoral spines, coracoids, hyomandibular, and so on, have clear annulations. All of these bony pieces are extracted from fresh specimens. Any form of preservation should be avoided. These bones are thoroughly dried before being examined under reflected light to provide a complete picture of annulations.

The visible rings of the vertebral centra are sometimes useful in determining the age of a fish. 0.7% pepsin in 0.2% HCl is used to remove the tissue attached to the body of the centrum. Under a microscope, the annuli and growth rings on the face of the centrum can be studied.

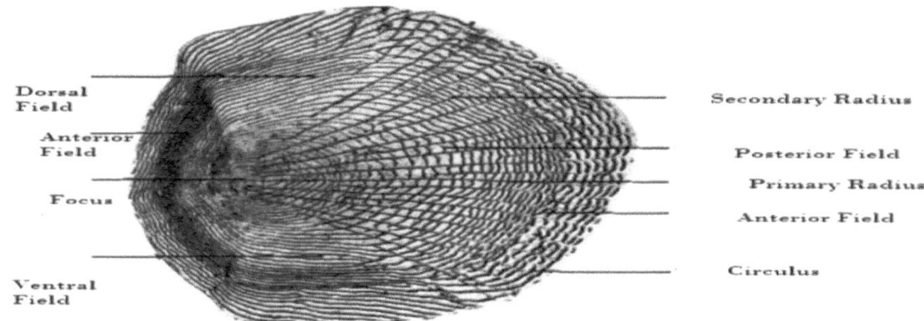

Fig. 35.5: Cycloid Scale (from Van Oosten, 1957).

When opercular bone is used, it is taken from a fresh species and boiled in water for a few minutes to remove any tissue that has become attached to it. The bone is treated for 15 minutes with 50% hydrogen peroxide before being washed and dried in the sun for three days. The annuli of a fish can be studied to determine its age.

35.2.4 Fin Ray method

Thin slices or transverse sections of a pectoral fin's marginal ray are cut into 3-5 thicknesses. The section is immersed in glycerine and viewed through a binocular. Age is determined by comparing 1-2% of indistinct annuli of a fin ray to 15-20% of annuli on the scales.

35.2.5 RNA-DNA ratio method

Protein synthesis rate is measured as an indicator of growth rate (Fig. 35.6). The amount of DNA in a cell is constant, but the amount of RNA varies during protein synthesis. As a result, the DNA-RNA ratio indicates the rate of protein synthesis in a tissue sample. In fish, the ratio correlates with weight gain. However, comparisons can only be made between individuals of the same species.

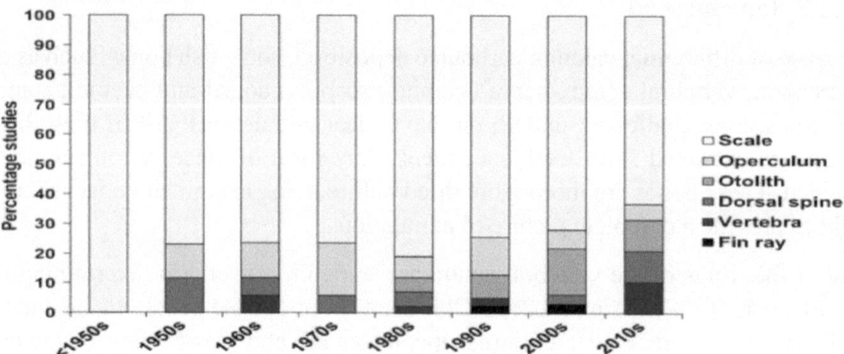

Fig. 35.6: Pattern of age determination (from https://journals.plos.org)

6. C^{14} uptake method: Scales are removed from a live fish and incubated in a C^{14}-glycine-containing medium. After about 4 hours of incubation, the rate at which C^{14}-glycine is incorporated into the scale is measured using -radiations from scales. Increased incorporation of C^{14} by the scales, as measured by the scintillation counter, indicates a faster rate of fish growth.

35.3 Growth of Fishes

Growth can be defined as a bioenergetic process of change in an organism's length and weight over time. It indicates both the individual's and the population's health. The rate of growth varies between fish species and also between species living in different environments. The length and weight of a fish are the two parameters that show its growth. Based on the relationship between two fundamental parameters, the following types of growth are known:

1. **Long-term change**: It includes growth in length.
2. **Seasonal change**: It includes growth in weight.

The growth of fish may be of the following types:

a. **Absolute growth**: It is the highest growth of the fish from the embryonic to the senility period.
b. **Allometric growth**: It is the lopsided growth of various patterns. For example: several fish species grow more in length than the width or weight.
c. **Isometric growth**: It shows an inclination to three perpendicular axes of the cube system.
d. **Relative growth**: It is the comparison from one life period to another.

35.4 Factors Affecting the Growth of Fishes

Several factors such as availability of food, temperature, photoperiod, dissolved oxygen, ammonia, salinity, stages of development of fishes, the density of fishes, crowding, diseases and completions influence the rate of growth of fishes.

1. **Temperature**: Temperature affects the optimum consumption of food. Maximum growth of *Salmon* is reported at 15^{0}C. In many fishes, growth slows from December to February.
2. **Dissolved oxygen**: It is temperature-dependent parameter. It appears that if the dissolved oxygen falls below a certain level, the fish is deprived of the extra energy required for growth.
3. **Ammonia**: The high concentration of ammonia slows down the growth rate of fish.
4. **Population density**: It influences the growth rate due to competition for available food. Higher density slows down the rate of growth while lower one increases it.
5. **Availability of food**: It is temperature dependent. Growth is rapid during warmer months when plenty of food is available and slows down during winter.
6. **Maturity of fish**: A young fish grows more compared to a mature fish.

35.5 Methods of Growth Determination of Fishes

35.5.1 Age Method

The larvae of the fish are placed in a rearing tank and left for 2-3 seasons. The periodic samples are collected, and the growth rate curve is calculated by measuring growth. Used fish are marked or tagged and released back into the wild.

Fish marking can be accomplished by clipping the fin rays, spray painting the skin, or placing fluorescent rings on the scales or bones. Certain dyes, such as chromium oxide, fast blue, lead acetate, or latex, can be injected into the fish to help identify it after recapturing it.

Fish are recaptured at regular intervals to determine the rate of growth over time. Other methods may be used to validate the value.

35.5.2 Length-frequency Method

Seasonal breeding fishes are made up of distinct stocks. Each fish in the sample is measured and an average is calculated after obtaining a sample that is a fair representation of the population. A distinct peak corresponding to each age group is recorded. Following these peaks through the seasons provides a clear picture of the growth. Peterson pioneered the technique.

Prerequisites

1. The fish should be as large as possible.
2. The sample should be collected within a specific time frame.
3. The sample must be representative.

Limitations

1. It applies to fish with a long life span but at a young age.
2. Individual growth rate variations should be identical.

35.5.3 Corroborative Methods

A. **Agreement with size frequency distribution**: The size of a fish is determined corresponding to the various lengths and the mean length of each year is calculated. This is an identical method.
B. **Seasonal record of zone formation**: If opaque and translucent zones are laid down in one year, a close examination of the growing periphery indicates the period of each zone formation throughout the year.
C. **Observations of the fishes of known age**: Fishes of known age are kept in the aquarium and their structures are examined periodically.

35.6 Relationship Between the Structure used and the Size of the Fish

The structure e.g., otolith grows along with the size of the fish; a relationship between the two can be established. Uniform growth is known is also known as isogenic growth. If the growth is constant and uniform, then:

$$L_x \quad - \times L_y$$

Where

L_x = Growth of the body of the fish

L_y = Final length of the fish

l_x = Length of the annual zone and

l_y = Final length of a structure (e.g., otolith)

The growth not proportional to the size of the fish is regarded as heterogenic growth.

The relationship follows the law of the following compound interest:

$$L \propto l$$

or $$L = kl$$

or $$\text{LogL} = \text{logk logl}$$

Where, k = Constant coefficient

35.7 Relationship between the body length and scale length

The following formula may be used to depict the relationship

logL = loga + b logS

Where, L = body length of the fish

a = intercept on the straight line on the axis of ordinate

b = slope

S = scale length

35.8 Determination of fish length

Measuring the radius from the focus of the scale to each annulus, the length of the fish at each year can be determined using the following formula:

$$Ln = a + \frac{(L + a)(Dn)}{Dr}$$

Where

Ln = calculated length of at n year

L = length of fish at the time of capture

Dn = distance from the focus to the nth annulus

a = intercept that often approximates the fish length at the time of scale formation

36

Fish-Migration

36.1 Introduction

Migration refers to the long-distance movement of fish from one location to another in search of food or breeding. It involves a temporary or permanent absence from one's home range and the establishment of a residence in another. It is one of the most remarkable and fascinating characteristics of fish. It is the mass movement of fish from one body of water to another. It entails longer-duration movements of schools of fish than those seen during normal daily activities.

36.2 Definition

It has been defined variously:

1. According to Heape (1931), migration is a type of movement in which migrants are compelled to return to the region from which they migrated. He also used the term impel to describe biological necessity.
2. According to Thompson (1952), migration is a seasonal movement that compels a return to the starting point. These fish have a variety of morphological, behavioural, and physiological adaptations.
3. According to Jones (1968), migration is a type of movement in which migrants are compelled to return to the region from which they migrated.
4. According to Baker (1978) Migration is the act of moving from one spatial unit to another. The return journey is not restricted.
5. According to Dingle (1980), Migration is a specialised behaviour evolved for the displacement of an individual in space.
6. According to Lack "Migration" is defined as "the regular, seasonal, large-scale, long-distance movement of a population twice a year between a fixed breeding and non-breeding area".

Article 64 of the United Nations Convention on the Law of the Sea (UNCLOS) defines highly migratory species (HMS). The list includes: tuna and tuna-like species (albacore or longfin tuna (*Thunnus alalunga*), bluefin tuna (*Thunnus thunnus*), bigeye tuna (*Thunnus obesus*), skipjack tuna (*Katsuwonus pelamis*), yellowfin tuna (*Thunnus albacares*), blackfin tuna (*Thunnus atlanticus*), little tunny (*Euthynnus alletteratus*), southern bluefin (*Thunnus maccoyii*), and bullet (*Auxis rochei*)) etc.

Wang et al. (2019) discovered additional potential Myxovirus resistance (Mx) genes in rainbow trout. Mx proteins are members of a GTPase family that aid in viral immunity and rainbow trout (*Oncorhynchus mykiss*) have been found to have three distinct Mx genes that aid in viral defense in both environments.

36.3 Examples of Migratory Fish

1. *Anguilla anguilla* (the European eel)
2. *Anguilla australis* (the Australian eel)
3. *Anguilla japonica* (the Japanese eel)
4. *Anguilla rostrata* (the American eel)
5. *Clupea harengus* (the herring)
6. *Gadus morhua* (the cod)
7. *Gasterosteus aculeatus* (the three-spined stickleback)
8. *Hilsa ilisha* (Hilsa)
9. *Lampetra fluviatilis* (the Lamprey)
10. *Oncorhynchus* spp. (the Pacific salmon)
11. *Petromyzon marinus* (the Lamprey)
12. *Pleuronectes platessa* (the flat fish)
13. *Salmo* spp. (the salmon)
14. *Scomber* spp. (the mackerel).
15. *Thunnus thynnus* (the Tunas)

36.4 Types

Based on different characteristics, migration has been classified variously (Fig. 36.1):

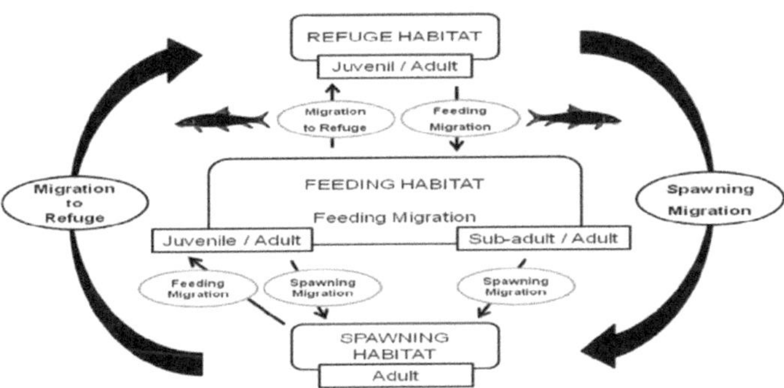

Fig. 36.1: Scheme of fish migration (from Lucas and Baras, 2001)

36.4.1 Based on the direction of water movement

1. Vertical migration: It is a daily migration of fish from the deep to the surface for food, protection, and spawning. Bull sharks (*Carcharhinus leucas*), for example, move vertically downward to greater depths in search of food.
2. Horizontal migration refers to movement from upstream to downstream. Example: Swordfish.
3. Latitudinal migration: It refers to the climatic movement of fish from north to south and vice versa. Swordfish, for example, migrate north in the spring and south in the autumn.
4. Shoreward migration: It is the movement of fish from water to land. However, this is only a temporary relocation. Eels, for example, migrate from one pond to another via moist meadow grass.

36.4.2 Based on the direction of water flow

1. Denatant migration: The movement along with the water current. e.g., pelagic eggs.
2. Contranatant migration: The movement against the flow of water, as in all adult fishes.

36.4.3 Based on purpose or need

1. Alimental migration, also known as feeding migration is movement in search of food and water when food sources are depleted. Example: Salmons, cods and swordfish.
2. Gametic or spawning migration: Movement during the breeding season in search of reproduction. Example: American eel and European eel.
3. Seasonal or climatic migration: Movement in search of a more secure and suitable climate.
4. Osmoregulatory migration: Movement in search of osmoregulation or water and salt balance.
5. Juvenile migration: Larval migration from spawning grounds to the feeding habitats of their parents.

36.4.4 Based on the river body (Myer, 1949)

1. Diadromous migration: According to Myers (1949), movement of known 444 species between river and sea (Fig. 36.2). It may be one of the following types:

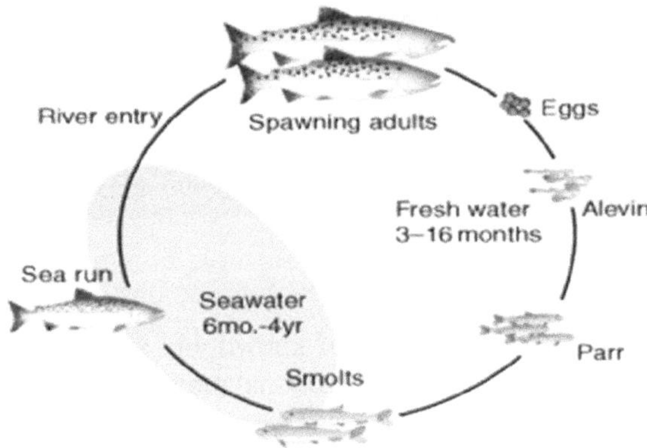

Fig. 36.2: Diadromous migration (from https://www.sciencedirect.com)

(a) Anadromous migration: Migration of known 147 species from sea to the river for spawning (Fig. 36.3a). e.g., Sturgeon (*Acipenser*), Salmon (*Salmo solar*), Hilsa (*Hilsa ilisha*), Shad, Sea lamprey (*Petromyzon marinus*), Pacific salmon (*Oncorhynchus nerka*), Striped bass (*Morone saxatilis*).

(b) Catadromous migration: Migration of 73 known species of fish from the river to the sea for spawning (Fig. 36.3d). e.g., European or freshwater eels (*Anguilla Anguilla, Anguilla vulgaris*) and American eels (*Anguilla rostrata*).

(c) Amphidromous migration: Migration of 224 known species of fish from the river to sea and vice versa, not necessary for breeding (Fig. 36.3c). e.g., gobies. Fish of orders Clureiformes and Mugiliformes show diadromous, anadromous, catadromous and amphidromous migration.

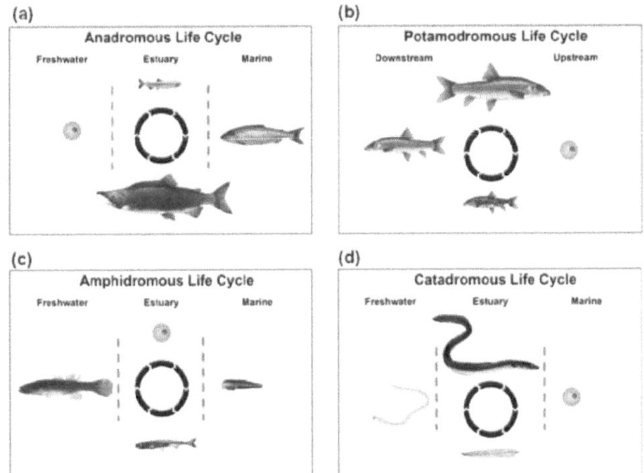

Fig. 36.3: Different types of migration (from https://www.researchgate.net)

2. Potanodromous migration: It is the freshwater migration of fish from one habitat to another for feeding or spawning (Fig. 36.3b). e.g., carp, trout and catfish.
3. Oceanodromous migration: It is the migration of fish within the sea in search of suitable feeding and spawning grounds. e.g., mackerels (*Scomber*), Tunas (*Thunnus*) and Herrings (*Clupea*).

36.4.5 Based on the movement of fishes

1. Drifting migration: This is a passive movement of known 444 species of fish along with water currents that can result in directional movement if the water movement is only one way. Example: American eel and European eel.
2. Random migration: A movement that results in a uniform distribution or aggregation.
3. Dispersal movement: This is a random locomotory movement of fish from a consistent habitat to a random direction.
4. Swimming movement: This is the fish-oriented movement toward or away from a stimulus source.
5. Oriented migration: It includes swimming in a particular direction. It may be:
   (a) Towards or away from the stimulus source.
   (b) A slant to the imaginary line that connects them to the source of stimulation.

36.5 Periodicity

Migration may be daily, monthly, seasonal, yearly, biannually, or longer; for example, Pacific salmon (*Oncorhynchus nerka*) may spend several years at sea before returning to spawning grounds. *Petromyzon marinus* larvae may spend several years in the mud before metamorphosis and migration to the sea. The eel (*Anguilla anguilla*) can spend up to 20 years feeding in freshwater before migrating to the sea to spawn. Thus, periodicity varies between species and even within species, and it appears to be caused by a variety of biotic and abiotic initiating factors (Fig. 36.4).

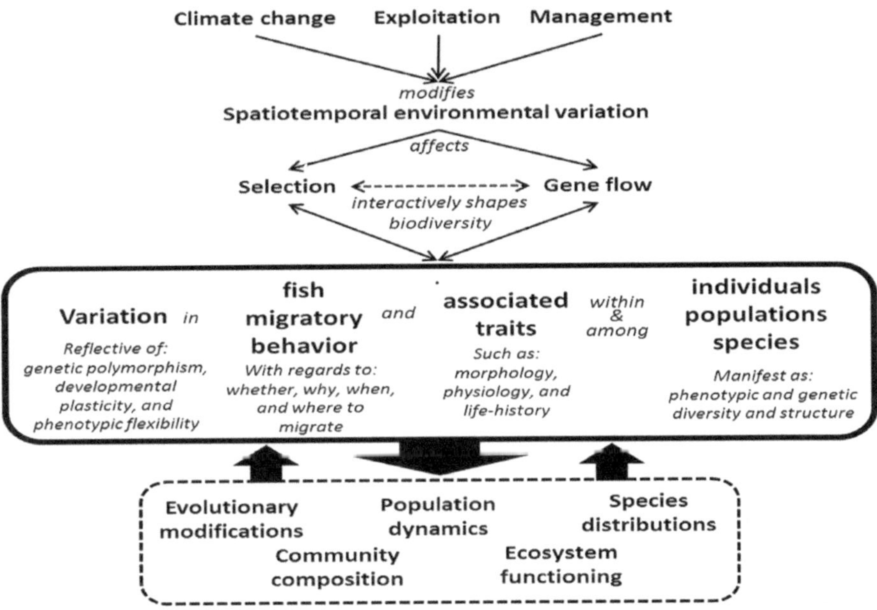

Fig. 36.4: Periodicity of migration (From Tamario et al, 2019)

36.6 Techniques/Methods of Study

Marking and tagging techniques, followed by recapture are used to study various aspects of fish migration, such as pattern, direction, and speed of movement. To study fish migration, one of the following methods is used:

1. Fish fin clipping is a quick and easy way to identify a limited number of individuals.
2. Removal of the fin as a marking technique is no longer useful due to fin regeneration or loss due to accidents or other causes.
3. Metal tools can be used to tag fish.
4. A fluorescent dye that becomes embedded in the scales of fish can be sprayed on them. Only UV radiation makes it visible.
5. Tetracycline has the potential to be used as a marker. It is given orally, by injection or by immersion because it does not affect fish growth. It can be seen through a fluorescent microscope.
6. Some water-soluble radioisotopes can be used as markers by immersing the fish or administering them orally.
7. A small piece of magnetised stainless steel wire (10.25mm) is injected into the nose of the fish. To demonstrate the presence of magnetised wire, adipose tissue is removed.
8. To identify the fish, various types of tags, such as small metallic clips, discs and celluloid have been developed.

External tags: (a) Mutilation: adipose fin missing (clipping or punching). (b) Peterson (1894) Discs: the most successful of all tags. This tag is made up of two half-inch celluloid or plastic discs that are attached to the fish with a pin or wire. (c) Carlin Darter Tag: plastic disc with steel wire. (d) Visible Implant Elastomer (VIE) Tag: injected as a liquid, solidifies and becomes transparent. (e) Floy Tag: adult migratory fish with a T-bar hook that interlocks with the skeleton (by gun).

Internal tags: (a) Radio Tag: sends radio signals in shallow conductivity water. (b) Sonar Tag: This is a hydrostatic tag. (c) Coded-wire Tag: attached to the snout or neck and detected by a metal detector.

These procedures require the catching, handling and releasing of fish which may cause death due to stress etc. Hence, tags are good markers for a short period.

The study of meristic and meristematic characters of scales and otoliths is also used to study fish migration. Besides, the following methods are also used:

1. Analysis of fishery statistics
2. Biochemical methods
3. Direct observation and
4. Echo-ranging techniques

36.7 Examples

36.7.1 Migration of Salmon

Atlantic salmon (*Salmon salar*) spawns during November and December. It represents an example of anadromous migration (Fig. 36.5). They ascend the rivers and stop feeding during this time. The silvery colour is replaced by a dull-reddish brown tint, and the skin on the back thickens. Males become spotted with red, orange, and large black spots (redfish), while females darken to become blackfish. After deciding on a suitable spawning location, the salmon form pairs and build a shallow, saucer-like nest where the spawning occurs. The fertilised eggs sink and the females cover them loosely with fine gravel. Following the process, the fish return to the sea to repeat the cycle.

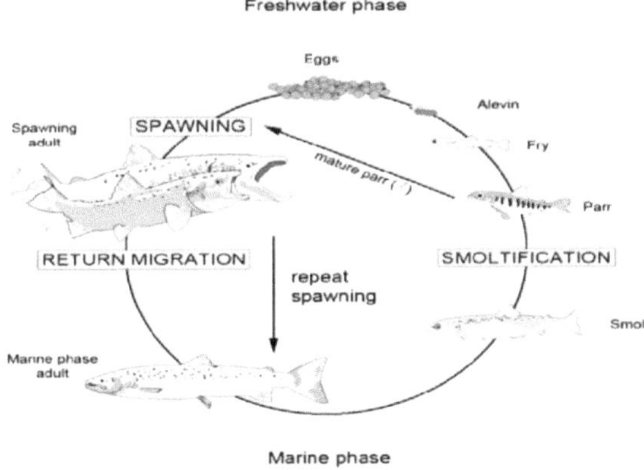

Fig. 36.5: Migration of Atlantic Salmon (from https://link.springer.com)

The king salmon of North America's Pacific coast travels approximately 3600 kilometres from sea to river to spawn. However, after spawning, they all perish. Young salmon that hatch in freshwater are unable to return to the sea until salt-secreting glands develop. Salmon migration is most likely caused by changes in environmental conditions. Hasler and Larson (1954) believed that a fish's sense of smell is important in locating its home grounds because different streams have different odours.

36.7.2 Migration of Eel

J. Schmidt (1922) described eel migration to explain catadromous migration (Fig. 36.6). Eels have two morphologically distinct phases in their life history:

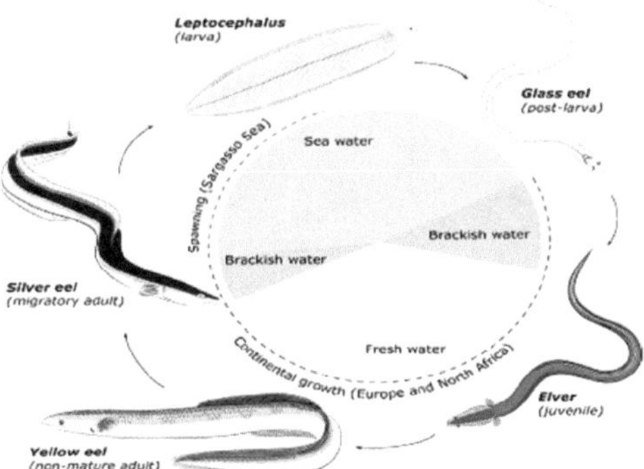

Fig. 36.6: Migration of Eel (from https://onlinelibrary.wiley.com)

(a) Yellow eel or feeding and spawning eels.

(b) Metallic silver eel or breeding eel.

During the autumn season, yellow eels transform into silver eels and prepare to migrate to spawning grounds. During this transformation, the yellow eel stops feeding, its lips become thinner, its alimentary canal shrinks, its pectoral fins become more pointed, its eyes become greatly enlarged, its snout becomes sharper, its gonads fully mature, and its yellow coloration is replaced by a silvery one. Their length ranges from 1.5 to 1.65m.

The silver eel has mature gonads and a narrowed alimentary canal. It travels down rivers to reach the sea. The eel travels approximately 4800-6400 kilometres. After the spawning process is complete, the parent dies. Tucker (1959) proposed that adult European eels die before spawning and that eggs are laid by American forms.

The eggs hatch into tiny, transparent larvae called Leptocphali. These are flat, leaf-like creatures with long, needle-like teeth for hatching. They begin their journey home and eventually metamorphose into elevers. They are approximately 8-9cm long at this stage. When the elevers reach about 3 years of age, they begin ascending the rivers in shoals in search of a suitable resting place. Elvers feed and grow here for years before becoming yellow eels. When the yellow eels reach maturity, they begin to migrate.

36.8 Adaptations or Factors Affecting Fish Migration

Many fish travel long distances to reach their destination, encountering various obstacles and environmental conditions along the way. Various biological, chemical, and physical factors influence fish migration, as described below (Fig. 36.7):

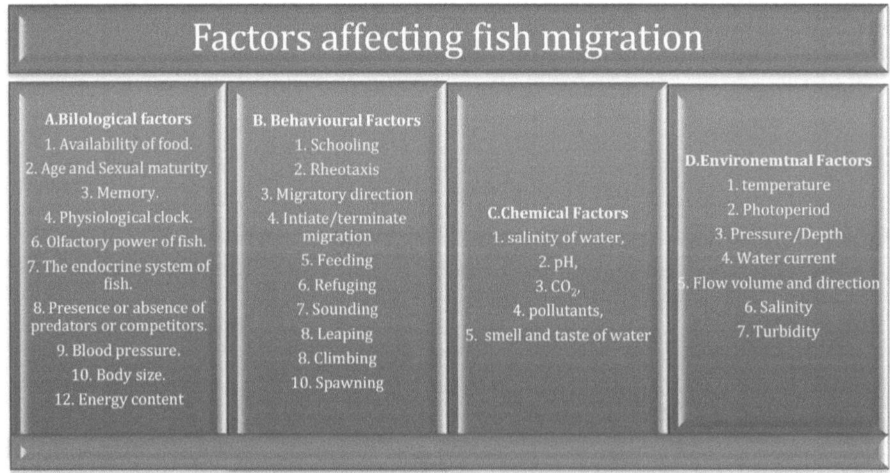

Fig. 36.7: Factors affecting fish migration

36.8.1 Biological factors

These are: 1. Availability of food. 2. Sexual maturity. 3. Memory. 4. Physiological clock. 6. Olfactory power of fish. 7. The endocrine system of fish. 8. Presence or absence of predators or competitors. 9. Blood pressure.

(a) Availability of food: It is one of the most important factors causing large-scale migration of many fish species in search of feeding areas.

(b) Smell and memory appear to guide fish during migration as well. Salmon have been shown to return to the same area of the river as adults for spawning and where they hatch and develop. Hasler and Larson (1954) supported the theory that olfaction is a determining factor in fish home range location.

(c) Endocrine system of fish: Thyroid and inter-renal activity increases during migration. Hoar demonstrated that thyroxin treatment of young salmon increased their tendency to remain in a fixed position concerning the bottom. Hormones (prolactin) play a direct or indirect role in migration.

(d) Orientation: Hasler (1970) and Able (1971) reviewed the concept of orientation (1980). A fish may recognise its home site through direct sensory stimulation, such as vision or olfaction (Homing or Piloting). Navigation is the process by which an animal finds its way to a specific location. During migration, a fish can use the sun to help it find its way.

(e) The stage of maturation of the gonads as well as the state of the endocrine systems is an important factor governing migration.

36.8.2 Chemical factors

These include salinity of water, pH, CO_2, pollutants, the smell and taste of water etc as discussed below:

1. Salinity of water: Most freshwater fishes are relatively intolerant of salinity changes (Stenohaline) and do not migrate in large numbers, limiting themselves to freshwater only. However, a few species, such as *Slamon, Anguilla, Hilsa, Gasteroesteus* and *Fundulus* can adapt to salinity changes (euryhaline) and migrate on a large scale. As a result, it appears to be an important factor, and fish movement is restricted within the range of their salinity tolerance.
2. pH: Violent pH changes are caused by chemicals, effluents, or sewage containing a high concentration of acidic and alkaline compounds, which may cause fish to migrate.
3. CO_2: High CO_2 concentrations can reflect fish. The CO_2 gradient may be guiding salmon.
4. Smell and taste of water: Some species have been reported to use water currents as orientation cues. The fish's olfactory sense is thought to play an

important role in locating the parent stream, and it can detect the smell of the parent stream from a long distance away.

36.8.3 Physical factors

Temperature, light intensity and photoperiod, orientation, bottom materials, water depth, pressure, current, tides and turbidity are important physical factors that influence migration:

1. Temperature: The warmer seawater in the summer encourages salmon to migrate to the sea. Similarly, rising river temperatures cause spawning upstream.
2. Light intensity and photoperiod: Light intensity and duration affect migration. Some fish are attracted to light and can be trapped by shining it in the right places. Sturgeons and lampreys (*Petromyzon marinus*) migrate at night. During the full moon, herrings (*Clupea harengus*) migrate.
3. The direction of movement of fishes is greatly influenced by the water current. Eggs and fry are passively transported to their feeding grounds by the current. After spawning, spent *Salmon* are carried downstream by river currents to the sea.

All of these factors interact in various ways and a few or several of them including homing tendency and sense of direction may be responsible for creating a strong urge to migrate.

4. Homing instincts: These include:

(a) Aquarium, stream, tank experiment suggests that topographical clues are used by fishes to recognise home range.

(b) In some, Gunnings (19590 showed that blind fish use another sensory channel (probably acoustic lateralis system) for homing.

(c) Idler (1961) found active components in the home stream water which are volatile, liable and natural.

(d) Hora (1965) showed that water from different sources produces notably different brass wave patterns in *Salmon* based on the electro-olfactory graphic studies.

(e) Halser (1966) gives 'olfactory theory' as a migratory guide.

(f) It has been seen by experimentation that plants are capable of imprinting their individual achromatic properties to water. These properties are detectable by the olfactory organs of fishes.

These factors initiate migration and guide the fish in finding its way and locating its destination.

36.9 Importance

Migration is an adaptation in which abundance is preferred as discussed below:

1. Spawning or nursery grounds may not have enough food to maintain both mature and immature members of a large population.
2. A better egg and larvae survival should lead to a greater number of spawners on particular grounds.
3. Results in precise timing or environmental conditions of spawning behaviour.
4. It helps to separate spawning, nursery and feeding grounds.
5. It protects from predators
7. It survives extreme climatic conditions
8. It increases genetic diversity
9. It is an adaptation for survival and existence.

Unit-V Threats to Fish Population

37

Effect of Climate Change on Fishes

37.1 Introduction

Temperature describes the kinetic energy of the gases that make up air. As gas molecules move more quickly, air temperature increases. Among several environmental changes, increased air temperature affects the water cycle by increasing water temperature, which causes changes in local weather, water quality and river bodies (Gossiaux *et al.*, 2019). This is because the water temperature is one of the most important factors for physicochemical properties of water quality (Szumiriska *et al.*, 2020), aquatic habitats (Duggals *et al.*, 2018), spawning rates (Besson *et al.*, 2016) and fish growth (Collas *et al.*, 2019) (Fig. 37.1).

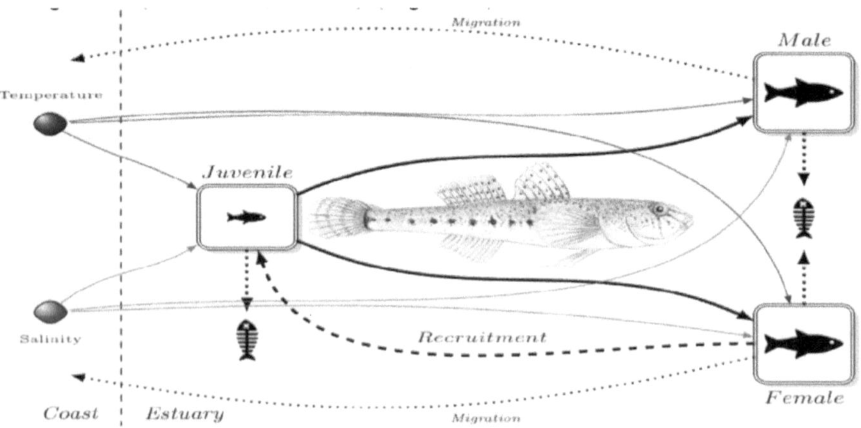

Fig. 37.1: Effect of Climate Change on Fishes (https://climefish.eu)

The average global temperature of 13.73⁰C has risen by 0.08°C per decade since 1880, and the rate of warming over the past 40 years is more than twice by 0.18°C per decade since 1981. With this increase, the present global temperature becomes 15.25⁰C. In the late 20th century, global warming has been occurring since industrialization and numerous unprecedented climatic changes have been observed for the first time in decades or centuries (Stocker *et al.*, 2013). The temperature drops by 5-6.5°C/km with rising height in the lower part of the atmosphere but cools by 10-13°C than a place at the same latitude of the sea level.

If global temperature rises by 5°C, almost 2/3rd of fish species could be eradicated by 2100AD. An increase of 3.2°C in global mean temperature would threaten less than 50% of the habitat of fish. It has been estimated that the east coast of India will reduce by 25% in 25 years and will result in a cumulative loss of US$17 billion. Over the next 40 years, the temperature in the Indian seas is expected to rise by 1-3°C. The oceans are projected to acidify; sea levels and currents are expected to change.

Climate change has a significant negative effect on society (Deepananda and Macusi, 2012). Climate change is a key factor controlling ecosystem structure and the distribution and abundance of aquatic organisms (Jena and Gopalakrishnan, 2012). The climate change impact on fish depends on the magnitude of change and the sensitivity of particular species or ecosystems. The impact of temperature shifts due to climate change on aquatic organisms will affect their biological functions, as most of them are poikilothermic. The report of the Intergovernmental Panel on Climate Change (IPCC, 2007) recognizes that human-induced warming temperature in lakes and rivers poses serious threats to various fish species and fish culture production (Ficke *et al.*, 2007; Cheung *et al.*, 2010).

Fish spend their entire lives in water and play an important role in aquatic biogeochemical processes, ecosystem structuring, and functioning via food web links (Fricke *et al.*, 2020). Because of their high vulnerability to climate change-induced environmental changes, wild fish diversity and global fisheries face an uncertain future (Free *et al.*, 2019).

In many areas, various fish species are living at the upper end of their thermal range and ecological models demonstrate that there would be significant losses of temperate fish species as climate change may lead to a reduction of fish habitat and diversity (Mohseni *et al.*, 2003). In temperate regions, coldwater fishes are important ecological indicators for climate change as they are very sensitive to changes in water temperature and other environmental conditions. The Himalayan coldwater fish species may be at the highest risk of global warming as many of them are endangered.

37.2 Climate Change and Fishes

37.2.1 Water temperature

Water temperature depends upon climate, sunlight and depth. The optimum water temperature ranges between 20-32°C. Water temperature is a limiting factor controlling the distribution of fish. The metabolic rate of fish doubles for every 7.78°C rise in temperature. In general, temperatures less than 16.7°C and more than 39.5°C are fatal to fish. Fishes may be:

1. Air-breathing fishes: *Anabas testudineus, Channa punctatus, Clarias batrachus, Heteropneustes fossilis* (39-41°C).
2. Coldwater fishes: The body temperature of coldwater fishes depends upon the temperature of the water body in which the fish live. yellow perch 0-20°C
3. Eurythermal fishes: These fishes display great variability in their tolerance to temperature. *Tilapia mossambica* (25-35°C) and *Anguilla japonica* (20-28°C).
4. Indian catfishes: 30°C
5. Indian major carps: *Labeo rohita, Catla catla* (18-32.8°C)
6. Stenothermal fishes: These fishes can tolerate a narrow range of temperature variations. Goldfish, Brook trout (*Salivelinus fontinalis*) (13-18°C).
7. Warm water fishes: 25-32°C

An increase in water temperature leads to the following changes in fishes

1. An increase in metabolic rate and earlier maturation by which respiration will increase which needs more dissolved oxygen.
2. Changes in the feeding habits
3. Changes in the size of the fish.
4. Production of more toxins by algal blooms
5. The toxins can stress or kill fish by clogging their gills or decreasing dissolved oxygen in the water.
6. Water temperature gradually increased from 25°C to 32°C over several hours and did not harm fish.

Studies have shown a direct relationship between metabolic rates and water temperature. This occurs as many cellular enzymes are more active at higher temperatures (Wetzell, 2001). Most living organisms perform their activities in a temperature range of 4-45°C. Moreover, according to Van't Hoff's rule, 10°C increases in water temperature will approximately double the rate of physiological function. Furthermore, temperatures above 35°C can begin to denature, or break down enzymes, reducing metabolic function ((Dash, 2006; Bennett and DeSanto, 2011). Enzyme action is temperature-dependent. Usually, most enzymes are inactivated above 35-40°C, while below 10°C enzymatic activities are at a minimum (Dash, 2006).

37.2.2 Salinity

The concentration of dissolved salt in water is called salinity. Its amount increases because of the increase in density with the increase in water temperature. Fish living in freshwater (1mg/l) cannot tolerate increased salinities (Senese, 2010).

On the other hand, *Hypophthalmichthys molitrix* and *Chanos chanos* show normal survival and growth at a salinity concentration of 8000 and 32000 mg/l respectively.

1. Euryhaline fishes: These fishes have a great tolerance to salinity. It constitutes 3-5% of all fish species. Example: *Poecilia sphenops*, Salmon, Eel, Herring.
2. Stenohaline fishes: These fishes have a narrow tolerance to salinity. Most fishes are stenohaline. Example *Carassius auratus*.

Fishes are highly sensitive to changes in salinity. Since warmer water can hold more salt and other molecules than cold water; it can have a higher salinity. In Polar Regions, changes in salinity affect ocean density more than temperature changes. Because saltwater is heavier, the density of the water increases and the water sinks (Hogan, 2012).

37.2.3 Turbidity

Turbidity is the haziness of fluid because of suspended solid dirt and other particles caused by large numbers of individual particles generally invisible to the naked eye. Turbidity can increase water temperature because suspended particles absorb more heat. Climate change leads to a doubling of turbidity and harms fish by reducing food supplies, degrading spawning beds and affecting gill formation (EPA, 2012).

Turbidity during heavy floods increases to 2000-5000mg/l in the river Ganga. Very high turbidity like 175000 mg/l and above causes fish mortality by gill clogging. High turbidity increases water temperature due to the particles absorbing sunlight. Higher temperatures of water result in less oxygen content, leading to hypoxic conditions (EPA, 2012).

37.2.4 pH

pH measures the acidity or basicity of aqueous or other liquid solutions. pH range of 5-10 is generally acceptable for fish culture (Wurts, 2012). Carps may survive at pH 10.8. But, more than 8.5 seems unproductive and more than 11 is fatal to fish. Due to rapid global warming, the pH will have dropped dramatically by 2100. But this does not mean that water becomes more acidic at higher temperatures. A solution is considered acidic if there is an excess of H^+ over OH^- (EPA, 2012). Ocean acidification occurs when the pH of water drops, due to the absorption of CO_2 from the atmosphere. Due to low pH, the solubility of $CaCO_3$ is reduced and respiration is affected at pH less than 5.

pH of freshwater varies from 6.5 to 10.5 (Dash, 2006). Very high pH (greater than 9.5) or very low pH (lower than 4.5) values are unsuitable for most aquatic organisms. Fish begin to die when pH falls below 4.0 (EPA, 2012). Aquatic organisms are extremely sensitive to pH levels below 5 and may die at these low pH values. High pH levels (9-14) can harm fish because ammonia will turn to toxic ammonia at a pH of more than 9 (Kumar and Pun, 2012).

37.2.5 DO

Dissolved oxygen (DO) is the amount of oxygen present in the intermolecular space of the water. DO is produced as a waste product of photosynthesis (Eissa and Zaki, 2011). Its value is 14.62 mg/l at 0°C and 7.04 mg/l at 35°C. DO of less than 1 mg/l leads to the death of fish and more than 15 mg/l causes gas bubble disease in fish. A concentration of 5 mg/L DO is optimum for fish health. Most species of fish are distressed when DO falls to 2-4 mg/L. Decreased dissolved oxygen also pushes fish to shallow water or fish can drown. *Ctenopharyngodon idella* and *Carrasius auratus* die at DO of 02-0.6 and 0.1-2.0 mg/l respectively (Pandey and Shukla, 2005)

The solubility of oxygen decreases with increasing temperature (10.15 mg/l at 15^{0}C to 7.1 mg/l at 35^{0}C). As a result, DO levels in a river are higher during the winter months than during summer. The change in dissolved oxygen with temperature is small compared with the larger change in corrosion rates. Warm water holds less dissolved oxygen than cool water and may not contain enough dissolved oxygen for the survival of different species of aquatic life.

37.2.6 BOD

Biological Oxygen demand (BOD) is a measure of the amount of oxygen required to remove waste organic matter from water during decomposition by aerobic bacteria. The amount of oxygen that can dissolve in water (DO) depends on temperature. The bacteria require oxygen for this process so BOD is high at this location. BOD of less than 50 mg/l is optimum for water bodies (Hernandez *et al.*, 2010).

Increased water temperature will speed up bacterial decomposition and result in higher BOD levels up to 50°C. Warmer water usually will have a higher BOD level than colder water. As water temperature increases, the rate of photosynthesis by algae and other plants in the water bodies also increases. The impact of rising water temperature on BOD depends on the interplay of the DO saturation and the self-purification rate of the water body (Chapra *et al.*, 2021).

37.2.7 Free CO_2

Free CO_2 is present in water as a dissolved gas. Air contains 0.03% FCO_2, which is essential for photosynthesis (Dash, 2006). It dissolves in water 200 times more easily than oxygen. The solubility of CO_2 goes down as water temperature goes up! This results in its increase with the increase of water temperature. Its value ranges from 0-16 mg/l in freshwater bodies. Less than 5mg/l of free CO□ supports the fish population. More than 20mg/l of free CO□ may infer oxygen utilization by the fish. Most fishes may survive in water containing up to 60mg/l of free CO_2 provided DO concentrations are very high (Pandey and Shukla, 2005). Surface water normally contains less than 10mg/l of free CO_2, while some groundwater may exceed that concentration.

The IPCC (2013) estimates that by the year 2100 average mean temperature is to increase by 4.8°C over the 1996–2005 average, and that CO_2 levels will reach ~1000 ppmv.

37.2.8 Total hardness

Total hardness is a measurement of the mineral (calcium, magnesium, iron, aluminium, zinc, manganese, strontium) in water that is irreversible by boiling and expressed as $CaCO_3$ equivalent. Swingle (1967) suggested that a total hardness of 50mg/l be a dividing line between soft and hard water. The solubility of most salts increases with the increase in water temperature. More specifically, total hardness is determined by the concentration of multivalent cations in water. Total hardness of 40-400mg/l is optimum for water bodies and less than 5mg/l leads to the eventual death of fish. The hardness of at least 20mg/l is maintained for optimum growth of fish.

37.2.9 Total alkalinity

Total alkalinity (TA) is the measure of water's ability to neutralize acids. Alkaline compounds like OH^- and CO_3^{-2} eliminates H□ ions from the water, which lowers the acidity of the water and results in a higher pH. Higher temperature slightly increases the CO_3^{-2} to HCO_3^- ratio. At the same time, the H^+ concentration increases slightly which causes a slight drop in pH (Dash, 2006). Total alkalinity of 40-150mg/l is optimum for water bodies Fish farmers maintain at least 20mg/l of total alkalinity for catfish production and 80-100 mg/l for hybrid bass production.

37.2.10 TDS

Total dissolved solids (TDS) comprise inorganic salts, principally calcium, magnesium, potassium, sodium, bicarbonates, chlorides, and sulphates and some small amounts of organic matter that are dissolved in water. As water temperature increases, the TDS of water also increases; for each 1°C increment, TDS may rise by 2–4% indirectly. Fish farmers maintain 650-725mg/l of TDSs for fish production. It can dehydrate the skin of fish which may be fatal.

37.2.11 Chloride

Chloride is common in most natural water and is most often found as a component of salt (sodium chloride) or in some cases in combination with potassium or calcium. Its value ranges from 1-100mg/l in freshwater. As water temperature increases, chloride also increases (Dash, 2006). Modelling showed that climate change impacts chloride concentrations in a variety of ways (EESC, 2019):

(1) An increasing occurrence of low river flows from summer to autumn reduces the dilution of the chloride that is emitted to the Rhine with a constant load thereby increasing its concentration;

(2) Increased open water: increased chloride alters reproduction and increased mortality.

37.2.12 Nitrate

Nitrate is formed naturally when nitrogen combines with oxygen or ozone. Studies of soil processes suggest climate change is likely to lead to increased nitrate leaching from the soil under future climate scenarios. Its amount increases with the increase in water temperature (Pandey and Shukla, 2005). Nitrate of less than 100 mg/l is optimum for water bodies. Fish become more susceptible to disease and inhibit the ability to reproduce. The growth of fish is also affected.

37.2.13 Phosphate

The phosphate increases with the increase in water temperature (Pandey and Shukla, 2005). Phosphate of 0.05-0.2 mg/l is optimum for water bodies. High phosphate is detrimental to the fish that they harbour (Wang *et al.*, 2022).

37.2.14 Ammonia

Ammonia is a colourless, pungent gaseous compound highly soluble in water. The range of ammonia is from 0.3 to 1.3 mg/l for most water bodies (Wurts, 2012; EPA, 2013). The toxicity of ammonia increases as temperature and pH increase (Pandey and Shukla, 2005). In lakes, ammonia did not reach a lethal level because of the low densities of fish.

37.3 Effect of Climate Change

According to the study, climate change will force 23% of shared fish stocks to move from their historical habitats and migration routes by 2030 AD, if nothing is done to halt greenhouse gas emissions. By the end of the century, that number could rise to 45%. However, a recent study noted global warming to have pushed fish populations towards colder waters in the north and south poles. Hundreds of fish species and shellfish will be forced to migrate northwards to escape the effects of climate change, putting global fisheries at risk.

38

Population Dynamics

38.1 Introduction

To fishery management, understanding the abundance of a given species population is critical to the success of that species' fishery. In this context, a fish population is a naturally occurring stock of a race of fish that occupy a specific area of space and act as a biological unit. The density of the fish population, on the other hand, is the population size per unit of space. Therefore

$$\text{The density of the fish population} = \frac{\text{The Population size}}{\text{Area or volume of space}}$$

When expressed in terms of the number of fish in the stock or the weight of the stock per unit of an aquatic ecosystem or unit area of a water body, density is known as abundance. Therefore

$$\text{Theabundanceoffishpopulation} = \frac{\text{Number of fishes in the stock or weight of the stock}}{\text{Unit of an aquatic ecosystem or area of water body}}$$

The fluctuation in the abundance of fish populations is known as fish population dynamics (Fig. 38.1).

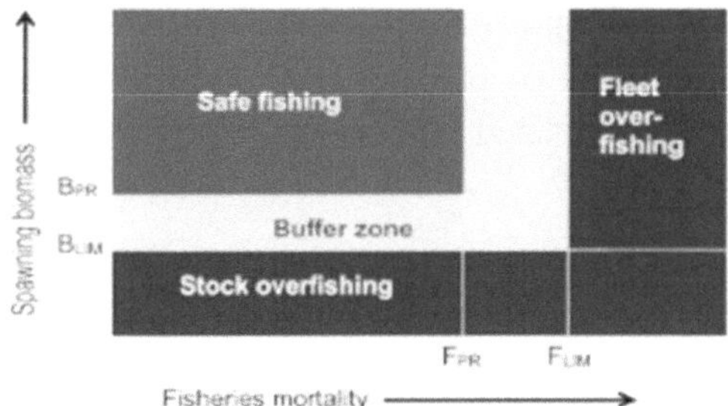

Fig. 38.1: Population Dynamics in Fishes (from https://en.wikipedia.org)

For effective fisheries management, a solid understanding of fish population dynamics is required. Models are typically developed as part of population dynamics research. Using numbers and equations to build a model forces the investigator explicitly to the hypothesised processes that influence fish population size. Therefore, population models should be regarded as hypotheses for management and future research. Rather than speculating that low recruitment is limiting fish abundance in a population, building a model can highlight the need to estimate recruitment trends to assess the impacts.

38.2 History

The first principle of population dynamics is most likely Malthusian's exponential law. The demographic work of Benjamin Gompertz and Pierre François Verhulst refined and adjusted the Malthusian demographic model, which dominated the early nineteenth century. Richards (1959) proposed a more general model, with the models of Gompertz, Verhulst, and Ludwig von Bertalanffy covered as special cases of the general formulation.

38.3 Measures of Fish Density in Population Dynamics

1. **Density-dependent requirement:** The abundance of adult stock in a population influences population requirement. If the adult density is high, the requirements may be high. The relationship between adult density and subsequent demand may not always be valid. The number of eggs laid is proportional to adult density.
2. **Density-dependent growth:** The slower the rate of growth, the higher the population density. The growth rate may be faster at low density.
3. **Density-dependent mortality:** Natural mortality is proportional to population size. Natural mortality is unaffected by density in general.
4. **Virtual population analysis (VPA):** It is a cohort modelling technique used in fisheries science to reconstruct historical fish numbers at age using data on individual deaths each year. This death is usually divided into two categories: catch by fisheries and natural mortality. It is an examination of fish population size that has not been directly measured but is assumed to have been a certain size in the past to support observed fish catches and assumed mortality due to non-fishery related causes (Fig. 38.2).

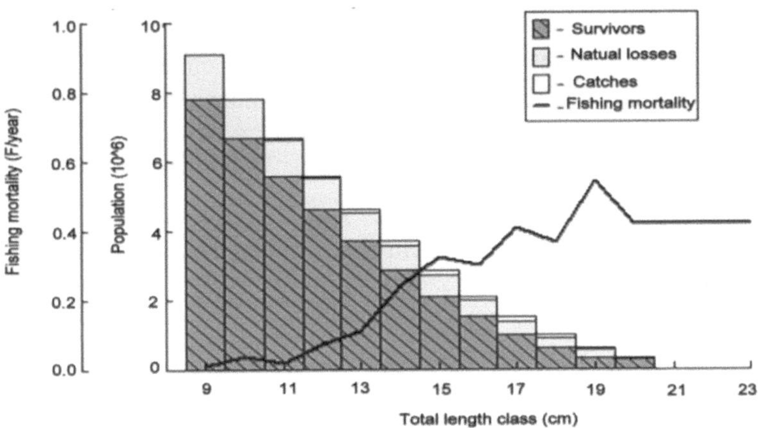

Fig. 38.2: Virtual population analysis in fish (from https://www.researchgate.net)

5. **The maximum sustainable yield (MSY):** It is the maximum catch that can be taken from a fishery stock indefinitely. Under the assumption of logistic growth, it will be exactly at half a species' carrying capacity, as population growth is greatest at this stage. It is typically greater than the maximum sustainable yield.
6. **Harvest Control Rule** (HCR): Some fisheries use this model to predict acceptable levels. This formalises and summarises a management strategy that can actively adapt to new information. It has some direct control and describes how management intends to control the harvest to some indicator of stock status.

   It may be of the following types:

   (A) Constant catch and

   (B) Constant fishing mortality
7. **Biological overfishing:** It occurs when mortality has reached a point where the stock biomass is growing at a negative marginal rate. Fish are being taken out of the water at such a rapid rate that stock replenishment through breeding is slowing. If replenishment continues to slow for an extended period, replenishment will reverse and the population will decrease.
8. **Economic** or **bioeconomic overfishing:** Overfishing is defined as a situation of negative marginal growth of resource rent. Fish are being taken out of the water at such a rapid rate that the growth in fishing profitability is slowing. If this continues for an extended period, profitability will suffer.

38.4 Relationship Between Fish Catch, Mortality and Population

The basic relation for population dynamics is the BIDE (Birth, Immigration, Death and Emigration) model is shown as:

$N_1 = N_0 + B - D + I - E$

Where, N_1 = the number of individuals at time 1, N_0 = the number of individuals at time 0, B = the number of individuals born, D = the number of mortality, I = the number of individuals immigrated and E = the number of individuals emigrated between time 0 and time 1. While immigration and emigration can be present in wild fisheries, they are usually not measured.

The relationship between fish catch, mortality and population may be expressed by a formula as given below:

$P = C + M + S$

Where, P = fish population, C = fish catch, M = natural mortality of fish and S = fish survival.

The natural mortality of fish is much higher in an unexploited population than in an exploited population. While the natural mortality rate may remain constant, actual mortalities may be greatly reduced after fishing. Fishing allows for more food and space per person. It would imply that natural mortalities are likely to be very low in intensively exploited fish populations. In such cases

$P = C + S$

In this way, the total annual catch may prove an index of the population size.

38.5 Estimation of Population

Various methods have been applied computation of the size of an population entire or a selected portion. Some important methods are discussed below:

1. **Direct method:** It includes counting all the fishes in a population. It is the simplest and the best method for accuracy.
2. **Indirect methods:** The following are indirect methods:

(A) **Age-frequency method:** The following formula is used to estimate the fish population.

$$P = \frac{C}{r}$$

Where, r = ratio of all deaths to population.

(B) **Correlated population method (Saville method of egg count):** It seems to be the best method. The following formula is used to estimate the fish population.

$$P = \frac{r_n E}{e}$$

$$r_n = \frac{N_{\text{書}} + N_{\text{京}} + N_i}{N_i}$$

Where, r_n = the ratio between the total number of individuals that exceed a specified size or age and the number of mature females. E = the total number of eggs laid down by the mature fish. $N_♂$ = the total number of mature male fish, $N_♀$ = the total number of mature female fish, N_i = the total number of immature fish exceeding a certain age or size.

Generally, for better estimation, the standard error is also calculated.

(C) **Deluty regression method**: It has theoretical value and has found little application in real practice since it is based on fish catches.

38.6 Fluctuation in the Abundance of Fish Population

The total stock of a species of fish varies from year to year and from region to region. Aside from anthropogenic activities such as fishing and pollution, these fluctuations in abundance are caused by natural factors. This variation is thought to reflect the biological characteristics of a particular fish as well as its adaptation to changing ecology. Survival reflects the fish's adaptability, while mortality reflects ecological adversity.

The following biological and physical factors are to blame for the fish population fluctuations:

38.6.1 Biological factors

(1) **Ethological factors:** Experimental evidence shows that epinephrine released by certain fish species, such as carp and gouramis, reduces the abundance of other fish species.

(2) **Inter-specific interactions in the food chain:** The population abundance of any one stock of fish is generally influenced by interactions with other fish species in an aquatic ecosystem. Temperature, food and predators are regarded as the most important factors in this context.

(3) **Inter-specific interactions on the spawning ground:** The population abundance of any one stock of fish is generally influenced by the presence of other fish species. For example, many fish species occupy the spawning grounds of *Salmon*.

(4) **Intra-specific interactions:** It is caused by an overpopulation of a single fish species. Overpopulation reduces food availability, resulting in a decrease in fish population abundance.

(5) **Natality versus mortality:** If natality equals mortality in any case, stock abundance remains stationary. If natality is high, the fish population grows; if it is low, fish population shrinks.

38.6.2 Physical factors

(1) **Temperature:** Temperature tolerance varies between fish. When there is a sudden cold spell, a large number of marine fish are killed in shallow

water. Temperature variations may also have an impact on fish spawning and survival, reducing abundance.

(2) **pH:** Many fish can tolerate a pH range of 6.5—8.5. When the pH falls below 6.5, the water becomes acidic, and the fish become restless, resulting in fish death and a decrease in the fish population. When the pH of water exceeds 8.2, free CO_2 does not exist.

(3) **Dissolved oxygen:** DO tolerance varies greatly between fish. Warm-water fish can withstand lower DO concentrations than cold-water fish. The normal DO tolerance range for various fishes is 0.5—5.0 mg/L as the minimum requirement. Sudden depletion of DO levels below 0.5 mg/L, or even super-saturation, may result in mass fish killing.

(4) **Free CO_2:** High CO_2 concentrations in aquatic bodies are harmful to fish. Any CO_2 concentration above 20 mg/L harms fish survival. High CO_2 concentrations are accompanied by low DO concentrations.

(5) **Salinity:** Most fish can only tolerate a narrow range of salinity. When salinity rises to 15ppt in a freshwater ecosystem, it becomes detrimental to fish survival. Similarly, when salinity rises from 30—35ppt to 70—100ppt, mass mortality in marine fish may occur.

(6) **$CaCO_3$:** The presence of $CaCO_3$ in water causes alkalinity. When the $CaCO_3$ concentration is low, the water is less biologically productive than when the $CaCO_3$ concentration is high.

38.7 Basic Fisheries Models

38.7.1 Yield-per-Recruit Models

These models are typically used to assess the risk of overfishing (i.e., fishing at a level that reduces the maximum yield per recruit). For these models, several formulations have been proposed. We used a simple formulation developed by Botsford and Wickham (1979), which was summarised by Walters and Martell (2004). The method employs Botsford incidence functions, which estimate a fish population's abundance and biomass per recruit.

38.7.2 Catch-at-Age Models

The ability to estimate population size improves fish population management. To assess the impact of fishing on population, an estimate of absolute abundance can be compared to the catch. The abundance of spawning stock is estimated to determine whether the abundance is low enough to limit recruitment. Prey and predator population levels can used to estimate food resources for predators or the impact of predation on prey species. When combined, analyses that predict population abundance provide a wealth of information about mortality, recruitment, and growth that relative abundance (i.e., CPUE) data cannot.

39

Fish Pathology

39.1 Introduction

Like all other animals, fish are also subjected to certain diseases. This is one of the most important problems confronting a fish culturist. Diseases at times may take the form of epidemics. The occurrence and magnitude of infections and diseases are closely related to the sanitary conditions prevalent in the water and also the general health of the fish.

39.2 Classification of Fish Diseases

(A) Based on the nature of causative agents, the fish disease may be of the following types:

(a) Parasitic diseases: Causative agents of this concern may be of the following types based on their taxonomic aspects:

I. Viral diseases: 1. Carp pox, 2. Chinook disease, 3. Channel catfish virus (CCV) disease, 4. Epizootic ulcerative syndrome (EUS), 5. Infectious pancreatic necrosis (IPN), 6. Infectious haemopoeitic necrosis (IHN), 7. Lymphocystis disease virus (LCDV), 8. Koi herpesvirus (KHV), 9. Salmon pox, 10. Spring Viremia of carp (SVC) or Infectious dropsy of carp (IDC), 11. Viral haemorrhage septicemia (VHS), 12. Viral Nervous Necrosis (VNN)

II. Bacterial diseases: 1. Septicemia, 2. Cottonmouth disease (CMD), 3. Cold Water Disease (CWD), 4. Columnaris, 4. Dropsy, 5. Eye diseases, 6. Fin and tail rot, 7. Furunculosis, 8. Gill disease, 9. Kidney disease in salmonids, 10. Mouth fungus, 11. Protrusion of scale, 12. Red Pest, 13. Scale Protrusion, 14. Tuberculosis, 15. Ulcer, 16. Vibriosis.

III. Fungal diseases (Mycoses): 1. Cotton wool Disease (CWD or Saprolegniasis), 2. Ichthyosporidiasis, 3. Exophialasis, 4. Branchiomycosis (Gill rot), 5. Reeling disease (Ichthyophonosis), 6. Dermatomycosis, 7. Aspergillomycosis, 8. Shrimpmycosis, 9. Skin and Kidney disease of salmonids.

IV. Protozoan diseases: 1. Chilodonella, 2. Diplostomosis, 2. Hexamita, 3.

Glugea and Henneguya, Gut blocking, 4. Knot disease, 5. Myxosporidisis, 6. Neon Tetra disease, 7. Sliminess of fishes (Costiasis), 8. White spot or Ich (Ichthyopthiriasis), 9. Whirling disease.

V. Helminthic diseases: 1. Acanthocephaliasis, 2. Black spot (Diplostomiasis), 3. Buccphaliasis, 4. Cestodiasis, 5. Dcatylogyrosis, 6. Gyrodactylosis, 7. Hirudinosis, 8. Myxosporodiosis, 9. Nematodiasis, 10. Trichodinosis.

VI. Arthropod diseases:

VII.Molluscan diseases:

(b) Non-parasitic disease:

I. Ailmental diseases: 1. Algal bloom, 2. Asphyxiation, 3. Chemical changes in water, 4. Physical changes in water, 5. Deficiency of oxygen, 6. Indigestion, Constipation and Dysentery, 7. Pox, 8. Acidosis, 9. Alkalosis.

II. Nutritional diseases: 1. Pinheads, 2. Deficiency of carbohydrates, 3. Lipoid hepatic degeneration disease (LHDD), 4. Deficiency of protein, 5. Vitaminosis A, 6. Deficiency of Thiamine (B_1), 7. Deficiency of Riboflavin (B_2), 8. Deficiency of Niacin (B_3), 9. Deficiency of Pantothenic acid (B_5), 10. Deficiency of Pyridoxine (B_6), 11. Deficiency of Biotin (B_7), 12. Deficiency of Inositol (B_8), 13. Deficiency of Folic acid (B_9), 14. Deficiency of Choline, 15. Deficiency of Cyanocobalamine (B_{12}), 16. Deficiency of Ascorbic acid (Vitamin C), 17. Deficiency of Vitamin D, 18. Deficiency of Vitamin E and 19. Deficiency of Vitamin K.

39.3.1 Viral Diseases

Fish can be infected by virus representatives from most of the virus families. Rhabdoviridae, Birnaviridae, Herpesviridae (Alloherpesviridae), Iridoviridae *(e.g. Ranavirus, Iriovirus, Chloridovirus, Lymphocystivirus, Megalocystivirus),* Reoviridae, Orthomyxoviridae and Retroviridae cause several diseases in fishes. Their infection may occur in the viscera, muscle, skin or fin. Such fishes are generally treated by Sulphanilamide, Sulpadizine, Sulphamerazine etc. The following are the common viral diseases in fish:

1. **Carp pox: It is** one of the oldest recognized fish diseases. It is caused by Cyprinid herpesvirus-1 (CyHV-1) in *Cyprinus carpio,* carps and koi. Its lesions are smooth and raised and may have a milky appearance. The gills of the infected fish show serious tissue damage.
2. **Chinook disease:** It occurs in *Oncorhynchus tshawytscha* (Chinook salmon) fingerlings due to infection with the virus. The diseased fishes show exophthalmus, a distended abdomen, and a dull red area on the dorsal surface anterior to the dorsal fin. The liver, spleen, kidney, gills and heart become pale.

3. **Channel catfish virus (CCV) disease:** The disease is caused by a DNA virus measuring about 1000. Affected fish hangs vertically in the water, showing spiral swimming and haemorrhage in visceral organs.
4. **Epizootic ulcerative syndrome (EUS) or Red Spot:** It is a cancer of fish caused by *Rhabdovirus*. It appeared in 1988 in India although known in Australia since 1972.

   It affects a variety of fish species including *Labeo rohita, Catla catla, Cirrhinus mrigala, Labeo calbasu, Cyprinus carpio, Ctenopharyngodon idella, Hypophthalmichthys molitrix, Anabas testudineus, Clarias batrachus, Heteropneustes fossilis, Amphipnous cuchia, Wallago attu, Oxygaster bacalia, Mugil subviridis, Mugil cephalus, Liza bornensis, Etroplus suretensis, Puntius* sp., *Channa* sp., *Mastacembalus* sp., *Glossogobius* sp., *Trichogaster* sp. *and Gadusia* sp.
5. **Infectious pancreatic necrosis (IPN) of *Salmon*:** It is the first fish virus isolated in endemic in most parts of Northern America, Europe and Japan. The virus contains RNA and measures about 700. It easily affects young fish. The diseased fishes show damaged intestine and pancreas; looks darker and swim in a spiral manner.

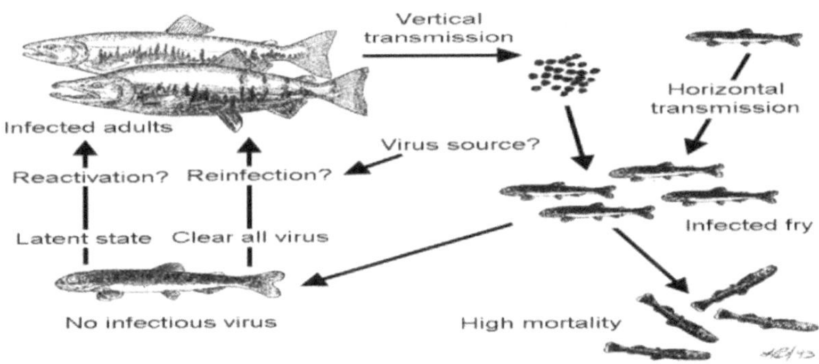

Fig. 39.1: Infectious pancreatic necrosis (from https://www.cabidigitallibrary.org)

6. **Infectious haemopoetic necrosis (IHN) of *Salmon*:** IHN was observed for the first time in trouts in British Columbia (Canada) in 1967. It is a bullet-shaped novirhabdovirus. It is a common name for the disease Orgeon sockeye disease (OSD) and Sacramento river Chinook disease (SRCD). It is confined to North America and Japan. The virus is transmitted by contact with infected fish, feeding on infected carcasses, exposure to water from an infected source and surface of eggs from infected stock. It is not generally observed at a temperature exceeding 15^{0}C.
7. **Lymphocystis disease virus (LCDV):** Woodcock (1904) identified this disease in fish. It is a common chronic disease caused by *Lymphocystivirus*

(Irido Virus). It affects the skin and fins of fish. Although it is serious, it only disfigures the fish. Fishes kept in aquariums or outside in ponds are susceptible to this disease. Tumor formation is an important characteristic of this viral disease. There is no cure for this virus.

8. **Koi herpesvirus (KHV):** It was first recognized in 1996 and caused by Cyprinid herpesvirus-3 (CyHV-3). It is widespread in the US and considered endemic. It causes clinical disease in koi and common carp *Cyprinus carpio*, *Carassius auratus* and *Ctenopharyngodon idella*.
9. **Salmon pox:** It affects the skin of wild and farmed *Salmon* and causes light-coloured worts.
10. **Spring Viremia of carp (SVC) or Infectious dropsy of carp (IDC):** It is caused by an RNA virus, *Rhabdovirus carpio* measuring 1200. The affected fish (carps, silver carp, grass carp, koi carp, and crucian carp) show darker colour, distended abdomen, loss of balance and haemorrhage in skin and gills. This virus is shed in the feces and urine, as well as the gill and skin mucus of infected fish.
11. **Viral haemorrhage septicemia (VHS) of rainbow trout or *Egtved*:** The disease is caused by a bullet-shaped virus confined to Europe. The virus contains RNA and measures about 500. It affects the different age groups of fish. The diseased fish show anaemia, pale discolouration of gills, swelling of eyes and belly, oedema of muscles, inflammation of the intestine and haemorrhage in most tissues. Sometimes there are sores on the skin, protrusion of the anus and seldom movement. The kidney and liver may also be affected.
12. **Viral Nervous Necrosis (VNN)**: It is caused by *Betanodavirus* in tropical, sub-tropical or cold-temperature fishes. *Betanodavirus* more commonly causes disease and mortality in larval stages. Abnormal swimming behavior, muscle tremors and hyperinflation of the swim bladder symptoms are generally because of *Betanodavirus* attack of the nervous system.

39.3.2 Bacterial Diseases

Bacteria cause many diseases in fish. Generally, fish possess great resistance to Gram-negative bacteria (such as *Aeromonas, Citobacter, Edwardsiella, Flavobacterium, Pseudomonas* and *Vibrio*) and Gram-positive bacteria (*Streptococcus, Lactococcus, Enterococcus, Vagococcus and Kocuria rhizophila*), but those weakened by unfavourable conditions or injured or infected by other parasites are highly susceptible to bacterial diseases. *Aeromonas, Vibrio* and *Streptococcus* more commonly affect freshwater, marine and ornamental fish respectively. These diseases are treated by Sulphanilamide, Sulpadizine, Sulphamerazine etc. Recently, Roccal (Alkyldimethybenzyl-ammonium chloride) and PMA (Pyridyl mercuric acid) are also recommended. The following are the common bacterial diseases in fish:

1. Septicemia: It is caused by *Aeromonas hydrophila* and affects the blood of *Catla catla, Labeo rohita, Cirrhinus mrigala* and Catfish. This disease shows ulceration of the skin, distended abdomen and inflamed fin and fin bases

2. Cotton mouth disease (CMD: It is caused by *Fexibacteria* and *Enterobacteria* having fungus-like tufts around the mouth. The disease is common in salmonids.

Treatments: (i) Use of 10 mg/l Chloramphenicol at 2-5 days intervals and 0.1-0.3 mg/l Furance for a long-term bath.

3. Coldwater Disease (CWD): It is caused by *Flavobacterium psychrophilia.* In infected fish, the caudal peduncle darkness and the caudal fin becomes frayed and eroded. According to Wood (1979), the transmission of the disease is associated with the presence of carrier fish, vertical transmission of the disease to the offspring may also be a possibility.

4. Columnaris: It is caused by *Chondococcus columaris* and *Cytophaga columnaris.* It is associated with a low DO level. Initially, the fish shows discoloured patches on the body. These patches spread and cause loss of scales and eventual death of fish.

Treatments: (i) Treatment with 1mg/l $CuSO_4$ solution. (ii) Apply Terramycin with artificial food (iii) Dip fish for 2 minutes in 1:2000 $CuSO_4$ solution or 10-30 seconds in 1:15000 Malachite green.

5. Dropsy: It is one of the most feared diseases in carp culture. It is caused by *Aeromonas hydrophilus* and *Pseudomonas punctatus.* Initially, it is caused by the virus but is further complicated by a secondary infection of bacteria.

Symptoms: (i) Due to the accumulation of yellow-coloured liquid in the viscera, the belly swells up considerably. (ii) Protrusion of scales and exophthalmic conditions are common. (iii) Apart from the inflammation of the intestine, discolouration of the liver is observed. (iv) When an ulcer appears on the skin, the fin lengthens and deformation of the backbone occurs. (v) Fish respond by jumping.

Treatments: (i) Treat with Chloromycetin (60mg/gallon) for 3-7 days. (ii) In acute cases, the liquid is first removed from the belly before treatment. (iii) To check epizooties, disinfect the pond with 1mg/l $KMnO_4$ solution for 2 minutes. (iv) Removal and destruction of fish followed by draining, drying and disinfecting the pond with quicklime. (v) Resistant bacteria may be killed by Chloromycetin, Streptomycin or Oxytetracycline along with supplementary food or by injecting the severely injured fish.

6. Eye diseases: It is an epidemic disease caused by *Aeromonas liquifaciens, Aeromonas punctatus* and *Staphylococcus aureus.* It is common in medium and large-sized *Catla catla.* The cornea of the eye becomes vascularized and later becomes opaque and the eyeballs get decayed.

Symptoms: (i) Infection of the eye, optic nerve and brain. (ii) Degeneration of eyeball. (iii) Cornea becomes red and later opaque.

Treatments: (i) Disinfect pond with 0.1 mg/l $KMnO_4$ solution. (ii) Boil with Chloromycetin (8-10mg/l) for 1 hour for 2-3 days.

7. Fin and tail rot: It is common in Indian major carp particularly when they are reared under bad sanitary conditions. The infection is caused by many *Myxobacteria* like *Haemophilus piscium, Aeromonas puncatus, Aeromonas hydrophilus, Pseudomonas fluorescence, Pseudomonas granulose, Mycobacterium* etc.

Symptoms: (i) Putrefaction of tail or other fins. (ii) Fins become torn and gradually destroyed. (iii) A white line appears on the fins as early infection. (iv) White lines move towards the base of the fins as late infection. (v) Infection may spread onto the body and involve connective tissues.

Treatments: (i) In early infection, use 1:2000 $CuSO_4$ solution for 2 minutes. (ii) A bath in dilute Acriflavine or Phenoxethol. (iii) In late and serious infection, affected parts are surgically removed and the wounds are disinfected by the touch of 1% $AgNO_3$ solution. (iv) Keeping the fish in 1:25000 $K_2Cr_2O_7$ for 7 days. (v) Treat with O_3 for 1 hour, repeat it 3-4 times daily for 10 days. (vi) Treat with Neomycin, Oxytetracycline or Chloromycetin (60mg/gallon) for 3-7 days.

8. Furunculosis: It is one of the most important bacterial diseases caused by *Aeromonas salomonicida* in trout and *Salmon*. The bacterium is rod-shaped and 2-3μ in length.

Fig. 39.2: Furunculosis in *Salmon* (from https://www.flickr.com)

Symptoms: (i) Formation of swellings containing a bloody and pus-like substance that may be discharged upon bursting of boils. (ii) Severe infections cause death within a few days.

Treatments: (i) Treat it with a 1:4000 formaldehyde solution for 1 hour. (ii) Treat with 1:50000 Roccal for 1 hour. (iii) Treatment with PMA (Pyridyl mercuric acid) in a 1:500000 ratio for 1 hour. (iv) Drain the pond and treat it with slaked lime. (v) Severely infected fish is treated by supplying food containing Introfurans or

Sulphonamides. (vi) Disinfect with a solution of 0.015% Merthiolate or 0.1855 Acriflavin or 1% Providone-iodine.

9. Gill disease: It is caused by *Myxobacteria* in salmonids under stress. The fish showed colonization of damaged gills.

Treatments: Treating with Neomycin or Chloromycetin may be useful.

10. Kidney disease in salmonids: It is caused by *Corynebacterium* and *Renibacterium salmoninarum*. It is more prevalent in fish in soft water. Diseased fishes show the darkness of the skin, gills are pale in colour and swelling of the kidney. Treatment by applying antibiotics is successful.

11. Mouth fungus (rot): It is caused by a rod-shaped, gram-negative bacterium, *Flavobacterium columnare*, it is especially common in newly imported fish. Treat with Maracyn.

12. Protrusion of scale: It is characterised by some red spots on the skin and projecting and falling off scales. Dip in 5mg/l $KMnO_4$ solution for 2 minutes is used to treat diseased fishes.

13. Red Pest: It is characterised by the appearance of bloody streaks on the body, fins and/or tail in fishes. In severe infection, these streaks could lead to ulceration and possibly followed by fin and tail rot with the tail and/or fins falling off.

Treatments: (i) Treat the tank with a disinfectant and clean it. (ii) To disinfect, use Acriflavine (Trypaflavine) or Monacrin (mono-amino-acridine) using a 1 ml/l of 0.2% solution 1 ml/l. (iii) Do not feed a lot of the fish during the treatment.

14. Scale Protrusion: Protruding scales without body bloat. An effective treatment is to add an antibiotic to the food. With flake food, use about 1% of antibiotics like Chloromycetin (Chloramphenicol) or Tetracycline.

15. Tuberculosis: It is caused by *Mycobacteria chelonian, Mycobacterium marinum, Mycobacterium piscium* and *Mycobacterium tuberculosis. Nocarida asteroids* cause similar chronic lesions in salmonids. The diseased fishes show rot of tail and fin, waxy coat and ulcer on the body, nodules in internal organs and loss of weight and appetite. These disease symptoms are mainly shown in the summer season.

Treatments: (i) Use of 1:2000 $CuSO_4$ solutions for 3-4 days. (ii) Keeping the fish sterilized with $KMnO_4$ or soaked lime. (iii) Use of Kanamycin.

16. Ulcer: It is caused by *Haemophilus piscium* reported mainly in major carps and salmonids. The diseased fishes may be carried by asymptomatic adults.

Symptoms: (i) Presence of ulcers on the body. (ii) Ulcers gradually increase in size exposing muscles.

Treatments: (i) Acutely infected fishes should be segregated. (ii) Dip in 1:2000 $CuSO_4$ solution for 3-4 days.

17. Vibriosis: It is caused by comma-shaped *Vibrio anguillarum* generally at a temperature above 11°C. It multiplies in the skin of salmonids and eels but the toxins produced act on the circulating blood to cause severe anaemia.

Treatments: (i) Dip in $CuSO_4$ solution. (ii) Application of antibiotics with artificial food.

39.3.3 Fungal Diseases (Mycoses)

Fungi infect various developmental stages of fish. They grow well in non-slimy areas and their filaments grow into sub-cutaneous tissue to cause the death of the fish. Many preventive and control methods are employed to control fungal growth in fish such as:

(i) Treatment of pond with 150-200 kg/hectare quicklime. (ii) Bathing of fish with 500mg/l $CuSO_4$ solution for 1-2 minutes. (iii) Use of artificial food containing Terramycin, Sulphamethalin, Erythromycin, Thiocyanate or Calomel. (iv) Keeping infected fish alternatively in bath water containing 0.5ml of HCHO solution in 1 litre water and Malachite green. (v) Use of *Argentum nitricum* (Q).

1. Cotton wool fungus (Saprolegniasis): It is mainly a secondary infection of *Saprolegnia diclina* and *Achyla hoferi*, seen after damage to the fish integument. George *et al.* (1998) reported that lesions grow surface of the skin, they usually do not penetrate deeply into the muscle. The area of skin and gill damage determines the severity of the disease. The pathogen attacks usually occur at low temperatures.

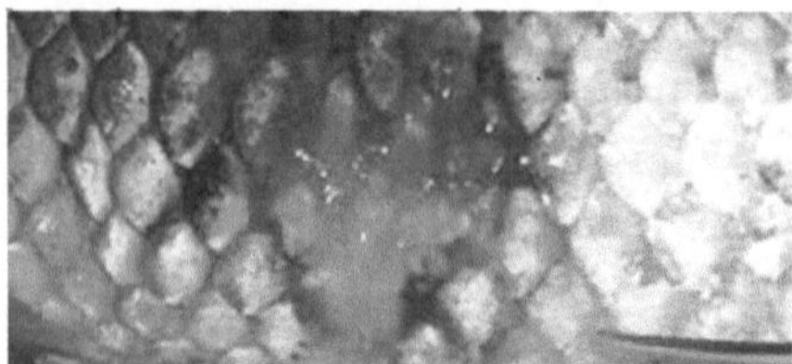

Fig. 39.3: Saprolegniasis in fish (from www.adfg.alaska.gov)

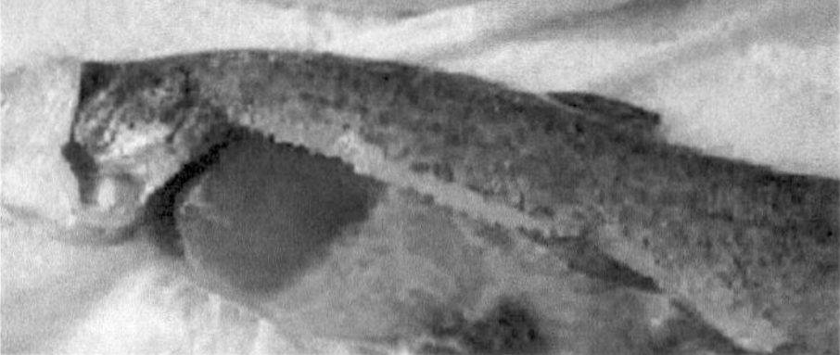

Fig. 39.4: Ichthyosporidiasis in fish (from units.fisheries.org)

Symptoms: (i) Gray or white patches on the skin/gills. (ii) May become brown/ green (later stage) as they trap sediment. (iii) *Saprolegnia* is established as small, focal infections that then spread rapidly over the body or gills.

Treatment: (i) Fish are removed from the water. (ii) Use the 100mg/ litter strong malachite green solution to clean the lesion and apply a waterproof cream.

2. Ichthyosporidiasis: Gustafson and Rucker (1956) reported that *Ichthyosporidium* is an endoparasitic fungus. It primarily attacks the kidneys and liver, but it spreads everywhere else.

Symptoms: The fish may become sluggish, lose balance and eventually show external cysts or sores.

Treatment: 1% Phenoxethol solution added to food or Chloromycetin added to the food has also been effective.

3. Exophialasis: *Exophiala salmonis* and *Exophiala psychrophila* hyphae that are septated, irregular in width and branched (Robert, 1989). Both fungal diseases infected many species of fish.

Symptoms: (i) Fish become darker and lethargic, with erratic and abnormal swimming behavior. (ii) Round yellow to white granulomas are present in visceral organs like the liver, kidney and spleen.

4. Gill rot (Branchiomycosis): It is caused by *Branchiomyces sanguinis* and *Branchiomyces demigrans* affecting the gill tissues of freshwater fish. *Branchiomyces sanguinis* grows in the blood vessels but *Branchiomyces demigrans* is found in the parenchymal tissue of the gills. Fish become weak in movement. There is respiratory distress in infected fish and do not swallow the air.

Fig. 39.5: Branchiomycosis in fish (from https://ilovepets.co)

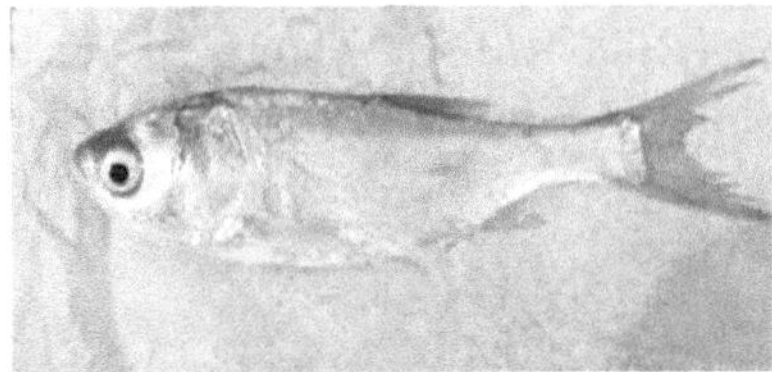

Fig. 39.6: Aspergillomycosis in fish (from http://pu.edu.pk)

Symptoms: (i) Gill filaments become reddish and subsequently fade. (ii) Drop leaving exposed cartilaginous supporting skeleton. (iii) Diseased fish die due to suffocation.

Treatments: (i) Difficult to save the fish from dying of disease. (ii) Introduction of freshly cooled water, prevention of pollution, discontinuation of artificial feeding. (iii) Dip the fish in 3-9% NaCl solution for 5-10 minutes or 5ppm $KMnO_4$ solution. (iv) Add 1-2g $CuSO_4$ m^3 of water.

5. Reeling disease (Ichthyophonosis): It is caused by *Icthyophonus hoferi, Penicillium, Phoma* and *Aphanomyces*. It is spherical or oval, yellowish brown and has granulated cytoplasm. This disease is transmitted through skin abrasions of damaged gills. There are no eternal signs. Fish with advanced infection have rough or granulomatous skin. These nodules are found in the heart, liver, kidney and brain and are filled with cellular debris and fungus. Infection of the swim bladder, which led to damage, fish lie on the bottom of the pond and die.

6. Dermatomycosis: It is characterized by the growth of thin threads on skin or fins. If fungal growth is abundant, it resembles a tuft of cottonwood. The fungus attacks only fishes that have been wounded or whose resistance has been weakened by other parasites or bad conditions. It is caused by various members of the class Oomycetes. In the case of *Saprolegnia* attack, the hyphae grow into sub-coetaneous parts and cause the death of the surrounding tissue and finally death of the fish.

Table 39.1: Group of fish and fungal patheogens

Sl.No.	Order	Fungal pathogens
1.	Leptomialales	*Leptomitus*
2.	Pernosporales	*Pythium*
3.	Saprolegniales	*Achyla, Aphanomyces, Dictyuchus, Leptogenia, Saprolegnia, Thranstotheca*

Symptoms: Fish becomes weak and lethargic and dies of blindness, deformities of bones and inflammation of the liver and intestine.

Treatments: (i) Take out the fish from the water; touch the affected part with 1:10 iodine solution or mercurochrome, 1% $K_2Cr_2O_7$ solution, 1:2000 $CuSO_4$ solution or 1:1000 $KMnO_4$ solution. (ii) Keep the fish in a tank containing 1:25000 $K_2Cr_2O_7$ solution for 7 days. (iii) Dip the fish in 3% NaCl solution until distress sign. (iv) The use of Phenoxethol destroys the fungus rapidly. (v) Use of 1:200000 malachite solution.

7. Aspergillomycosis: It is a cutaneous disease caused by *Aspergillus niger*, *Aspergillus falvus*, *Aspergillus fumigates* and *Aspergillus terreus* generally in *Tilapia* and *Hypophthalmichthys molitrix*.

Treatments: (i) Use of Isavuconazole, Itraconazole, Posaconazole, Voriconazole. (ii) Use of Caspofungin, Micafungin. (iii) Use of lipid Amphotericin formulations.

8. Shrimpmycosis: It is caused by *Lagenidium callinectes* and *Sirolpidium*.

9. Skin and Kidney disease of salmonids: It is caused by *Scolecobasidium*. This fungus is found in soil but occasionally invades salmonids.

39.3.4 Protozoan Diseases

Protozoans cause numerous external and internal diseases, particularly in young fishes. At least thirty-four freshwater fish species under nine orders in Bangladesh were retrieved from the mentioned articles and found 48 species of protozoan parasites under 19 genera (Kabita *et al*, 2020).

1. Chilodonelliasis: *Chilodonella uncinata* is the most dangerous parasite and can cause mass mortalities, especially in overstocked ponds and aquaria. It is between 30-60 microns in diameter and the adult has a heart-shaped appearance.

Symptoms: (i) Opaqueness of the skin, opened gill cover, especially between the head and dorsal fin. (ii) Affected gills are destroyed, quickly killing the host. (iii) Classic signs of flashing and rubbing may hold their fins clamped against their body and appear listless. (iv) Fishes hang at the pond surface and gasp for air in severe infestations.

Treatment: (i) Treatments include Malachite Green, Formalin, Potassium Permanganate and Salt baths at 3%.

2. Costiasis (Sliminess of the skin of fish): It is a fatal disease caused by *Costia necatrix, Bodomonas rebae, Trichidina indica* and *Chilodon* in ponds where fishes live densely in water with a low pH and poor condition food. *Costia necatrix* has a groove leading to the gullet and bears two rows of flagella. The parasites live in large numbers on the skin, fins and gills of *Labeo rohita, Catla catla* and *Cirrhinus mrigala. Trichidina indica* infects mostly major carp. *Chilodon* causes the slimness of the body of carp and catfishes. *Tilapia* is a very resistant fish against the attack of *Costia necatrix.*

Fig. 39.7: (a) *Costa necatrix* **(b)** *Chilodonella uncinata*

Symptoms: (i) The symptoms are the appearance of grey-blue film on the skin which turns to red patches in severely affected cases. (ii) The infected fish becomes weak; loss of appetite occurs and finally dies.

Treatments: (i) They can be treated with 3% common salt for 10 minutes. (ii) Use of 1: 4000 formalin solutions for 1 hour. (iii) Use of 1:1500 acetic acid solution. (iv) A solution of Methylene blue, $KMnO_4$ and $CuSO_4$ can also be used. (v) Overcrowding should be avoided and the pH of the water should be checked regularly.

3. Ichthyophthiriasis Disease (Ich, Ick and White Spot Disease)**:** It is caused by a ciliate, *Ichthyophthirius multifilis*. The parasite is 0.2-1.0mm long, spherical or oval and covered by fine cilia arranged in rows. The young parasites grow between the epidermis and dermis and after becoming large fall to the bottom of the pond. It is considered to be the most devastating protozoan disease in fish.

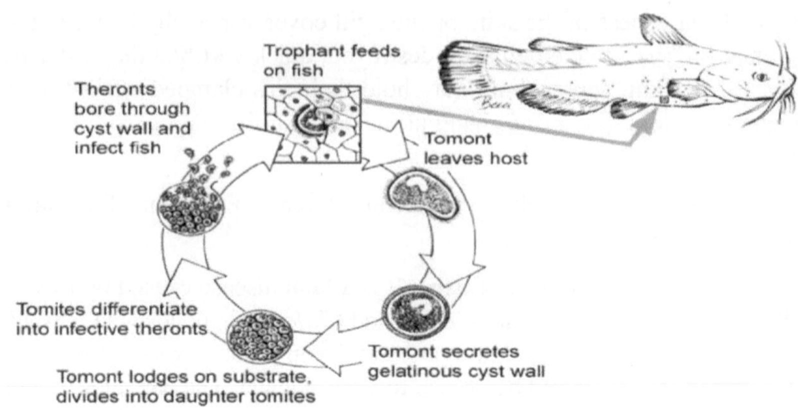

Fig. 39.8: Ichthyophthiriasis in fish (from https://www.cabidigitallibrary.org)

Symptoms: (i) Each spot (0.5-1.0mm in diameter) is a small bladder containing many parasites. (ii) Infected fish develop small white spots on the skin, fins and gills. (iii) Fish respond by jumping in the water and rubbing their body against the objects in the pond. (iv) The cyst geminates many young ones by multiplication. (v) Respiration gets affected and they finally die.

Treatments: (i) Dip treatment in 1.5 ppm of Malachite green or 10 ppm of Acriflavine. (ii) Treatment with 3% salt solution, 1:4,000 formalin and 1:100000 Quinine hydrochloride kills parasites in 7-14 days. (iii) Apply 1:500000 methyl blue in ponds to kill parasites (iv) Heat with Mepacrine (Atabrine) hydrochloride. (v) Disinfect the pond with quicklime 300-500 kg/ha/annum in 2-3 installments.

4. Marine White Spot Disease: It is caused by *Cryptocoran irritans*. $CuSO_4.5H_2O$ appears to be the most effective treatment to date. If the *Cryptocaryon* strain affecting your fish is tolerant of low salinity, Chloroquine and salinity, as well as formalin, have also proven to have some effectiveness.

5. Milky fish disease: It is caused by *Henneguya* possessing tadpole-shaped cysts. The infected fishes show whitening of the skin.

6. Trichodinasis: It is caused by *Trichodina indica* in generally young Indian major carp. It seems species-specific disease. The parasite measures about 40-60μ and shows rotating movement.

Symptoms: (i) Infected fish comes to the water surface and rubs its body on the margin of the pond. (ii) Both the gills and skin of fish secrete more slime.

Treatments: (i) They can be treated with 3% common salt for 10 minutes. (ii) Use of 1: 4000 formalin solution for 1 hour. (iii) Use of 1:1500 acetic acid solution.

7. Ulcer in skin: It is caused by *Myxobolus* sp. in the skin of common carp.

8. Velvet Disease (gold-dust, rust and coral disease): It is caused by

Piscinoodinium pillulare, is a yellowish, parasitic dinoflagellate that infects the skin and gills of *Amyloodinium* (a marine fish) and *Oodinium* (a freshwater fish). The parasite can multiply very rapidly in aquaria and kill fish. The disease gives infected organisms a dusty, brownish-gold color. There are commercial fish medications available that contain copper formulations.

9. Whirling Disease: It is caused by *Myxosoma cerebralis* and *Ceratomyxa* in rainbow trout and salmonids respectively.

Symptoms: (i) Pancreatic necrosis, lesions and disintegration of the cartilaginous skeletal support. (ii) Rapid tail-chasing type of whirling when the fish is frightened or trying to feed. (iii) The typical symptoms usually appear 1-2 months after exposure to the disease.

Treatments: (i) If the pond contains all infected fish, it is better to destroy them by deep burial. (ii) The pond is cleaned thoroughly and disinfected with calcium cyanamide, quick lime or sodium hypochlorite.

39.3.5 Helminth Diseases

Dhole *et al*, (2010) reported that four cestodes, two trematodes (flukes) and one nematode affect the fish. Both monogenetic (*Dcatylogyrus, Diplozoon* and *Gyrodactylus*) and digenetic (*Cryptocotyle* and *Diplostomum*) affect fishes.

(A) Disease caused by flatworms

1. Gyrodactylosis: It occurs in trout and carp due to a monogenetic, external and viviparous trematode, *Gyrodactulus elegans* (skin fluke). It can be seen within the body of fish in various stages of development in the culture ponds. The parasite lives on the fins and body of fish.

Symptoms: (i) Affected surface becomes covered by bluish slime due to increased mucous secretion. (ii) In heavy infection, colour becomes pale and skin slimy. (iii) Fins frayed and gradually torn. (iv) Fishes rub their body against the sides or bottom of the aquarium.

Treatments: (i) Place the fish in 1:500 CH_3COOH for 1-2 minutes. (ii) Place the fish in 1:2000 NH_3 for 5-10 minutes. (iii) Treat the fish in 1:4000 HCHO for 2-3 days. (iv) Treat the fish with Mosoten and Neguran in 1% dip for 1-2 minutes. (v) Dip treatment for 3-5 minutes in 5% NaCl solution. (vi) Left the fish for 2-3 days in the water tank containing 10 ppm or Acriflavin of 2-4 mg/l of Methylene blue.

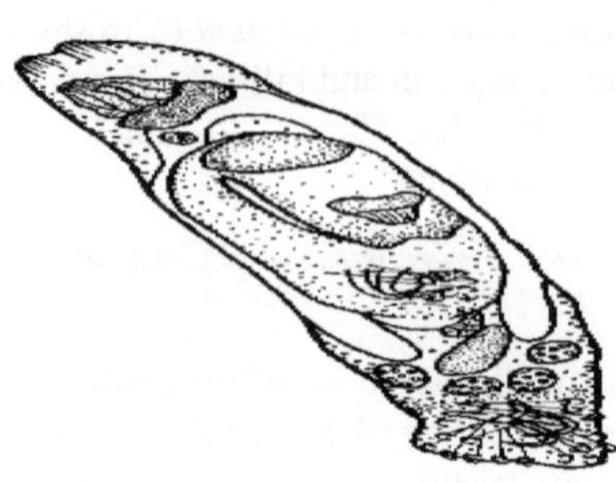

Fig. 39.9: *Gyrodactylus*

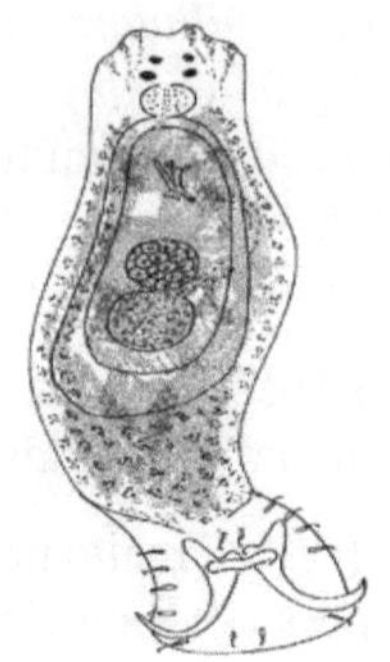

Fig. 39.10: *Dactylogyrus*

2. Dactylogyrosis (Gill flukes): It is caused by a monogenetic, external and oviparous trematode parasite, *Dactylogyrus exitensis, Dactylogyrus vastator* and *Dactylogyrus lamellatus* in carps. The parasites are found on the body, fins and gills.

Symptoms: (i) The parasites start appearing in the ponds during the rainy season. (ii) The most infected size group is 61-100 mm. (iii) Infected fishes rest near the surface of the pond margin, swim slowly and feel suffocation. (iv) The infected fish show slim, dropping and folding fins and pale gills.

Treatments: (i) Alternative baths with 1:2000 CH_3COOH and 2% NaCl is effective. (ii) 10 ppm of $KMnO_4$ bath for 1-2 hours and 5 ppm in the pond may give good results. (iii) Bromex – 50 (0.18mg/l) and Dylox (0.25mg/l) are effective to control the disease.

3. Diplostomiasis (black spot) of freshwater fishes: It is caused by a digenetic parasite, *Diplostomum* (= *Diplostomulum*) and *Neodiplostomum*. It is prevalent in some rearing and stocking ponds.

Symptoms: (i) Development of small brown or black spots on several parts of the body and fins. (ii) Spots are present in the cutis and underlying muscles. 3. The cysts are surrounded by an accumulation of pigment cells. (iv) The cyst (0.8-4.0mm diameter) contains *Postodiplostomum cuticula*. (v) The adult stage of the parasite lives on the heron.

Treatments: (i) Bath in a solution of picric acid 2-7:100000 for 1 hour.

4. Black spot of marine fishes: It is caused by *Cryptocotyle lingua*. Cercaria larva is the infective stage of the fishes

Symptoms: (i) Metacercaria is formed in the skin and the host fish lays down a readily visible black capsule around them. (ii) The life cycle depends upon a multiplication stage within *Littorina littorea*.

Treatments: (i) Destruction of secondary host.

5. Diplozooniasis (Gill rot): It is caused by a monogenetic trematode, *Diplozoon paradoxum* (twin-worm). The parasite affects *Cyprinus, Carassius, Phoxinus, Rhodeus, Rutilus, Gobio* etc. Another trematode affecting the gill of fish includes *Bucephalus* (Ox-head worm).

Symptoms: (i) The parasites are found on the gills and feed on their blood. (ii) The two individuals of the parasites are always found attached permanently together in the form of a cross.

Treatments: Removal of the fish by treating a dilute solution of Methylene blue. (ii) Use of dilute solution of picric acid.

6. Clinostomumiasis (yellow grub): It is caused due to a trematode 5-6mm long parasite, *Clinostomum* (yellow grub) in many freshwater fishes.

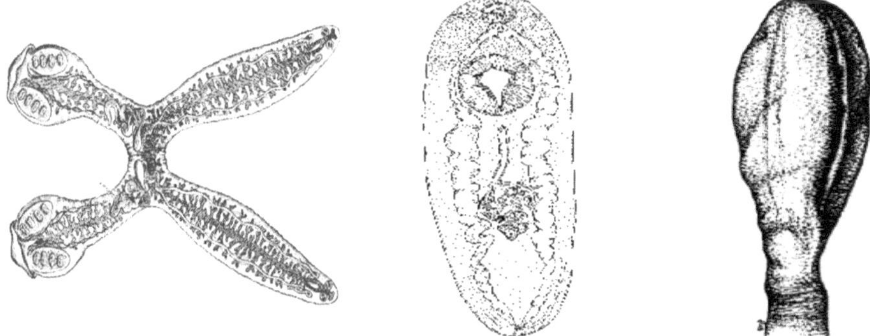

Fig. 39.11: *Diplozoon* **Fig. 39.12:** *Clinostomum* **Fig. 39.13:** *Diphyllobothrium*

Symptoms: (i) The parasite is enclosed in a small cyst lying just beneath the skin. (ii) The cysts are in the form of small, creamy nodules on the head, body and fins of fishes. (iii) The cyst contains many metacercariae of the adult fluke. (iv) The metacercaria reach the birds when infected fish is eaten up. (v) The eggs are shed into water and hatch miracidia which penetrate a snail. (vi) The cercaria comes out of the body of the snail and penetrates below the scale of the fish to form the cyst.

Treatments: (i) Careful opening of the cyst by a scalpel and removal of parasite. (ii) Disinfection of wound by mercurochrome. (iii) Removal of the snail from the body.

7. Cestodiasis: It is caused by many species of cestodes. *Diphyllobothrium latium* is a common fish tapeworm, whose larvae are found on the viscera of fishes. When abundant the larvae may the cause death of fish. *Bothriocephalus gowkongensis* infects grass carp and silver carp. *Ligula intestinalis* is another cestode infecting cyprinid fishes.

Symptoms: (i) Dull blue colour of fish. (ii) Swollen parts of the alimentary canal. (iii) Sometimes gall bladder is also affected.

Treatments: (i) Bath in 4mg/l of $KMnO_4$ solution, 3-5% of NaCl solution, 500mg/l $CuSO_4$ for 1-2 minutes or 250mg/l of HCHO. (ii) Treatment with 65-70mg/l of Malachite green for 10-30 seconds.

(B) Diseases caused by roundworms:

1. **Nematodiasis**: The most commonly reported nematode parasites are *Cucullaunus, Camallanus, Neocamallanus, Heliconema, Proleptus* and *Paracamallanus. Philometra* and *Camallanus* infect tropical fishes.

Symptoms: Their infection causes much harm.

Treatments: Use of Piperazine and Cadmium compounds. (ii) Dip fish in 3-100000 of picric acid for 45-60 minutes.

(C) Diseases caused by spring-headed worms:

1. **Acanthocephaliasis**: It is caused by *Acanthocephalus, Balbosoma, Centrorhynchus, Corynosoma, Echinorhynchus, Pompharhynchus* and *Rhodinorhynchus* etc. They are typical intestinal parasites of fish.

Symptoms: Penetrated intestine of fish.

Treatments: Use of Parachlorometa xylenol.

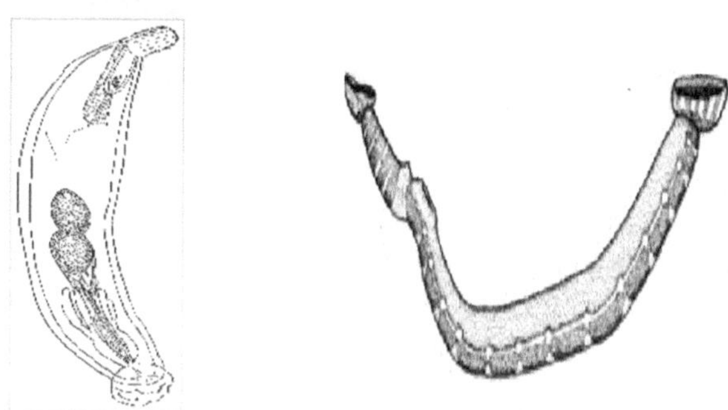

Fig. 39.14: *Acanthocephalus* **Fig. 39.15:** *Piscicola*

(D) Diseases caused by segmented worms:

1. **Hirudiniasis**: It is caused by parasites of Rhynchobdellidae and Gnathobdellidae. *Piscicola* is a typically segmented parasite of fish. They mainly infect major carp and feed on the blood of fish. Its wound may be affected by the fungus causing more harm to the fish.

Symptoms: (i) Irritates fish. (ii) Abnormal movement.

Treatments: (i) Place the fish in 2-5% NaCl solution for 10 minutes. (ii) Dip in 1:1000 CH_3COOH solution with 1:10000 $KMnO_4$ solution. (iii) Use 5mg/l Gammexane. (iv) Dip in 5-15 seconds in Lysol solution.

39.3.6 Arthropod Diseases

Crustaceans among the arthropods are chief ectoparasites causing skin diseases in fish as given below:

1. *Argulus* (Carp louse): The parasite remains attached to the skin by two means of its two suckers. The young parasites can attack fish.

Symptoms: (i) Stunted growth. (ii) Loss of scales. (iii) Appearance of a red spot at the sign of infection.

Treatments: (i) Removal of parasites with forceps. (ii) Paralysis of parasites with strong NaCl solution. (iii) Addition of the required amount of $KMnO_4$ in the pond. (iv) Segregation of infected fish and treatment with 1:1000 acetic acid for 5 minutes followed by NaCl solution for 1 hour.

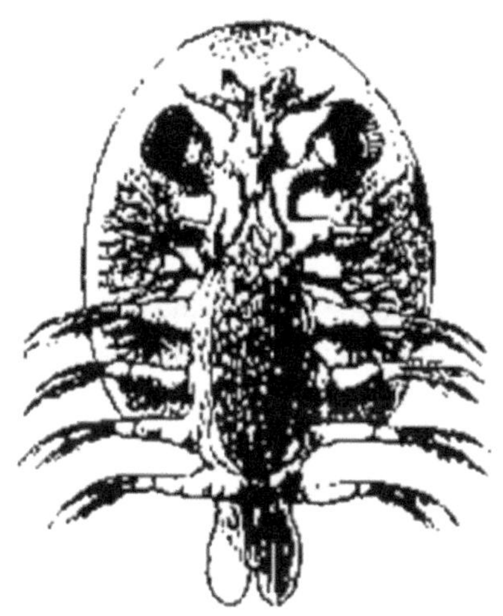

Fig. 39.16: *Argulus*

Fig. 39.17: *Lernea*

2. *Lernea* (Anchor worm): The parasite resembles a worm due to the absence of appendages. It affects goldfish, Indian carp, silver carp and trout. The nauplius larvae of the parasite swim freely and grow till they meet another fish

Symptoms: (i) Wounds and holes in the body. (ii) The body is later infected by bacteria and fungus. (iii) Fish die from secondary infection. (iv) The anterior region of the worm is buried deep into fish skin. (v) Only the posterior region remains visible. (vi) Red sores on the body.

Treatments: (i) Died and felled off parasites may be removed by forceps. (ii) Remove infected fish from the pond with 0.1% $KMnO_4$ solution for 2-3 minutes. (iii) A thin film of castor oil over the surface of the pond prevents infection. (iv) A half-hour bath in 500mg/l HCHO solution, 5 minute bath in 1-1000 glacial acetic acid solution followed by an hour bath in 1% NaCl solution. (v) Drained and dried ponds may be disinfected by using Lindane of 8ml/1000 litre.

3. *Ergasilus*: This looks like a normal copepod. It is generally found attached to the gills of fishes. Some other crustaceans include species of *Caligus, Pseudocycnus, Clavellisa* and *Lernaenicus.*

4. *Salmincola* (gill-maggot): It is commonly seen in salmon and sea trout of young size.

Treatments: (i) Use of Benzene hexachloride of 0.5mg/l.

39.3.7 Molluscan Diseases

1. Glochidium larva: It causes heavy infection in fishes. It may lead to the death of fish.

Symptoms: (i) Damage of gill arches and filaments. (ii) Appearance of tumours.

Treatments: (i) Mechanical removal of parasites.

39.3.8 Non-Parasitic Diseases

1. **Asphyxiation:** It is caused by stress and supersaturation of dissolved oxygen.

Symptoms: (i) Bubbles appear beneath the skin, around the eyes, on the fins, in the blood capillaries and alimentary canal. (ii) Fish swims at an angle of 45^0 and leads downward. (iii) The fish shows suffocation, (iv) Mouth widely open, gill opercula raised and gills spread wide apart.

Treatments: Transfer the fish or change the environment.

2. **Deficiency of oxygen:** There is anaemia and increased fat in the body.

Treatments: Keeping the concentration of dissolved oxygen constant or within a range essential to fish.

3. **Physical changes in water**: Changes in depth, temperature, turbidity, light etc. constitute the most important parameters responsible for proper growth. These changes result in mortality.

4. **Chemical changes in water**: Free CO_2, alkalinity, pH, dissolved salt etc. are responsible for the maintenance of the proper balance of water. Changes in these parameters are responsible for the mortality of fish.

5. **Algal bloom**: It occurs in ponds, especially in the summer season. The spread of algae all over the surface of the water is noticed. It obstructs sunlight.

Symptoms: (i) Lowering of the percentage of dissolved oxygen in the water. (ii) Chocking of gills.

Treatments: (i) Algae are removed by beating the water surface with bamboo sticks. (ii) Fishes may be transferred to another pond.

6. **Indigestion, Constipation and Dysentery**:

Symptoms: (i) Listlessness and queer movements. (ii) The fish remain stationary in mid-water. (iii) Scales are erected and the stomach becomes swollen. (ii) Rectum becomes red. (iii) Appearance of mucous in large quantities in the gut.

Treatments: (i) Stop the hard food supply to affected fish.

7. **Pox**: Occurrence of small milky white spots which gradually increase in size. Later, the spot becomes solid and fishes become weak.

Treatments: Treat with 5% $KMnO_4$ solution daily for 5 minutes.

8. **Acidosis**: Most fish live at pH 7-8. In acidosis, the pH of water goes down below 5.5 drastically due to the reduction of calcium salts or the release of humic acids from the soil.

Symptoms: (i) Fish show very rapid swimming movements and a tendency to jump out of the water.

9. **Alkalosis**: Alkalosis **comes about when** the pH of freshwater becomes highly alkaline (e.g. 9.2-10.8),

Symptoms: (i) The effects on fish may include: death, damage to outer surfaces like gills, eyes, and skin and an inability to dispose of metabolic wastes.

39.3.9 Nutritional Diseases

Fishes raised in fish farms require high-quality food and hence may become uneconomic if they develop diseases. Standard artificial diets tend to eradicate nutritional diseases. Nutritional diseases are more important in the pisciculture programme. It can be attributed to under-nutrition (deficiency), over-nutrition (excess) or mal-nutrition (improper balance) of components present in the food available.

1. **Pinheads**: It is the enlarged head and slender body of fishes as well as their starved larvae in the nurseries. It occurs due to an insufficient supply of food.
2. **Deficiency of carbohydrates**: Depress digestion; enlarged livers. Sikoki disease in carp.
3. **Lipoid hepatic degeneration disease (LHDD):** The liver of fish becomes yellow-brown. Anaemia follows with gills-looking pallor. Goiter is caused by iodine deficiency. Dicalcium phosphate deficiency causes scoliosis in carp.

4. **Deficiency of protein:** Reduced growth rate and body deformities
5. **Vitaminosis of A**: Deficiency of vitamin A causes undergrowth, blindness and haemorrhagic kidney and keratomalacia, Expthalmos, ascites and odema in fishes. Conversely, excess vitamin A results in squamous metaplasia, splenomegaly, hepatomegaly and necrotic caudal fin.
6. **Hypovitaminosis of thiamine (B_1)**: poor appetite, muscle atrophy, loss of equilibrium in trout, odema and poor growth.
7. **Hypovitaminosis of Riboflavin (B_2)**: corneal vascularisation, cloudy lens, hemorrhagic eye, photophobia, dim vision, incardination, discoloration, poor growth and anemia.
8. **Hypovitaminosis of Nicotinic acid (B_3):** Loss of appetite, photophobia, swollen gills, reduced coordination and lethargy.
9. **Hypovitaminosis of pantothenic acid (B_5):** It causes the clubbing of secondary lamellae while gill filaments show hyperplasia and inappetence. The affected fish show respiratory troubles and anaemia. Loss of appetite, necrosis and scarring, cellular atrophy and poor growth.
10. **Hypovitaminosis of pyridoxine (B_6):** It affects larval growth and causes lesions in nerves. The affected fish show gasping of opercular folds. Nervous disorders hyperirritability, anaemia serous fluid.
11. **Hypovitaminosis of Biotin (B_7):** Blue slime patch, loss of appetite, muscle atrophy, fragmentation of erythrocytes, skin lesions and poor growth.
12. **Hypovitaminosis of Inositol (B_8):** Fin necrosis anaemia, distended stomach, skin lesions and poor growth.
13. **Hypovitaminosis of Folic acid (B_9):** Poor growth, lethargy, the fragility of the caudal fin, Dark colouration, macrocytic anaemia, decreased appetite.
14. **Hypovitaminosis of Choline:** Anaemia, hemorrhagic kidney and intestine, poor growth.
15. **Hypovitaminosis of B_{12}:** Erratic haemoglobin level, erythrocyte counts and cell fragmentation.
16. **Hypovitaminosis of ascorbic acid:** It brings spinal deformities due to abnormal skeletal development and poor wound healing. Lordosis and scoliosis eroded the caudal fin, deformed the gill operculum and impaired collagen formation.
17. **Hypovitaminosis of D:** Necrotic appearance in the kidney.
18. **Hypovitaminosis of E:** Exophthalmia, distended abdomen, anemia with reduced RBC numbers and haemoglobin content. Accumulation of ceroid in fish liver.
19. **Hypovitaminosis of K:** Mild cutaneous hemorrhages due to the ineffectiveness of blood clotting.

Prevention: The following measures are generally applied to prevent nutritional diseases in fish. (i) Use of good quality vitamin-rich food. (ii) Food should lack a lot of fat and salt. (iii) Controlled distribution of supplementary food. (iv) No over overfeeding. (v) Fish should not be fed if the climate is either very hot or cold.

Unit-VI Ecology and Importance of Fish

40

Deep-Sea Fishes

40.1 Introduction

India has an 8,118-kilometer-long coastline that stretches through nine Maritime States (Gujarat, Tamil Nadu, Kerala, West Bengal, Maharashtra, Odisha, Andhra Pradesh, Karnataka, and Goa) and four Union Territories (Daman and Diu, Puducherry, Lakshadweep Islands, and Andaman and Nicobar Islands). The country also has a Continental Shelf of 0.53 million km^2 and an Exclusive Economic Zone (EEZ) of 2.02 million km^2. The EEZ's annual potential yield is estimated to be 3.93 million tonnes of fish. Marine Fisheries has played an important role in the Indian Fisheries Sector. Gujarat, Tamil Nadu, Kerala, West Bengal and Maharashtra have the highest marine catches. According to the National Bureau of Fish Genetic Resources (NBFGR) database in Lucknow, 1,518 fish species are from the marine environment.

The estimated marine fish landings from all maritime states and two union territories on the Indian mainland for the year 2021 were 3.05 million tonnes, an increase of about 11.8% over the landings in 2020 (2.73 million tonnes), but still 14.4% less than the landings in 2019, the last pre-pandemic year (3.56 million tonnes).

The most important Marine Fisheries are classified as follows:

1. **Surface-water Fish (Pelagic)**: Sardines, Anchovies, Ribbonfish, Mackerel, Seer fish, Tuna, etc.
2. **Mid-water Fish (Pelagic)**: Bombay duck, Cobia, Silver Bellies, Horse Mackerel, etc.
3. **Bottom-water Fish (Demersal)**: Perches, Catfish, Pomfrets, Flatfish, Eels, etc.

40.2.1 Sardines

Sardines are one of India's most important commercial pelagic schooling clupeid fish. The Indian Oil Sardine is one of the most regionally restricted *Sardinella* species, only found in the northern Indian Ocean. Out of the 14 species, the Oil

Sardine (*Sardinella longiceps*) (Fig. 40.1) is the most important single-species fishery, accounting for approximately 15% of total marine fish catches. The highest number of oil sardine landings occurs during the post-monsoon season, followed by the monsoon and pre-monsoon seasons.

Fig. 40.1: *Sardinella*

The other 13 species, known as "Lesser Sardines," include all species of Sardines (*Sardinella* spp.) other than the Indian Oil Sardine and account for 8-9% of total marine fish catch. They are known to live primarily in near-shore waters up to a depth of 25-30 m. *Sardinella gibbosa, Sardinella fimbriata, Sardinella sirm, Sardinella albella, Sardinella dayi, Sardinella sindensis, Sardinella clupeoides, Sardinella melanura* and *Sardinella jonesi* are the nine species of Lesser Sardines found in Indian water, listed in order of abundance. *Sardinella dagi, Sardinella fimbriata,* rainbow sardine (*Dussumiera*), and white sardine (*Kowala*) are also important *Sardinella* species. Lesser Sardine is found in four states: Goa, Karnataka, Kerala, Tamil Nadu, Pondicherry, Andhra Pradesh, and Orissa.

40.2.2 Anchovies

Stolephorus, Engraulis mystax, Engraulis purina, Thrissocles, Thryssa, Setipinna, and Coilia dussumieri are small clupeid fish with silver-colored longitudinal stripes running from the head to the base of the caudal fin (Fig. 40.2). They are found in a variety of locations throughout the world's oceans but are most common in temperate waters and are rare or absent in very cold or very warm seas. Anchovies account for about 6% of India's total pelagic fish landings. In Indian water, 28 species of anchovies have been identified. The three coastal states of India that support 95% of the country's average annual catch of anchovies are Andhra Pradesh, Tamil Nadu, and Kerala on the east coast and Kerala on the west coast. Anchovies primarily consume zooplankton. The fishing season is divided into two parts: January to May and September to November, with the latter being the peak season for anchovy catches.

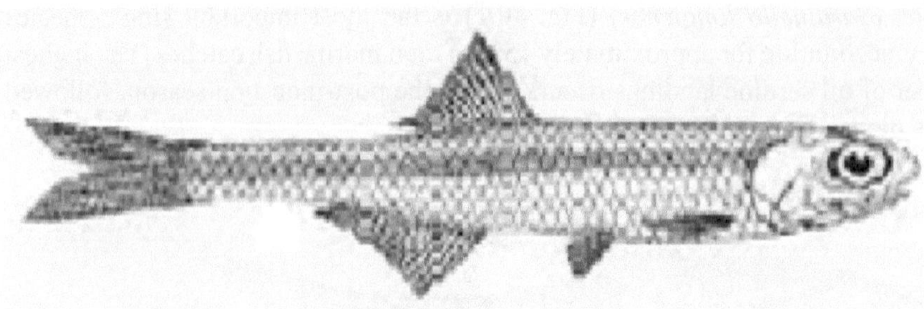

Fig. 40.2: *Stolephorus*

Hilsa ilisha, Albula vulpes, Elops saurus, Megalops cyprinoides and *Chanos chanos* are also members of this group

40.2.3 Mackerels

Mackerel is a common name for several species of schooling epipelagic fish in the Scombridae family. The Indian Mackerel (*Rastrelliger kanagurta*), Indian Chub Mackerel (*Scomber indicus*), *Rastrelliger brachysoma* and *Rastrelliger faughni* are all found in Indian water (Fig. 40.3). They can be found in both temperate and tropical seas, mostly along the coast or offshore in the ocean. Mackerels have longitudinal bands on the upper half of their bodies, dorsal and anal finlets, keels on the caudal peduncle, and a deeply forked caudal fin. It accounts for about 20% of the country's total marine catches. The majority of mackerel fishing occurs along the western coast of Tamil Nadu, Andhra Pradesh, and Orissa.

Fig. 40.3: *Rastrelliger kanagurta*

Considering the great economic importance of the Mackerel, the Government of India has established a Research Station at Karwar, Maharashtra.

40.2.4 Seer fishes

Another group of commercially important fishes in Indian coastal waters is the Seer fishes, which include Narrow-barred Spanish mackerel (*Scomberomorus commersonii*), Indo-Pacific King mackerel (*Scomberomorus guttatus*) and *Scomberomorus interruptus*.

40.2.5 Tunas

Scombridae tunas are among the largest, most specialised, and commercially important of all fish. They are found in temperate and tropical oceans all over the world and contribute significantly to global fishery production. Tunas are unique among fish because their body temperature is several degrees higher than the ambient water temperature and they have a high metabolic rate, which allows them to grow at an extraordinary rate. They have streamlined bodies and come in a wide range of sizes, colours, and fin lengths.

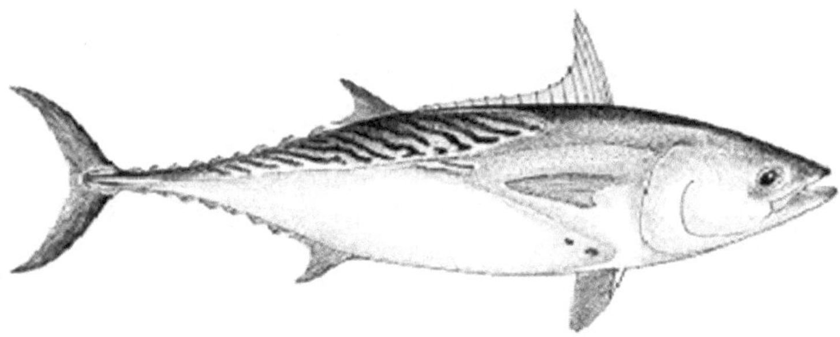

Fig. 40.4: *Euthynnus affinis*

The Indian Ocean accounts for approximately 19% of global tuna catches. With 100,000 tonnes of Coastal Tunas and 82,000 tonnes of Oceanic Tunas are available for exploitation, the water of the Andaman Islands appears to have the world's richest tuna stocks (Fig. 40.4). Tuna stocks in Lakshadweep are estimated to be between 50,000 and 90,000 tonnes. Tunas account for approximately 85% of total marine fish landings in the Lakshadweep Islands. Little Tunna (*Euthynnus affinis*), Frigate Tuna (*Auxis thazard*), Oriental Bonito (*Sarda orientalis*), Yellowfin Tuna (*Thunnus albacares*), Bigeye Tuna (*Thunnus obesus*), Skipjack Tuna (*Katsuwonus pelamis*), and others are commonly found in Indian water.

40.2.6 Bombay duck

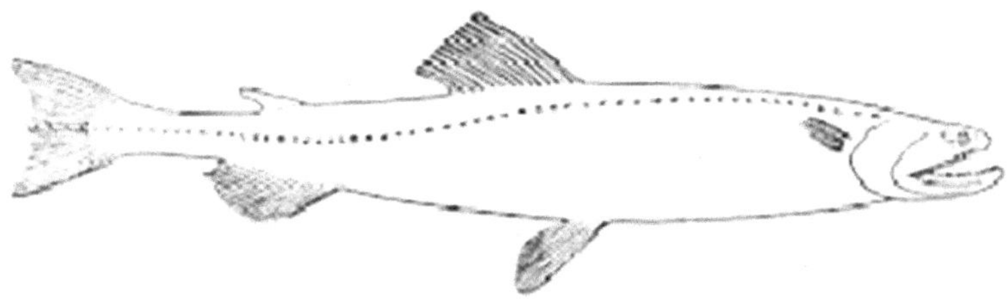

Fig 40.5: *Harpodon mirochir*

The Bombay duck (*Harpodon nehereus*) is a single euryhaline species of lizard fish found in the mid-water column (Fig. 40.5). It is a white, soft, and delicate fish with predatory habits that contribute to a major fishery along Maharashtra, Gujarat, West Bengal, Sunderbans, and Andhra-Orissa coasts and occurs sporadically along the east coast. It accounts for about 10% of the total marine catch. It is caught using a distinctive bag-net known as a 'dol,' which is used against tidal currents. It is a popular food fish along India's northwestern coast; it is sold fresh, dried or salted.

40.2.7 Cobia

Cobia (*Rachycentron canadum*) is found throughout the water column and is found in warm-temperate to tropical waters of the West and East Atlantic, the Caribbean, and similar zones in India (Fig. 40.6), Australia, and Japan's Indo-West Pacific region. In search of food, the fish congregates at reefs, shipwrecks, harbour buoys, and other submerged structures. In search of prey, juveniles may enter estuaries and mangroves. It is dark brown and young ones have alternate black and white horizontal stripes. It can grow to be 2 m long and weigh 50 kg. It is a fast-growing, high-value marine fish that is farmed in Sea Cages in India.

Fig 40.6: *Rachycentron canadum*

CMFRI standardised cobia seed production and farming technology, which is being successfully adopted by fish farmers. The success of the 'Cobia Farming Association' in the Palk Bay region, Rameswaram, and subsequent adoption by many fishermen groups in Tamil Nadu, Kerala, Karnataka, and Goa demonstrated that the technology is commercially viable.

40.2.8 Flatfishes

Flatfishes (Pleuronectiformes) are deep-sea bony fishes with oval-shaped and flattened bodies that have both eyes on one side. Flatfish can blend in by imitating the colour of the sea floor. Halibut, Flounder, Sole, Turbot, Plaice, Dab, and other species are among them. Flatfishes are important marine fishery resources in India, particularly along the West Coast, particularly the North Kerala and South Kannada coasts. Although there are approximately 91 species of flatfish reported in Indian waters, only a small number contribute to large catches, and only the Malabar Tongue Sole (*Cynoglossus macrostomus*) contributes to a regular fishery of significance (Fig. 40.7). Other species such as Large Scale Tongue Sole (*Cynoglossus macrolepidotus*), Indian Halibut (*Psettodes erumei*), Large Tooth Founder (*Pseudorhombus arsius*), etc. are also found in small quantities.

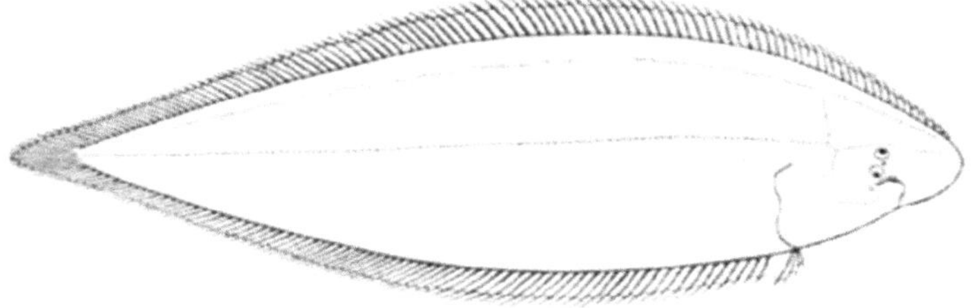

Fig 40.7: *Cynoglossus macrostomus*

Table 40.1: Other marine fishes and Their Contributions to fish catch

Sl.No.	Name of the group	Examples	Contributions
1.	Elasmobranches	*Aetobatus, Aetomyalus, Carcharhinus, Dasyatis, Galeocerdo, Herbius, Pristis, Rhinobatis, Rhincodon, Rhinoptera, Rhynchobatus, Scoliodon, Sphyrna, Stegostoma, Trygon.*	5-7% of total marine catches.
2.	Jewfish (Scianids)	*Johinus dussumieri, Otolithoides brunneus, Otolithes, Pseudosciana diacanthus.*	3.5-4.0% of total marine catches.
3.	Catfishes	*Osteogeneiosus militaris, Plotossus anguillaris, Plotossus canius, Tachysurus caelatus, Tachysurus dussumeieri, Tachysurus jelia, Tachysurus maculatus, Tachysurus sona, Tachysurus thalassinus, Tachysurus tennispinis,*	3-4% of total marine catches.
4.	Ribbon fish (Hair tail)	*Trichiurus gangeticus, Trichiurus haumela (Trichiurus lepturus), Trichiurus internedius, Trichiurus muticus, Trichiurus savala. Trichiurus pantulli*	2.5% of total marine catches.
5.	Perches	*Ambassis serranus, Acanthurus, Gerres, Lates calcarifer, Lethrinus, Lutianus, Therapon, Platax.*	1.3-1.5% of total marine catches.
6.	Eels	*Anguilla bengalensis, Muraenesox cinereus, Muraenesox talabonoides.*	0.3-0.5% of total marine catches.

41

Coldwater Fishes

41.1 Introduction

Fishes adapted to live at a temperature ranging from 10-20^{0}C constitute coldwater fish. Upland water at high altitudes in the mountains and spring water at low elevations in temperate regions remain cooler than the rest, allowing coldwater fisheries to thrive in these areas. Such bodies of water, which include several hill streams, pools, lakes and reservoirs, are abundant in the Himalayan and Deccan plateau regions of peninsular India. These are fed by either melting snow or springs in the north or rain on the Deccan plateau. The bodies of water in the hills are distinguished by their low temperature, high oxygen content, low carbon dioxide, and high velocity. As a result, the ecological conditions of hill streams differ significantly from those of plains streams.

41.2 Characteristics

The coldwater fishes are distinguished by:

1. The coldwater fishes adapted to live below 10^{0}C to 20^{0}C temperature.
2. Their mouth is adapted for rasping food with rocks, pebbles etc.
3. Their oxygen requirement for intensive feeding is about 10-11mg/l.
4. Their reparation is suppressed at an oxygen level of 4-5mg/l.
5. Their spawning varies with temperature.
6. They are adapted to freezing point temperatures.
7. They are very active and agile having great power of locomotion, clinging and burrowing habits.
8. They can thrive in hard water lacking organic contents but abundant oxygen.
9. They drop feeding intensity at a temperature of 4-7^{0}C.
10. They hatch at 9^{0}C and depend upon water temperature.
11. They tend to like shady places and a bottom that is sandy and stony.

41.3 Coldwater Lakes In India

The following are the most important coldwater lakes in India:

Table 41.1: List of some important lakes of India

Sl.No.	Name of lake	Name of State	Sl. No.	Name of lake	Name of State
1.	Bhimtal	Uttaranchal	2.	Dal	Kashmir
3.	Devikolam	Kerala	4.	Elephant	Kerala
5.	Kishensar	Kashmir	6.	Kodaikanal	Tamilnadu
7.	Nainital	Uttaranchal	8.	Ooty	Tamilnadu
9.	Renuka	Himachal	10.	Sattal	Uttaranchal
11.	Shevaray	Pradesh Tamilnadu	12.	Walar	Kashmir

41.4 Coldwater Hatcheries

In recent years, there has been a growing recognition in India for the development of coldwater fisheries, as coldwater production is negligible in comparison to total inland catch. One potential source is the total hatcheries of Kashmir (Harwan), from which brown trout have been transplanted to upland water in Jammu, Kashmir, Kullu, Shimla, Kangra, Nainital, Shilong, and Arunachal Pradesh. Additional hatcheries are being built in Avalanche (Nilgiri), Eravikulam, and Rajamallay (Kerala).

ICAR-Directorate of Coldwater Fisheries Research, Bhimtal (2022) is working towards sustainable development of coldwater aquaculture, management and conservation of the hill stream fishes.

41.5 Coldwater Fishes

These water resources harbour 272 fish species belonging to 21 families and 76 genera in the country of which 203 are recorded from the Himalayas and 91 from the Deccan Plateau (Sehgal, 1999). Out of these, a maximum of 255 species are recorded from the North-East Himalayas, 203 from the west and central Himalayas and 91 from the Deccan plateau (Ali, 2010). The Government of India has established the National Research Centre on Coldwater Fisheries (NRCCWF). A large number of indigenous fish species and several other exotic fishes thrive well in the mountain water:

A. Indigenous coldwater fishes: e.g., Mahseer, Snow trout, Indian hill trout etc,

B. Exotic coldwater fishes: Exotic fishes found in hill streams of India mainly include trout, mirror carp, tench and carp etc.

41.5.1 Mahseer

The Himalayan Mahseer is an important game fish. It grows to its greatest size and abundance in mountain streams, preferably rocky ones. In hot weather, the Rocky

Mountains provide positive security against river dwindling. Crabs and shellfish are abundant and provide a significant portion of their diet. Their juveniles can be found in the hill stream's cleaner water. During the monsoon, when the rivers are swollen with rain water, mahseer migrates downstream and ascends the rivers to spawn. Some of the most important mahseer species are listed below:

Fig. 41.1: *Acrossocheilus hexagonolepis*

1. *Acrossocheilus (Neolissochilus) hexagonolepis* (Mc Clelland) or Copper or Chocolate mahseer (Fig. 41.1): This is distributed in Upper India, Assam and the Cauvery River. It attains a length of over 60cm. it has a hexagonal shape of its scale and thin lips. It is adapted for living in ponds.

2. *Tor khudree* (Skyes): It is found in Odisha and throughout peninsular India. It attains a length of about 1.3m and its head is as long as the depth of the body.

3. *Tor mosal* (Skyes): It is found in Mountain Rivers on the foothills of the Himalayans, Kashmir, Assam and Sikkim (Fig. 41.2). It is charcaterised by its head, which is more or less equal to the depth of the body.

4. *Tor mosal mahanadieus*: It resembles mosal mahsser except its distribution. It is found in Mahanadi. Its eyes are small and higher than the depth of the body.

5. *Tor mussulah* (Skyes): It inhabits the freshwater of peninsular India and grows about 1.2m in length. Its head is much shorter than the depth of the body.

Fig. 41.2: *Tor mosal*

6. *Tor nelli* (Day): It is found in the Tunga Bhadra and Krishna Rivers of peninsular India. Its head is much shorter than the depth of the body.
7. *Tor progencius* (Mc Clelland): It is found in the rivers of the Himalayas in Assam.
8. *Tor putitora* (Ham.) or golden or common Himalayan mahseer: It can be found from Kashmir to Darjeeling in the Himalayas. It has a large head and is 2.75cm long. The fish breeds three times a year, during the winter, May-June, and July-September, according to Khan (1930). However, only two breeding seasons were reported for this fish by Sehgal *et al.* (1971). Spawning takes place in shallow rocky areas of streams, and the developmental stages feed on plankton. It has created the majority of the lakes in the Kumaun Hills. It also serves as an independent fishery for the rivers Beas and Sutlej and their tributaries.
9. *Tot tor* (Ham.): It occurs all along the Himalayan foothills from Kashmir to Assam in the upper reaches of the rivers Narmada and Tapti in Madhya Pradesh and Punjab, provide an important fishery in these areas. It grows to 1.5m in length and has a head that is shorter than the depth of the body. From July to September, this fish has a long breeding season. It lays eggs in batches, and after the last batch is laid, the ovary begins spawning preparations for the next season. The juveniles eat insects, but the adults eat plants.
10. Dark mahseer (*Naziritor chelynoides*) (Mc clelland): It is found in India and Nepal.
11. Wayanad mahseer (*Neolissochielus wynaadensis*) (Day)*:* It is collected from Wayanad (Kerala) and Kodagu (Karnataka).
12. *Crossocheilus periyarensis* (Menon and Jacob): This species is only found in Periyar River in Kerala, India.
13. *Osteobrama belangeri:* As of 2022, it is found in the wild only in Myanmar. The extirpation from Manipur was caused by dam building, habitat degradation and the introduction of alien species which caused the populations to fragment.
14. *Chagunius chagunio:* It inhabits wetlands in Bangladesh, India, Nepal and Bhutan

    It may be inferred that there has been a considerable decline in the mahseer fishery.

41.5.2 Snow trouts

They are represented by *Schizothorax* and *Schizothoraichthys*.

1. *Schizothorax:* It has a suctorial lip. It is found in the snow-fed streams of Assam, Eastern Himalayas, Nepal, Sikkim, Kashmir and Kumaon hills of the Himalayas. The following species of *Schizothorax* are known:
   (i) *Schizothorax richardsoni* (Fig. 41.3)
   (ii) *Schizothorax plagiostomus* and
   (iii) *Schizothorax molesworthi.*
2. *Schizothoraichthys:* It lacks a suctorial lip. It has the following species:
   (i) *Schizothoraichthys esocinus* (Haeckel)*:* It is found in Kashmir and Ladakh.

Fig. 41.3: *Schizothorax richardsoni*

   (ii) *Schizothoraichthys progastus* (Mc clelland)*:* It is found in the hill streams of the Ganges of Hardwar and Darjeeling.
   (iii) *Schizothoraichthys kumauoneus* (Menon)*:* It is found in Nainital.

41.5.3 Indian hill trouts

They are represented by four species of *Barilius* and various other genera of lesser importance as summarized in the following table (Fig. 41.4):

Fig. 41.4: *Barilius bola*

Table 41.2: Some important genera of Indian hill trouts

Sl. No.	Fish	Distribution	Maximum size
1.	Trout barb or Indian trout (*Barilius bola,* Ham.)	Rivers of bills and pools of Arunachal Pradesh, Assam, Bay of Bengal, Bihar, Haryana, Himachal Pradesh, North East India, Uttar Pradesh, West Bengal and Orissa.	30cm 15cm
2.	*Barilius bendelisis* (Ham.)	Stream and pool in Assam, Bihar, Punjab, Shimla, UP, Nilgiri and West Bengal.	
3.	*Barilius bendelisis chedra* (Ham.)	River Beas	152cm
4.	*Barilius gatensis* (Valenciennes)	Coorg, Mysore, Western Ghats, Travancore, Malabar, Nilgiri	
5.	*Barilius vagra* (Ham.)		
6.	*Barilius shacra* (Ham.)	Koladyne basin, India	
7.	*Glyptothorax stolickae* (Steindechner)	Punjab, Himalaya's snow-fed stream	
8.	*Glyptothorax pectinopterus* (Mc clelland)	Asia Minor to South East Asia	
9.	*Glyptothorax brevipinnis* (Hora)	River Ken, Central India.	91cm
10.	*Garra lamta*	Hill streams	
11.	*Garra gotyla*	Snow fed rivers	
12.	*Labeo dero*	Rapids and pools of Assam, Darjeeling, Eastern Himalayas, Punjab, UP, Western	
13.	*Labeo dyocheilus* (Fig. 41.5) *Labeo gonius*	Himalaya Assam, Nepal, Himalayas hill streams	
14.	*Bangana dero*		
15.		Himalayan foothills in India, Nepal, China and Bangladesh	

Fig. 41.5: *Labeo dyocheilus*

41.5.4 Trouts

Trouts are thought to be native to mountain streams. However, Danish and French Salmonid forms have a good reputation for having the largest population and the highest return per acre of cultivation.

Trout hatcheries have also been established in India, including Avalanche in Northern Ireland, Rajamally in Kerala, Achaibal, Laribal and Herwan in Kashmir, Katrain and Barot in Himachal Pradesh, Kaldayani and Talwari in Uttar Pradesh, Bimdilla in Arunachal Pradesh, and Shilling in Meghalaya. Some of the most important trouts are as follows:

1. *Slamo gairdneri gairdneri* (Richardson) or Rainbow trout or Steel Head (Fig. 41.6): It is indigenous to North American Pacific water. It was brought to India in 1907. It feeds quickly on artificial food and can withstand high temperatures and oxygen-depleted water. Their incubation period is shorter, and their growth rate is quicker. In three years, they grow to be 41-50cm long and weigh about 5.5kg. It is primarily a river fish, but it can also be grown in confined water, either naturally or artificially. They lay their eggs in shallow pools. They eat plankton, but adults eat meat.

Fig. 41.6: *Slamo gairdneri gairdneri*

   The rainbow trout of Nilgiris and Kodai hills constitute one of the important fisheries of Tamil Nadu but are poorly represented by the Kumaun and Nepal hills.

2. *Salmo truta fario* (Linnaeus) or brown trout: It is indigenous to Central and Western Europe. It is less adaptable than rainbow trout and has only been found in the streams and farms of Kashmir and the river Beas in Punjab. At the bottom, it feeds on crustaceans and large living prey. It reaches a maximum length of 46.5cm. During the breeding season, the fish migrate upstream to spawn on gravel-bedded shallows in fast-flowing water. They are more timid and demanding than other trouts.
3. *Oncorhynchus nerka* (Walbaum) or golden rainbow trout or Ko Kanee.
4. Tiger trout (*Salmo trutta* × *Salvelinus fontinalis*): It is a sterile, intergeneric hybrid of the brown trout and the brook trout.
5. *Peninsular hilltrout (Lepidopygopsis typus):* It is endemic to Kerala, India where it is only known from the Periyar River and Periyar Lake.

41.5.5 Tench (*Tinca tinca*) or Doctor Fish

It arrived in India in 1874 from Europe, Asia Minor, and Western Siberia (Fig. 41.7). Although they were introduced into Ooty Lake, they thrived best in the Nilgiris. They grow slower than carp and reach a maximum length of 50cm. It has small scales on its body, a dark golden-greenish body, and a short tail. It symbolises a type of bottom fish. Their fingerlings eat crustacean plankton, but the adults eat anything. From February to April, they breed. Sunkesula Fish Farm in Andhra Pradesh, India has successfully bred these.

Fig. 41.7: ***Tinca tinca***

41.5.6 Mirror carp

They are native to China. They were introduced into the Nilgiris in 1939 and regarded as ideal fish for coldwater fisheries. In 1947, these were introduced in the Kumaun hills and in 1954, the various lakes of that region. These have also been introduced in the warm water of the plain where they bred successfully. They are carnivorous and fast-growing fish and are also cultured in composite fish culture. They breed in confined water and grow up to 100cm. They are represented by:

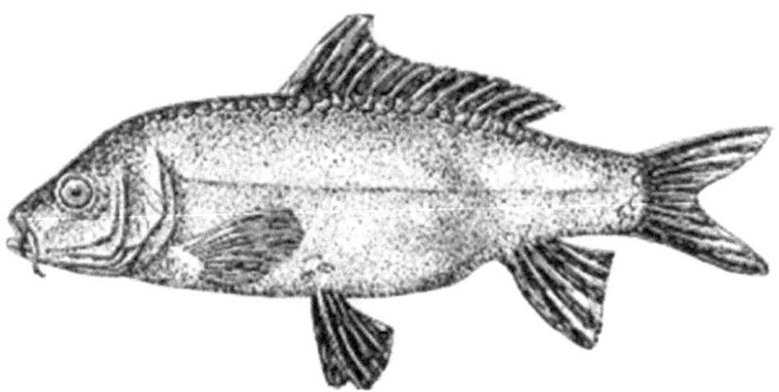

Fig. 41.8: *Cyprinus carpio specularis*

a. *Cyprinus carpio communis* (sacle carp),

b. *Cyprinus carpio specularis* (mirror carp) (Fig. 41.8) and

c. *Cyprinus carpio nudus* (leather carp).

41.5.7 Crucian carp (*Carassius carassius*) and golden carp (*Carassius aurattus*)

These are natives of the UK. In 1874, they were introduced along with tenches. It was successfully bred in Ooty Lake, Nilgiris Water and Sunkesula Lake. It can survive in ponds having low dissolved oxygen concentration. It grows slowly. The fish is scarcely appreciated.

41.5.8 Koi or Amur carp (*Cyprinus rubrofuscus*) and pond loach or Dojo loach or Oriental weather fish (*Misgurnus anguillicaudatus*, Cantor)

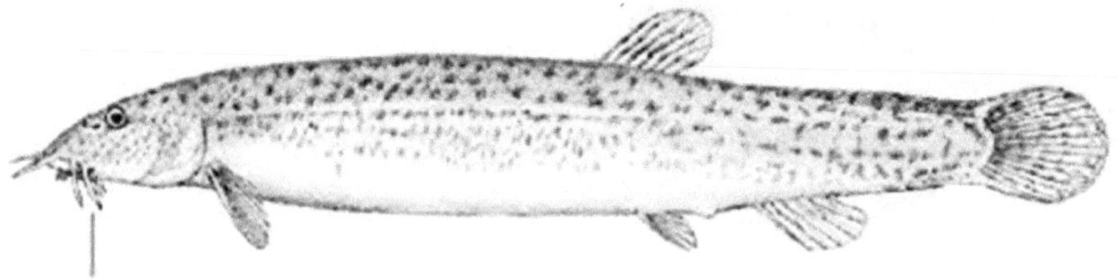

Fig. 41.9: *Misgurnus anguillicaudatus*

These are also considered to be cold-water fish because of their ability to survive at very low temperatures, but their physiological optimal temperatures are 32°C and 26-28°C respectively (Fig. 41.9).

42

Exotic Fishes

42.1 Introduction

Several fish species including their seeds, eggs and spores have been imported either occasionally or artificially and introduced into the water of our nation. These fishes are not native to our country and are called exotic fishes. Based on utility, these fishes include food fishes, game fishes and larvicidal fishes. In other words, the fish species that are not of Indian origin but transplanted into Indian water for cultural purposes are called exotic fishes. Exotic fish are an alien species that are non-indigenous having their origin in another country.

The introduction of exotic fishes to Indian water can be traced as having more than a century-old history. While the country was under British rule, such fisheries were possibly introduced as a means for recreational fisheries. During 1870-1947, 9 species of temperate food carp as exotic fishes namely Tench or Doctor fish (*Tinca tinca*), Crucian carp, wild goldfish or golden carp (*Carassius carassius*), Common carp, European carp, Mirror carp or German carp *(Cyprinus carpio),* Goramy (*Osphronemus goramy*), Cichlid *(Gambusia affinis),* Cichlid (*Labistes reticulus*), Brown trout (*Salmo truta*) and rainbow trout (*Salmo gairdneri* or *Onchorhynchus mykiss*) were introduced. The post-independence India witnessed the introduction of 8 exotic fish namely *Cyprinus carpio,* Grass carp or witnessed the (*Ctenopharyngodon idella*), Silver carp (*Hypophthalmichthys molitrix*), Tawes (*Puntius javanicus*), *Tilapia* (*Tilapia mossambica*), *Salvelinus fontinalis, Onchorhynchus nerka* and *Salmo salar*.

During the last several decades, over 300 species of exotic fish have been brought into India for experimental aquaculture, sport fishing and mosquito control.

Table 42.1: Some aspects of important exotic fishes

Name of fish	Home	Year	Destination	Purpose
Salmo trutto fario	The U.K.	1863	Nilgiris (Tamil Nadu)	As a game fish
Carassius carrasius	The U.K. England,	1870	Otacamund or Ooty lake (Tamil Nadu)	As a food fish
Tinca tinca	The U.K. England, Europe	1870, 1874	Otacamund and Nilgiris (Tamil Nadu)	As a food fish
Salmo trutto fario	The U.K. England	1899, 1901	Harwan (Kashmir)	As a game fish
Salmo gairdneri (*Onchorhynchus mykiss*)	Sri Lanka, Germany, The U.K.	1904,1907, 1912	Harwan (Kashmir)	As a game fish
Lebistes reticulatus	South Africa	1908		As a larvicidal fish
S. gairdneri	Sri Lanka, Germany, New Zealand	1909-1910	Nilgiris (Tamil Nadu)	As a game fish
Salmo trutta	England, The U.K	1901, 1909-1938	Munnar High Range (Kerala)	As a game fish
Osphronemas gorami	Java And Mauritius	1916	Tamil Nadu	As a food fish
Gambussia affinis holobrooki	Italy	1928	Cuttack (Orissa)	As a larvicidal fish
Salmo gairdneri gairdneri	The U.K. Sri Lanka	1938-1940	Munnar High Range (Kerala)	As a game fish
Cyprinus carpio	Sri Lanka, Bangkok	1939, 1947	Nilgiris (Tamil Nadu), Nainital, Kumaon and Bangalore	As a food fish
Salmo gairdneri	The U.K	1941	Munnar High Region (Kerala)	As a game fish
Tilapia mossambica	Bangkok, Sri Lanka	1952	Tamil Nadu and Kerala	As a food fish
Hypophthalmichthys molitrix	Hongkong, Japan	1959	Cuttack	As a food fish
Ctenopharyngodon idella	Hongkong, The U.K.	1959	Cuttack	As a food fish
Salmo salar	Canada	1960-1970	Harwan/Laribal (Kashmir)	As a game fish
Salvelinus fontinalis	U.S.A	1963	Harwan (Kashmir)	Aquaculture
Onchorhynchus nerka	Japan	1968	Nilgiris (Tamil Nadu)	As a game fish
Salmo trutta	Japan	1968	Nilgiris (Tamil Nadu)	As a game fish

Name of fish	Home	Year	Destination	Purpose
Japanese Rainbow Trout	Japan	1968	Nilgiris (Tamil Nadu)	As a game fish
Tiger Trout (Brown Trout X Eastern Brook Trout)	Japan	1968	Nilgiris (Tamil Nadu)	As a game fish
Albino Rainbow	Japan	1968	Nilgiris (Tamil Nadu)	Aquaculture
Lake trout (Lake Trout X Brown Trout)	Canada	1968	Harwan/Laribal (Kashmir)	Aquaculture
Puntius javonicus	Indonesia	1972		As a food fish
Rainbow trout Selective quick Mann	Islet of (The U.K.)	1984	Kokernag (Kashmir)	Aquaculture
Pangasius sutchi	Bangladesh, Thailand	1994-1995	West Bengal	Aquaculture

(Source: National Bureau of Fish Genetic Resources, Lucknow)

42.2 Account of some important exotic fishes

42.2.1 Green tench or Doctor fish (*Tinca tinca,* Linnaeus, 1758)

Tinca tinca is a freshwater and brackish water fish native to Eurasia, ranging from Western Europe, including the British Isles, east into Asia as far as the Ob and Yenisei Rivers (Fig. 42.1). It is also discovered in Lake Baikal. It prefers slow-moving environments, such as lakes and lowland Rivers. Tench are most commonly found in still waters with clay or muddy substrate and plenty of vegetation. It can tolerate low oxygen concentrations found in water where even carp cannot survive. Tench are bottom feeders that prefer animals such as chironomids on the bottom of eutrophic waters and snails in well-vegetated water. Breeding occurs when the sticky green eggs can be laid. Spawning occurs in the summer and can produce up to 300,000 eggs. However, spawning in Nilgiris coldwater occurs from February to April. Growth is rapid, and fish can reach 0.11 kg (0.25 lb) in their first year. Tench are edible and can be used in recipes that would normally call for carp, but they are rarely eaten these days.

Fig. 42.1: *Tinca tinca*

42.2.2 Crucian carp, wild goldfish, English carp or golden carp (*Carassius carassius,* Linnaeus, 1758)

Crucian carp is indigenous to Europe, from which it has spread to Japan, Borneo, China, Indonesia, and India (Fig. 42.2). It is a freshwater fish that thrives in ponds and lakes due to its ability to breed in confined water. These eat various types of plankton. It can grow to a maximum length of 45 cm. The eggs are 0.8mm in diameter and sticky. These hatch alongside aquatic plants. It can breed with *Cyprinus carpio*. As a result, it is necessary to avoid its introduction into ponds containing *Cyprinus carpio*.

Fig. 42.2: *Carassius carassius*

42.2.3 European Carp or Common Carp (*Cyprinus carpio*)

This fish is found all over the world. Based on the texture of the skin and size of the scales, *Cyprinus carpio* has been divided into 3 main varieties (Fig. 42.3). They are:

Fig. 42.3: *Cyprinus carpio*

A. Scale carp, Eurasian carp or European carp (*Cyprinus carpio* var. *communis*) body completely covered with smaller scales.
B. Mirror carp (*Cyprinus carpio* var. *specularis*) with large size shiny scale over the body.
C. Leather carp or naked carp (*Cyprinus carpio* var. *nudus*) bodies almost without any scales.

Mirror carp was introduced to India from Sri Lanka in 1939. The German strain of this fish was introduced in 1946 at the Bhowali hatchery in Uttarakhand for stocking the Kumaon lakes. The species was introduced into Ooty Lake and has since become well-established in Nilgiri water. Because common carp were not reproducing freely in India's tropical waters, a shipment of scale carp was brought from Bangkok to Cuttack in August 1957. Later, it found a home in Kashmir lakes, where it was invaded and heavily infested to the exclusion of all other local species, particularly the *Schizothorax* (Sehgal, 1989). Because of its bottom-feeding omnivorous habitat, it competes for space and food with the indigenous *Cirrhinus mrigala, Clarias batrachus* and scampi (*Macrobranchium rosenbergii*).

In the absence of adequate food, it burrows for food, causing damage to the pond dyke sand and making the water turbid, reducing natural productivity by suppressing phytoplankton production. It has recently been recorded from almost all major rivers as a result of pond and reservoir overflow. Instances of common carp have been reported to be causing the decline of reservoirs such as *Cirrhinus* spp. (Valsankar, 1987) and Krishnaraja sagar (Sugunan, 1995). A study on the catch composition of carp availability along the Ganga stretches revealed that the population of Common carp has increased while the availability of Gangetic carp has decreased (Singh et al, 2010). It spawns twice a year, from January to March, and again from July to August. It is a good fish to grow in conjunction with other Indian carp.

42.2.4 Tilapia (*Tilapia mossambica*)

Fig. 42.4: *Tilapia mossambica*

Tilapia a native and hardy South African fish, was introduced to India by CMFRI in 1952 (Fig. 42.4). A few fingerlings were brought to Madras the same year. In 1953, a few fingerlings were brought to the CIFRI centre in Cuttack for a detailed investigation. The adult fish is herbivorous and grows quickly, reaching 9cm in two months. It breeds at 15-60-day intervals throughout the year. It is a riverine fish that also lives in estuaries. It did not do well in ponds from the start, as stunted populations appeared rather quickly, and the impact was found to be quite severe on major carp, pearl spot and milkfish. It is grown alongside *Channa striatus* and *Channa marulius*. It can survive in a variety of salinities and can be cultured in both coastal and paddy fields. *Tilapia* mono-sex culture is profitable because they grow in size rather than reproducing at an early stage. It cannot survive at a temperature of less than 10^0C.

42.2.5 Silver carp (*Hypophthalmicthys molitrix*)

This fish was introduced in Cuttack and is now found throughout India (Fig. 42.5). It is a non-predatory fish that primarily feeds on phytoplankton but also consumes rice bran and flour. It is a fast-growing fish that can weigh up to 5 kg. It is commonly grown with *Catla catla, Labeo rohita* and *Cirrhinus mrigala*. It produces well when mixed with *Catla* in a 2:1 ratio.

Fig. 42.5: *Hypophthalmichthys molitrix*

42.2.6 Grass carp (*Ctenopharyngodon idella*)

Fig. 42.6: *Ctenopharyngodon idella*

The grass carp was introduced in the year 1959 primarily for controlling submerged vegetation (Fig. 42.6). However, due to its fast growth rate soon it became an integral part of composite fish culture. It is generally cultured with *Catla catla, Labeo rohita* and *Cirrhinus mrigala*. It attains a weight of 5-7 kg in three years. The Grass carp feed on *Hydrilla verticillata*, but it does not consume *Eichhornia crassipes*, *Pistia stratiotes* and *Salvinia molesta*. Its daily requirement is as much as its body weight and so it is a problem to provide it with a large number of preferred weeds. On one hand, controlling the weeds brings in circulation the nutrients locked up by the weeds and produces valuable fish protein; on the other hand, it impacts the survival of those fishes and prawns that hide in the weeds to escape the predators, especially the murrels. The negative impacts on the environment by grass carp are alteration of water quality including an increase in turbidity, reduced dissolved oxygen, and an increase in plant nutrients. Removal of aquatic vegetation may alter the invertebrate community, thereby influencing species abundance and composition of fishes. Reduction in bluegill populations potentially attributed to increased turbidity, distribution of bluegill nesting, or silt-covered eggs.

42.2.7 Thai catfish (*Pangasius sutchi*)

It has been introduced illegally and bred in hatcheries (Fig. 42.7). It is an obligatory air-breathing shark catfish that uses a swim bladder. It plays an important role in Asian Aquaculture and commercial fishing. It is an exotic fish introduced in India during the mid-'90s. It is also known as an iridescent shark, due to its typically attractive appearance during the juvenile stage. Owing to its fast growth of about 1 kg in 3 months, the fish is already established as a profitable species for aquaculture in West Bengal and Andhra Pradesh.

Fig. 42.7: *Pangasius sutchi*

It is reported that it invades the natural water. The fish was found to be infected with at least 6 species of myxozoan parasites in Malaysia. Whatever may be the reason for its limited area of cultivation particularly, the coastal belt; the impact of this exotic catfish still requires studies.

42.2.8 Salmonids

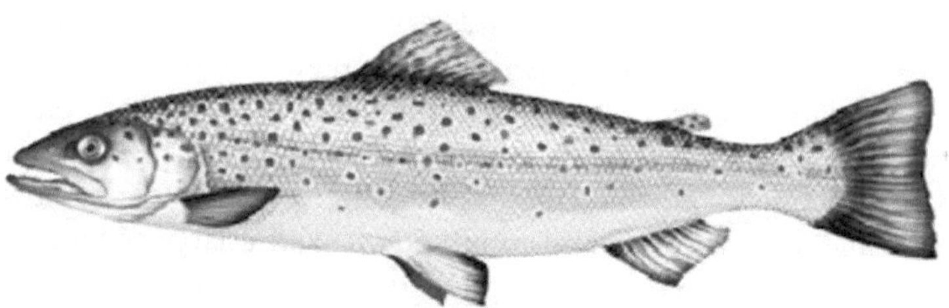

Fig. 42.8: *Salmo trutta fario*

The salmonid family truly belongs to the temperate water of the Northern Hemisphere. In India, the introduction of salmonids was largely confined to the Himalayan states of Kashmir, and Himachal Pradesh and comprised brown trout, (*Salmo trutta fario*) in 1899 and rainbow trout (*Salmo gairdneri*) in 1912 (Fig. 42.8). Post Independence additions are Eastern Brook Trout (*Salvelinus fontinalis*) and the land-locked salmon (*Salmo salar*). The Kokane salmon (*Onchorynchus nerka*), was also added in 1968 in Tamil Nadu. It has been found to displace local snow trout species.

Both brown trout (*Salmo trutta fario*) and rainbow trout (*Oncorhynchus mykiss*) are now being regularly stocked in different river stretches in Himachal Pradesh, Jammu and Kashmir as well as Uttaranchal. However, there is no organized study to assess the impact of this kind of unplanned introduction of trouts in open water. Whatsoever may be the impact of such introductions, the recent mass mortality in rainbow trout in Himachal Pradesh during 2002 possibly on account of severe problems has indeed drawn the attention of the scientists calling for quarantining needs and ecological concerns. Sehgal, 1989 has reported that the introduction of *Salmo trutta fario* has eradicated the local cold water spp. It breeds in winter from December to February when the temperature is 2.5^{0}C.

42.2.9 Silver Barb (*Puntius gonionotus*)

Fig. 42.9: *Puntius gonionotus*

The fish has gained popularity in West Bengal on account of its fast growth rate but was never considered a weed-eating fish and a competitor of grass carp. Experiments on its compatibility with the Indian major carps have shown it to be affecting the growth and production of Rohu (Fig. 42.9).

42.2.10 Black carp (*Mylopharyngodon piceus*)

It is stocked in composite carp culture to control molluscs and it is found to grow fast and attain a weight of 4 kg. Due to this the culture of indigenous *Labeo calbasu* and *Pangasius pangasius* is being neglected (Fig. 42.10).

Fig. 42.10: *Mylopharyngodon piceus*

42.2.11 Gaurami (*Osphronemus goramy*)

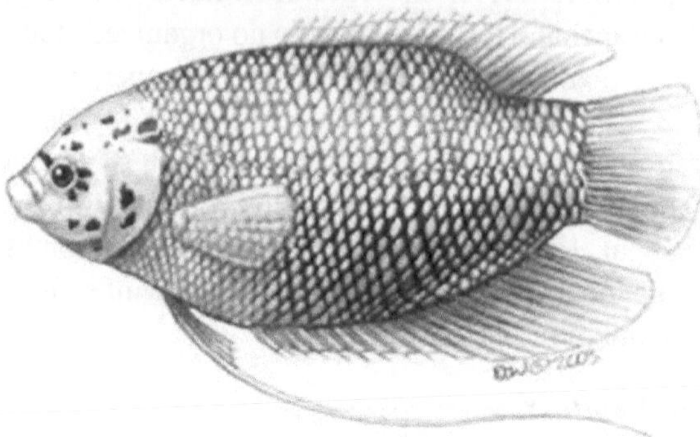

Fig. 42.11: *Osphronemus goramy*

Gaurami is an excellent food fish because of its good flavor. It has been introduced in ponds and streams of Mumbai, Bengal, Orissa and other states (Fig. 42.11). Its fries are insectivorous but adults are herbivorous. This is not a fast-growing fish. They may survive in water with low oxygen content because these possess accessory sir breathing organs. It does not survive temperatures below 15^{0}C. It builds oral nests with a circular mouth.

43

Game Fishes

43.1 Introduction

Game fish (also known as sport fish or quarry) are freshwater or saltwater fish that are pursued by recreational anglers or fished with a rod and line from clear running water (Hora 1943). Some game fish (such as salmon and tuna) are commercially targeted. In other words, game fish are fish caught for sport or subsistence. Trophy fish are game fish that are larger and heavier than the average for the species.

The World Record Game Fishes, published annually by the International Game Fish Association (IGFA), keeps records on 400 different species all over the world. The records are divided into categories, with separate records for juniors, tackle and line used, fly fishing, and location. The International Game Fish Association is also in charge of organising world saltwater championship tournaments.

Salmon, trout, and char are considered "game fish" in the United Kingdom.

Other than Salmonids, popular freshwater fish are referred to as "coarse fish" or "rough fish". Examples of coarse fish are bream (*Abramis brama*), barbel (*Barbus barbus* or *Tor barbus*), chub (*Squalius cephalus*), dace (*Leuciscus leuciscus*), ide (*Leuciscus idus*), gudgeon (*Gobio gobio*), pike (*Esox lucius*), roach (*Rutilus rutilus*), rudd (*Scardinius erythrophthalmus*) and tench (*Tinca tinca*). They are native to the British Isles. There is no taxonomic basis for the distinction between coarse and game fish. Coarse fish have scales that are generally larger than game fish scales.

Some popular game fish have been introduced and stocked throughout the world. Rainbow trout, for example, can be found almost anywhere there is a suitable climate, from their native range on North America's Pacific Coast to the mountains of southern Africa, and is now considered one of the worst invasive species.

Game fish such as barracuda (*Sphyraena barracuda*), mullet (*Mugil cephalus*), perch (*Perca fluviatilis* and *Perca flavescens*), tuna (*Thunnus thynnus*), marlin (*Makaira nigricans*), sailfish (*Istiophorus platypterus*) and seerfish (*Scomberomorus commerson*) can be found off the coasts of India. The Andaman and Nicobar Islands are frequently referred to as the "best game fish destination" in the world. Since 1993, the government has permitted "sea game fishing," particularly along the Andaman Islands.

43.2 Types

1. Atlantic salmon (*Salmon solar*)
2. Brown trout or sea trout (*Salmo trutta*)
3. Char (*Salvenius*)
4. Grayling (*Thymallus thymallus*)
5. Kokanace (*Oncorhynchus nerka*)
6. Mahseer (*Tor* and *Acrossocheilus*)
7. Rainbow trout (*Salmo irridius, Salmo gairdnerii gairdnerii, Oncorhynchus mykiss*)

Anglers' favourite fish varies according to geography and tradition. Some fish are sought after for their food value, while others are sought after for their fighting abilities or the difficulty of successfully luring the fish to bite the hook.

Tuna, tarpon, grouper, and billfish (sailfish, marlin, and swordfish) are popular saltwater bony fish. Other predatory fish such as sharks, barracuda, and dolphin fish are occasionally pursued.

Many anglers in North America target common snook, redfish, salmon/trout, bass, northern pike and muskellunge, walleye/sauger, sturgeon, and several catfish species.

Panfish are the smallest fish sought by younger anglers because they can fit whole into a standard cooking pan. Crappies, perch, rock bass, bluegills, and other sunfish are examples (Centrarchidae).

43.2.1 Freshwater Mahseer

Mahseer is a large cyprinid fish that is endemic to Asia and can be found in Afghanistan, Pakistan, India, Sri Lanka, Nepal, Bhutan, Myanmar, Thailand, China, Laos, Cambodia, Vietnam, Indonesia, and Malaysia. In its report on fisheries, the National Committee on Agriculture (1976) stated that "there was a general decline in the mahseer fishery due to indiscriminate fishing of brood and juvenile fish and the adverse effect of river-valley projects".

They are popular cultural icons of economic, recreational, and conservation interest in many of these countries, and serve as a "freshwater game fish". Currently, approximately 17 species of Tor are recognised as valid, many of which are of significant conservation concern. Of the 17 species, 14 have been assessed for conservation status by the International Union for Conservation of Nature (IUCN) in their Red List of Threatened Species.

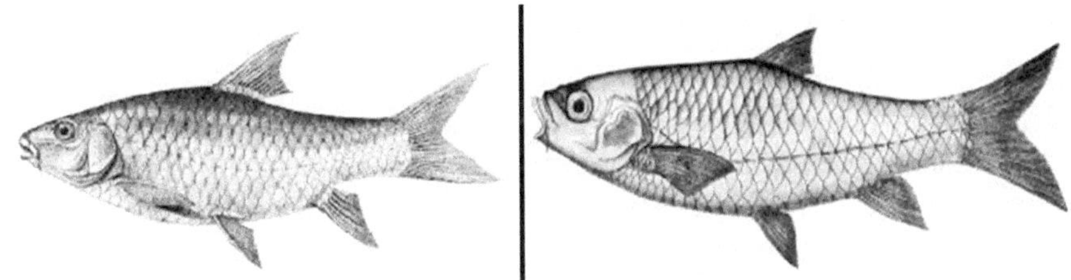

Fig. 43.1: *Tor khurdee*

Fig. 43.2: *Tor tor*

In India primarily seven valid species are listed as valid species by IUCN, the yellow or Deccan mahseer (*Tor khudree*) (Fig. 43.1), dwarf mahseer (*Tor kulkarni*), Malabar mahseer (*Tor malabaricus*), the copper mahseer (*Tor mosal*), the golden or the yellow-finned Himalayan mahseer (*Tor putitora*), hump-backed mahseer (*Tor remadeviae*) and deep-bodied or red-finned mahseer (*Tor tor*) (Fig. 43.2).

43.2.2 Indian Major Carps

Three species of Indian carp are the most economically important. They are: catla (*Catla catla*), rohu (*Labeo rohita*) and mrigal (*Cirrhinus mrigala*). These fishes' natural habitats are the rivers and backwaters of Northern India, Pakistan, and Burma. In addition, rohu can be found in rivers in India and Nepal. These large carp have also spread to many other areas and countries. Though not as widely accepted as common and Chinese carps, which are more adaptable and tolerant of a wider temperature range, Indian major carps are still the most important cultured fish species in India, Pakistan, and Burma.

Of lesser importance are: calbasu (*Labeo calbasu*), reba (*Cirrhinus reba*), white carp (*Cirrhinus cirrhosa*), bata (*Labeo bata*) (Fig. 43.3) and fringe-lipped carp (*Labeo fimbriatus*). In addition to these invasive carp, which are almost always present in every river system and are also considered game fish are Big Head Carp (*Hypophthalmichthys nobilis*) (Fig. 43.4), Silver Carp (*Hypophthalmichthys molitrix*), Common Carp (*Cyprinus carpio*) and Grass Carp (*Ctenopharyngodon idella*).

Fig. 43.3: *Labeo bata*

Fig. 43.4: *Hypophthalmichthys nobilis*

43.2.3 Indian Freshwater Catfish

A catfish is a diverse group of ray-finned fish that range in size and behaviour from the three largest species, the Mekong giant catfish from Southeast Asia, the wels catfish from Eurasia, and the Piraba from South America, to detritivores and even a parasitic species, the candiru (*Vandellia cirrhosa*) (Fig. 43.5). There are armour-plated and scale-less varieties. Contrary to popular belief, not all catfish have prominent barbel. Siluriformes are distinguished by features of the skull and swim bladder.

Fig. 43.5: *Vandellia cirrhosa*

Goonch, Mulley, Asian Red Tail, Long whiskered catfish (*Sperata aor*) also known as Indian shovelnose catfish and Giant river catfish are important game catfish (*Sperata seenghala*) (Fig. 43.6).

Fig. 43.6: *Sperata seenghala*

43.2.4 Indian Snakeheads

Snakeheads are freshwater fish from the Channidae (=Ophiocephalidae) family that are native to Africa and Asia. Long dorsal fins, large mouths, and shiny teeth distinguish these elongated, predatory fish. They breathe air through their gills, allowing them to travel short distances over land. When they get older, they develop supra-branchial organs, which are a primitive form of a labyrinth organ.

Fig. 43.7: *Channa diplograma*

The game fish species of snakeheads also known as murrels in India are bullseye snakehead (*Channa marulius*), spotted snakehead (*Channa punctata*), striped snakehead (*Channa striata*) and the great snakehead or Malabar snakehead (*Channa diplograma*) (Fig. 43.7).

43.2.5 Other Species of Freshwater Game Fish

Other Game fish in the freshwater are Indian Mottled Eel, Knife Fish, Spiny back eel, Pearl Spot, other *Labeo* species, Snow Trout and Indian Trout.

Fig. 43.8: *Hypselobarbus dobsoni* **Fig. 43.9:** *Piaractus brachypomus*

A large number of *Hyspelobarbus* are considered good game fish including *Hypselobarbus dobsoni* also known as Krishna carp and Puliches (Fig. 43.8).

The other invasive game fish of India are *Tilapia*, Pacau (*Piaractus brachypomus*) (Fig. 43.9), Brown Trout and Rainbow Trout.

44

Ornamental Fishes

44.1 Introduction

Ornamental fish are a diverse and colourful species benefiting millions of aquarium enthusiasts. The term "ornamental fish" was first coined in the early 19th century to describe fish kept solely for decorative purposes by ancient civilizations such as the Sumerians and Egyptians.

They are known for their shapes, sizes, vibrant colours, intricate patterns and distinct behaviours with hobbyists and collectors. They are widely used to distinguish between fish kept for food or other practical purposes.

Care is required to ensure their health and well-being. Their fishing can be a rewarding hobby that offers relaxation and enjoyment to people of all ages. They embrace creating and maintaining an appropriate aquarium environment, which includes proper filtration, heating, lighting and water parameters. Enthusiasts frequently enjoy decorating their aquariums with plants, rocks, and decorations to provide a visually appealing and stimulating environment for their fish.

44.2 Commercial Freshwater Ornamental Fishes

44.2.1 Angel fish: *Pterophyllum scalare* are very popular and a favorite among hobbyists as they are beautiful and look graceful in the aquarium (Fig. 44.1). In Asia, there are 25 varieties now with unique colour patterns. They may be single-coloured ones like black, silver and gold or multi-coloured pattern ones such as leopard, striped or zebra and lace-like, mottled or marble, half black etc. Black veil tail, diamond, ghost, blushing, golden marble, pearl scale and koi are the other popular varieties in Asia. They originate from the Amazon basin and Guyana River where the water bodies have been densely overgrown by aquatic plants. They prefer an environment with thick aquatic vegetation.

Fig. 44.1: *Pterophyllum scalare*
(from https://www.researchgate.net*)*

Fig. 44.2: *Barbus tetrazona*
(from https://nas.er.usgs.gov)

44.2.2 Dwarf chameleon: *Badis badis* is a native of India, Thailand and Myanmar. It changes its colour quite often depending on mood and environment, especially during courtship and breeding (Fig. 44.2). *Badis* has three subspecies, *Badis badis badis* found in the Indian sub-continent from the northeastern side, *B. badis siamensis* of Thailand and *B. badis burmanicus* from Myanmar.

44.2.3 Barbs: They comprise more than colourful 400 species in their entire range, which extends from Africa, Asia and Europe to Central China, the Philippines and the East Indies. About 150 species of barbs are of ornamental value, comprising mostly wild-caught varieties. *Barbus tetrazona* (Tiger barb), *Barbus conchonius* (Rosy barb) and *Puntius titteye* (cherry barb) are the most common barbs in the aquarium industry. *Puntius denisonii*, a wild-caught variety of barb is mainly found in Kerala and Karnataka.

44.2.4 Catfish (*Corydoras*): They are natives of Brazil, Uruguay, northern Argentina, Venezuela, Peru and Colombia. The common species found in the hobby market are *Corydoras ambiacus*, *C. agaassizii*, *C. leucomelas*, *C. schwa rtzi*, *C. punctatus*, *C. parallelus*, *C. pulcher* and *C. ornatus*.

44.2.5 Danio: They are small and lively fish native to the Indian peninsula, Sri Lanka, Pakistan, Thailand, Myanmar, Malaysia and Indonesia. They are found in different habitats from boulder-strewn mountain torrents to small pools in dry zone streams. There are more than 12 species reported today of which *Danio malabaricus* (Pearl Danio), *D. albolineata* and *Brachydanio rerio* (zebra fish) are common in the hobbyist market. *Brachydanio rerio* is mainly found in Kerala and Karnataka.

44.2.6 Discus (*Symphysodon sp*): It is said to be the queen of the aquarium, native to South America and found in slow-flowing streams with alkaline water. The wild forms are the blue discus (*S. aequifa sciatus haraldi*), and the green discus (*S. aequifasciata*), Heckel discus (*S. discus*) and the brown discus (*S. aequifasciatus axelrodi*) (Fig. 44.3). The wild blue discus is distributed in the rivers of Peru and Brazil while green discus is found mainly in the Peruvian Amazon. The blue discus has a wide variety of blue colouration with some having blue stripes on the body, head and fins. The green discus has also a varying degree of colour ranging from yellowish green to olive green and solid green to light brown and has green stripes and red spots on the sides of the body.

Fig. 44.3: *S. aequifa sciatus haraldi* *(from https://www.shutterstock.com)*

Fig. 44.4: *Epalzeorhynchos bicolor* *(from https://www.aqa.cz)*

44.2.7 Eels (Spiny eels or swamp eels): They are distributed in India, Pakistan, Sri Lanka, Bangladesh, Nepal, Myanmar, Thailand, Malaysia, Vietnam and Indonesia. *Macrognathus aculeatus* (spotted spiny eel) and *Mastacembalus armatus*, (zigzag eel) are of much importance as far as the ornamental fish market is concerned. *Macrognathus aculeatus* is commonly referred to as the peacock or spotted spiny eel. These are mainly found in West Bengal, Tamil Nadu, Kerala and Maharashtra.

44.2.8 Fighter (*Betta splendens*): Siamese fighting fish is a popular aquarium fish native to the Mekong basin of Laos, Cambodia, Vietnam and Thailand. The colour patterns of this species range from vivid red, blue, bright green, purple and white and black to cream with red fins. Apart from coloration, the finnage also has varieties like veil tail, crown tail, half moon, butterfly, double tail etc. The males of their kind and fight ferociously until death, hence the name of the fish.

44.2.9 Freshwater sharks: They are different types of minnows belonging to *Balantiocheilos, Epalzeorhynchos, Labeo* and *Luciosoma* (Fig. 44.4). These minnows are widely distributed over Southeast Asia, the Malay Archipelago, Indonesia, and parts of the Middle East and Africa. *Epalzeorhyncho* has four types of freshwater sharks, *E. bicolor, E. frenatus, E. kalopterus*, and *E. munense*. Two species of *Labeo*, *Labeo chrysophekadion* and *L. cyclorhynchus* are popular in the aquarium trade. The genus, *Luciosoma* has five species, of which *L. siplopleura* is popular in the hobby. Bala shark, as popularly known, is a short form of its Latin name, *Balantiocheilus melanopterus*, which is native to Malaysia, Thailand, Cambodia and Indonesia (Sumatra and Borneo).

44.2.10 Glassfish: They comprise mostly fish with transparent bodies. They originate from Asia, especially Pakistan, India, Nepal, Bangladesh, Myanmar and Thailand. There are four species commonly referred to as glassfish in the trade: *Chanda ranga, C. lala, C. nama* and *C. baculis*.

44.2.11 Goldfish (*Carassius auratus*): It is considered to be the most popular and attractive pet fish among all ornamental fishes, due to its many variations such as colour, fin shape, size and body structure. Though similar in appearance to carp (*Cyprinus carpio*), goldfish lack barbels and a dark spot at the base of each scale. The goldfish is the most common carnivorous aquarium fish and one of the oldest and best-known fish in the industry.

44.2.12 Gourami or labyrinth fish: It is a native of western and southern Africa, northeastern, southeastern, south and Southeast Asia. The Dwarf Gourami or Khosti (*Colisa lalia*), occurs naturally in freshwater ponds, streams and the paddy fields of northeastern India and Bangladesh, and is a most popular among the Gouramis. Important varieties include The Honey Gourami or Dwarf Gourami (*Colisa chuna*), Indian Gourami (*C. fasciata*), Thick-lipped Gourami (*C. labiosa*), Three spot gourami (*Trichopodus trichopterus*), Pearl Gourami (*T. leeri*), Snakeskin (*T. pectoralis*), Moonlight Gourami (*T. microlepis*) and the Kissing Gourami (*Helostoma temmincki*) (Fig. 44.5).

Fig. 44.5: *Trichopodus trichopterus* (*from* https://www.researchgate.net)

Fig. 44.6: *Hyphessobrycon herbertaxelrodi (from https://www.istockphoto.com)*

44.2.13 Guppy, millions of fish or rainbow fish (*Poecilia reticulata*): It is one of the most popular freshwater peaceful livebearer aquarium fish species after goldfish. The origin is from Central America to Brazil and the West Indies and is a prolific breeder. Depending on the variations in the tail fins, they are also known as round tail, spear tail, fan tail, veil tail, pin tail etc.

44.2.14 Loaches (*Botia*): There are around 40 species of loaches known today. They are natives of Thailand, India, Pakistan, China, Bangladesh and some Indonesian Islands. They are probably the most diverse group of fish in the hobby, both in pattern and behaviour.

44.2.15 Molly: Black molly (*Poecilia sphenops*) is a native of Central America, from Mexico to Columbia; sail fin molly (*P. latipinna*) is from Southeastern North America, from the Carolinas through the Gulf coast of Southern Mexico while Lyre tail molly (*P. velifera*) is found in the streams of Southern Mexico. Molly is a very familiar fish in the aquarium. About 25 strains of molly are being cultured and traded in the market, black molly being well-known.

44.2.16 Oscar (*Astronotus ocellatus*), tiger Oscar or marble cichlid or the velvet cichlid: They are native to South America, particularly Peru, Colombia, Brazil and French Guiana.

44.2.17 Platy (*Xiphophorus maculatus*): It is native to the east coast of Central America and Southern Mexico. It is a livebearer and is most popular among aquarium fishes. Several different colour variations have been developed so far such as red, yellow, orange, blue and white. Three species as Southern Platy (*Xiphophorus maculatus*), Variatus Platy (*X. variatus*) and Swordtail Platy (*X. xiphidium*) are commonly available.

44.2.18 Pufferfish (*Carinotetraodon*): *Carinotetraodon lorteti*, *C. salivator*, *Monotetrus travancoricus*, *Chonerhinos amabilis*, *C. nefastus*, *C. modestus*, *C. remotus*, *C. asellus*, *Colomesus asellus*, and *C. psittacus*. *Monotetrus travancoricus* popularly known as Malabar or dwarf pufferfish, is a native of India and is considered the smallest puffer in the trade. *T. fluviatilus*, known as the Ceylon puffer, is a native of Sri Lanka, India, Bangladesh, Myanmar and Borneo.

44.2.19 Severum: These are widespread throughout the northern Amazon Basin and Guyana and are commonly referred to as 'poor man's discus". There are several different color variations available, including the "Gold," Green, Brown, and "Peruvian Green" strains. The most common species in trade include *Heros severus* and *Heros efasciatus*.

44.2.20 Shark catfishes (*Pangasius*): Important species belonging to this group are *Pangasius pangasius*, *P. nieuwenhuisii*, *P. humeralis*, *P. lithostoma*, *P. kinabatanganensis*, *P. macronema* and *P. pleurotaenia*. *Pangasius pangasius* is distributed throughout India.

44.2.21 Snakehead (*Channa* or *Ophiocephalus*): 28 species of *Channa* are known today, *C. bleheri*, *C. burmanica*, *C. gachua*, *C. micropeltes*, *C. lucius*, and *C. orientalis*, are well known in the aquarium trade. *C .bleheri* is widely known as the rainbow snakehead because of its body colouration. It originates from the upper region of northeastern India, particularly in the Brahmaputra river basins of Assam. *C. gachua* or dwarf snakehead found in rainforest streams of Sri Lanka. They are also found in Southeastern Iran, Eastern and Western Pakistan and Southern China.

44.2.22 Swordtail (*Xiphophorus hellerii*): It is a native of North and Central America stretching from Veracruz, Mexico, to northwestern Honduras. They have been interbred to produce all kinds of interesting colors and different types of finnage. Some of the more common varieties of the Swordtail are Red, Red Wag, Red Tux, Painted, Neon Green, Marigold (and wag), Pineapple, Black, Red Twin bar, Sunset, and Gold Tuxedo.

44.2.23 Tetra: Tetra includes several small species of freshwater fish from Africa, Central America and South America belong to the biological family

Characidae. There are ten varieties of tetras known as Neon tetras (*Paracheirodon innesi*), Cardinal tetra (*Paracheirodon axelrodi*), Black neon (*Hyphessobrycon herbertaxelrodi*), Black tetra (*Gymnocorymbus ternetzi*), Bleeding heart tetra (*Hypessobrycon erythrostigma*), Bloodfin (*Aphyocharax anisitsi*), glow light tetra (*Cheirodon erythrozonus*), Head and tail light tetra (*Hemigrammus ocellifer*), Lemon tetra (*Hyphessobrycon pulchripinnis*), Serpae tetra (*Hyphessobrycon callistus*) (Fig. 44.6). They are peaceful and pose no threat to any other fish in the aquarium and are ideal for a community aquarium with other fish of similar disposition.

44.3 Commercial Marine Ornamental Fishes

44.3.1 Marine Angelfish: They are favourite aquarium fishes, especially when small. It includes *Centropyge bicolor* (Bicolor angelfish), *C. eibli* (Blacktain angelfish), *C. multispinis* (Dusky angelfish), *Chaetodontoplus melanosoma* (Black- velvet angelfish), *Pomacanthus annularis* (Blue ring angelfish), *P. imperator* (Emperor angelfish) and *P. semicirculatus* (Semicircle angelfish).

44.3.2 Blennies: The blennies are small carnivorous with drab colours which harmonise with their environment but may be brilliantly coloured. *Andamia reyi* (Suckerlip blenny), *Blenniellia periophthalmus* (Bluedashed rock skipper), *Cirripectes castaneus* (Chestnut eyelash blenny), *C. filamentosus* (Filamentous blenny), *C. perustus* (Flaming blenny), *C. polyzona*, *C. quagga* (Squiggly blenny), *C.stigmaticus* (Red streaked blenny) *C. variolosus* (Red speckled blenny), *Ecsenius midas* (Persian blenny), *Enchelyurus kraussii* (Krauss' blenny), *Exallias brevis* (Leopard blenny), *Salarias fasciatus* (Jewelled blenny) (Fig. 44.7). Blennies are found in tropical and subtropical waters in the Atlantic, Pacific and Indian Oceans; some species are also found in brackish and even freshwater environments.

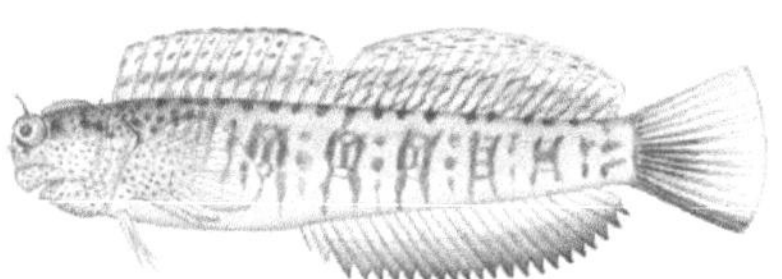

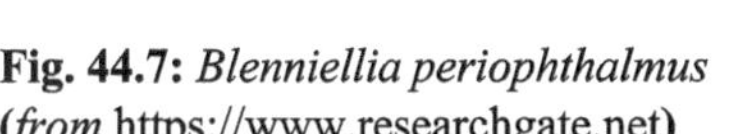

Fig. 44.7: *Blenniellia periophthalmus* (*from* https://www.researchgate.net)

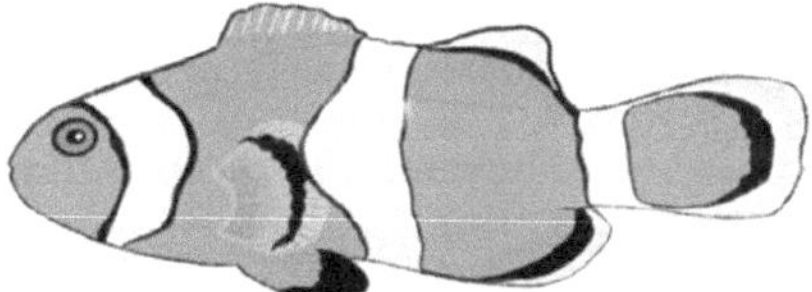

Fig. 44.8: *Amphiprion ocellaris* (*from* https://urbanfishkeeping.com)

44.3.3 Butterflyfish: These are found among tropical reefs around the world but are concentrated in the Indo-Pacific oceanic region. They move about with a flitting, darting motion and are very brightly coloured. The major species include four-eyed butterflyfish (*Chaetodon capistratus*), spotfin butterflyfish (*C. ocellatus*) and Feather-fin bullfish (*Heniochus acuminatus*).

44.3.4 Cardinal fish: Cardinal fishes are red-coloured and occur in a diversity of habitats, but individual species are restricted to relatively narrow ecological zones. Most prefer caves and crevices of rock or coral reefs. At dusk, they emerge from these retreats for nocturnal feeding. A firm, succulent fish that flakes easily when cooked. They are a deepwater species, commonly caught off the east coast of the North Island at depths between 300 and 800m.

44.3.5 Cleaner wrasse: The blue streak cleaner wrasse is known to clean balaenopteridae, chondrichthyans, homaridae, octopodidae, and dermochelyidae.

44.3.6 Clownfishes: They are hardy, coloured and in good demand in the marine aquarium trade. Clownfishes usually live in association with the sea anemones. Clownfishes exude a mucous substance that protects them from the stings of the anemone; when alarmed they immediately take shelter among the tentacles of the anemone, but never straying far. Clownfishes swim with an odd rising and falling waddling, wriggling motion which has earned them the name of clownfishes. Major clownfish species are *Amphiprion ocellaris* (Clown anemone fish), *A. percula* (Orange clownfish), *A. bicinctus* ((Two-band anemone fish), *A. chrysogaster* (Mauritian anemone fish), *A. ephippium* (Saddle anemone fish), *A. frenatus* (Tomato clownfish), *A. nigripes* (Maldive anemone fish), *A. polymnus* (saddleback clownfish), *A. sebae* (Sebae anemone fish) (Fig. 44.8).

44.3.7 Damsel fishes or demoiselle: They are a large group together with Amphiprion (Clownfish) and Chrominae (Chromis). These are relatively small, primarily tropical marine fishes found in the Atlantic and Indo-Pacific oceans. Many species are brilliantly coloured, often in shades of red, orange, yellow or blue. Common damselfish available include white tail humbug Damselfish (*Dascyllus aruanus*), Lemon Damsel (*Pomacentrus moluccensis*), Yellow tail blue Damselfish (*Chrysiptera parasema*), Allen's Damselfish (*Pomacentrus alleni*), Yellow Belly Damselfish (*Chrysiptera hemicyanea*), Blue Devil Damselfish (*Chrysiptera cyanea*), Blue/Green Chromis (*Chromis viridis*) (Fig. 44.9).

44.3.8 Gobies: They are small colourful species, usually solitary, sedentary, bottom-dwelling fishes. *Amblyeleotris gymnocephala* (masked shrimp goby), *Mahidolia mystacina* (Flagfin prawn goby), *Paragobiodon echinocephalus*, (Redhead goby), *Periophthalmus argentilineatus* (Barred mud skipper), *P.barbarus* (Atlantic mud skipper), *Priolepis eugenius* (Noble goby), *P. inhaca* (Brick goby), *Trimma annosum* (Grey beared pygmy goby), *T. winterbottomi* (Winterbottom's goby), *Valenciennea muralis* (Mural goby), *V. sexguttata* (Sixspot goby), *V. strigata* (Blueband goby) (Fig. 44.10). It can be found in mangrove ecosystems and mudflats of East Africa and Madagascar east through the Sundarbans of Bengal, Southeast Asia to Northern Australia, southeast China, and southern Japan to Samoa and Tonga Islands.

Fig. 44.9: *Chrysiptera parasema* (from https://www.aquariumbcn.com)

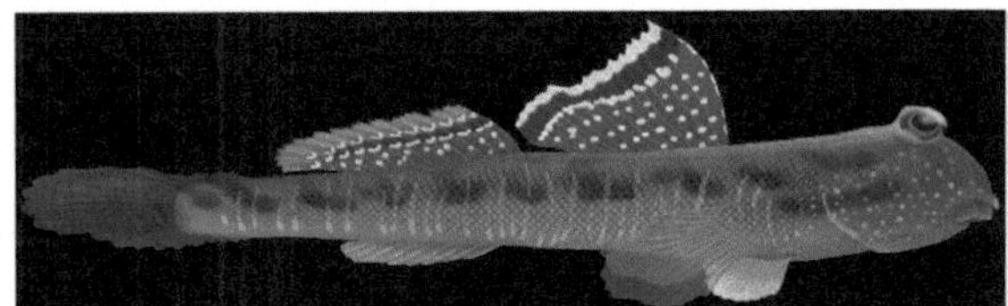

Fig. 44.10: *Paragobiodon echinocephalus* (from https://fishesofaustralia.net.au)

44.3.9 Lionfish: It is a longstanding showstopper in home aquaria, flourishing invasive species in U.S. Southeast and Caribbean coastal waters. *Pterois* is venomous marine fish native to the Indo-Pacific. It is characterized by conspicuous warning coloration with red or black bands and ostentatious dorsal fins tipped with venomous spines.

44.3.10 Moorish idol (*Zanclus cornutus*): It is a marine ray-finned fish found on reefs in the Indo-Pacific region.

44.3.11 Parrot fish: With roughly 95 species, this group's largest species richness is in the Indo-Pacific.

44.3.12 Rabbitfish or spine-foots: These are perciform fishes of 29 species in the genus, *Siganus*. In some now obsolete classifications, the species having prominent face stripes—colloquially called fox faces in the genus *Lo. Chimaera monstrosa*, also known as the rabbitfish or ratfish, is a northeast Atlantic and Mediterranean species of cartilaginous fish.

44.3.13 Sand-smelt Fish: It is a common pelagic fish in the northeastern Atlantic from the Danish straits, where it is rare, and Scotland to the Canary Islands and the western Mediterranean Sea (*Atherina hepsetus*).

44.3.14 Scorpionfish: The world's most venomous species. It is a large group with hundreds of members. They have a type of "sting" in the form of sharp spines coated with venomous mucus.

44.3.15 Sea anemones: These are soft-bodied, primarily sedentary marine animals resembling flowers. They are the largest, most numerous, and most colourful in warmer seas. Some live in brackish water. The common species include Bubble-tip anemone (*Entacmaea quadricolor*), Condy seaanemone (*Condylactis gigantea*), delicate sea anemone (*Heteractis malu*) and long tentacled anemone (*Macrodactyla doreensis*).

44.3.16 Sea Stars: The largest population of starfish lives in the Indian and Pacific oceans. A sea star can generate lost arms and nearly its entire body if at least a portion of the sea star's central disc remains. The major species include ochre sea star (*Pisaster ochraceus*), reef sea star (*Stichaster australis*) and Crown-of-thorns starfish (*Acanthaster planci*).

44.3.17 Sergeant (Unicon) Fish, píntano or damselfish: It grows to a maximum length of about 22.9cm. The Indo-Pacific sergeant (*Abudefduf vaigiensis*) is also known as the Sergeant Major (Fig. 44.11).

44.3.18 Squirrel fish: Squirrel fish is a well-known dish in Jiangsu cuisine, originally from Suzhou. It is prepared by deboning and carving a mandarin fish into an ornamental shape similar to a squirrel and then deep-frying it in batter before dousing it in sweet and sour sauce. Squirrelfish are edible fish found throughout the tropics. They have spiny fins and rough, prickly scales; some also have a sharp spine on each cheek.

44.3.19 Triggerfish: Often marked by lines and spots, they inhabit tropical and subtropical oceans throughout the world, with the greatest species richness in the Indo-Pacific. The species recorded are *Balistapus undulatus* (Orange lined triggerfish), *Balistes vetula* (Queen triggerfish), *Balistoides conspicillum* (Clown triggerfish), *B. viridescens* (Titan triggerfish), *Melichthys niger* (Black triggerfish), *Odonus niger* (Redtoothed triggerfish), *Psuedobalistes flavimarginatus* (Yellow margin triggerfish), *P. fuscus* (Yellow-spotted triggerfish), *Rhinecanthus aculeatus* (Black bar triggerfish), *R. rentangulus* (Wedge tail triggerfish), (*Sufflamen chrysopterum* (Halfmoon triggerfish) and *S. fraenatum* (Masked triggerfish).

44.3.20 Trunk (Box) fish: It can be found in reefs throughout the Caribbean, as well as the southwestern Atlantic Ocean. The species gets its name from the black spots it has covered over its whitish or yellow-golden body.

Fig. 44.11: *Abudefduf vaigiensis* (from https://www.dreamstime.com)

Fig. 44.12: *Arowana* (from https://www.petco.co)

44.3.21 Vastu fish: The Asian *Arowana*, also called the dragon fish, according to Vastu and Feng Shui, is a symbol of good luck, health and prosperity (Fig. 44.12).

45

Larvivorous Fishes

45.1 Introduction

Mosquitoes of various species breed in ponds, swamps, gutters, and other bodies of stagnant water, carrying parasites that cause malaria, filaria, dengue fever, chikungunya, Japanese encephalitis, Zika and yellow fever. Various approaches have been used to control mosquito populations to reduce the incidence of vector-borne diseases, including chemical, physical, and biological measures. In all types of habitats, biological control against mosquito vectors is safe, friendly, and cost effective. Mosquito biological control entails introducing natural enemies such as parasites, disease organisms, and predatory animals into the environment.

Fish are one of the most important vertebrate groups for men, influencing their lives in a variety of ways. They are a rich source of food and a means of overcoming man's nutritional difficulties. In addition to being an important source of food, several fish species contribute to mosquito biological control by feeding on larvae. These types of fish are known as larvivorous fish.

Atkins reported the larvicidal nature of fish as early as 1901. It was followed by Raj (1916), Hora (1927), Nair (1938), Mukherjee (1953), and Singh et al. (1977) observe various features of larvicidal fishes. In disease control policy documents, the WHO includes biological control of malaria vectors by stocking ponds, rivers, and water collections near where people live with larvivorous fish.

Ghosh and Dash (2007) reported 315 larvivorous fish species from seven genera. In Andhra Pradesh, Ramarao (2014) identified 58 species of larvivorous fish. Similarly, Krishna et al. (2016) reported 29 species of larvivorous fish from Lake Kolleru (Andhra Pradesh), divided into six orders, 14 families, and 20 genera. According to Das et al. (2018), 9 species with widespread distribution are abundantly available in rivers, ponds, streams, water canals, and tanks in Jharkhand, indicating their sustainability and ability to tolerate a wide range of ecological conditions. Gogoi and Biswas (2021) recently reported the activities of two ornamental fishes, *Channa bleheri* and *Channa stewartii*, as mosquito larvae biocontrol agents.

45.2 Definition and Features

Fish are a natural enemy of mosquito eggs and larvae, and their use as a means of biological control has been recognized since ancient times. According to Job (1940), a larvicidal or larvivorous fish should possess the following features:

1. A surface feeder and carnivorous.
2. Breed freely in confirmed water.
3. Hardy to withstand transport to distant places.
4. No food value but compatible with regionally existing fish fauna.
5. Small-sized to be capable of moving freely among weeds.
6. Surviving capacity in oxygen-deficient, deep as well as shallow water.
7. They should be able to withstand transportation for long distances.
8. They should be difficult to catch so that they may escape from their natural enemies.
9. They should possess adaptations in jaw mechanism in capturing and feeding upon mosquito larvae.

The eggs, larvae and adults of mosquitoes form excellent food for several species of fish, but small fish prefer mosquito larvae and feed on them throughout their lives.

45.3 TYPES

45.3.1 Based on the native places

1. Exotic larvivorous fishes: Goldfish (*Carassius auratus*) (Fig. 45.1), *Carassius idella,* Cichlids (*Lebistes reticulates* and *Gambusia affinis*) etc.

Fig 45.1: *Carassius auratus*

2. Indigenous larvivorous fishes: *Anabas testudineus, Aphanius, Aplocheilus panchax, Badis, Barilius bendelisis, Barilius vagra, Colisa fasciatus, Danio rario, Etroplus suratensis, Etroplus maculatus, Mugil or Rhinomugil cephalus, Notopterus notopterus, Oxygaster (Chela) bacaila, Oryzias mlanostigma, Puntius sophore, Rasbora daniconius, Macropodus cupanous, Glossogobous giuri, Panchax, Therapon jarbua,* etc.

45.3.2 Based on the efficiency of mosquitocidal activity

Jacob (1953) and Hora and Mukerjee (1953) divided larvivorous fishes into the following six groups:

1. Typical surface feeder or primary mosquito fish: They are the best or highly efficient for anti-malarial work. e.g., *Aplocheilus panchar, Gambusia affinis* etc.
2. Surface feeder: They eat mosquito larvae. They are less efficient for anti-malarial work. e.g., *Lebistes reticulatus, Oryzias, Aphanius* etc.
3. Sub-surface feeder: They are larvicidal to a considerable extent. e.g., *Amblypharyngodon mola, Carassius, Danio, Esomus, Rasbora daniconius* etc.
4. Column feeder: They feed on larvae when getting a chance. e.g., *Puntius sophore, Anabas testudineus, Chanda, Colisa fasciatus, Macropodus* etc.
5. Large-sized edible fish: They feed on larvae. e.g., *Labeo rohita, Catla catla, Cirrhinus, Mugil cephalus* etc.
6. Carnivorous and Predatory fishes: their fry feed on larvae but young and adults are destructive for other fishes. e.g., *Wallago attu, Channa punctatus, Mystus vittatus, Notopterus notopterus* etc.

45.4 Role of Larvivorous Fishes

Biological control of mosquitoes with the help of suitable larvivorous species is the most practical and cheapest method. With proper instructions, the method can be practicised by the poor villagers also. The relative utility of different larvivorous fishes includes the following:

1. *Aphanius dispar*: This is an effective destroyer of mosquito larvae. The male is larger than the female.
2. *Badis badis*: It is small-sized beautifully coloured fish growing to 5-6cm in length. They are reported to feed on mosquito larvae.
3. *Barilius bendelisis, Barilius vagra*: They are considered effective larvicidal forms in hill streams. They are omnivorous fish that prefer larvae of mosquitoes (Fig. 45.2).

Fig 45.2: *Barilius bendelisis*

4. Carp minnow (*Puntius sophore, Puntius ticto, Punitus chola, Puntius conchonius*): Several species are found in ponds, lakes and other water bodies. They feed on the eggs and larvae of mosquitoes. They are useful as larvicidal forms. They are hardy fishes so they can withstand unfavorable conditions for a long time and can be easily transported. They have little food value.
5. Chalhawa or Chilwa (*Oxygaster bacaila, Oxygaster gora*) (Fig. 45.3): They are found in rivers and ponds. Their larger species are used as food. Their mouth is turned upward as an adaptation. They are of great utility in mosquito control. Their small varieties feed on mosquito larvae throughout life.
6. Cichlid (*Lebistes reticulatus*): It was brought to India in 1908 from South Africa and is flourishing well in ponds and tanks of South India.
7. Murrels (*Channa punctatus, Channa marulius, Channa gachua* and *Channa striatus*): Their young ones feed on mosquito larvae.
8. Climbing perch (*Anabas testudineus*): This is an important food fish. This is reported to be a useful larvicidal form and of great commercial value.
9. *Colisa fasciatus*: This is a beautifully coloured, hardy fish of small size. They are reported to be good larvicidal forms off confined water. They have food values. It breeds in ponds and ditches in confined water.

Fig 45.3: *Colisa (Trichogaster) fasciatus*

10. *Danio rario:* It is a small-sized (5.0cm) fish found in plains. This has some value in mosquito control in plains.
11. *Esomus danricus*: This is an aquarium fish found in shallow ponds, lakes, paddy fields and ditches. They feed on the eggs and larvae of mosquitoes. This can be employed on a large scale for malaria control. They have no food value.
12. *Etroplus suratensis, Etroplus maculatus*: Their young ones feed on larvae and destroy them by pecking. Their adult stages feed on algae. They can live in freshwater as well as in brackish water. They have food value (Fig. 45.4).

Fig. 45.4: *Etroplus suratensis*

13. Goldfish (*Carassius auratus*): This may be used for mosquito control. However, due to their high cost, their use on a large scale is not possible.
14. Guppy (*Poecilia reticulata*) (Fig. 45.5): it is an exotic fish introduced in India in 1910. The temperature range suitable for breeding is from 24^{0}C to 34^{0}C. It can survive in water with a pH ranging from 6.5 to 9.0. However, it cannot survive in cold water. A single fish eats 80 to 100 larvae in 24 hours.

Fig. 45.5: *Poecilia reticulate*

15. Indian topminnow (*Aplocheilus panchax, Aplocheilus lineatus*): This brackish water, hardy fish is regarded as an important larvicidal fish. This can be living in freshwater as well as in brackish water. It grows to a maximum length of 9.0cm.

16. *Laubuca (=Chela) laubuca*: These are larvicidal fish not suitable for transport to long distances.
17. *Macropodus cupanus*: It is a small hardy fish of the Malabar Coast. It breeds in ponds, ditches, paddy fields and any shallow water body. This is considered useful for mosquito control.
18. Moy (*Notopterus notopterus*): This is considered useful for mosquito control. The young stages of this fish feed on mosquito larvae.
19. Mullet (*Mugil or Rhinomugil cephalus*) (Fig. 45.6): They live in freshwater as well as brackish water. Their young ones destroy mosquito larvae but cannot be used for extensive malaria operations.

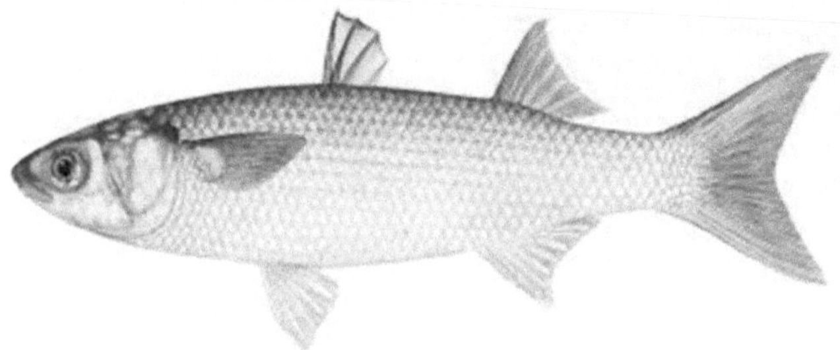

Fig. 45.6: *Mugil cephalus*

20. *Oryzias melanostigma:* This is a delicate allied of *Aplocheilus*, and is a very useful larvicidal form of confined water. It cannot be transported over long distances (Fig. 45.7).
21. *Rasbora daniconius:* This fish has a wide distribution. This is regarded as the effective larvicidal form but is not suitable for extensive use.
22. *Therapon jarbua*: This is useful for destroying mosquito larvae in brackish water.
23. Topminnow (*Gambusia affinis*): It has been brought from Italy in 1928. It measures about 6.25-7.50cm in size. This has been extensively used for mosquito control in several countries. This is successfully used for malaria control as it may consume 100-300 larvae in 24 hours. The optimum temperature for reproduction ranges from 24^{0}C to 34^{0}C but the fish can survive at freezing temperatures. The most suitable pH of water is between 6.5 and 9.9. A single female may produce between 900 and 1200 offspring during its lifespan (Fig. 45.9).

Fig. 45.7: *Aplocheilus panchax*

Fig. 45.8: *Gambusia affinis*

24. Translucent fish (*Ambasis (=Chanda) nama, Ambasis ranga, Chanda baculis*): They are small-sized transparent surface feeder fish. They mainly feed on Cyclops and are useful in controlling guinea worms. They can be used for mosquito control.

The actual control work must be carefully planned while taking into account the ecological features of the area to be treated. A suitable species can be introduced directly into the water or after removing any dense surface vegetation. Before introducing larvivorus fish, any predaceous fish in the water should be removed.

According to the account above, young forms of all fishes found in the plains consume mosquito larvae. As a result, increased pisciculture aids in mosquito control. However, the utility of mosquito control via fish is limited to permanent and semi-permanent bodies of water. Areas inaccessible to fish in temporary collections of water are suitable and safe places for mosquito breeding. As a result, using larvivorous fishes seldom solves the problem and must be supplemented by other methods. Because most anti-malarial chemicals are toxic to fish, the two methods must be mutually adjusted.

This is the simplest and cheapest method of mosquito control, and it can be practised by poor villagers with the proper instructions. To control malaria and other diseases, suitable species can be introduced into bodies of water, but the ecological conditions must be carefully studied.

46

Common Food Fishes

46.1 Introduction

Since humans settled and flourished adjacent to the aquatic bodies and therefore fishes might have attracted their attention. The increasing population has threatened mankind variously amongst which, the most important one is the food problem. Fish being a valuable and easily available source of food can be used to overcome the food problem to a certain extent. Indian rivers are the major source of irrigation systems, drinking water and fish as food. The fish industry of West Bengal and Orissa is some 1500 years old and in West Bengal, every family traditionally has at least one small or moderate-sized pond. Except for some sharks, caught for food, the majority of food fishes in India are teleosts.

As per the database of the National Bureau of Fish Genetic Resources (NBFGR), Lucknow, 877 fish species are from freshwater habitats in India.

The most common food fishes of India may be arranged as:

(i) Major carp: *Catla catla, Cirrhinus mrigala, Labeo rohita.*

(ii) Minor carp: *Cirrhinus reba, Cirrhinus chaudhryi, Labeo bata, Labeo calbasu, Labeo gonius.*

(iii) Catfishes: *Bagarius bagarius, Eutropichthyes vacha, Mystus aor, Mystus cavasius, Mystus seenghala, Ompak (=Callichorus) bimaculatus, Pangasius pangasius, Rita rita, Silonia silondia, Wallago attu.*

(iv) Exotic Carp: *Cyprinus carpio, Ctenopharyngodon idella, Hypophthalmichthys molitrix, Hypophthalmichthys nobilis.*

(v) Feather backs: *Notopterus chitala, Notopterus notopterus.*

(vi) Herrings: Species of *Clupea, Engraulis purava, Gadusia chapra, Hilsa (=Clupea) ilisha,* Gangetic anchovy *(Setipinna phasa),*

(vii) Live fishes: *Channa (=Ophiocepholus) punctatus. Channa gachua, Channa marulius, Channa striatus, Channa melasoma, Channa argus, Channa orientalis, Channa maculata, Clarias batrachus, Heterpneustes fossilis.*

(viii) Mullets: *Mugil cephalus, Mugil cursula, Rhinomugil corsula, Sicamugil cascasia.*

(ix) Perch: *Anabas testudineus, Ambassis ambassis, Chanda nama, Sciaena coitor.*

(x) Garfishes and Half beaks: *Xenentodon (Belone) cancila, Hemiramphus gorakhpurensis,*

(xi) Eels: *Anguilla bengalensis.*

(xii) Miscellaneous: *Botia dario, Colisa fasciatus, Garra lamta, Rasbora trilineata, Macrognathus aral, Mastacembelus armatus, Mastacembelus pancalus, Muraena robusta, Nandus nandus, Puntius sophore, Puntius ticto, Salmo gairdneri, Schizothorax, Tor (Barbus) tor, etc.*

46.2.1 Carp

Carp account for more than 85% of total aquaculture production in India. Out of the 266 carp species available in the Indian region, approximately 34 are economical and produced primarily through capture fisheries, while less than 10 are produced through both cultures and capture fisheries in the country. India is known as "Carp Country" because carp have been cultured in the country since ancient times and is a popular delicacy.

The Carp are native to the Indus-Ganges River Systems/ Indo-Gangetic Plains of India, referred to as the Gangetic Carp / Indian Major Carp (IMC), comprising *Catla, Rohu* and *Mrigal* contribute 60% of total Carp production. The Carp that were introduced from other countries are referred to as Exotic Carp such as Silver Carp, Grass Carp and Common Carp.

Besides the Major Carp, there are also smaller Carp often referred to as Minor Carp such as Reba (*Cirrhinus reba),* Bata (*Labeo bata*), Fringe-lipped carp (*Labeo fimbriatus*), Calbasu (*Labeo calbasu*), white carp (*Cirrhinus cirrhosus*) and Cauvery carp (*Labeo kontius*).

1. *Catla catla* (Hamilton, 1822)

Katla or Catla, also known as the large Indian carp, Bengal carp, or Bhakura, is a well-known freshwater fish found primarily in rivers and lakes in Assam and North India (Fig. 46.1). It has a large and broad head, as well as an upturned, wide mouth with overlapping lips. It feeds on the surface and in the middle of the water. It is a very strong fish that can jump over the fishing net if it is caught inside the net. It grows best in water temperatures ranging from 25 to 32^{0}C. It is the most important aquacultured fish in India, along with rohu and mrigal. The fish is available all year long. The fish has a conservation status of least concern (IUCN 3.1) and a mean survival rate of 79.64%.

Fig. 46.1: *Catla catla*

A mature Katla can weigh up to 2 kg. It is a popular oily fish. The fish is high in essential fatty acids and protein. Katla has a healthy (0.7) 6 to 3 ratio. This fish has a moderate Hg level, making it safe to eat. It has a high food value and delicious flesh. It is another source of income in Bihar and West Bengal. It can grow to be up to 1.82m long and weigh up to 38.6 kg.

Table 46.1: Nutritional values of some freshwater fishes

Sl. No.	Name of fish	Protein (%)	Energy (%)	Minerals (%)	Calcium (mg/kg)	Iron (mg/kg)	Phosphorus (mg/kg)	Carbon (%)	Fat (%)	Moisture (%)
1.	*Catla catla*	19.5	111	1.5	530	0.9	235	2.9	2.4	73.7
2.	*Channa sps.*	11.2	-	-	140	0.54	200	-	2.3	-
3.	*Cirrhinus mrigala*	14.5	98	1.5	350	1.1	280	3.2	**0.8**	75.0
4.	*Cirrhinus reba*	15.2	50	-	260	0.64	-	-	4.5	73.4
5.	*Clarias batrachus*	15.0	86	1.3	210	0.7	290	4.2	1.0	78.5
6.	*Heteropneustes fossilis*	**22.8**	124	1.7	670	2.3	**650**	**6.9**	0.6	**68.0**
7.	*Hilsa ilisha*	21.8	**273**	**2.2**	180	2.1	280	2.9	**19.4**	53.7
8.	*Labeo rohita*	16.6	97	**0.4**	650	1.0	175	4.4	1.4	76.7
9.	*Labeo bata*	14.3	89	2.0	**790**	1.1	200	2.2	2.5	79.0
10.	*Labeo calbasu*	14.7	**76**	1.3	320	0.8	380	2.0	1.0	**81.0**
11.	*Notopterus chitala*	18.6	108	1.0	**180**	**3.0**	250	3.1	2.3	75.0
12.	*Pangasius pangasius*	**14.2**	161	1.0	**180**	**0.5**	**130**	**1.7**	10.8	72.3
13	*Puntius javanicus*	14.9	53	-	210	0.45	-	-	4.7	73.6

2. White carp (*Cirrhinus mrigala*) (Hamilton, 1822)

Mrigala, Nain, Yerramosu, Mirrgah, or Mori is similar in shape to Rohu but slightly more slender. It is found in the countries of India, Pakistan, Bangladesh, Nepal, and Burma (Fig. 46.2). Its pectoral, pelvic, and anal fins are orange with black tips. It also has two small barbels on it. Its only surviving wild population is in the Cauvery River, making it vulnerable according to the IUCN. The depth is approximately equal to the length of the head. The scales on the body are cycloid.

Fig. 46.2: *Cirrhinus mrigala*

3. *Labeo rohita* (Hamilton, 1822)

Rohu, Rui, Ruee, Tambada masa, Rohiti, Ruhi or Roho also known as freshwater carp fish is a major carp found in ponds, lakes, rivers and streams across Northern as well as Central India and widely utilized in aquaculture (Fig. 46.3). It has a triangular head, a ventral pointed mouth, and fringed lips. The enormous silver-coloured rohu is a significant freshwater aquacultured species that is widely consumed in India and is considered a delicacy in Madhya Pradesh. The fish has a conservation status of least concern (IUCN 3.1) and a mean survival rate of 89.49%.

Fig. 46.3: *Labeo rohita*

The fish can weigh up to 2kg and is mostly eaten for its meat. It contains a lot of fatty acids. It is inexpensive and tasty, making it a popular choice among the general public! Adult fish can grow to be 45 kg in weight and 2 m in length. Rohu is a popular fish in the Indian market because it is readily available and inexpensive. The fish is herbivorous and bottom feeder, and it is available all year. Rohu is a staple in Andhra Pradesh, Bihar, and West Bengal, and it can be found in a variety of dishes.

46.2.2 Catfishes

They have prominent barbels that resemble the whiskers of a cat. Although catfish are typically found in faster-flowing rivers and streams, some catfish species have adapted to living in shallow salt-water environments, while others live underground in caves. Because they are negatively buoyant, most catfish are bottom feeders. Air-breathing catfishes such as Magur and Singhi live in shallow waters, can withstand low oxygen levels, and are marketed as "live fishes"; they are sold live and command a higher price.

4. Walking catfish *(Clarias batrachus)*

The Magur is a walking catfish that is found in India and Southeast Asia. It is an omnivorous species that is prized in the Indian states of Assam, Maharashtra, and Uttar Pradesh (Fig. 46.4). This fish prefers slow-moving, often stagnant waters such as those found in ponds, swamps, streams, and rivers, as well as flooded rice paddies. The fish has a conservation status of least concern (IUCN 3.1). When it is invasive, it is considered harmful. It has up to 18.22% protein content.

Fig.46.4: *Clarias batrachus*

5. Singhi (*Heteropneustes fossilis*)

The Asian stinging catfish, also known as the fossil cat, is an air-sac catfish that can be found in India, Bangladesh, Pakistan, Nepal, Sri Lanka, Thailand, Myanmar, and Bhutan (Fig. 46.5). It has also been introduced into Iran's Tigris River Basin. The fish has a low conservation concern (IUCN 3.1) and a stable population. Because their pectoral fins contain a venom gland, the fish may be considered one of the most dangerous kept by aquarists. This is an important food fish because it is high in protein, iron, and calcium.

Fig.46.5: *Heteropneustes fossilis*

6. Asian Striped Dwarf Catfish (*Mystus tengara*)

Fig.46.6: *Mystus tengara*

Tengra (Tengna) is a small omnivorous catfish used in *Tangra Macher Jhal* recipes from Bengal (Fig. 46.6). Tengra fish are mostly found in rivers in the Indian states of Bihar, Odisha, Chhattisgarh, and Bengal. The fish is available all year long. It measures 18.0 cm in total length. It is both ornamental and edible. It is a delicious food that is high in protein and can help with calcium deficiency.

7. *Rita rita*

Rita rita is a bagrid catfish that can be found throughout southern Asia (Fig. 46.7). Afghanistan, Bangladesh, India, Myanmar, Nepal, and Pakistan have all reported it. It can reach a length of 150 cm. It is fished commercially for human consumption. It's a slow-moving, bottom-dwelling catfish. It prefers muddy water to clear water and lives in rivers and estuaries. It also prefers quiet eddies in backwaters. It is an omnivorous catfish that feeds primarily on mollusks. It also consumes small fish, crustaceans, and insects, as well as decaying organic matter.

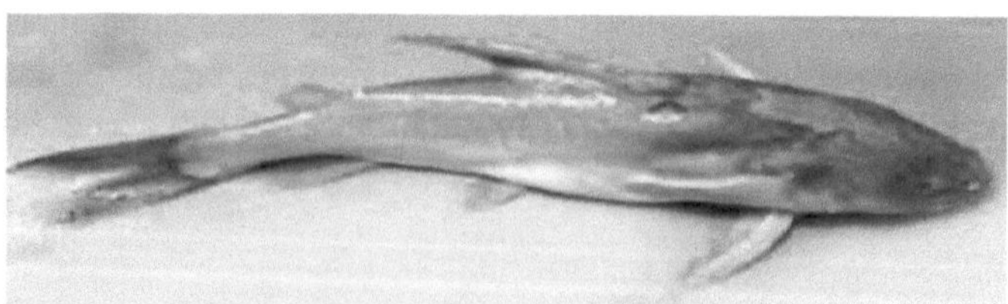

Fig. 46.7: *Rita rita*

8. Freshwater shark or Boal (*Wallago attu*)

Wallago attu (Helicopter catfish or Wallago catfish) is fast-growing catfish with high nutritional value (Fig. 46.8). It is also popular as a sport fish and has recently made an appearance in ornamental fish markets. It was once found in India, Bangladesh, Pakistan, Sri Lanka, Nepal, Afghanistan, Indonesia, Myanmar, Thailand, Vietnam, and Cambodia. The fish is classified as vulnerable (IUCN 3.1).

Fig. 46.8: *Wallago attu*

9. Pabdah Catfish (*Ompak pabda*)

Ompok belongs to the family Siluridae found in lakes and large rivers throughout South and Southeast Asia (Fig. 46.9). They are listed as near threatened by IUCN.

Fig. 46.9: *Ompak pabda*

10. The Pangas Catfish or Basa Fish (*Pangasius pangasius*)

It is a shark catfish native to the fresh and brackish water of Bangladesh, India, Myanmar, and Pakistan (Fig. 46.10). It has also been introduced to Cambodia and Vietnam. It can live in salt concentrations of around 0.7—1% and alum water which can be tolerated at temperatures of around 30°C. It shows potamodromous migration within the streams or rivers. Phytase supplementation significantly improved weight gain, protein efficiency, apparent net protein utilization, and feed conversion. Eating *Pangasius* nowadays is safe.

Fig. 46.10: *Pangasius pangasius*

46.2.3 Featherbacks or Knifefish

These fish are adapted to flowing conditions and are found in deep and clear waters in rivers, beels, reservoirs, and ponds. The Bronze Featherback has been observed swimming in brackish water. They are carnivorous and predatory fish that feed on aquatic insects, mollusks, prawns, and small fishes, as well as insects and tender roots of aquatic plants in their early stages of development. Despite the presence of a large number of intramuscular spines, they are high in nutritional value and command a higher market price. It has six different species.

11. Indian knifefish or Chitala (*Notopterus chitala*)

It has been found in the Brahmaputra, Indus, Ganges, and Mahanadi River basins in Bangladesh, India, Nepal, and Pakistan (Fig. 46.11). The IUCN lists them as near threatened. This species is revered in Hinduism as one of Lord Vishnu's avatars; in the first episode, "Matsya," Narayana was born as a golden knifefish to kill the demon.

Fig. 46.11: *Notopterus chitala*

12. Bronze knifefish, Asian Knifefish or Ghost knifefish (*Notopterus notopterus*)

It is native to South and Southeast Asia. Although it is mostly found in freshwater, it has been observed in brackish water (Fig. 46.12). It is currently the only member of the genus *Notopterus*, but as defined, it is most likely a species complex. The fish has a conservation status of least concern (IUCN 3.1).

Fig. 46.12: *Notopterus notopterus*

46.2.4 Small Indigenous Fish Species (SIFS)

They are defined as fish that can reach a mature size of 25-30 cm. Rivers and tributaries, floodplains, ponds and tanks, lakes, beels, streams, lowland areas, wetlands, and paddy fields are all home to them. In India, approximately 450 of the 877 native freshwater fish species are SIFS. The North East Region has the greatest diversity of SIFS in freshwaters, followed by the Western Ghats and Central India. Approximately 62 SIFS have been classified as food fish, while 42 species have been classified as ornamental fish. Mola, Climbing Perch, Barbs and Bata include some cultivable SIFS.

13. Koi or Climbing perch (*Anabas testudineus*)

The climbing perch (*Anabas testudineus*) is a freshwater amphibious fish (Fig. 46.13). The fish, a labyrinth fish native to Far Eastern Asia, lives in freshwater systems in Pakistan, India, and Bangladesh. The climbing perch is an important food fish in certain parts of South and Southeast Asia. The fish has a conservation status of least concern (IUCN 3.1). This is a tough fish that is popular in the fishing industry.

Fig. 46.13: *Anabas testudineus*

14. Spot fin Swamp Barb, Stigma Barb or Pool Barb or Chiddu (*Puntius sophore*)

It is a commercially important Assamese freshwater and brackish water fish with both food and ornamental value (Fig. 46.14). It is one of the most crucial SIFS.

The profiling and role of bioactive molecules from fish are extremely important in defence mechanisms. The fish has a conservation status of least concern (IUCN 3.1).

Fig. 46.14: *Puntius sophore*

15. Silver Hatchet Chela (*Chela cachius*)

Fig. 46.15: *Chela cachius*

The silver hatchet chela is a danionin fish in the Cyprinidae family. It is endemic to South Asia and can be found in Pakistan, India, Bangladesh, and Myanmar (Fig. 46.15). The fish has a conservation status of least concern (IUCN 3.1). It is a commercially important freshwater Cyprinidae for both subsistence and artisanal fisheries.

46.2.5 Snakeheads

The snakeheads (*Channa punctatus, Channa marulius* and *Channa gachua*) are members of the freshwater fish, native to parts of Africa and Asia (Fig. 46.16). They live in swampy areas and have gills that allow them to breathe air. They can survive out of water for up to four days if they are wet, and have been observed wriggling up to 400 metres on wet land to other bodies of water. They have two air chambers (suprabranchial cavities) that develop from the pharynx and are lined by vascular epithelium. They take in air and function like lungs.

Snakeheads feed on plankton, aquatic insects, and mollusks in their early stages before becoming predatory and cannibalistic as they mature. Snakehead meat is tasty, nutritious, and has high pharmaceutical values. Snakehead also contains all of the essential amino acids for wound healing, particularly glycine, which is required for the formation of human skin collagen.

Fig. 46.16: *Channa punctatus*

16. Pink Perch

Fig. 46.17: *(Nemipterus japonicas)*

Rani fish (Pink Perch) is a popular freshwater game fish in India (Fig. 46.17). The fish is available all year long. A perch is typically 30 cm / (0.45 kg) or less in length, with anything over 40 cm / (0.91 kg) considered a great catch. The fish is pink, small in size, and has a mild flavour. The fish is referred to as lean because it contains only 5% body fat. If protein is your only goal, this fish is an excellent choice.

17. *Ailia coila*

The Gangetic ailia (Kajuli) is common in large rivers and their associated water systems. This species is critical to India's indigenous commercial fisheries. The fish is available all year long. This species can reach a length of 30cm.

46.2.6 Tilapias

Tilapias are a type of "Cichlid" fish that are native to Africa. Tilapia farming in ponds was introduced in Central African countries after WWII and quickly spread to most tropical and subtropical countries around the world, earning them the moniker "international fish." Although the majority of Tilapia's natural resources are found primarily in Africa, Asia accounts for nearly 80% of the global Tilapia aquaculture production of approximately 5.0 million metric tonnes. Tilapias are considered the most important aquaculture species of the twenty-first century, and they are commercially cultured in 100 countries worldwide, ranging from extensive to super-intensive scales.

Tilapia (Cichlid) is a freshwater fish found in shallow water, rivers, and lakes, but they can also be found in brackish water in India. It is also one of the most popular fish in the world and one of the most important species in aquaculture after carp and salmon.

The Pearl Spots are another name for the Karimeen fish. It is a type of Green chromide or 'Cichlid fish' found in Kerala's backwaters. Green chromide is native to the Indian Subcontinent's coastal areas, where it lives in brackish water habitats such as river deltas. It can grow to be 20-40cm long.

46.2.7 Coldwater Fishes

Coldwater fishes occupy an important place amongst the freshwater fishes of India. Coldwater fisheries deal with fisheries activity in water where the temperature of water ranges from 5 to 20 degrees centigrade. The gills of cold water fish are greatly reduced and the gill opening is smaller in size for adaptation to cold temperatures. Important coldwater fishes of India are Mahseers such as *Tor putitora, T. tor, T. khudree, T. mosal,* Snow Trout such as *Schizothorax richardsonii* and *Schizothoraichthys esocinus*, Mountain Trout such as *Barilius vagra, B. bendelisis,* other fish such as *Glyptothorax sp., Garra sp.,* etc.

18. *Tor tor* (Hamilton, 1822) or *Barbus tor* (Day)

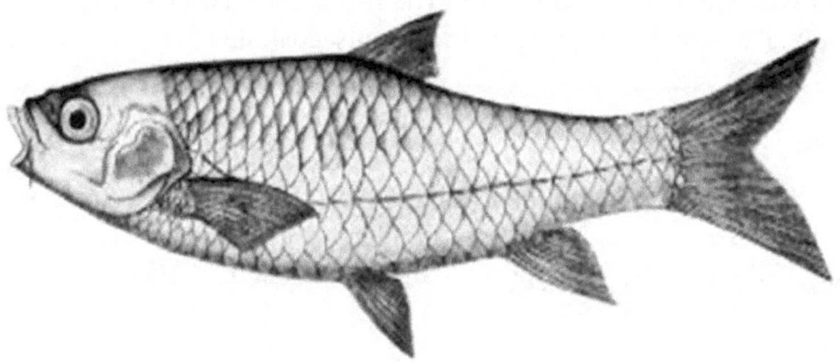

Fig. 46.18: *Tor tor*

The *Mahseer, Mahasher, Nahram, Mahasol, Putitor, Khadchi, Mesta, Peruval, Meruval* or the *Golden Mahseer* is a type of popular game fish, freshwater sport and food fish (Fig. 46.18). Mahseer is majorly present in *Wayanad, Kali River* and also in *Sarda River* and the *Himalayan Rivers*. Out of the total of 47 types of *Mahseer species* in existence in the world, India provides habitat to 15 of them!

46.2.8 Other Fishes

19. *Hilsa ilisha*

Fig. 46.19: *Hilsa ilisha*

Hilsa shad (Ilish) is a famous freshwater and brackish water fish in India, particularly in West Bengal, Odisha, Tripura, Assam, and Andhra Pradesh. In Andhra Pradesh and Bengal, fish is a popular dish (Fig. 46.19). Pulasa fish from the Godavari River in Andhra Pradesh is the tastiest and most expensive of all the popular fish in India. It is the national fish of Bangladesh. It contains a 0.1 high ratio of 6 to 3. This delicious fish has a slice of tender meat.

A popular saying in Andhra Pradesh says

"Pustelu ammi ayina Pulasa tinocchu"

It means that it is always worth eating Pulasa Fish even if you have to sell the *Mangalsutra.*

20. Rawas (Indian Salmon)

Rawas is one of the most loved and popular edible fish. Rawas is widely available in India and is famous for its pink to orange meat with a mild flavor. It is specifically an oily fish, which means it consists of oil almost all over its body. A salmon fillet, which is a boneless piece of fish is said to have around 30% of the oil. This oil is though great for your health containing Omega 3, Vitamin A, and Vitamin D.

The Trout fish is a type of freshwater fish and is closely related to *Salmon fish* and *Char fish* and can be found in the Himalayan Region of India. The Himalayan Rivers of the Indian subcontinent are rich in *German Brown* and *Rainbow Trout.* Fishing in these Himalayan Rivers is quite restricted and requires a special permit from the state government with a policy of *'catch and release'*. A normal adult rainbow trout generally weighs about 2–3 kg, while its maximum length is 120 cm total and maximum weight is 25.4 kg and 11 years of age.

21. Bangda (Indian Mackerel)

Indian Mackerel or Bangda belongs to the list of staple fishes of India. It is a saltwater fish found in the Indian Ocean and the seas around it. The fish is oily and therefore is a great source of protein and Omega 3.

22. Surmai (King Fish/Seer Fish) (*Scomberomorus guttatus*)

The fish is extremely popular as well as an expensive one. It is most popular in South and Central India. Surmai is categorically mackerel and is rich in protein, Omega 3 and other vitamins.

23. Pomfret

Pomfret is an exotic fish that is extremely enjoyed in India. It falls under the category of butterfish vastly found in South Asia including the Indian Ocean (Fig. 46.20). The fish is not oily and has white meat which is highly delicious. Silver, white and black pomfret are popular in India.

Fig. 46.20: *Pomfret*

24. Prawn

Fig. 46.21: Prawn

The Indian prawn is one of the main commercial prawn species in the world. It is found in the Indo-West Pacific from eastern & south-eastern Africa, through India, Malaysia & Indonesia to southern China & northern Australia (Fig. 46.21). Both Prawns and shrimp can be found in salt water as well as fresh water. Both are almost the same in taste. They just differ in size as prawns are bigger than shrimps.

47

Economic Importance of Fishes

47.1 Introduction

Fish are one of the most important groups of vertebrates and have always been a significant species in man's life, influencing it in a variety of ways. Fish are highly valued in human civilization for their food, cooking, and nutritional value. They have significant economic, nutritional, medicinal, industrial, aesthetic and religious values, and they employ millions of people worldwide. Several species are sought after as luxury foods in high-end restaurants. They help to ensure food security in many parts of the world by serving as a valuable supplement to varied and nutritious diets. Fish consumption has numerous health, nutritional, environmental, and social benefits over other terrestrial animal meats. Doctors recommend eating fish once a week. It should be heated to temperatures above 70°C.

Fish has been used as a source of dietary protein by humans throughout history. Historically and currently, the majority of fish harvested for human consumption has come from catching wild fish. Fish is estimated to provide about one-sixth of the world's protein. This proportion is significantly higher in some developing countries and regions that rely heavily on seafood. Similarly, fish has been linked to the primary industries as well as related food, feed, pharmaceutical production, and service industries.

Millions of people suffer from hunger and malnutrition, and fish are a rich source of food that can help people overcome their nutritional challenges. Fish, in addition to being an important source of food, provide us with several byproducts.

The importance of fish can be discussed under the following headings.

47.1.1 Useful Activities

(I) Fish as Food for Humans: Fish flesh has high protein, low fat and cholesterol, almost zero carbohydrate, high omega-3 fatty acids, high iodine, and high Ca, Mg, Zn, Fe, Cu, Mn, K, Na, P, F, Br and vitamins A, B, and D. Above all, it is easily digestible by men. These polyunsaturated fatty acids can help lower cholesterol levels in the blood, lowering the risk of a heart attack. The flesh of fish is white, contains approximately 13 to 30% protein, and has a food value of 300 to 1600 calories per pound. Fisheries supply approximately 35% of the world's animal protein. Fish protein is superior to egg protein, milk protein, and beef protein. The nutritive values of preserved fish are lower than those of fresh ones

Fish edible tissues are significantly larger than chicken, pig, and sheep/goat edible tissues. For example, about 65% of the raw weight of fish is consumed, compared to 50% of chicken and pigs and 40% of sheep and goats. Fish has a high biological value (HBV), net protein utilized (NPU) and protein efficiency ratio (PER) of 80, 74 and 35 compared to 74, 76 and 3.2 of other meat.

Fish meat, according to Pottinger and Baldwin, contains at least 5 to 10 essential amino acids (lysine, arginine, histidine, leucine, isoleucine, valine, threonine, methionine, phenylalanine and tryptophan) as well as a significant amount of phosphorus. Caver eats the eggs of certain fish, such as Russian sturgeon.

The major food fishes of India include

(a) Freshwater species

(i) Major carp: *Catla catla, Cirrhinus mrigala, Labeo rohita.*

(ii) Minor carp: *Cirrhinus reba, Labeo bata, Labeo calbasu, Labeo gonius.*

(iii) Catfishes: *Bagarius bagarius, Eutropichthyes vacha, Mystus aor, Mystus cavasius, Mystus seenghala, Ompak (=Callichorus) bimaculatus, Pangasius pangasius, Rita rita, Silonia silondia, Wallago attu.*

(iv) Exotic carps: *Cyprinus carpio, Ctenopharyngodon idella, Hypophthalmichthys molitrix, Hypophthalmichthys nobilis.*

(v) Feather backs: *Notopterus chitala, Notopterus notopterus.*

(vi) Herrings: Species of *Clupea, Engraulis purava, Gadusia chapra, Hilsa (=Clupea) ilisha,* Gangetic anchovy *(Setipinna phasa),*

(vii) Live fishes: Species of *Channa (=Ophiocepholus), Clarias batrachus, Heterpneustes fossilis.*

(viii) Mullets: *Mugil cephalus, Rhinomugil corsula, Sicamugil cascasia.*

(ix) Perch: *Anabas testudineus, Ambassis ambassis, Chanda nama, Sciaena coitor.*

(x) Garfishes and Half beaks: *Xenentodon (Belone) cancila, Hemiramphus gorakhpurensis,*

(xi) Eels: *Anguilla bengalensis.*

(xii) *Miscellaneous: Botia dario, Colisa fasciatus, Garra lamta, Rasbora trilineata, Macrognathus aral, Mastacembelus armatus, Mastacembelus pancalus, Muraena robusta, Nandus nandus, Puntius sophore, Puntius ticto, Salmo gairdneri, Schizothorax, Tor (Barbus) tor, etc.*

(b) Brackish water species: *Chanos chanos, Eluitheronema tetradactylum, Etroplus surantensis, Lates calcarifer, Mugil cephalus, Polynemus indicus, Therapon, etc.*

(c) Marine species

i. Anguilla anguilla, Anguilla bengalensis,

ii. *Chanos chanos (Chanos salmoneus),*

iii. *Dussumieria,*

iv. *Eutroplus suratensis, Exocoetus obtusirostris, Exocoetus poecilopterus, Exocoetus volitanus,*

v. *Formigo (Stromateus) niger,*

vi. *Harpodon neherius,*

vii. *Hilsa (=Clupea) ilisha, Hilsa kanagurta, Hilsa toli,*

viii. *Lates calcarifers, Lactarius lactarius (=Lactarius delicalutus), Lutianus argentimaculata, Lutianus lutjanus, Lutianus malabarius, Lutianus sangunea,*

ix. *Muraenesox,*

x. *Otolihes argentius, Otolihes brunneus, Otolihes maculatus.*

xi. *Pampus argentus, Polynemus paradiscus, Polynemus indicus, Polynemus heptadactylus, Pristis clavata, Psettodes erumei,*

xii. *Rastrelliger brachisoma,*

xiii. *Sciaena dussumieri, Scoliodon, Sphyrnaea, Stromateus cinereus, Sardinella (=Clupea) longiceps, Sardinella albella, Sardinella clupeoides, Sardinella dayi, Sardinella fimbriata, Sardinella gibbosa, Sardinella melanura, Sardinella pacifica, Sardinella perforate, Sardinella sidensis, Sardinella sirm, Setipinna (=Engraulis) breviceps, Setipinna taty, Sillago sihama, Scomber microlepidotus (=Rastrelliger kanegurta), Scomberomorus (=Cybium) commersonii, Scomberomorus guttatum, Scomberomorus kuhli, Strometeus sinensis,*

xiv. *Tilapia mossambicus, Trichiurus lepturus, Thunnus albacores, Thunnus alalvinga, Thrissocles boelema, Thrissocles dussumieri, Thrissocles hamiltoni, Thrissocles kammalensis, Thrissocles malabaricus, Thrissocles mystax, Thrissocles setirostris, Thrissocles purava, Thynunus thunini, Trygon zugei,* etc.

The cartilaginous sharks, skates and rays are also used as human food in several countries. They are consumed by the poorer classes who live along the seacoasts of India and Sri Lanka. The canned meat of sharks is sold commercially under the name "gray fish." In Southeast Asian countries, shark fins are dried and boiled to get a gelatinous material used in soups.

Fresh fish meat is usually cooked for human consumption. However, large quantities are refrigerated, salted, smoked, canned or pickled. Today, fisheries of the world carry on business worth several hundred thousand rupees annually and also employ thousands of people.

The refined oils extracted from the livers of fish are rich in vitamins A and D.

(II) Fish as Food of Cattle: The scrap from canneries, as well as entire fishes that are not relished by man, are dried and ground in the mill. This is called fish meal and is used as artificial food for poultry, pigs and cattle. Fish meal is produced in several states like Maharashtra, Andhra Pradesh, Tamil Nadu, Bengal and Kerala chiefly from sardines, mackerels, ribbon fish, etc.

The fish is first cooked in large pots containing a sufficient quantity of water, on fire on steam. The cooked material is then pressed to remove moisture and dry in the sun on suitable platforms. The resultant product is then stored and if preserved in airtight containers after sterilisation retains its nutritive value for a long time.

The fish meal contains about 50-70% protein and a high percentage of calcium phosphate so it is very valuable for cattle and poultry. The manufacture of fish meals can be undertaken as a cottage industry requiring little expenditure.

(III) Fish for Disease Control: It is possible to control diseases like malaria, yellow fever and other dreadful diseases like malaria, yellow fever and other dreadful diseases that are spread by mosquitoes in tropical countries. Larvivorous fish eat mosquito larvae. Top minnows (*Gambusia affinis*), *Trichogaster, Chela, Puntius, Barilius, Danio, Colisa, Rasbora, Esomus, Ambassis, Aplocheilus, Barbus, Panchax*, etc., are the major larvivorous fish. Other larvivorous species of *Brachydanio, Barilius, Chela*, etc. help in the biological control of dengue, malaria, filaria, encephalitis, etc. Similarly, herbivorous species like grass carp (*Ctenopharyngodon idella*), tilapia (*Oreochromis*), silver barb (*Puntius gonionotus*) (Fig. 47.1) etc. are used to control aquatic weeds and vegetation.

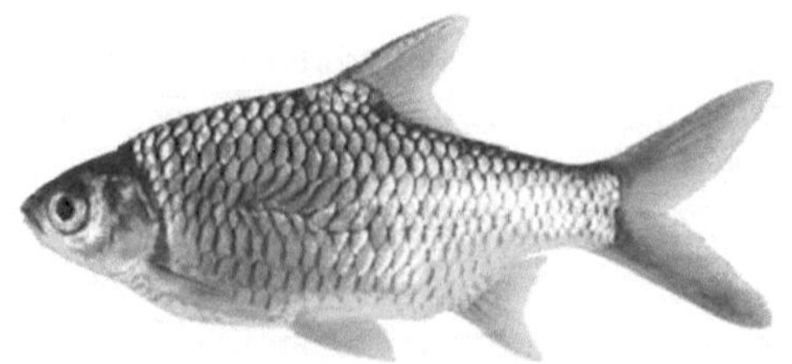

Fig. 47.1: *Puntius gonionotus*

Fig. 47.2: *Pterophyllum*

(III) Aesthetic Quality: In aquariums, a significant number of fish are bred for their beauty and graceful movements. *Macropodus, Trichogaster,* goldfish (*Carassius*) and angelfish (*Pterophyllum*) (Fig. 47.2) are important aquarium fishes.

(IV) Trade and Industry: The fishing industry is prevalent even in the time of Jesus Christ. It is the 2nd biggest industry in Korea. Some countries like Portugal, Canada, Norway and Holland export more fish than they import while countries like the USA, Germany and France import more fish than they export. Fishes are a part of the national dishes of Angola, Portugal, Japan, Norway and Sweden. China, Morocco, Bangladesh, India, Indonesia and the Philippines account for 73% of the total fish production in the world (FAO, 1997). Total global production of Fish was 7.5 million tonnes in 1986 and it increased to 8.6 million tonnes in 1996.

(V) Use of Fish in Medicine: Fish are used to extract Insulin. When Insulin of lower concentration is required for treatment, Fish Insulin can be used. Besides fish are generally low in sugar content and they can be relished by Diabetic patients. This way it can be an important component of the diet with an emphasis on healthy eating.

Certain fishes and their bye-products contribute to useful Ayurvedic and Unani medicines for the treatment of duodenal ulcers, skin diseases, night blindness, general weakness, loss of appetite, colds, coughs, bronchitis, asthma, tuberculosis, etc.

a. Asthma: Children who eat fish are less likely to develop asthma.
b. Brain and eyes: Fish rich in Omega-3 fatty acids can contribute to the health of the brain and the retina. The IQ level of children whose mothers consumed about 340g of fish per week during pregnancy was found higher than non-fish eaters. Similarly, breastfed babies whose mothers eat fish have better eyesight, perhaps due to the Omega-3 fatty acids transmitted in breast milk.
c. Cancer: The Omega-3 fatty acids in fish reduce the risk of cancers by 30 to 50 percent, especially in the oral cavity, oesophagus, colon, breast, ovary and prostate. The cartilage of sharks is used in the preparation of medicine for cancer.
d. Cardiovascular disease: Eating fish every week reduces the risk of heart disease and stroke by reducing blood clots and inflammation, improving blood vessel elasticity, lowering blood pressure, lowering blood fats and boosting good cholesterol.
e. Dementia: Elderly people who eat fish or seafood at least once a week may have a lower risk of developing dementia, including Alzheimer's disease.

f. Depression: People who regularly eat fish have a lower incidence of depression. Depression is linked to low levels of Omega-3 fatty acids in the brain.

g. Prematurity: Eating fish during pregnancy may help reduce the risk of delivering a premature baby.

(VI) Recreational Value: Fishing is an important outdoor activity for a lot of people. *Tor tor, Catla catla, Bagarius bagarius* are some of the species used for this purpose. Scales of garpike (*Lepidosteus*) are used for jewelry and novelties. From the scales of some fish is secured a pigment whose water suspension is known as pearl essence. It is used in the manufacture of artificial pearls in Europe, especially in France. The most common sports fishes of Nepal are Sahar (*Tor tor*), Asla (*Schizothorax telchitta*) and rainbow trout (*Oncorhynchus mykiss*).

(VII) Ornamental Value: Fish aquariums and displays kept in the house and workplace offer an ornamental value. Common aquarium fishes are goldfish (*Carassius auratus*), angelfish (*Pterophyllum*), swordtail guppy (*Xiphophorus*), minnow (*Gambusia affinis*), siamese fighter (*Betta splendens*), paradise fish (*Macropodus*), *Hemigrammus, Aphyocharax*, loach (*Botid*), *Trichogaster, Tilapia*, etc. Goldfish are cultured and not found in nature and the Japanese have produced several curious artificial varieties. Pet shops nowadays stock many kinds of fish for hobbyists and scientists.

The common ornamental fish species kept in an aquarium in Nepal are goldfish (*Carassius*), gourami (*Colisa fasciatus*), zebrafish (*Brachi danio*) (Fig. 47.3), guppy (*Poecilia*), fighting fish (*Betta splendens*) (Fig. 47.4), koi (*Cyprinus carpio*) etc.

Fig. 47.3: *Brachi danio*

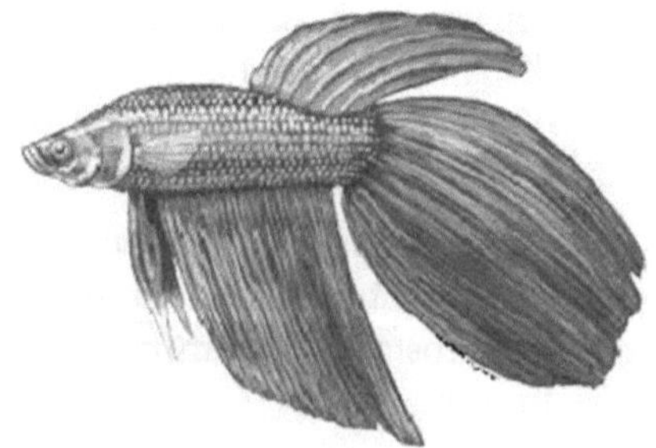

Fig. 47.4: *Betta splendens*

(VIII) Fishes in Rural Development: Small-scale fish farming can be done in rural areas to improve the life standard of rural fishermen. This can also help in income generation.

(IX) Impact on Migration: Jobs generated through fishing like production, processing, transport and marketing can be expected to control, to some extent, the drift of rural people to urban areas.

(X) For Social Benefit: Fish farming in a community aims at the priorities of the community. It helps to attain the knowledge of human, economic and socio-infrastructural development.

(XI) Employment Opportunities: Crop-farming farmers nowadays produce fish side by side with their crops. This creates increased farm activity with minimal cost and more employment per unit of capital invested. More fish farming projects have the advantage of being more widely distributed geographically, and locally owned, enabling improved income distribution among the population.

(XII) Contribution to GDP: Fishing in India is a major sector contributing 1.07% of its total GDP. Its fishing sector supports the livelihood of over 28 million people in the country, especially within marginalized and vulnerable communities. India is the third largest fish-producing country in the world accounting for 7.96% of the global production and the second largest producer of fish through aquaculture, after China. The total fish production during the financial year 2020-2021 is estimated at 14.73 million metric tonnes. According to the National Fisheries Development Board, the Fisheries Industry generates export earnings of Rs 334.41 billion. Centrally sponsored schemes will increase exports by Rs 1 lakh crore. According to the Ministry of Fisheries, Animal Husbandry, Dairying, fish production increased from 7.52 lakh tonnes in the years 1950–1951 to 125.90 lakh tonnes in the years 2018–2019, a 17 times increase.

(XIII) Scientific Study: Fishes have considerable use as experimental animals, especially in the fields of Genetics, Embryology, Animal Behaviour and Pharmacology. Certain fishes such as *Latimeria* (Fig. 47.5) and Dipnoans have anatomical features of great zoological interest. Fishes like dogfish (*Scoliodon*) (Fig. 47.6), perch (*Perca*) and carp (*Labeo*), etc., are dissected for anatomical study in zoological laboratories. Research in Ichthyology is conducted for the benefit of fisheries and mankind.

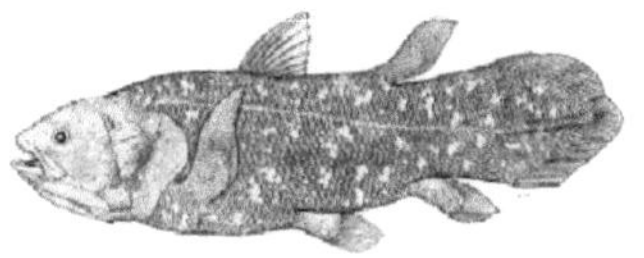

Fig. 47.5: *Latimeria chalumnae*

Fig. 47.6: *Scoliodon laticaudous*

47.1.2 Harmful Fishes

(I) destructive Activities: All the cartilaginous fishes are predaceous and feed chiefly on large quantities of crabs, lobsters, squids and other valuable marine animals. They cause great harm by destroying eggs, young and even adults of our food fishes. Sharks are extremely dangerous in the sea and injure fishermen and damage their nets, steal baits, devour captured fish and drive away squids and fish to be netted. The damage done by these sharks is several hundred thousand rupees.

(II) Injurious Activities: Some larger sharks and swordfish may capsize small boats and injure or even kill fishermen. All sharks are a menace to bathers and skin divers in shallow waters. The piranhas of Central and South America have very strong teeth and are dangerous even to men.

Some fishes like cartilaginous electric rays (*Torpedo*) (Fig. 47.7) have electric organs which give powerful shocks to swimmers and fishermen. An extreme case of specialisation is the bony electric eel (*Electrophorus electricus*) of the Amazon. It is nearly 1.5 metres long and the posteroventral four-fifth part of the body is occupied by the electric organ. It can generate electricity up to 600 volts in potential with a maximum output of 1000 watts. Carnivorous fishes eat away the larvae of useful insects.

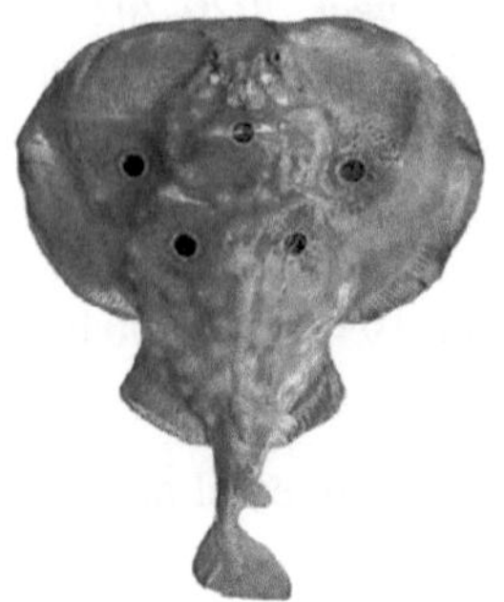

Fig. 47.7: *Torpedo*

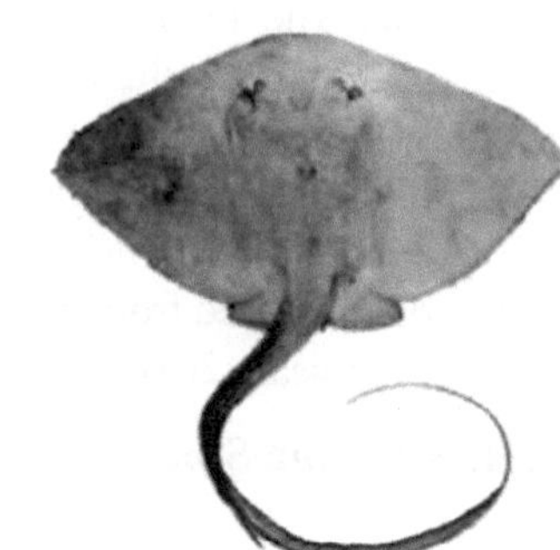

Fig. 47.8: *Trygon*

Fig. 47.9: *Diodon*

Fig. 47.10: *Tetradon*

(III) TOXIC ACTIVITIES: Poisonous glands are found in sting rays (*Trygon*) (Fig. 47.8) and eagle rays (*Myliobatis*). They can inflict painful wounds, sometimes fatal by poisonous stings or spines. Poisonous glands are also found in dogfish (*Squalus*), *Heterodontos, Chimaera*, porcupine fish (*Diodon*) (Fig. 47.9), globe fish (*Tetrodon*) (Fig. 47.10), scorpion fish, etc. These are also capable of causing wounds by their spines or spiny opercula. The flesh of some tropical fishes (e.g., *Tetrodon*) is also poisonous and may prove fatal to man. Among freshwater teleosts, *Clarias batrachus, Heteropneustes fossilis*, introduce poison by the pectoral spine.

48

Multiple Choice Questions

1.Class Placodermi includes

(a) Acanthodi (b) Coccostei

(c) Pterichthys (d) All.

2. Division Siluri lacks

(a) ***Channa*** sp. (b) ***Clarias*** sp.

(c) ***Mystus*** sp. (d) ***Wallago*** sp

3. Agnatha appeared in .... period

(a) Craboniferous (b) Devonian

(c) Ordovician (d) Silurian

4. Which of the following suffix is added to the family?

(a) formes, (b) idae,

(c) ini, (d) oidei

5. Placodermi appeared in .... period

(a) Craboniferous, (b) Devonian,

(c) Ordovician, (d) Silurian

6. Which of the following is known as a blue shark?

(a) ***Carcharinus***, (b) ***Carcharodon***,

(c) ***Cetorhinus***, (d) ***Stegoistoma***

7. Which of the following is not a dipnoan fish?

(a) ***Lepidosiren***, (b) ***Neoceratodus***,

(c) ***Notopterus***, (d) ***Protopterus***

8. Which of the following is known as a 'living fossil'?

(a) ***Lepidosiren***, (b) ***Neoceratodus***,

(c) ***Notopterus***, (d) ***Protopterus***

9. The scales present in ***Latimeria*** is?

(a) Ctenoid, (b) cycloid,

(c) Placoid, (d) Rhomboid

10. Gonopodium, a copulatory organ of some fishes is a derivative of .... fin
    (a) Anal, (b) Caudal,
    (c) Dorsal, (d) Ventral
11. Major carps breed in..........
    (a) Estuarine water, (b) Marine water,
    (c) Running freshwater, (d) Stagnant freshwater
12. Which of the following is the most common form of locomotion in fishes?
    (a) Anguilliform, (b) Carangiform,
    (c) Ostraciform, (d) All
13. In general, fish travel on average the ...... times the length of their body
    (a) 5, (b) 10,
    (c) 15, (d) 20
14. ***Hilsa ilisha*** is a/an .... fish
    (a) Amphidromous, (b) Anadromous,
    (c) Catadromous, (d) Oceandromous
15. Sharpey's fibres are present in which of the following scales?
    (a) Ctenoid, (b) cycloid,
    (c) Placoid, (d) Rhomboid
16. Whitish/silvery appearance of fish is due to the presence of .... pigments?
    (a) Erythrophores, (b) Iridophores,
    (c) Melanophores, (d) Xanthophores
17. Which of the following fish is used to study the bones of fishes?
    (a) ***Labeo rohita***, (b) ***Scoliodon***,
    (c) ***Wallago attu***, (d) All
18. Which of the following is the largest opercular bone of fishes?
    (a) ***Interopecular***, (b) ***Opercular***,
    (c) ***Preopercular***, (d) All
19. Ribs are characteristics of
    (a) Bony fishes, (b) Cartilaginous fishes,
    (c) Dipnoi, (d) Placodermi
20. The number of rays in the anal fins of fishes ranges from
    (a) 50-60, (b) 60-70,
    (c) 70-80, (d) 80-90

21. Which of the following is the 'L' shaped bone in the pectoral girdle of fishes
    (a) Cleithrum, (b) Coracoid,
    (c) Post-temporal, (d) Pre-temporal
22. Which of the following fish is a plankton eater?
    (a) ***Catla catla***, (b) ***Cirrhinus reba***,
    (c) ***Hilsa ilisha***, (d) All
23. Which of the following fish is a column feeder?
    (a) ***Labeo bata***, (b) ***Labeo calbasu***,
    (c) ***Labeo gonius***, (d) ***Labeo rohita***
24. Maximum value of Gastrosomatic Index is observed during .... phase?
    (a) Migratory, (b) Post-spawning,
    (c) Pre-spawning, (d) Spawning
25. The rasping organs in fishes are made up of .... like gill rakers?
    (a) Hard, (b) Long,
    (c) Teeth, (d) All
26. Maximum number of gill pairs has been reported in?
    (a) ***Bdellostoma***, (b) ***Myxine***,
    (c) ***Petromyzon***, (d) ***Scoliodon***
27. Pseudobranchs are found in?
    (a) ***Amia*** sp. (b) ***Channa*** sp.
    (c) ***Clarias*** sp. (d) ***Wallago*** sp
28. Opercular lungs are found in?
    (a) ***Anabas*** sp. (b) ***Channa*** sp.
    (c) ***Heteropneustes*** sp. (d) All
29. Which of the following fish is known as a bichir?
    (a) ***Catla catla***, (b) ***Cirrhinus reba***,
    (c) ***Hilsa ilisha***, (d) ***Polypterus***
30. Transitory and rudiment air bladder is present in
    (a) Bony fishes, (b) Cartilaginous fishes,
    (c) Dipnoi, (d) Placodermi
31. The largest bone of Weberian ossicles is
    (a) Claustrum, (b) Intercalarium,
    (c) Scaphium, (d) Tripus

32. The tripus is derived from ribs of ……… vertebra
    (a) 1st, (b) 2nd,
    (c) 3rd, (d) 4th
33. The first pair of aortic arch is known as
    (a) Mandibular, (b) Hyoid,
    (c) Pulmonary, (d) All
34. Parental behaviour in fish is controlled by
    (a) Diencephalon, (b) Mesencephalon,
    (c) Rhombencephalon, (d) Telencephalon
35. Zero nerve has been reported in
    (a) Fishes, (b) Reptiles,
    (c) Birds, (d) Mammals
36. Ogawa (1961) reported the marine teleostean kidney is divided into ……… groups
    (a) I, (b) III,
    (c) V, (d) VII
37. Ogawa (1961) reported the freshwater teleostean kidney is divided into ……… groups
    (a) I, (b) III,
    (c) V, (d) VII
38. The kidneys of freshwater teleosts are ………… to body size
    (a) Large, (b) Very large,
    (c) Small, (d) Very small
39. Most of the fishes are
    (a) Asexual, (b) Bisexual,
    (c) Unisexual, (d) All
40. The testes of fishes differentiated into …….. compartments
    (a) Interstitial, (b) Lobular,
    (c) Both, (d) None
41. Resting germ cells of fishes are produced during ……… phase
    (a) Maturing, (b) Resting,
    (c) Spawning, (d) Post-spawning
42. Maximum value of Gonadosomatic Index is observed during ……… phase?
    (a) Migratory, (b) Post-spawning,
    (c) Pre-spawning, (d) Spawning

43. ***Myliobatis*** is a/an .... fish
(a) Oviparous, (b) Ovoviviparous,
(c) Viviparous, (d) None
44. The fishes which deposit their eggs on sandy places are known as
(a) Lithophils, (b) Phytophils,
(c) Psammophils, (d) Pelagophils
45. The cleavage in fish are generally
(a) Discoidal, (b) Holoblastic,
(c) Merobalstic, (d) Periblastic
46. The pit organ is a component of ....system
(a) Neuromast, (b) Nervous,
(c) Excretory, (d) Digestive
47. Ampullae of Lorenzini have .... chambers
(a) 4-5, (b) 8-9,
(c) 12-13, (d) 16-17
48. Mating and egg-laying in fishes is controlled by
(a) Oxytocin, (b) Vasopressin,
(c) Both, (d) None
49. The corpuscles of Stannius have been reported in
(a) Fishes, (b) Reptiles,
(c) Birds, (d) Mammals
50. The maximum number of the corpuscles of Stannius has been reported in
(a) ***Catla catla***, (b) ***Channa*** sp.
(c) ***Clarias*** sp. (d) ***Labeo*** sp
51. Which of the following are not characteristic endocrine glands in fishes?
(a) The corpuscles of Stannius, (b) Hypophysis.
(c) Urohypophysis. (d) The ultimobranchial glands
52. The calcium level in the blood of fish is controlled by
(a) The corpuscles of Stannius, (b) Hypophysis.
(c) Urohypophysis. (d) The ultimobranchial glands
53. The hormone responsible for the production of light in some fish is
(a) Luciferin, (b) Oxytocin,
(c) Pepsin, (d) Vasopressin

54. The world's smallest fish is

(a) ***Catla***, (b) Gobies,

(c) Rohu, (d) Magur

55. Who is known as the "Father of Ichthyology?"

(a) Aristotle (b) Berg

(c) Muller (d) Peter Arditi

56. Which of the following period covers the "Late modern period in Ichthyology?"

(a) 1833-1858 (b) 1858-1880

(c) 1851-1931 (d) 1881-1920

57. Which of the following period is known as the "golden age of fishes?"

(a) Craboniferous (b) Devonian

(c) Ordovician (d) Silurian

58. ***Petromyzon*** is variously known as

(a) Lamprey (b) Lampern,

(c) Sand Pride (d) All

59. In fishes, which of the following endocrine glands originated from endostyle?

(a) Pituitary (b) Thyroid,

(c) Adrenal (d) Gonads

59. Jaw suspension may be ……… in cartilaginous fishes

(a) Amphistylic (b) Holostylic

(c) Hyostylic (d) All

60. Which of the following is known as the nurse shark?

(a) ***Alopias***, (b) ***Cetorhinus***

(c) ***Scoliodon***. (d) ***Stegostoma***

61. Which of the following is known as the nurse shark?

(a) ***Alopias***, (b) ***Cetorhinus***

(c) ***Scoliodon*** (d) ***Stegostoma***

62. ***Chimaera*** is variously known as

(a) King of Herrings (b) Rabbitfish,

(c) Ratfish, (d) All

63. Bony fishes have …… pairs of aortic arches

(a) 2, (b) 4,

(c) 6, (d) 8

64. Which of the following is known as Burnett Salmon?
    (a) ***Lepidosiren*** (b) ***Neoceratodus***
    (c) ***Notopterus*** (d) ***Protopterus***
65. Lolach is the common name of?
    (a) ***Lepidosiren*** (b) ***Neoceratodus***,
    (c) ***Notopterus*** (d) ***Protopterus***
66. Squarehead catfish is the common name of?
    (a) ***Chaca*** (b) ***Heteropneustes***,
    (c) ***Mystus*** (d) ***Wallago***
67. ***Pterois*** is variously known as
    (a) Scorpionfish (b) Rockfish,
    (c) Turkey fish (d) All
68. Leaf fish is the common name of?
    (a) ***Badis*** (b) ***Nandus***
    (c) ***Pristolepis*** (d) All
69. Bubble nest is produced by which of the following fish?
    (a) ***Betta*** (b) ***Micropodus***
    (c) ***Macropodus*** (d) ***Trichogaster***
70. Pomfret is the common name of which of the following fish?
    (a) ***Harpadon*** (b) Perch,
    (c) ***Stromateus*** (d) Tuna
71. Maximum number of families of freshwater fishes is reported in .... Region
    (a) Australian (b) Ethiopian
    (c) Neotropical (d) Oriental
72. Which of the following month is recommended for adding fish to the diet?
    (a) April (b) May
    (c) June (d) July
73. The book written by which of the following authors is used to identify the fish of Eastern Up and Bihar
    (a) Aristotle (b) Berg
    (c) Shukla and Pandey (d) Srivastava GJ
74. Barbels are used to identify which of the following order?
    (a) Cypriniformes (b) Holocephaliforms
    (c) Mugiliforms, (d) Siluriformes

75. X-ray fish is the common name for?

(a) Perch (b) *Polypterus*

(c) *Pristolepis* (d) *Pristella*

76. Blue Revolution is related to

(a) Cattle production (b) Fish production

(c) Milk production (d) Pig production

77. Common carp, grass carp and silver carp are varieties of ...... found in Madhya Pradesh.

(a) Cattle (b) Fish

(c) Grass (d) Pig.

78. 'Mahseer', the state fish of Madhya Pradesh belongs to the genus

(a) *Torpedo* (b) *Tor*

(c) *Tinca* (d) *Taenia*

79. In paddy fields, which of the following fish is cultured?

(a) Rohu (b) Catla

(c) Channa (d) All

80. **Which fish shows dorsal fin modified into suckers?**

(a) *Neoceratodus* (b) *Echeneis*

(c) *Protopterus* (d) *Notopterus*

81. ***Matsya*** was the first of ten primary avatars of ....

(a) ***Brahma*** (b) ***Vishnu***

(c) ***Mahesha*** (d) All

82. Who is regarded as the father of the Blue Revolution?

(a) Eric Godwin Silas (b) Gopal Ji Srivastava,

(c) Hiralal Chaudhury (d) V.G. Jhingran

83. Which of the following is the largest order of fish?

(a) Cypriniformes (b) Cyprinodontiformes

(c) Clupeiformes (d) Perciformes

84. Which of the following is the warm-blooded fish?

(a) Featherback (b) Herrings

(c) Opah (d) Shad

85. The number of 2n chromosomes of ***Labeo rohita*** is

(a) 23 (b) 26

(c) 46 (d) 52

86. The number of caudal vertebrae in a typical teleost fish is
    (a) 16-17 (b) 21
    (c) 37-38 (d) 57
87. A spiral valve takes 20 turns in
    (a) Actinopterygii (b) Brachiopterygii
    (c) Crossopterygii (d) All
88. Which of the following has a maximum number of species?
    (a) ***Lepidosiren*** (b) ***Neoceratodus***
    (c) ***Notopterus*** (d) ***Protopterus***
89. Which of the following is a living fossil?
    (a) ***Lepidosiren*** (b) ***Neoceratodus***
    (c) ***Notopterus*** (d) ***Protopterus***
90. Which of the following show(s) parental care?
    (a) ***Lepidosiren*** (b) ***Neoceratodus***
    (c) ***Protopterus*** (d) More than one
91. Chromaffin tissues of dipnoans resemble the .... of higher animals
    (a) Adrenal (b) Parathyroid
    (c) Pituitary (d) Thyroid
92. The median fins of a fish include
    (a) Anal fin (b) Caudal fin
    (c) Dorsal fin (d) All
93. Which of the following fish generally lack pelvic fins?
    (a) Burrowing fish (b) Eels
    (c) Pipefish (d) All
94. Adipose fin is a modification of
    (a) Anal fin (b) Caudal fin
    (c) Dorsal fin (d) Pectoral fin
95. Epithelial cells of the fish epidermis are also known as
    (a) Common cells (b) Keratocytes
    (c) Malpighian cells (d) All
96. Goblets cells of the fish epidermis are generally absent in
    (a) Lamprey (b) Paddlefish
    (c) Mudskipper (d) All

97. Merkel cells are a type of .... in fish epidermis
   (a) Club cells, (b) Neurons
   (c) Paraneurons (d) Synapses
98. Which scales of fish are more primitive
   (a) Cycloid (b) Ctenoid
   (c) Ganoid (d) Placoid
99. Which scales of fish are more advanced
   (a) Cycloid (b) Ctenoid
   (c) Ganoid (d) Placoid
100. Spinoid scales are modification of
   (a) Cycloid (b) Ctenoid
   (c) Ganoid (d) Placoid
101. Elasmoid scales are not found in
   (a) Elasmobranchs (b) Extinct lungfish
   (c) Teleosts (d) Zebrafish
102. Cyvloid and ctenoid scales originated from .... scale
   (a) Elasmoid (b) Ganoid
   (c) Leptoid (d) Placoid
103. Which of the following fish is not selected to describe the skeletal system of a fish?
   (a) ***Heteropneustes*** (b) ***Mystus***
   (c) Rohu (d) ***Wallago***
104. The number of bones in the skeletal system of a fish is generally
   (a) 100 (b) 125
   (c) 150 (d) 175
105. Which of the following fish lost their ability to swim but move with their fins?
   (a) Butterflyfish (b) Parrotfish
   (c) Trunkfish (d) All
106. Fish provide the least resistance when moving through viscous water due to their
   (a) Fins (b) Gills
   (c) Skin (d) Streamlined body
107. The maximum speed of a fish reported in
   (a) Blenny (b) Bonito
   (c) Swordfish (d) Sea trout

108. Which type of swimming is reported in ***Rohu***

(a) Anguilliform (b) Carangiform
(c) Ostraciform (d) All

109. Nociceptors are related to

(a) Acceleration (b) Equilibrium
(c) Pain (d) Touch

110. The ear of a fish contains

(a) External ear (b) Internal ear
(c) Middle ear (d) All

111. Macula neglecta is related to

(a) Hear the sound (b) Excel the body
(c) sensation of touch (d) Feeling of pain

112. Ampulla of Lorenzini contains ……… ampullary sacs

(a) 5-6 (b) 6-7
(c) 7-8 (d) 8-9

113. The type of pseudobranch present in ***Calta*** is

(a) Covered (b) Free
(c) Glandular (d) All

114. Pseudobranch is absent in

(a) ***Channa*** (b) ***Mystus***
(c) ***Wallago*** (d) All

115. The number of pairs of gill pouches in different fish groups varies between

(a) 5-7 (b) 5-9
(c) 5-12 (d) 5-14

116. The spiracular pouch is generally absent in

(a) Elasmobranchs (b) Teleostomi
(c) Dipnoi (d) Latimeria

117. Pillar cells are the characteristics features of

(a) Elasmobranchs (b) Teleostomi
(c) Dipnoi (d) *Latimeria*

118. Hemibranchs are the characteristics features of

(a) Elasmobranchs (b) Teleostomi
(c) Dipnoi (d) *Latimeria*

119. holobranchs are the characteristics features of
    (a) Elasmobranchs (b) Teleostomi
    (c) Dipnoi (d) *Latimeria*
120. The arborescent organ is the characteristic feature of
    (a) ***Anabas testudineus*** (b) ***Channa*** sps.,
    (c) ***Clarias batrachus*** (d) ***Heteropneustes***
121. Delicate vascular outgrowth of pelvic fins acts as respiratory function in
    (a) ***Lepidosiren*** (b) ***Neoceratodus***
    (c) ***Protopterus*** (d) More than one
122. Which of the following is also known as the fish maw?
    (a) Gills (b) ABO
    (c) Swim bladder (d) Pharynx
123. Which of the following is the smallest bone of Weberian apparatus?
    (a) Claustrum (b) Intercalarium
    (c) Scaphium (d) Tripus
124. Relative gut length (RLG) of herbivorous fishes varies between
    (a) 0.2-2.5 (b) 6-8
    (c) 8-20 (d) None
125. Which of the following fish is a column feeder?
    (a) ***Labeo bata*** (b) ***Labeo calbasu***
    (c) ***Labeo gonius*** (d) ***Labeo rohita***
126. In carnivorous fish, gill rakers are
    (a) Long (b) Hard
    (c) Teeth-like (d) All
127. Sinu-atrial aperture is absent in
    (a) ***Channa striatus*** (b) ***Clarias batrachus***
    (c) ***Mystus aor*** (d) ***Wlallago attu***
128. Adenohypophysis of fish is divided into ....parts
    (a) Pro- (b) Meso-
    (c) Meta (d) All
129. Growth hormone is secreted by which part of the adenophypophysis
    (a) Pro (b) Meso
    (c) Meta (d) All

130. Oxytocin in fish controls
   (a) Reproduction (b) Metabolism
   (c) Osmoregulation (d) All
131. Melatonin in fish controls
   (a) Reproduction (b) Metabolism
   (c) Osmoregulation (d) Rhythm
132. TRH has been shown to stimulate the secretion of
   (a) GH (b) PL
   (c) ACTH (d) All
133. Brockmann bodies are related to .... in fish
   (a) Pituitary (b) Thyroid
   (c) Adrenal (d) Islets of Langerhans
134. Ultimobranchial gland is not found in
   (a) Cyclostomata (b) Elasmobranchs
   (c) Teleostomi (d) Dipnoi
135. Urotensin, secreted by Urohypophysis is of the following ...... types
   (a) Two (b) Four
   (c) Eight (d) Sixteen
136. Corpuscles of Stannius is the characteristic of
   (a) Cyclostomata (b) Elasmobranchs
   (c) Teleostomi (d) Dipnoi
137. Maximum number of Corpuscles of Stannius is found in
   (a) ***Channa striatus*** (b) ***Clarias batrachus***
   (c) ***Mystus aor*** (d) ***Wlallago attu***
138. Which fish has two pupils in each eye?
   (a) ***Benthobates*** (b) ***Dialommus***,
   (c), ***Gigantura*** (d) ***Periophthalamus***
139. Choroid plexus is absent in
   (a) Cyclostomata (b) Elasmobranchs
   (c) Teleostomi (d) Dipnoi
140. Cones are absent in
   (a) Cyclostomata, (b) Elasmobranchs
   (c) Teleostomi (d) Dipnoi

141. Tapetum lucidum is the characteristic of
(a) Cyclostomata (b) Elasmobranchs
(c) Teleostomi (d) Dipnoi

142. Most of the electric fish is found in
(a) Africa (b) India
(c) South America (d) More than one

143. Maximum number of electric organs is found in
(a) *Electrophorus* (b) *Gymnarchus*
(c) *Malapterurus* (d) *Torpedo*

144. Fish of which of the order does not have bioluminescent organ
(a) Cypriniformes (b) Clupeiformes
(c), Lophiformes (d) Perciformes

145. Photoprotein may act in place of .... in fish
(a) Luciferin (b) Luciferase
(c), Lysine (d) All

146. Tubular testes in fish are found in which of the following order
(a) Autheriniformes (b) Clupeiformes
(c), Lophiformes (d) Perciformes

147. The resting phase of testes in fish spans from
(a) January to March (b) April to June
(c), July to August (d) August to September

148. ***Cyprinus carpio*** shows .... condition
(a) Monomorphic (b) Temporary dimorphic
(c), Permanent dimorphic (d) All

149. Mature female fish are larger than male ones in
(a) ***Catla catla*** (b) ***Cirrhinus mrigala***
(c) ***Heteropneustes fossilis*** (d) ***Labeo rohita***

150. Length of embryogenesis of indigenous major carp varies from .... days
(a) 24-30 (b) 60-70
(c), 150-180 (d) 260-770

151. Which of the following develops from endoderm?
(a) Para-thyroid (b) Pituitary
(c), Thymus (d) Thyroid

152. Which of the following methods is not generally used in the age determination of fish?

(a) Bone (b) Fin-ray

(c), Otolith (d) Scale

153. Most of the known species of fish show which of the following migration

(a) Amphidromous (b) Anadromous

(c), Diadromous (d) Potamodromous

154. Freshwater migration of fish from one water body to another is known as

(a) Amphidromous (b) Anadromous

(c), Diadromous (d) Potamodromous

155. Which antibiotic is used to study fish migration?

(a) Amoxicillin (b) Amphotericin B,

(c), Azithromycin (d) Tetracycline

156. Which range of temperature is fatal to fish?

(a) <6.7 - >29^{0}C (b) <16.7 - >39^{0}C,

(c), <26.7 - >49^{0}C (d) <36.7 - >59^{0}C

157. At which DO, there is fish mortality

(a) <1ppm (b) <2ppm,

(c), <3ppm (d) <4ppm

158. The relationship between fish catch, mortality and population is

(a) P = C+M+S (b) P = M+S+C

(c) P = S+C+M (d) P = M+C+S

159. Vasodentine is present in..........scale

(a) Ctenoid (b) Cycloid

(c), Ganoid (d) Placoid

160. Which of the following is not a fungal disease?

(a) Branchiomycosis (b) Exophialasis

(c) Ichthyosporidiasis (d) Mouth fungus

161. Which of the following is a protozoan disease?

(a) Branchiomycosis (b) Exophialasis

(c) Ichthyosporidiasis (d) Myxosporidiasis

162. Which of the following is also known as a red spot?

(a) EUS (b) IHN

(c) IPN (d) KHV

163. Red pest disease of fish is caused by?
(a) Virus (b) Bacteria
(c) Algae (d) Fungi

164. The maximum contribution of marine fish catch comes from
(a) Cyclostomata (b) Elasmobranchs
(c) Teleostomi (d) Dipnoi

165. Coldwater fish tolerate temperature range of
(a) 5-10^{0}C (b) 10-15^{0}C
(c), 10-20^{0}C (d) 10-25^{0}C

166. The first exotic fish to India is
(a) ***Cyprinus carpio***, (b) ***Puntius javonicus***
(c), ***Salmo trutto fario*** (d) ***Tinca tinca***

167. In India, the number of native freshwater fish species as SIFS is
(a) 42 (b) 62
(c) 450 (d) 877

168. The range of protein content of fish is
(a) 13-15% (b) 13-20%
(c) 13-25% (d) 13-30%

169. Fish meat contains ……. essential amino acids
(a) 0-5 (b) 5-10
(c) 10-20 (d) 15-20

170. Which of the following fish aesthetic value
(a) Goldfish (b) Silverfish
(c) Starfish (d) Butterflyfish

171. Fish has an HBV of
(a) 35 (b) 74
(c) 80 (d) All

172. Which of the following tools identify genetic variations in fish?
(a) DNA fingerprinting (d) ELISA
(b) RNA sequencing (c) Western blotting

173. Which of the following is the first transgenic fish approved for human consumption?
(a) AquaAdvantage Salmon (b) GloFish
(c) Salmon Express (d) Trout Titan

174. Which of the following is not a DNA marker for fish genetics research?
    (a) GLPs (b) RAPDs
    (c) Microsatellites (d) SNPs
175. Which of the following is a brackishwater fish?
    (a) Asian seabass (b) Grey mullet
    (c) Milkfish (d) All
176. Which fish disease does not any cure?
    (a) Branchyomycosis (b) Icthyophonus
    (c) sparolegniosis (d) All are curable
177. 15-50 ppm of formalin treats
    (a). Dropsy (b) Hexamitiasis
    (c) Icthyobadiasis (d) Velvet

Answers

1(d), 2(a), 3(c), 4(b), 5(b), 6(a), 7(c), 8(b), 9(b), 10 (a), 11(a), 12(b), 13(b), 14(b), 15(d), 16(b), 17(d), 18(b), 19(a), 20(d), 21(a), 22(d), 23(d), 24(b), 25 (d), 26(a), 27(a), 28(d), 29(d), 30(b), 31(d), 32(c), 33(a), 34(d), 35(a), 36(c), 37(b), 38(b), 39(c), 40(c), 41(d), 42(d), 43(b), 44(c), 45(c), 46(a), 47(c), 48(c), 49(c), 50(c), 51(b), 52(d), 53(a), 54(b), 55(d), 56(d), 57(b), 58(d), 59(b), 60(d), 61(d), 62(d), 63(b), 64(b), 65(a), 66(a), 67(d), 68(d), 69(a), 70(c), 71(d), 72(a), 73(d), 74(d), 75(d), 76(b), 77(b), 78(b), 79(d), 80(b), 81(b), 82 (c), 83(d), 84(c), 85(d), 86(a), 87(c), 88(d), 89(b), 90(d), 91(a), 92(a), 93(d), 94(c), 95(d), 96(d), 97(c), 98(c), 99(b), 100(b), 101(a), 102(c), 103(a), 104(c), 105(d), 106(d), 107(c), 108(b), 109(c), 110(b), 111(a), 112(d), 113(a), 114(d), 115(d),116(a), 117(b), 118(a), 119(b), 120(c), 121(a), 122(c), 123(a), 124(c), 125(d), 126(d), 127(a), 128(d), 129(b), 130(c), 131(d), 132(d), 133(d), 134(a), 135(b), 136(c), 137(b), 138(b), 139(c), 140(c), 141(b), 142(d), 143(b), 144(a), 145(b), 146(a), 147(d), 148(b), 149(c), 150(b), 151(b), 152(b), 153(c), 154(d), 155(d), 156(b), 157(a), 158(a), 159(c), 160(d), 161(d), 162(a), 163(b), 164(b), 165(c), 166(c), 167(c), 168(d), 169(b), 170(a), 171(c). 172(a), 173(a), 174(a), 175(d), 176(b), 177(c).

References

Abangan AS, Kopp D, Faillettaz R. Artificial intelligence for fish behavior recognition may unlock fishing gear selectivity. *Frontiers in Marine Science.* 2023;10:1010761. doi: 10.3389/fmars. 2023.1010761.

Abbas HH, Authman M, Zaki MS, Mohamed GF. Effect of seasonal temperature changes on thyroid structure and hormones secretion of white grouper (*Epinephelus aeneus*) in Suez Gulf. *Egyptian Life Science Journal,* 2012;9:700-705.

Abdulhadi H. Some comparative histological studies on alimentary tract of tilapia fish (*Tilapia spilurus*) and sea bream (*Mylio cuvieri*). *Egyptian Journal of Aquatic Research,* 2005;31:387–397.

Abdussamad EM, Gopalakrishnan A, Mini KG, Sukumaran S, Divya PR, Retheesh TB, Muhammed AA, Dipti NV, Akhil AR, Thomas T, Jacob KD. Description of a new species of queenfish, *Scomberoides pelagicus* from Indian seas. *Journal of Environmental Biology*, 2022;43:105-114.

Aboim AN. L'organe interrenal des cyclostomes et des poissons. *Portugaliae Acta Biological*, 1946;4:353-383.

Agrawal N, Singh HS. On a known and three unknown monogenetic *Urocleidus* Mueller, 1934. Pranikee 1982;3:22–34.

Ahuja G, Nia SB, Zapilko V, Shiriagin V, Kowatschew D, Oka Y, Korsching SI. Kappa neurons, a novel population of olfactory sensory neurons. *Science Reporter.* 2014;4:4037.

Akagawa I, Tsukamoto Y, Okiyama M. Sexual dimorphism and pair spawning into a sponge by the filefish *Brachaluteres ulvarum* with a description of the eggs and larva. *Japanese Journal of Ichthyology*, 1995;41:397-407.

Ali S. *Mountain fisheries in India: Exploration, exploitation and sustainable utilization.* In: P. C. Mahanta & Debajit Sarma (ed.) Coldwater fisheries management. DCFR, Bhimtal. pp.187-198; 2010.

Alikunhi KH *Fish culture in India.* Farm Bulletin 1957;20:1-144.

Almada VC, Santos RS. Parental care in the rocky intertidal: a case study of adaptation and expatiation in Mediterranean and Atlantic blennies. *Rev Fish Biol Fisheries* 1995;5:23–37.

Aparicio S, Chapman J, Stupka E, Putnam N, Chia JM, Dehal P, Brenner S. Whole-genome shotgun assembly and analysis of the genome of *Fugu rubripes*. *Science,* 2002;297(5585):1301–1310.

Atherton JA, McCormick MI. Active in the sac: Damselfish embryos use innate recognition of odours to learn predation risk before hatching. *Animal Behaviour* 2015;103:1–6.

Atta KI. Morphological, anatomical and histological studies on the olfactory organs and eyes of teleost fish: *Anguilla anguilla* in relation to its feeding habits. *Journal of Basic and Applied Zoology.* 2013;66:101–108.

Authman MMN, Zaki MS, Khallaf EA, Abbas HH. 2015. Use of Fish as Bio- indicator of the Effects of Heavy Metals Pollution. *Journal of Aquatic Research and Development*, 2015;6:328.

Baber BG. Researches on the minute structure of the thyroid gland. *Philosophy of Transaction of Royal Society London*. 1881;172:577-608.

Baecker, R. Jber die Nebennieran der Teleostier. *Z. Mikrosk-Anat. Forsch*. 1928;15:204-273.

Bagchi A, Jha P. Fish and Fisheries in Indian Heritage and Development of Pisciculture in India. *Reviews in Fisheries Science*. 2011;19(2):85-118.

Baker KF. Heterotopic thyroid tissue in fishes. I. The origin and the development of heterotopic tissue in platyfish. *Journal of Morphology*, 1958;103: 91-134.

Baker RR. The evolutionary ecology of animal migration. Holmes and Meier, New York. 1978

Bandyopadhyaya J. 2007. *Class and Religion in Ancient India. Anthem Press. p. 136.*

Barlow GW. A test of appeasement and arousal hypotheses of courtship behaviour in a cichlid fish, *Etroplus maculatus*. *Zeitschrift für Tierpsychologie* 1970;27:779–806.

Beamish RJ, Chilton DE Preliminary evaluation of a method to determine the age of sablefish (*Anoplopoma fimbria*). *Canadian Journal of Fisheries and Aquatic Sciences* 1982;39:277- 287.

Bean TH. Descriptions of five new species of fishes sent by Prof. A. Dugles from the province of Guanajuato, Mexico. *Proceedings of US National Museum*. 1887;10:370–375

Belcher WR Fish Symbolism and Fishing in Ancient South Asia. Walking with the Unicorn-Jonathan Mark Kenoyer Felicitation. Archaeo Press. 27-41,1991.

Belsare DK. On the evolution of testicular endocrine tissues in some teleosts. *Z. Mikrosk. Anat. Forsch., Leipzig* 1973;87:610-618.

Benton MJ. Vertebrate Paleontology, Blackwell, 3rd edition.2005.

Berg LS. A classification of fish-like vertebrates. Bulletin de l'Académie des Sciences de l'U.R.S.S. Classe des Sciences mathématiqueset naturelles, 1937.

Berg LS. System der rezenten und fossilen Fischartigen und Fischr. VEB, *Veriag der Wissenschaften*, Berlin. (German translation of 2nd Russian edition), 1958.

Berg LS: Classification of fishes, both recent and fossil, vol. 5 (part 2): *Trudy Zoologiceskogo Instituta, Akademii Nauk SSSR*, Moskva, Leningrad; 1940.

Berkes F, Coldings J, Folke C. Rediscovery of Traditional Ecological Knowledge as Adaptive Management. *Ecological Applications*, 2000;10:1251-1262.

Berkes F. Rethinking community-based conservation. *Conservation Biology*, 2004;18:621–630.

Bernatchez L, Wellenreuther M, Araneda C, Ashton DT, Barth JMI, Beacham TD, Withler RE. Harnessing the power of genomics to secure the future of seafood. *Trends in Ecology & Evolution,* 2017;32(9):665–680.

Berquist RM, Gledhill KM, Peterson MW, Doan AH, Baxter GT, Yopak KE, Kang N, Walker HJ, Hastings PA, Frank LR. The Digital Fish Library: Using MRI to Digitize, Database, and Document the Morphological Diversity of Fish. *PLoS ONE*, 2012;7(4):e34499.

Bertelsen E, Marshall NB. The Mirapinnati, a new order of fishes. *Dana Reproduction*. 1956;42:1–35.

Berthelot C, Brunet F, Chalopin D, Juanchich A, Bernard M, Noel B, Alberti A. The rainbow trout genome provides novel insights into evolution after whole genome duplication in vertebrates. *Nature Communications,* 2014;5:3657.

Bertin L, Arambourg, C. Super-ordre des téléostéens. In *Traité de Zoologie* (Ed. P.-P. Grassé). 1958;13: 2204–2500. Paris: Masson.

Besson M, Aubin J, Komen H, Poelman M, Quillet E, Vandeputte M, van Arendonk JAM, de Boer IJM. Environmental impacts of genetic improvement of growth rate and feed conversion ratio in fish farming under rearing density and nitrogen output limitations. *Journal of Clean Production*, 2016;116:100-109

Betancur-R R, Broughton RE, Wiley EO, Carpenter K, Lopez JA, Li C, et al. The tree of life and a new classification of bony fishes. *PLoS Currents Tree of Life*. 2013.

Bhagwat SA, Rutte C. Sacred groves: potential for biodiversity management. *Frontiers in Ecology and the Environment*, 2006;4:519–524.

Billard WW. Stages and rates of normal development in the holostean fish, *Amia calva. Journal of Experimental Zoology*, 1986;238:337–354.

Biswas C, Chakraborty S, Munilkumar S, Gireesh-Babu P, Sawant PB, Chadha NK. Effect of high temperature during larval and juvenile stages on masculinization of common carp (*Cyprinus carpio,* L) *Aquaculture*, 2021;530:735803,

Blanton ML, Specker JL. The hypothalamic-pituitary-thyroid (HPT) axis in fish and its role in fish development and reproduction. *Critical Review of Toxicology*. 2007;37(1-2):97-115.

Bleeker P. Enumeratio specierum piscium hucusque in Archipelago indico observatarum, adjectis habitationibus citationibusque, ubi descriptiones earum recentiores reperiuntur, nec non speciebus *Musei Bleekeriani Bengalensibus, Japonicis, Capensibus Tasmanicisque. Acta Society of Science,* Indo-Neerl. v. 6: i–xxxvi + 1–276, 1859.

Bonnefoy Y. *Asian Mythologies*. University of Chicago Press. 1993.

Botsford, L. W., and D. E. Wickham. 1979. Population cycles caused by inter-age, density-dependent mortality in young fish and crustaceans. Pages 73–82 in E. Naybr and R. Hartnoll, editors. Cyclic phenomena in marine plants and animals. Proceedings of the 13th European marine biology symposium. Permagon Press, New York.

Bradbury RH, Young PC. The effects of a major forcing function, wave energy, on a coral reef ecosystem. Mar Ecol Prog Ser 1981;5:229–241.

Breder CM. The locomotion of fishes. ZoolNY, 1926;4:159–297.

Bridge TW, Haddon AC. Contribution to the anatomy of fishes II. The air-bladder and Weberian ossicles in the siluroid fishes. *Philosophical Transactions of the Royal Society B: Biological Sciences*, 1893;184:65–333.

Bridge TW, Haddon AC. Contribution to the anatomy of fishes. I. The airbladder and Weberian ossicles in the Siluridae. *Proceedings of the Royal Society London*, 1889;46:309–328.

Bruce BC. Menon AGK, 1921–2002, *Copeia*, 2003;(4):922.

Brzeski VJ, Doyle RW. A morphometric criterion for sex discrimination in *Tilapia*. In The Second International Symposium on Tilapia in Aquaculture (Pullin,R. S. V., Bhukaswan,T., Tonguthai, K. &Maclean, J.L., eds), pp. 1988:439–444.Manila: ICLARM Conference Proceedings 15.

Bulger RE. Trump BF. Renal morphology of the English sole (*Parophrys vetulus*). *American Journal of Anatomy*, 1968;123:195226.

Bushnan JS. *Ichthyology*, Memoir of Hippolito Salviani, in: William Jardine *The Naturalist's Library* 1853;**35**:17–43.

Calamia MA. A methodology for incorporating traditional ecological knowledge with geographic information systems for marine resource management in the Pacific Ocean, 1999. http://www.spc.int/Coastfish/*publications*/bulletins/traditional-management/212-traditional-information-bulletin-10.html,

Carrier JC, Pratt HL, Castro JL. Reproductive biology of elasmobranchs. In J. C. Carrier, J. A. Musick, & M. R. Heithaus (Eds.), Biology of Sharks and their Relatives (pp. 269e286). Boca Raton, FL: CRC Press, 2004.

Carrizo SF, Smith KG, Darwall WRT. Progress towards a global assessment of the status of freshwater fishes (Pisces) for the IUCN Red List: application to conservation programmes in zoos and aquariums. International Zoo Yearbook, 2013;47:46–64.

Casselman SJ, Schulte-Hostedde AI. Reproductive roles predict sexual dimorphism in internal and external morphology of Lake Whitefish, *Coregonus clupeaformis*. *Ecology of Freshwater Fish*, 2004;132:217-222.

Cavin L. A new Clupavidae (Teleostei, Ostariophysi) from the Cenomanian of Daoura (Morocco). C. R. Acad. Paris, 1999;329(9):689-695,

Cawthorn DM, Steinman HA, Witthuhn RC. DNA bar-coding reveals a high incidence of fish species misrepresentation and substitution on the South African market. *Food Research International*, 2012;46:30–40.

Chacko PI, Zubairi ARK, Krishnamurthy B. The radii of the scales of *Hilsa ilisha* (Hamilton) as an index of growth and age. *Current Science*, 1948;5:158-159.

Chapman DW. Food and space as regulators of salmonid populations in streams, *American Naturalist*, 1966;100:345–357.

Chauhan T, Rajiv K. Molecular markers and their applications in fisheries and aquaculture. *Advances in Bioscience and Biotechnology*, 2010;1:281-291.

Chavin W. Adrenal histochemistry of some freshwater and marine teleosts. *General and Comparative Endocrinology*, 1966;6:183-194

Chavin W. Pituitary-adrenal control of melanization in the xanthic goldfish, *Carassius auratus* L. *Journal of Experimental Zoology, 1956*;133:1-45.

Chen X, Zhong L, Bian C, Xu P, Qiu Y, You X, Wang M. (2016). High-quality genome assembly of channel catfish, *Ictalurus punctatus*. *Giga Science,* 2016;5(1):39.

Cheung WW. Lam L, Sarmiento YWY, Kearney JL, Watson K, Zeller R, Pauly D. Large-scale redistribution of maximum catch potential in the global ocean under climate change. *Global Change Biology*, 2010;16:24–35.

Claes JM, Ho HC, Mallefet J (2012) Control of luminescence from pygmy shark (*Squaliolus aliae*) photophores. *Journal of Experimental Biology*, 2012;215(10):1691-1699.

Conte MA, Gammerdinger WJ, Bartie KL, Penman DJ, Kocher TD. A high quality assembly of the Nile Tilapia (*Oreochromis niloticus*) genome reveals the structure of two sex determination regions. *BMC Genomics,* 2017;18(1):341. https://doi. org/10.1186/s12864-017-3723-510.1186/s12864-017-3723-3725.

Corwin JT, Northcutt RG. Auditory centers in the elasmobranch brain stem: deoxyglucose autoradiography and evoked potential recording. *Brain Research*. 1982;236:261–273.

Corwin JT. Audition in elasmobranchs, in Hearing and Sound Communication in Fishes. W.N. Tavolga, A.N. Popper, and R.R. Fay, Eds., Springer-Verlag, New York, 81–105, 1981.

Craig-Bennett A. The reproductive cycle of the three-spined stickleback, *Gasterosteus aculeatus*, Linn. Philosophy of Transactions of Royal Society of London, 1931.

Dan SS. Age and growth of the catfish *Tachysurus tenuispinis* Day. *Indian Journal of Fish*, 1980;27(1&2):226-235

Dandekar P. India›s community fish sanctuaries protect wild fish and rivers. *World Rivers Review*, 2011;26:1–7.

Darlington, P. J. (1957). *Zoogeography: the geographical distribution of animals*. New York : Wiley.

Darwin CR. On the Origin of Species by Means of Natural Selection, Or, the Preservation of Favoured Races in the Struggle for Life. London: John Murray, Albemarle Street; 1859.

Das MK, Rao MR Kulsreshtha AK. Native larvivorous fish diversity as a biological control agent against mosquito larvae in an endemic malarious region of Ranchi district in Jharkhand, India. *Journal of Vector Borne Diseases,* 2018;55:34-41.

Das SM. The scales of freshwater fishes of India and their age distribution and systematics. *Proceedings of First All India Congress of Zoology,* 1959;Part III, 621-628

Das TC. The Cultural Significance of Fish in Bengal (continued), *Man in India*, 12, pp. 96-115,1952.

Das TC. The Cultural Significance of Fish in Bengal, *Man in India*, 11, pp.275-303, 1931.

Davey JW, Blaxter ML. RAD Seq: next-generation population genetics. *Briefings in Functional Genomics*, 2011;9:416-423.

Davies MP, Sparks JS, Smith WL. Repeated and widespread evolution of bioluminescence in marine fishes. *PLOS ONE*, 2016;11(6):e0155154.

Day F. The fishes of India, being a natural history of the fishes known to inhabit the seas and freshwaters of India, Burma, and Ceylon. William Dawson & Sons, London, Pt IV, pp. i-xx, 553-778 and pis. 134-195, 1878.

Day, F. Great International Fisheries Exhibition, London, 1883. Catalogue of the exhibits in the Indian Section. London, 198 pp. (list of Day's own specimens, pp. 176-197), 1883.

Day, F. The fauna of British India, including Ceylon and Burma. Fishes, 1. Taylor & Francis, London, 548 pp, 1889.

De Smet W. Considerations on the Stannius corpuscles and the interrenal tissues of bony fishes, especially based on researches into *Amia*. *Acta Zoologica*. (Stockholm) 1962;43:201-219.

Deepananda KHMA, Macusi ED. Human disturbance on macrobenthic assemblages and the role of Marine Protected Areas in reducing its impact. *Philippine Agricultural Scientist*, 2012;95(1): 51-62.

Dhole J, Jawale S, Waghmare S, Chavan R. Survey of helminth parasites in freshwater fishes from Marathwada region, MS, India. *Journal of Fisheries and Aquaculture*. 2010;1(1):01-07.

Di Santo V, Bennett WA. Effect of rapid temperature change on resting routine metabolic rates of two benthic elasmobranchs. *Fish Physiology and Biochemistry*. 2011;37: 929–34.

Dijkgraff S. Hearing in Bony Fishes. Proceedings of the Royal Society of London. Series B, Biological Sciences. 152:946, A Discussion on the 'Ear' Under Water, 51-54,1960.

Dingle H. Ecology and evolution of migration. In: Gauthreaux SA, editor. *Animal Migration, Orientation, and Navigation*. New York: Academic Press, 1–101, 1980.

Drew JA. Use of Traditional Ecological Knowledge in Marine Conservation. *Conservation Biology*, 2005;19:1286-1293.

Dudley N, Higgins-Zogib L, Mansourian S. The links between protected areas, faiths, and sacred natural sites. *Conservation Biology*, 2009;23:568–577.

Dufossé L. Recherches sur les bruits et les sons expressifs que font entendre les poissons d' Europe et sur les organes producteurs de ces phénomènes acoustiques, ainsi que sur les appareils de l'audition de plusieurs de ces animaux. Annals of Science Natural. 1874.

Eid J, Fehr A, Gray J, Luong K, Lyle J, Otto G, Turner S. (Real-time DNA sequencing from single polymerase molecules. *Science,* 2009;323(5910):133–138.

Eschmeyer WN, Fricke R, Van der Laan R. Catalog of Fishes: Genera, Species, References. 2022.

European Economic and Social Committee (EESC). Proposal for a Regulation of the European Parliament and of the Council amending Regulation (EU) 2019/833 of the European Parliament and of the Council of 20 May 2019.

Feist SW, Longshaw M. Histopathology of fish parasite infections–Importance for populations. *Journal of Fish Biology*, 2008;73:2143–2160.

Ferreira HM, Reuss-Strenzel GM, Alves JA. Schiavett A Local ecological knowledge of the artisanal fishers on *Epinephelus itajara* (Lichtenstein, 1822) (Teleostei: Epinephelidae) on Ilhéus coast–Bahia State, Brazil. *Journal of Ethnobiology and Ethnomedicine,* 2014;10(51):1-15.

Ficke *AD*. Potential impacts of global climate change on freshwater fisheries. *Revolution of Fish Biology and Fisheries*. 2007.

Figueras A, Robledo D, Corvelo A, Hermida M, Pereiro P, Rubiolo JA, Gut M. Whole genome sequencing of turbot (*Scophthalmus maximus*; Pleuronectiformes): A fish adapted to demersal life. *DNA Research,* 2016;23(3):181–192.

Filleul, A. & Maisey, J. G. Redescription of *Santanichthys diasii* (Otophysi, Characiformes) from the Albian of the Santana formation and comments on its implications for Otophysan relationships. *American Museum of Novit*, 2004;3455:1-21.

Fish MP. The character and significance of sound production among fishes of the western North Atlantic. In *Bulletin of the Bingham Oceanographic Collection*, Yale University XIV, article 3, 1954.

Fonge BA, Tening AS, Egbe AE, Awo EM, Focho DA, Oben PM, Asongwe GA, Zoneziwoh RM. Fish (Arius heudelotii Valenciennes, 1840) as bioindicator of heavy metals in Douala Estuary of Cameroon. *African Journal of Biotechnology*, 2011;10(73):16581-16588.

Ford P. *Some observations on the corpuscles o£ Stannius*, p. 728-734. In A. Gorbman [ed.], Comparative Endocrinology. John Wiley and Sons, New York. 1959.

Fournie JW, Wolfe MJ, Wolf JC, Courtney LA, Johnson RD, Hawkins WE. Diagnostic criteria for proliferative thyroid lesions in bony fishes. *Toxicology and Pathology*, 2005;33:540-551

Franz V. Zur mikroscopischen Anatomie der Mormyriden". *Zoologisch Jahrbuch Abteilung für Anatomie und Ontogonie*. 1921;42:91–148.

Free CM, Thorson JT, Pinsky ML, Oken KL, Wiedenmann J, Jensen OP. Impacts of historical warming on marine fisheries production. *Science* 2019:363, 979–983.

Fricke R, Eschmeyer WN, Van der Laan R. (eds) Eschmeyer's Catalog of Fishes: Genera, Species. 2022.

Gadgil M. Conserving India's biodiversity: the societal context. *Evolutionary Trends in Plants*, 1991;5:3–8.

Gegenbaur C. *Vergleichende Anatomie der Wirbeltiere*. Bd. 1, S. 523, Leipzig, 1898.

Genten F, Terwinghe E, Danguy A. Atlas of Fish Histology Science Publishers, Enfield, NH, USA, 223, 2008

Germanà A, Montalbano G, Laurà R, Ciriaco E, Del Valle M, Vega J. S100 protein-like immunoreactivity in the crypt olfactory neurons of the adult zebrafish. *Neuroscience Letters*, 2004;371:196–198.

Gerringer ME, Dias AS, vonHagel AA, Orr JW, Summers AP,Farina S. Habitat influences skeletal morphology and density in the snailfishes (family Liparidae). *Frontiers in Zoology*. 2021;18:16.

Ghosh SK, Dash AP. Larvivorous fish in malaria control: A new outlook. *Transactions of Royal Society of Tropical Medicine and Hygiene*. 2007;101:1063-1064.

Gill EL, Watson DMS. An undescribed fish from the coal measures of Lancashire. *Annals of Magazine of Natural History*. 1923;9(11):465-472.

Gillis A, Tidswell O. Evolution: Origin of vertebrate gills. *Nature*, 2017;542 (7642):394. doi:10.1038/542394a. PMID 28230134.

Gills JA, Modrell MS, Northcutt G, Catania KC, Luer CA, Baker CVH. Electrosensory ampullary organs are derived from lateral line placodes in cartilaginous fishes. *Development*, 2012;139:3142–3146.

Gittleman JL.The phylogeny of parental care in fishes. *Animal Behaviour*, 1981;29(3):936-941.

Gogoi A, Biswas SP. Investigation on Larvicidal Efficacy of Two Native Ornamental Murrels of Assam under Controlled Condition. *Agricultural Science Digest*. 2021. **DOI:** 10.18805/ag.D-5404.

Goldberg RL, Downing PA, Griffin AS, Green JP. The costs and benefits of paternal care in fish: a meta-analysis. *Proceedings of the Royal Society B (Biological Sciences)*, 2020.

Gong G., Dan C, Xiao S, Guo W, Huang P, Xiong Y, Mei J. Chromosomal-level assembly of yellow catfish genome using third-generation DNA sequencing and Hi-C analysis. *Giga Science*, 2018;7(11), https://doi.org/10.1093/gigascience/giy120.

Goodrich ES. Studies on the Structure and Development of Vertebrates. Facsimile of the 1930 edition (1986). University of Chicago Press, Chicago, IL. 1930.

Goodrich ES. Vertebrata Craniata I. Cyclostomes and Fishes. In E. R. Lankester, *A Treatise on Zoology*, vol. IX. London, 1909.

Gorbman A. Thyroid function and its control in fishes. In: Fish physiology (eds. W.S. Hoar, D. J. Randall) II, p. 241–274. New York-London: Academic Press 1969

Gong B, Kanyuan D, Shao J, Ling J, Yingyi C. Fish-TViT: A novel fish species classification method in multi water areas based on transfer learning and vision transformer. 2023; Helion 9(6): e16761.

Graham JB. *Air Breathing Fishes: Evolution, Diversity, and Adaptation.* San Diego, CA: Academic Press, 1997.

Gray J. Studies in animal locomotion. I. The movement of fish with special reference to the eel. *Journal of Experimental Biology.* 1933;10:88–104.

Greenwood PH, Rosen DE, Weitzman SH, Myers GS. Phyletic studies of teleostean fishes, with a provisional classification of living forms. *Bulletin of American Museum Natural History.* 1966;131(4):339–456.

Greenwood PH. J. R. Norman's A History of Fishes. Third edition, xxv + 467 pages. New York: John Wiley and Sons. 1975.

Gregory WK, Raven HC. Studies on the origin and early evolution of paired fins and limbs, *Annals of New York Academy Science.* 1941;42:273–360.

Gupta N, Sivakumar K, Mathur VB, Chadwick MA. The 'tiger of Indian rivers': stakeholders' perspectives on the golden mahseer as a flagship fish species. *Area,* 2014b;46:389–397.

Gupta N, Raghavan R, Sivakumar K, Mathur VB. Freshwater fish safe zones: a prospective conservation strategy for river ecosystems in India. *Current Science*, 2014a;107:949–950.

Guraya SS. Recent advances in the morphology, histochemistry and biochemistry of steroid-synthesizing cellular sites in the non-mammalian vertebrate ovary. *International Review of Cytology*, 1976;44:365–409.

Hafeez MA. Light microscopic studies on the pineal organ in teleost fishes with special regard to its function. *Journal of Morphology.* 1971;134:281-313.

Hansen A, Zielinski BS. Diversity in the olfactory epithelium of bony fishes: development, lamellar arrangement, sensory neuron cell types and transduction components. *Journal of Neurocytology,* 2005;34:183–208.

Hara TJ, Zielinski B. Structural and functional development of the olfactory organ in teleosts. *Transactions of American Fish Society,* 1989;118(2):183–194.

Hara TJ. *Morphology of the olfactory (smell) system in fishes*. In: Farrell AP (ed) Encyclopedia of Fish Physiology. Academic Press, San Diego, 2011.

Hasler AD. Chemical ecology of fish, p. 219-234. In E. Sondheimer and J .13. Simeone (ed.) Chemical Ecology. Academic Press, New York, 1970.

Hawkins AD, Chapman CJ. Underwater sounds of the haddock, *Melanogrammus aeglifinus*. *Journal of the Marine Biological Association of the United Kingdom*, 1966;46:241–247.

Heape W. Emigration, migration and nomadism. London, England. Cambridge: Heffer and Sons, 1931.

Heather JM, Chain B. The sequence of sequencers: The history of sequencing DNA. *Genomics,* 2016;107(1):1–8.

Hebert PDN, Ratnasingham S, de Waard JR. Barcoding animal life: Cytochrome c oxidase subunit 1 divergences among closely related species. *Proceedings of the Royal Society B*, 2003;270:96-99.

Hedrick AV, Temeles EJ. The evolution of sexual dimorphism in animals: hypotheses and tests. *Trends in Ecology and Evolution*, 1989;4:136-138.

Hemmer-Hansen J, Therkildsen NO, Pujolar JM. Population Genomics of Marine Fishes: Next-Generation Prospects and Challenges. *Biological Bulletin,* 2014;227:117-132.

Herler J, Michaela K, Philipp M. Sexual dimorphism and population divergence in the Lake Tanganyika cichlid fish genus *Tropheus*. *Frontiers in Zoology*, 2010. http://www.frontiersinzoology.com/content/7/1/4.

Hernandez Jr, Frank J, Powers SP, Graham WM. Seasonal variability in ichthyo plankton abundance and assemblage composition in the northern Gulf of Mexico off Alabama. Aquadocs, 2010.

Hickman CPJr. Glomerular filtration and urine flow in the euryhaline southern flounder, *Paralichthys lethostigma,* in seawater. *Canadian Journal of Zoology*, 1968;46(3):427–437.

Hilborn R, Amoroso RO, Anderson CM et al, Effective fisheries management instrumental in improving fish stock status. *Biological Sciences*. 2010;117 (4) 2218-2224.

Hinegardner R, Rosen DE. Cellular DNA content and the evolution of teleostean fishes. *American Naturalist*. 1972;621–644.

Hoar WS, Randall DJ. (ed.) *Fish Physiology,* Vol. 7. *Locomotion*. New York: Academic Press. 1978.

Hogan C. Commensalism. retrieved from http://www.eoearth.org/view/article/171, 2012.

Holmes RL, Ball JN. The pituitary gland-a comparative account. Biological structure and function Cambridge University Press, 1974.

Holmgren N. Stensio EA. Kranium und Visceralskelett der Akranier Gyclostomen und Fische. Handb. vergl. Anat. 1936;4:233-500.

Honma *Y. et al.* Studies on the Japanese chars of the genus *Salvelinus. IV.* The caudal neurosecretory system of the Nikk-iwana, *Salvelinus leucomaenis pluvius* (Hilgendorf). *General and Comparative Endocrinology*. 1967.

Hora SL. Knowledge of the Ancient Hindus Concerning Fish and Fisheries of India. 1. Referencesto Fish in Arthasastra (ca. 300 B.C.), *Journal of the Royal [sic] Asiatic Society of Bengal. Science*, 1948b;14:10.

Hora SL. Animal life in torrential streams. *Journal of Bombay Natural History*, 1927;40C(32):111–126.

Hora SL. Evidence of distribution of fishes regarding rise in salinity of the river Hugli. *Current Science*, 1943;12:69-90.

Hora SL. Fish in the Ramayana', *Journal of the Asiatic Society. Letters*, 1952;18:63-69:pl1.

Hora SL. Knowledge of the Ancient Hindus Concerning Fish and Fisheries of India. 4. Fish in the Sutra and Smrti Literature', *Journal of the Asiatic Society. Letters*, 1953;19:63-77

Hora SL. Saraswati SK. Fish in the Jataka Tales, *Journal of the Asiatic Society. Letters*, 1955;21:15-30

Hora SL. Structural modifications of fishes of mountain torrents. *Records of the Indian Museum* XXIV: 46–58, 1922.

Howe K, Clark MD, Torroja CF, Torrance J, Berthelot C, Muffato M, Stemple DL The zebrafish reference genome sequence and its relationship to the human genome. *Nature,* 2013;496(7446):498–503.

Hubbs CL, Trautman MB. A revision of the lamprey genus *Ichthyomyzon*. Misc. Publ. *The University* of *Michigan Museum* of *Zoology*, 1937;35:1–109.

Hughes GM, Morgan M. The structure of fish gills in relation to their respiratory function. *Biological Review*. 1973;48, 419–475.

Hughes GM, Munshi, JSD. Scanning electron microscopy of the respiratory surfaces of *Saccobranchus* (= Heteropneustes) fossilis (Bloch.). *Cell and Tissue Respiration*, 1978;195:99–109.

Hughes GM, Dube SC, Munshi JSD. Surface area of the respiratory organs of the climbing perch, *Anabas testudineus* (Pisces: Anabantidae). *Journal of Zoology*, 1973;170:227–243.

Huxley TH. On the application of the laws of evolution to the arrangement of the vertebrata, and more particularly of the mammalia. *Proceedings of Zoological Society of London.* 1880;649–661

Idler DR, Bitners II, Schmidt PJ. II-ketotestosterone: an androgen for sockeye salmon. *Canadian Journal of Biochemistry and Physiology*, 1961;39:1737–1742.

IPCC. Impacts, Adaptation and Vulnerability Summary for Policy Makers. Intergovernmental Panel on Climate Change, Working Group II, Fourth Assessment Report, 16 pp. 2007.

Isaia J, Girard JP, Payan P. 1978. Kinetic study of gill epithelial permeability to water diffusion in the freshwater trout, *Salmo gairdneri:* effect of adrenaline. *Journal of membrane Biology. 1978;*41:337-347.

IUCN. *The IUCN Red List of Threatened Species. 2014.2.* https://www.iucnredlist.org.

Jarvik E. Aspects of vertebrate phylogeny. *In* Orvig, T. [ed.] Current problems of lower vertebrate phylogeny. Almqvist and Wiksell, Stockholm. 1968.

Jarvik E. On the structure of the lower jaw in dipnoans: with a description of an Early Devonian dipnoan from Canada, *Melanognathus canadensis* gen. et sp. nov. *Zoological Journal of Linnaean Society*. 1967;47: 155–183.

Jarvik E. Specializations in early vertebrates. *Annals of Society of Royal Zoology Belgium*. 1964.

Jena JK, Gopalakrishnan G, Aquatic biodiversity management in India. *Proceedings of the National Academy of Sciences, India Section B: Biological Sciences*, 2012:82:363-379.

Jenyns L. Fish. In: The zoology of the voyage of H. M. S. Beagle, under the command of Captain Fitzroy, R. N., during the years 1832 to 1836. Smith, Elder, and Co., London. Issued in 4 parts. i–xvi + 1–172, Pls. 1–29, 1842.

Jessen H. Die Crossopterygier des Oberen Plattenkalkes (Devon) der Bergisch-Gladbach-Paffrather Mulde (Rheinisches Schiefergebirge) unter Berucksichtigung von amerikanischem und europaischem Onychodus-Material. Arkiv for Zoologi 1966;18: 305-389.

Jhingran VG. Age determination of Indian major carps *Cirrhina mrigala* (Ham.) by mean of scales. Nature, London, 1957;179:468-469.

Jinbo U, Kato T, Ito M. Current progress in DNA barcoding and future implications for entomology. *Entomological Science*, 2011;14(2):107-124.

Jones BW. The cod and the cod fishery at Faroe. *Fishery Investment Ministry and Agriculture.* Fish. Food (G.B.) Ser. II, 1966;24(5): 1-32.

Jones EG. *The Thalamus*. Cambridge: Cambridge University Press, 2007.

Jones FC, Grabherr MG, Chan YF, Russell P, Mauceli E, Johnson J, White S. The genomic basis of adaptive evolution in three spine sticklebacks. *Nature*, 2012;484(7392):55.

Jones FRH, Marshall NB. The structure and functions of the teleotean swim bladder, *Biological Review*. 1953;28, 1683.

Jones R. A much simplified version of the fish yield equation. Doc. No. P. 21, presented at the Lisbon joint meeting of Int. Comm. Northwest Atl. Fish., Int. Counc. Explor. Sea, and Food Agric. Organ., United Nations. 8p. 1957.

Jones R. Appendix to the Report of the North West Working Group, Int. Counc. Explor. Sea 2p. 1968.

Jones R. The analysis of trawl haul statistics with particular reference to the estimation of survival rates, Rapp. P.-V. Reun. Cons, Perm, Int.Explor.Mer. 1956;140:30-39,

Jones S, Kumaran M. Fishes of the Laccadive Archipelago. *Nature Conservation and Aquatic Sciences Service*, Trivandrum, 760 pp, 1980.

Jordan DS. A classification of fishes including families and genera as far as known. Stanford Univ. Publ., Univ. Ser., Biol. Sci. v. 1923;3 (no. 2): 77–243 + i–x.

Jorgensen BR, Huss HH. Growth and activity of *Shewanella putrefaciens* isolated from spoiling fish. *International Journal of Food Microbiology*, 1989;9:51-62.

Jung B, Moritz ME, Berchtold JP. Fine structure and function of interrenal (adrenocortical) cells of dexamethasone-treated trout (*Salmo fario* L.). *Cell Tissue Respiration*, 1981;214:641-649.

Kabita FN, Bhuiyan AI, Jhinu JN. A Checklist on the Protozoan Parasites of Freshwater. Fishes of Bangladesh. *Bangladesh Journal of Zoology,* 2020;48(1): 21-35.

Kanagavel A, Raghavan R, Veríssimo, D. Beyond the general public: implications of audience characteristics for promoting species conservation in the Western Ghats Hotspot, India. *Ambio,* 2014;43:138–148.

Kanagavel, A., Pandya, R., Prithvi, A. Raghavan, R. Multi-stakeholder perceptions of efficiency in biodiversity conservation at limited access forests of the southern Western Ghats, India. *Journal of Threatened Taxa*, 2013;5:4529–4536.

Kandimalla V, Richard M, Smith F, Quirion J, Torgo L, Whidden C. Automated detection, classification and counting of fish in fish passages with deep learning. *Frontier of Marine Science.* 2022;8. doi: 10.3389/fmars.2021.823173

Katwate C, Pawar R, Shinde V, Apte D, Katwate, U. How long will social beliefs protect the pride of River Savitri? MIN, 2014;2:21–24.

Kaufman L, Liem KF. Fishes of the suborder Labroidei (Pisces: Perciformes): Phylogeny, ecology and evolutionary significance. *Breviora Museum of Comparative Zoology*, 1982;472: 1-19.

Kaufman LS. Effects of Hurricane Allen on reef fish assemblages near Discovery Bay, Jamaica. *Coral Reefs*. 1983;2:43–47.

Ke R, Mignardi M, Hauling T, Nilsson M. Fourth generation of next-generation sequencing technologies: Promise and consequences. *Human Mutation,* 2016;37(12):1363–1367.

Kerr JG. Textbook of embryology, Vol. 2, Vertebrata. MacMillan, London, 1919.

Khodadoust D, Ismail A, Zulkifli SZ Tayefeh FH. Short Time Effect of Cadmium on Juveniles and Adults of Java Medaka (*Oryzias javanicus*) Fish as a Bioindicator for Ecotoxicological. *Life Science Journal*, 2013;10(1):1857-1861.

Khoo G, Lim MH, Suresh H, Gan DKY, Lim KF, Chen F, Chan WK, Lim TM, Phang VPE. Genetic Linkage Maps of the Guppy (*Poecilia reticulata*): Assignment of RAPD Markers to Multipoint Linkage Groups. *Marine Biotechnology*, 2003;5:279-293.

Khoo G, Lim TM, Phang VPE. A Review of PCR-Based DNA Fingerprinting Using Arbitrary Primers in Tropical Ornamental Fishes of South East Asia. *Journal of Advanced Medical Research,* 2011;1:71-93.

Kierszenbaum AL, Tres LL. Histology and cell biology. An introduction to Pathology. Philadelphia: Elsevier. 2012;701.

Klimenkov IV, Sudakov NP, Pastukhov MV, Kositsyn NS. Adaptive changes in supporting cells of olfactory epithelium in fishes during their spawning period. *Limnology and Freshwater Biology,* 2020;4:795–796.

Kocher TD, Stepien CA. (eds). *Molecular Sytematics of Fishes*, Academic Press, San Diego, 1997.

Koepfli KP, Paten B, O'Brien SJ, Scientists GKC. The genome 10K project: A way forward. *Annual Review of Animal Biosciences,* 2015;3(3):57–111.

Kress JW, Garcia-Robledo C, Uriarte M, Erickson DL. DNA barcodes for ecology, evolution, and conservation. *Trends in Ecology and Evolution*, 2015;30:25-35.

Kreuzberg WG, Gross GW. General morphology and axonal ultrastructure of the olfactory nerve of the pike, *Esox lucinus. Cell Tissue Respiration*, 1977;181:443–457.

Krishna CH, Rao JCS, Veeraiah K. Diversity of larvivorous fish fauna in Lake Kolleru (AP), India. *International Journal of Fauna and Biological Studies,* 2016;3(3): 24-28.

Krönström J, Holmgren S, Baguet F, Salpietro L, Mallefet J. Nitric oxide in control of luminescence from hatchetfish (*Argyropelecus hemigymnus*) photophores. *Journal of Experimental Biology*, 2005;208(15): 2951-2961.

Kumar V, Sinha AK, Makkar HPS, Boeck GDe, Becker K. Phytate and phytase in fish nutrition. *Journal of Animal Physiology and animal Nutrition.* 2012; https://doi.org/10.1111/j.1439-0396.2011.01169.x.

Kuo SR, Lin HJ, Shao KT. Seaosnal changes in abundance and composition of the fish assemblage in Chiku Lagoon, southwestern Taiwan. *Bulletin of Marine Science*, 2001;68:85–99.

Lagler KF, Bardach JE, Miller RR, Passino DRM. Ichthyology. Wiley, New York, 1977.

LaPotin S, Swartz ME, Luecke DM, Constantinou SJ, Gallant JR, Eberhart JK, et al. Divergent cis-regulatory evolution underlies the convergent loss of sodium channel expression in electric fish. *Science Advancement,* 2022;8 (22), eabm2970.

Lauder GV, Anderson EJ, Tangorra J, Madden PGA. Fish biorobotics: kinematics and hydrodynamics of self-propulsion. *The Journal of Experimental Biology*, 2007;210: 2767-2780.

Lauder GV, Liem KF. The evolution and interrelationships of the actinopterygian fishes. *Bulletin of the Museum of Comparative Zoology.* 1983;150(95–197):103.

Leidy RA, Moyle PB. (Conservation status of the world›s fish fauna: an overview. In Conservation Biology for the Coming Decade (eds Fiedler, P.L. & Kareiva, P.M.), Chapman & Hall, New York, USA, 1997.

Li L, Stoeckert CJ, Roos DS. Ortho MCL: Identification of ortholog groups for eukaryotic genomes. *Genome Research*, 2003;13(9).

Li N, Bao L, Zhou T, Yuan Z, Liu S, Dunham R, Li Y, Wang K, Xu X, Jin Y, Zeng Q, Gao S, Fu Q, Liu Y, Yang Y, Li Q, Meyer A, Gao D, Liu Z. Genome sequence of walking catfish (*Clarias batrachus*) provides insights into terrestrial adaptation. *BMC Genomics*, 2018;19:952.

Li *C. et al,* Numerical investigation of fish exploiting vortices based on the Kármán gaiting model. *Ocean Engineering*. 2017.

Liao JC, Beal DN, Lauder GV, Triantafyllou MS. The Kármán gait: novel body kinematics of rainbow trout swimming in a vortex street. *Journal of Experimental Biology* 2003;206:1059–1073.

Liao JC. The role of the lateral line and vision on body kinematics and hydrodynamic preference of rainbow trout in turbulent flow. *Journal of Experimental Biology*. 2006;209:4077–4090

Lien S, Koop BF, Sandve SR, Miller JR, Kent MP, Nome T, Zimin, A. The Atlantic salmon genome provides insights into rediploidization. *Nature,* 2016;533(7602):200.

Lissmann H. Continuous Electrical Signals from the Tail of a Fish, *Gymnarchus Niloticus* Cuv, *Nature*, 1951;167: (4240):201–202.

Lofts B, Marshall AJ. *Cyclical changes in the distribution of the testis lipids of a teleost fish, Esox lucius*. *Quarterly Journal of Microscopic Science*, 1957;98:79-88.

Lorenzini, S. *Osservazioni intorno alle torpedini*. Florence, Italy: Per l'Onofri, 1678.

Lucas MC. Baras, E. Lucas MC, Baras E. Migration of freshwater fishes. Blackwell Science, Oxford. 2001.

Luther G. Observations on the biology and fisheries of the Indian mackerel *Rastrelliger kanagurta* (Cuvier) from the Andaman Islands. *Indian Journal of Fish*, 1973;20(2):425-

427.

Luther, G. 1973. The Dorab Fishery Resources of India. *Proceedings of the Syposium on Living Resources of The Seas Around India*, 11: 445-454.

Mackay KT. Further records of the stromateoid fish *Centro lophusniger* from the northwestern Atlantic, with comments on body proportions and behavior. *Copeia*, 1972:185-187.

MacKenzie S, Jentoft S. 11 - future perspective. In S. MacKenzie, & S. Jentoft (Eds.). *Genomics in aquaculture.* San Diego: Academic Press, 2016.

Mann RHK. The growth and reproductive strategy of the gudgeon *Gobio gobio* (L) in two hard-water rivers in Southern Englan. *Journal of Fish Biology*, 1980;17:163-176.

Margulies M, Egholm M, Altman WE, Attiya S, Bader JS, Bemben LA, Rothberg JM. Genome sequencing in micro-fabricated high-density pico-litre reactors. *Nature,* 2005;437(7057):376–380.

Marshall NB. Swim bladder structure of deep-sea fishes in relation to their systematic and biology. *Discovery Reproduction.* 1960;31:1-121.

Mateus CS, Stange M, Berner D, Roesti M, Quintella BR, Alves MJ. Salzburger W. Strong genome-wide divergence between sympatric European river and brook lampreys. *Current Biology*, 2013;**23**:R649–R650.

Matthew SS, Smith DC. Concentration of radioactive iodine by the thyroid gland of the parrot fish, *Sparisoma* sp. *American Journal of Physiology*, 1948;153:222-225.

Matthews SA. The thyroid gland of the Bermuda parrotfish, *Pseudoscarus guacamaia. Anatomical Records*, 1948;101: 251-263.

Matty AJ. Fish endocrinology. Timber Press, Portland, USA, 1985.

Maurer F. Schilddruse and Thymus der Teleostier. *Morphology of Jb*, 1886;11:129-175.

McCormick SD. Endocrine Control of Osmoregulation in Teleost Fish. *American Zoology*, 2001;41:781–794.

McElroy D, Chamberlain DA, Moon E, Wilson K.J. Development of *gusA* reporter gene constructs for cereal transformation: Availability of plant transformation vectors from the CAMBIA molecular genetic resource service. *Molecular Breeding,* 1995;1:27-37.

Mehinto AC, Martyniuk CJ, Spade DJ, Denslow NC. Applications of next-generation sequencing in fish ecotoxico genomics. *Frontiers in Genetics*, 2012;3:1-10.

Menon AGK. The external relations of Indian freshwater fishes with special reference to the countries bordering the Indian ocean, Journal of Asiatic Sciences, 1955;21:2:3138.

Menon MD. The use of bones other than otoliths in determining the age and growth rate of fishes. *J. du Cons.,* 1950;16(3):311-340.

Miya M, Holcroft NI, Satoh TP, Yamaguchi M, Nishida M, Wiley EO. Mitochondrial genome and a nuclear gene indicate a novel phylogenetic position of deep-sea tube-eye fish (Stylephoridae). *Ichthyology Research.* 2007;54(4):323–32.

Mogalekar HS, Jawakar P, Canciyal J. Fish diversity of rivers of Karnataka. *Journal of Inland Fishery Society of India,* 2016;48(1):56–83.

Mogdans J. Sensory ecology of the fish lateral-line system: Morphological and physiological adaptations for the perception of hydrodynamic stimuli. *Journal of Fish Biology*, 2019;95(1):53-72.

Mohanty RK, Mishra A, Panda DK, Patil DU. Water budget in gina carp prawn polyculture system: impacts on production performance, water productivity and sediment stack. *Aquatic Respiration.* 2016;47:2050–2060.

Mohseni O, Stefan HG, Eaton JG. Global warming and potential changes in fish habitat in U.S. streams. *Climate Change*, 2003;59: 289–409.

Mok HK. Nasal-sinus papillae of hagfishes and their taxonomic implications. *Zoological Studies* 2001;40:355–364.

Møller PR, Jones WJ *Eptatretus strickrotti* n. sp. (Myxinidae): first hagfish captured from a hydrothermal vent. *Biological Bulletin,* 2007;212:55–66.

Moses ST. Fish and Religion in Southern India. *Journal of the Mythic Society of Bangalore*, 1922-23;(13):549-54.

Moulton JM. The acoustical behavior of some fishes in the Bimini area. *Biological Bulletin*, 1958;114:357–374.

Mukerjee SK. An enumeration of the orchids of Ukhrul, Manipur. *Royal Botanic Garden Edinburgh*, 1953;21:149–154.

Mukherjee SK. Origin and distribution of *Saccharum*. *Botanical Gazette*, 1957;19:55- 61.

Muller,J. 1844. Uber den Ban und die Grenzen der Ganoiden undu iber das naturliche systemder Fische.Abhandl. Akad. Wiss. Berlin, phys.-math., 1844;201-204.

Munshi JSD, Olson KR, Ghosh TK, Ojha J. Vasculature of the head and respiratory organs in an obligate air-breathing fish, the swamp eel *Monopterus* (=*Amphipnous*) *cuchia*. *Journal of Morphology*, 1990;203:181-203.

Munshi JSD. Chloride cells in the gills of certain freshwater teleosts. *Quarterly Journal of Microscopic Science*, 1964;105:79-89.

Murray RW. Electrical sensitivity of the ampullae of Lorenzini. *Nature*. 1960;187:(4741):957.

Murray RW. The response of the ampullae of Lorenzini of elasmobranchs to mechanical stimulation. *Journal of Experimental Biology*, 1960;37:417– 424.

Nandi J. The structure of the interrenal gland in teleost fishes. *University of California Publication of Zoology*, 1962;65:129-212

Naseka AM, Tuniyev SB, Renaud CB. *Lethenteron ninae*, a new nonparasitic lamprey species from the north-eastern Black Sea basin (Petromyzontiformes: Petromyzontidae). *Zootaxa*, 2009;2198: 16–26.

Nautiyal P. Review of the art and science of Indian mahseer (game fish) from nineteenth to twentieth century: road to extinction or conservation? *Proceedings of the National Academy of Sciences, India*, 2014;84:215–236.

Nelson GJ. Observations on the gut of the Osteoglossomorpha. *Copeia* 1972;325–329.

Nelson JA. Breaking wind to survive: fishes that breathe air with their gut. *Journal of Fish Biology* 2014;84:554– 576.

Nelson JS, Grande T, Wilson MVH. 5th ed. Fishes of the World. Hoboken: John Wiley & Sons; 2016.

Nelson JS. Fishes of the World, 4th Edition. New Jersey: John Wiley & Sons Inc, 2006.

Nelson JS. Fishes of the world, First edition edn. Hoboken: John Wiley & Sons; 1976.

Nelson JS. Fishes of the world, Second edition edn. Hoboken: John Wiley & Sons; 1984.

Nelson JS. Fishes of the world. 3rd ed. Hoboken: John Wiley & Sons; 1994.

Nelson K. The evolution of a pattern of sound production associated with courtship in the characid fish, *Glanduloeauda inequalis*. *Evolution*, 1965;18:526-540.

Nikolsky GV. The Ecology of Fishes. (Translated by L. Birkett). Academic Press, London and New York, 1963.

NOAA (National Oceanic and Atmospheric Administration). National Centers for Environmental Information - U.S. Climate Normals (Data Products). 2022.

Noda M, VoVan T, Kusakabe I, Murakami K. *Agriculture Biology and Chemistry*, 1982;46, 1565.

Northcutt RG. Central auditory pathways in anamniotic vertebrates. In: Popper AN, Fay RR (eds) Comparative Studies of Hearing in Vertebrates. New York: Springer-Verlag, 1980;79–118.

O'Reilly P, Wright JM. The evolving technology of DNA fingerprinting and its application to fisheries and aquaculture. *Journal of Fish Biology*, 1995;47:29–55.

Ojha J, Rooj NC, Mittal AK, Munshi JSD. Light and Scanning Microscopic Studies on the effect of biocdal plant sap on the gills of a hill-stream fish, *Garra lamta*. *Journal of Fish Biology*, 1988; 34:165-170.

Oliveira, RF, Almada VC. Sexual dimorphism and allometry of external morphology in *Oreochromis mossambicus*. *Journal of Fish Biology*, 1995;46:1055-1064.

O'Reilly P, Wright JM. The evolving technology of DNA fingerprinting and its application to fisheries and aquaculture. *Journal of Fish Biology*, 47 (Supplement A): 1995;29-55.

Owen R. Anatomy of Vertebrates. Vol. 1: Fishes and Reptiles, Longmans, Green and Co., London, 1866.

Padilla DK, Williams SL. Beyond ballast water: aquarium and ornamental trades as sources of invasive species in aquatic ecosystems. *Frontiers in Ecology and the Environment*, 2004;2(3):131-138.

Pandey K, Shukla JP. Deleterious effect of Arsenic on the growth of fingerlings of freshwater fish, *Colisa fasciatus* (BI. Sch.). *Acta Pharmacology and Toxicology*, 1982;50:398-400.

Pandey K, Shukla JP. Fish and Fisheries; Rastogi Publications, Meerut; p504, 2005.

Pandey KC, Tewari S.K. A redescription of *DIplostomulum opthalml* Pandey, 1968; with a note on validity of certain other Indian species. *Uttar Pradesh Journal of Zoology*, 1984;4: 179-184.

Parra G, Bradnam K, Korf I. CEGMA: a pipeline to accurately annotate core genes in eukaryotic genomes. *Bioinformatics*, 2007;23(9):1061–1067.

Patterson C. Chanoides, a marine eocene otophysan fish (Teleostei: Ostariophysi). *Journal of Vertical Paleontology*, 1984;4(3):430-456.

Pereira LHG, Hanner R, Foresti F, Oliveira C. Can DNA bar-coding accurately discriminate mega diverse Neotropical freshwater fish fauna? *BMC Genetics*, 2013;14(20):1-14.

Peter RE, Crim LW, Goos HJ, Crim JW. Lesioning studies on the gravid female goldfish: neuroendocrine regulation of ovulation. *General Comparative Endocrinology*, 1978;35:391-401

Pickford GE, Atz JW. The physiology of the pituitary gland of fishes. p6l3p. *New York Zoological Society, New York.* 1957.

Pillai TVR.A morphometric and biometric study of the systematics of certain allied species of the genus *Barbus* Cuv. & Val., *Proceedings of National Institute of Science, India,* 1951;17 (5): 331–348.

Potter IC. *Mordacia praecox* n. sp., a nonparasitic lamprey (Petromyzonidae), from New South Wales, Australia. *Proceedings of the Linnaean Society of New South Wales*, 1968;92 (3).

Pough FH, Heiser JB, McFarl WN. *Vertebrate Life*. New York: Macmillan Publishing Co. 1989.

Power DM, Llewellyn L, Faustino M, Nowell MA, Björnsson BT, Einarsdottir IE, Canario AVM, Sweeney GE. Thyroid hormones in growth and development of fish. *Comparative Biochemistry Physiology*. C, 2001;130:447-459.

Prasenan P, Suriyakala CD. Fish species classification using a collaborative technique of firefly algorithm and neural network. *EURASIP Journal on Advances in Signal Processing,* 2022;116 https://doi.org/10.1186/s13634-022-00950-8.

Qasim SZ, Bhatt VG. Occurrence of growth zones in the opercular bones of the freshwater murrel, *Ophiocephalus punctatus* (Bloch). *Current Science*, 1964;33:19-20.

Quraishia SF, Panneerchelvam S, Zainuddin Z, Rashid NHA. Molecular Characterization of Malaysian Marine Fish Species using Partial Sequence of Mitochondrial DNA 12S and 16S rRNA Markers. *Sains Malaysiana*, 2015;44(8):1119-1123.

Raj BS. Notes on a collection of fish of Madras. *Records of the Indian Museum*, 1916;12(6):249-294.

Rakshit A, Paul A, Bhattacharjee S, Banik T, Saran R, Mandal B, Poddar D, Gangopadhyay K. Cytogenetic and molecular profiling of spotted snake head fish *Channa punctatus* (Bloch, 1793) from three districts (Nadia, Hooghly and north 24 Parganas) of west Bengal, India. *International Journal of Fisheries and Aquatic Studies*, 2015;3(1), 312-319.

Ramarao K. A study on larvivorous fish species efficacy of lower Manair Dam at Karimnagar, Andhra Pradesh, India. *Pelagic Research Library*, 2014;5(2):133-43.

Rao SR. A study of the otoliths of *Psettodes erumei* (Bl. & Schn.). *Proceedings of 21st Session of Indian Science Congress Association, Calcutta*, 1935;Part III, 319

Raschi W, Mackanos LA. Evolutionary trends in the peripheral component of chondrichthyan electroreceptors. *Archives of Biology*, 1987;98:163– 186.

Raschi W. Mackanos LA. The structure of the ampullae of Lorenzini in Dasyatis garouaensis and its implications on the evolution of freshwater electroreceptive systems. *Journal of Experimental Zoology*, 1989;2:101–111.

Regan CT. The classification of teleostean fishes of the order Ostariophysi. *Annals of Magazine National Hitt*. 1911;8:13–32.

Regan CT. Fishes. Pp. 305-329 in Encyclopedia Britannica, 14th edition, volume 9. Encyclopedia Britannica, Inc., New York, 1929.

Regan CT. On the anatomy and classification of the scombroid fishes. *Annals of Magazine Natural History*. 1909;8(3):66-75.

Regan CT. The Asiatic fishes of the family Anabantidae. *Proceedings of the Zoological Society of London*, 1910;767-787.

Regan CT. The classification of the percoid fishes. *Annals of Magazine Natural History*. 1913;8(12):111-145.

Renaud CB, Economidis PS. 2010. *Eudontomyzon graecus*, a new nonparasitic lamprey species from Greece (Petromyzontiformes: Petromyzontidae). *Zootaxa*, 2477: 37-48.

Retzius G. *Das Gehororgan der Wirbelthiere. Morphol.-Histol. Stud. I. Das Gehörorgan der Fische und Amphibien.* Samson & Wallin.Stockholm; 1881;1–200.

Reynolds JD, Goodwin NB, Freckleton RP. Evolutionary transitions in parental care and live bearing in vertebrates. *Philosophical Transactions of the Royal Society of London. Series B,* 2002;357:269–281. doi: 10.1098/rstb.2001.0930.

Rice AN, Farina SC, Makowski AJ, Kaatz IM, Lobel PS, Bemis WE, Bass AH. Evolutionary Patterns in Sound Production across Fishes. *Ichthyology & Herpetology,* 2022;110(1):1-12. https://doi.org/10.1643/i2020172.

Roberts TR. The Freshwater Fishes of Western Borneo (Kalimantan Barat, Indonesia). California, *Califonia Academy of Science*, 1989.

Romer AS. Fish origins-fresh or salt water? Deep-Sea Research, Suppl. 1955;3:261–280.

Romer AS. The Chafiares (Argentina) Triassic reptile fauna. XL Two new long snouted thecodonts, *Chanaresuchus* and *Gualosuchus*. *Breviora*, 1971;379:1-22.

Rosen D, Greenwood PH. Origin of the Weberian apparatus and relationships of the ostariophysan and gonorhychiform fishes. *American Museum Novitiates*, 1970;2468:1–49.

Rosen DE. Interrelationships of higher euteleostean fishes. In: Greenwood PH, Miles RS, Patterson C, editors. Interrelationships of fishes. London: Academic Press; 1973;397–513.

Roy B, Tandon V. Effect of root-tuber extract of *Flemingia vestita*, a leguminous plant, a scanning electron microscopy study. *Parasitological Respiration*, 1996;82:248-252.

Ruparel H, Bi LR, Li ZM, Bai XP, Kim DH, Turro NJ. Design and synthesis of a 3'-O-allyl photocleavable fluorescent nucleotide as a reversible terminator for DNA sequencing by synthesis. *Proceedings of the National Academy of Sciences of the United States of America,* 2005;102(17):5932–5937.

Saleh A, Sheaves M, Rahimi Azghadi M. Computer vision and deep learning for fish classification in underwater habitats: A survey. *Fish Fish.* 2022;23:977–999. doi: 10.1111/faf.12666.

Sand A. The function of the ampullae of Lorenzini, with some observations on the effect of temperature on sensory rhythms. *Proceedings of Royal Society of London*, B, 1938;125:524-553.

Sanger F, Nicklen S, Coulson AR. DNA sequencing with chain-terminating inhibitors. *Proceedings of the National Academy of Sciences,* 1977;74(12):5463–5467.

Santos RS, Hawkins S, Monteiro LR, Alves M, Isidro EJ. Marine research, resources and conservation in the Azores. *Aquatic Conservation.* 1995;5(4):311–354.

Sargent FJ, Leary TJ, Crewz DW, Kreuer CR. Scarring of Florida's sea grasses: assessment and management options. *Florida Department of Environmental Protection, FMRI Technical Report TR*, 1998;1:46.

Sarma D, Akhtar MS, Singh AK. *Coldwater Fisheries Research and Development in India. Aquaculture in India*, 2018;93-133.

Save-Soderbergh G. Preliminary note on Devonian stegocephalians from East Greenland. Meddr Greenland 1932;94, 1-107.

Schmidt J. The breeding places of the eel. *Philosophical Transaction of Royal of Society, B: Biological Science,* 1922;**211**: 179–208

Schmitz A, Bleckmann H, Mogdans J. Organization of the superficial neuromast system in goldfish, *Carassius auratus*. *Journal of Morphology,* 2008;269:751–61.

Sclater PL. On the general geographical distribution of the members of the Class Aves. *Linnean Society of London Zoological Journal* **2**: 130–145. Reprinted in Sterling 1974 and in Lomolino et at. 2004:118–133, 1858.

Sehgal KL, Shukla JP, Shah KL. Observations on the fishery of Kangra valley and adjacent areas with special reference to mahseer and other indigenous fishes. *Journal of Inland Fishery Society. India.* 1971;3:63-71.

Sehgal KL. Coldwater fish and fisheries in the Indian Himalayas: rivers and streams. p. 4163. In T. Petr (ed.) Fish and fisheries at higher altitudes: Asia. FAO Fish Technical Paper 385. FAO, Rome. 304 p. 1999.

Sehgal KL. The Ecology and Fisheries of Mountain Streams of N.W. Himalayas. Thesis for the award of D.Sc. Degree, Meerut University, 1988.

Seshappa G. Recent studies on age determination of Indian fishes using scales, otoliths and other hard parts. *Indian Journal of Fish,* 1999;46(1):1-11.

Shaw GE, Shebbeare EO. The Fishes of North Bengal. *Journal of Royal Asiatic Society of Bengal, Science* 1937;3(1):1–137.

Singh A, Singh JPN, Gupta RC. Lips and associated structure in *Aspidoparia morar*. *Recent Advance in Ecophysiology of Fishes,* 2009;92-99.

Singh B. Studies on biology of *Labeo dero* (Hamilton) from Nangal and Gobindsagar reservoirs. Ph.D. Thesis, Panjab University, Chandigarh. 1978.

Singh HR, Badola SP, Dobriyal AK. Geographical distribution list of Ichthyofauna of the Garhwal Himalaya with some New Records. *Journal of Bombay Natural History Society*, 1987;84:126-132.

Singh SK, Singh SK, Yadav RP. Toxicological and Biochemical Alterations of Cypermethrin (Synthetic Pyrethroids) Against Freshwater Teleost Fish *Colisa fasciatus* at Different Season. *World Journal of Zoology*, 2010;5(1): 25-32.

Sinha M. Fish genetic resources of the north-eastern region of India. *J. Inland Fish. Soc India.* 1994;26:1–19.

Sinha RK. Biodiversity conservation through faith and tradition in India: some case studies. *International Journal of Sustainable Development & World Ecology*, 1995;2:278–284.

Smarda P, Bures P, Horova L, Leitch IJ, Mucina L, Pacini E, Rotreklova O. Ecological and evolutionary significance of genomic GC content diversity in monocots. *Proceedings of the National Academy of Sciences of the U S A,* 2014;111(39):E4096–E4102.

Smith HW. The absorption and excretion of water and salts by elasmobranch fishes. *American Journal of Physiology,* 1931;98:296–310.

Srivastava GJ. Fishes of Eastern Uttar Pradesh, Vishwavidyalaya Prakashan, Varanasi, pp xxii +163, 1968.

Srivastava HC, Srivastava CBL Morphological variation in teleosts urophysis. *Proceedings of National Academy of Science, India,* 1984;54(8) 4, 288-297.

Stacey N, Karam J, Dwyer D. Speed C, Meekan M. Assessing Traditional Ecological Knowledge of Whale Sharks (*Rhincodon typus*) in eastern Indonesia: A pilot study with fishing communities in Nusa Tenggara Timur. Canberra: Charles Darwin University. 2008.

Stannius H. Uber Nebennieren bei Knochen fischen. *Achieves of Anatomical Physiology and Wiss Medicine*, 1839;6,97–101.

Stepien C, Kocher TD. Molecular systematics of fish: Chapter 1, Molecules and morphology in studies of fish evolution. *Academic Press. 1997.* https://www.utoledo.edu/nsm/lec/pdfs/moschp1.pdf

Strayer DL, Dudgeon, D. Freshwater biodiversity conservation: recent progress and future challenges. *Journal of the North American Benthological Society*, 2010;29:344–358.

Sugunan VV. Reservoir Fisheries in India. Fisheries Technical Paper No. 345. FAO, Rome, 1995.

Swingle HS. Standardization of Chemical Analysis for Water and Pond Mud. *FAO Fisheries Report*, 1967;4:397-421.

Szabo T, Kalmijn AJ, Enger PS, Bullock TH. Micro ampullary organs and a sub-mandibular sense organ in the freshwater ray *Potamotrygon. Journal of Comparative Physiology*, 1972;79:15-27.

Taberlet P, Coissac E, Pompanon F, Brochmann C., Willerslev E. Towards next-generation biodiversity assessment using DNA metabarcoding. *Molecular Ecology*, 2012;21:2045-2050.

Takahashi A, Kobayashi Y, Mizusawa K. The pituitary-interrenal axis of fish: a review focusing on the lamprey and flounder. *General and Comparative Endocrinology*, 2013;188:54-59

Takvam M, Wood CM, Kryvi H, Nilsen TO. Ion Transporters and Osmoregulation in the Kidney of Teleost Fishes as a Function of Salinity. *Frontier of Physiological and Scientific Aquatic Physiology*. https://doi.org/10.3389/fphys.2021;664588.

Talwar PK, Kacker RK. Commercial sea fishes of India *Handbook Zoological Survey of India* 1984;4:1–997

Tamario C, Sunde J, Erik PE, Tibblin P, Forsman A. Ecological and Evolutionary Consequences of Environmental Change and Management Actions for Migrating Fish. *Frontier of Ecology and Evolution Sec. Behavioral and Evolutionary Ecology*. 2019; https://doi.org/10.3389/fevo.2019.00271.

Tandon KK, Johal MS. Characteristics of larval mark and origin of radii. *Current Science*. 1993;64(7):524- 526.

Tarallo A, Angelini C, Sanges R, Yagi, M. Agnisola C, D'Onofrio G. On the genome base composition of telosts: The effect of environment and lifestyle. *BMC Genomics,* 2016;17, 173. https://doi.org/10.1186/s12864-016-2537-1.

Taverne, L. Description de l'appareil de Weber du téléosteen crétacé marin *Clupavus maroccanus* et ses implications phylogénétiques. *Belgium Journal of Zoology,* 1995;125(2):267-282.

Tavolga WN. Mechanisms of sound production in the ariid catfishes *Galeichthys* and *Bagre. Bulletin of American Museum and Natural History,* 1962;24:1–30.

Tester AL, Kendall JI, Milisen WB. Morphology of the ear of the shark genus *Carcharhinus*, with particular reference to the macula neglecta. *Pacific Science,* 1972;26:264–274.

Thapar R. *Asoka and the Decline of the Mauryas* (Oxford: Oxford University Press). 1961.

Thompson DW. *On growth and form*. Cambridge: University Press; (1116 pages), 1952.

Tibbits KJ. Non-parametric multivariate comparisons of soil fungal composition: Sensitivity to thresholds and indications of structural redundancy in T-RFLP data. *Soil Biology and Biochemistry,* 2008;40, 1601–1611.

Tillmar AO, Dell'Amico B, Welander J, Holmlund GA. Universal Method for Species Identification of Mammals Utilizing Next Generation Sequencing for the Analysis of DNA Mixtures. *PLoS ONE*, 2013;8(12):e83761.

Tinbergen N. "Courtship and Threat Display." in J. Huxley, A. Hardy and E.B. Ford. Evolution as a Process, London: George Allen and Unwin, 1954.

Tine M, Kuhl H, Gagnaire PA, Louro B, Desmarais E, Martins RS. Reinhardt R. European sea bass genome and its variation provide insights into adaptation to euryhalinity and speciation. *Nature Communications,* 2014;5:5770.

Trewavas E. Tilapias: taxonomy and speciation, R.S.V. Pullin and R.H. Lowe-McConnell (eds.). 1982.

Tricas TC, Sisneros JA. Ecological functions and adaptations of the elasmobranch Electrosense. In G. Emde, J. Mogdans, & B. G. Kapoor (Eds.), *The senses of fish* (308– 329). Dordrecht, the Netherlands: Springer.2004.

Trinkaus JP. Relation of blastoderm to periblast in epiboly of the Fundulus egg. *Anatomical Records*, 1950;108:586.

Tucker DW A new solution to the Atlantic eel problem. *Nature*. 1959; 183, 495-501.

Valborg H. Symbols of the Eternal Doctrine: From Shamballa to Paradise. Theosophy Trust Books. p. 313,2007.

Valenciennes A (1794-1865): Comité des travaux historiques et scientifiques. Retrieved 20 January *2011*

Van der Laan R, Eschmeyer WN. Fricke R. Family-group names of recent fishes. *Zootaxa,* 2014;3882(2):1-230.

Van Oosten J. The skin and scales. In The physiology of fishes 1: 207-244. Brown, M. E. (Ed.) New York: Academic Press. 1957.

Vingum R. How Fish Developed Electric Organs and What it means for Human Disease. *Trending Plants & Animals*. 2022.

Wakisaka N, Miyasaka N, Koide T, Masuda M, Hiraki-Kajiyama T, Yoshihara Y. An adenosine receptor for olfaction in fish. *Current Biology*, 2017;27(10):1437-1447.e1434.

Walters CJ, Martell SJD. Fisheries ecology and management. Princeton University Press, Princeton, New Jersey, 2004.

Walton TL. Coastal History Notes: Estero Island, Florida Sea Grant, University of Florida, Gainesville, 1960.

Wang J, Gaughan S, Lamer JT, Deng C, Hu W, Wachholtz M, Lu G. Resolving the genetic paradox of invasions: Preadapted genomes and postintroduction hybridization of bigheaded carps in the Mississippi River Basin. *Evolutionary Applications*. 2019; https://doi.org/10.1111/eva.12863.

Wang T. Lineage/species-specific expansion of the Mx gene family in teleosts: Differential expression and modulation of nine Mx genes in rainbow trout *Oncorhynchus mykiss*. *Fish and Shellfish Immunology*. 2019;90:413–430.

Wang Y, Zhou X, Xie B, Chen J, Huang L. Climate-induced habitat suitability changes intensify fishing impacts on the life history of large yellow croaker (*Larimichthys crocea*). *Authorea Preparation.* 2022;12(10):e9342. doi: 10.1002/ece3.9342

Wang YP, Lu Y, Zhang Y, Ning ZM, Li Y, Zhao Q, Zhu ZY. The draft genome of the grass carp (*Ctenopharyngodon idellus*) provides insights into its evolution and vegetarian adaptation (vol 47, pg 625, 2015). *Nature Genetics,* 2015;47(8), https://doi.org/10.1038/ng0815-962a 962-962.

Watson JM. The development of the Weberian ossicles and anterior vertebrae in the goldfish. Proceedings of Royal Society of London, Series B. 1939;127:452-472.

Weber EH. De aure et auditu hominis et animalium. Pars. I. De aure animalium aquatilium. Lipaiae. 1820.

Westoll TL. *On the evolution of the Dipnoi.* In Genetics, paleontology and evolution, Princeton Univ. Press, Princeton, 121-184,1949.

Westoll TS. The lateral fin-fold theory and the pectoral fins of ostracoderms and early fishes. In T. S. Westoll (Ed.), *Studies on fossil vertebrates* (pp. 180–211). London, UK: The Athlone Press, University of London. 1958.

Wetzel RG. Limnology: lake and river ecosystems, 3rd edition. Academic Press, San Diego, California. 2001.

Wiebe JP. The reproductive cycle of the viviparous surfperch, *Cymatogaster aggregata gibbons. Canadian Journal of Zoology*, 1968;46:1221–1234.

Williams JE, Macdonald CA, Deacon C, Williams, H. Weeks, G. Lampman, Sada DW, Prospects for recovering endemic fishes pursuant to the U.S. Endangered Species Act, *Fisheries*, 2005;30:24–29.

Wisby WJ, Hasler AD. Effect of olfactory occlusion on migrating silver salmon (*O. kisutch*). *Journal of the Fisheries Research Board of Canada*, 1954;11:472–478.

Wong LL, Peatman E, Lu, J, Kucuktas H, He S, Zhou CJ, Nanakorn U, Liu ZJ. DNA Barcoding of Catfish: Species Authentication and Phylogenetic Assessment. *PLoS ONE*, 2011;6(3):e17812.

Woodcock HM. Note on a remarkable Parasite of Plaice and Flounders. *Proceedings and Transactions of Liverpool Biological Society,* 1904;18: 143.

Wright PA, Wood CM. A new paradigm for ammonia excretion in aquatic animals: role of Rhesus (Rh) glycoproteins. *Journal of Experimental Biology*, 2009;212:2303-2312. doi: 10.1242/jeb.023085.

Wueringer BE, Tibbetts IR. Comparison of the lateral line and ampullary systems of two species of shovelnose ray. *Reviews in Fish Biology and Fisheries*, 2008;18:47– 64.

Wueringer BE, Tibbetts IR. Comparison of the lateral line and ampullary systems of two species of shovelnose ray. *Reviews in Fish Biology and Fisheries* 2008;18:47–64.

Würtz M. (ed.) Mediterranean Submarine Canyons: Ecology and Governance. Gland, Switzerland and Málaga, Spain: IUCN. 2012;216 pp.

Xu P, Xu J, Liu G, Chen L, Zhou Z, Peng W, Sun X. The allotetraploid origin and asymmetrical genome evolution of the common carp *Cyprinus carpio*. *Nature Communications,* 2019;10(1):4625.

Xu P, Zhang X, Wang X, Li J, Liu G, Kuang Y, Sun X. Genome sequence and genetic diversity of the common carp, *Cyprinus carpio*. *Nature Genetics,* 2014;46(11):1212–1219.

Xu W, Chen S. Genomics and genetic breeding in aquatic animals: Progress and prospects. *Frontiers of Agricultural Science and Engineering,* 2017;4(3):305–318.

You X, Bian C, Zan Q, Xu X, Liu X, Chen J, Wang J, Qiu Y, Li W, Zhang X. Mudskipper genomes provide insights into the terrestrial adaptation of amphibious fishes. *Natural Communication*. 2014;5:5594.

Yuan ZH, Liu SK, Zhou T, Tian CX, Bao LS, Dunham R. Comparative genome analysis of 52 fish species suggests differential associations of repetitive elements with their living aquatic environments. *BMC Genomics,* 2018;19(1):1–10.

Zaccone G, Lauriano ER, Gioele Capillo G and Kuciel M. Air- breathing in fish: Air- breathing organs and control of respiration: Nerves and neurotransmitters in the air-breathing organs and the skin. *Acta Histochemia*. 2018;120(7): 630-641.

Zhang ZH, Fu YQ, Zhang Z., Zhang XM, Chen SC. A comparative study on two territorial fishes: The influence of physical enrichment on aggressive behavior. *Animals*. 2021;11 (7). doi: 10.3390/ani11071868

Zielinski BS, Hara TJ. Olfaction. Fish physiology, Academic Press, New York, 2006.

Webliography

http://pu.edu.pk
http://spinops.blogspot.com
http://www.reformacja-pomorze.pl
https://alchetron.com
https://alchetron.com/Birkenia
https://allthatsinteresting.com
https://animaldiversity.org
https://animals.fandom.com
https://biologyeducare.com
https://bioone.org
https://blog.everythingdinosaur.com
https://climefish.eu
https://en.bdfish.org
https://en.wikipedia.org
https://etc.usf.edu
https://fishesofaustralia.net.au
https://fishionary.fisheries.org
https://ilovepets.co
https://in.pinterest.com
https://insaindia.res.in
https://journals.plos.org
https://link.springer.com
https://nas.er.usgs.gov
https://oceanadventures.co.za
https://onlinelibrary.wiley.com
https://prabook.com
https://prehistopedia.fandom.com
https://prehistoric-wiki.fandom.com/wiki
https://slideplayer.com
https://swarborno.com
https://twitter.com
https://urbanfishkeeping.com
https://vanderbrugghenfossils.org
https://www.aqa.cz
https://www.aquariumbcn.com
https://www.astroved.com
https://www.bioscience.com.pk
https://www.brainkart.com
https://www.brainkart.com/article
https://www.britannica.com

https://www.cabidigitallibrary.org
https://www.calacademy.org
https://www.daviddarling.info
https://www.deviantart.com
https://www.dreamstime.com
https://www.enchantedlearning.com
https://www.faunafondness.com
https://www.flickr.com
https://www.istockphoto.com
https://www.mdpi.com
https://www.notesonzoology.com
https://www.petco.co
https://www.pinterest.com
https://www.researchgate.net
https://www.sciencedirect.com
https://www.sciencephoto.com
https://www.semanticscholar.org
https://www.shutterstock.com
https://www.slideshare.net
https://www.wikidata.org/wiki
https://www.wikiwand.com
https://www.xwhos.com
twitter.com
units.fisheries.org
www.adfg.alaska.gov
www.basu.org.in
www.lkouniv.ac.in
www.old.amu.ac.in
www.prehistoric-wildlife.com
www.scuba.com

Fish Name Index

D

E

F

G

Subject Index

R

S

W

Y

Z